The Origins of Life

A subject collection from *Cold Spring Harbor Perspectives in Biology*

ALSO FROM COLD SPRING HARBOR LABORATORY PRESS

OTHER SUBJECT COLLECTIONS FROM COLD SPRING HARBOR PERSPECTIVES IN BIOLOGY

Cell–Cell Junctions
Generation and Interpretation of Morphogen Gradients
NF-κB: A Network Hub Controlling Immunity, Inflammation, and Cancer
Symmetry Breaking in Biology
The p53 Family

RELATED LABORATORY MANUALS

Molecular Cloning: A Laboratory Manual, Third Edition

OTHER RELATED TITLES

The RNA World

The Origins of Life

A subject collection from *Cold Spring Harbor Perspectives in Biology*

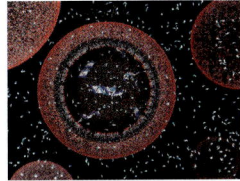

EDITED BY

David Deamer
University of California, Santa Cruz

Jack W. Szostak
*Howard Hughes Medical Institute,
Massachusetts General Hospital*

COLD SPRING HARBOR LABORATORY PRESS
Cold Spring Harbor, New York • www.cshlpress.com

The Origins of Life

A Subject Collection from *Cold Spring Harbor Perspectives in Biology*
Articles online at www.cshperspectives.org

All rights reserved
© 2010 by Cold Spring Harbor Laboratory Press, Cold Spring Harbor, New York
Printed in the United States of America

Publisher	John Inglis
Acquisition Editor	Richard Sever
Director of Development, Marketing & Sales	Jan Argentine
Project Manager	Barbara Acosta
Permissions Coordinator	Carol Brown
Production Editor	Kaaren Kockenmeister
Production Manager	Denise Weiss
Book Marketing Manager	Ingrid Benirschke
Sales Account Managers	Jane Carter and Elizabeth Powers
Cover Designer	Denise Weiss

Front cover artwork: Model of a simple protocell, consisting of a fatty acid based protocell membrane encapsulating nucleic acid molecules that constitute the protocell genome. The small molecules represent the nucleotide building blocks of the nucleic acid polymers, which are illustrated at various stages of replication. Many thanks to Janet Iwasa, the creator of the image.

Library of Congress Cataloging-in-Publication Data

The origins of life / edited by David Deamer, Jack W. Szostak.
    p. cm.
  ISBN 978-1-936113-04-0 (hardcover : alk. paper)
 1. Life–Origin. 2. Biomolecules. 3. Microorganisms I. Deamer, D. W.
II. Szostak, Jack W. III. Title.

  QH325.O69186 2010
  576.8'3–dc22

                                                               2010015162

10 9 8 7 6 5 4 3 2

All World Wide Web addresses are accurate to the best of our knowledge at the time of printing.

Authorization to photocopy items for internal or personal use, or the internal or personal use of specific clients, is granted by Cold Spring Harbor Laboratory Press, provided that the appropriate fee is paid directly to the Copyright Clearance Center (CCC). Write or call CCC at 222 Rosewood Drive, Danvers, MA 01923 (978-750-8400) for information about fees and regulations. Prior to photocopying items for educational classroom use, contact CCC at the above address. Additional information on CCC can be obtained at CCC Online at http://www.copyright.com/.

All Cold Spring Harbor Laboratory Press publications may be ordered directly from Cold Spring Harbor Laboratory Press, 500 Sunnyside Blvd., Woodbury, New York 11797-2924. Phone: 1-800-843-4388 in Continental U.S. and Canada. All other locations: (516) 422-4100. FAX: (516) 422-4097. E-mail: cshpress@cshl.edu. For a complete catalog of all Cold Spring Harbor Laboratory Press publications, visit our website at http://www.cshlpress.com/.

Contents

Preface, vii

Introduction and Overview, 1
David Deamer and Jack W. Szostak

Historical Development of Origins Research, 5
Antonio Lazcano

SETTING THE STAGE

Cosmic Carbon Chemistry: From the Interstellar Medium to the Early Earth, 21
Pascale Ehrenfreund and Jan Cami

The Hadean-Archaean Environment, 35
Norman H. Sleep

Earth's Earliest Atmospheres, 49
Kevin Zahnle, Laura Schaefer, and Bruce Fegley

COMPONENTS OF FIRST LIFE

Planetary Organic Chemistry and the Origins of Biomolecules, 67
Steven A. Benner, Hyo-Joong Kim, Myung-Jung Kim, and Alonso Ricardo

The Organic Composition of Carbonaceous Meteorites: The Evolutionary Story Ahead of Biochemistry, 89
Sandra Pizzarello and Everett Shock

Ribonucleotides, 109
John D. Sutherland

The Origin of Biological Homochirality, 123
Donna G. Blackmond

PRIMITIVE SYSTEMS

Bioenergetics and Life's Origins, 141
David Deamer and Arthur L. Weber

Contents

Mineral Surfaces, Geochemical Complexities, and the Origins of Life, 157
Robert M. Hazen and Dimitri A. Sverjensky

From Self-Assembled Vesicles to Protocells, 179
Irene A. Chen and Peter Walde

Membrane Transport in Primitive Cells, 193
Sheref S. Mansy

FIRST POLYMERS

Primitive Genetic Polymers, 207
Aaron E. Engelhart and Nicholas V. Hud

Closing the Circle: Replicating RNA with RNA, 229
Leslie K.L. Cheng and Peter J. Unrau

The Origins of Cellular Life, 245
Jason P. Schrum, Ting F. Zhu, and Jack W. Szostak

TRANSITION TO A MICROBIAL WORLD

Origin and Evolution of the Ribosome, 261
George E. Fox

Deep Phylogeny—How a Tree Can Help Characterize Early Life on Earth, 279
Eric A. Gaucher, James T. Kratzer, and Ryan N. Randall

Constructing Partial Models of Cells, 295
Norikazu Ichihashi, Tomoaki Matsuura, Hiroshi Kita, Takeshi Sunami, Hiroaki Suzuki, and Tetsuya Yomo

An Origin of Life on Mars, 305
Christopher P. McKay

Index, 313

Preface

THE QUESTION OF HOW LIFE FIRST APPEARED ON THE EARTH CONTINUES to fascinate both scientists and lay readers. A scientific database reveals over 31,000 papers related to this topic, and if one uses "origin of life on Earth" as key words to search a well-known online bookseller, perhaps a hundred or so books are available that deal directly with the science related to life's beginning. Why should we add another?

The reason is that the origin of life remains an open question, yet there is a growing optimism in the scientific community that we are getting closer to an answer. A number of new research themes that matured over the last decade have contributed to the burgeoning interest in life's origins, not the least of which is that for the first time in history we are searching for evidence of life on Mars. Results from the rovers show that Mars had liquid water over three billion years ago, at the same time that life began on Earth. A second independent origin would revolutionize our understanding of life as a universal phenomenon, not just confined to our planet.

The advent of systems biology and synthetic biology also changed the way we think about the origin of life. At some point in the pathway leading to life, there must have been a process by which molecular systems were encapsulated in cellular compartments. This understanding is now driving serious efforts to assemble artificial cells using the tools of synthetic biology, in a sense attempting to achieve a second origin of life that will tell us much about the first origin.

The editors wish to thank Richard Sever for initiating the process by which this book came to be, and Barbara Acosta for her tireless work to make it happen. We also thank all of the authors who agreed to contribute chapters. It is not easy to take time away from research to write a book chapter, but the authors agreed with us that this would be a valuable way to share their ideas. The result is a wonderful collection of expert articles that we hope will interest other scientists and guide the next generation of young investigators who are attracted to the problem of life's origin.

DAVE DEAMER AND JACK SZOSTAK
January 2010

Introduction and Overview

The past decade has seen a resurgence of interest, both public and scientific, in the Origin of Life. One of the most dramatic reasons for this is the discovery of hundreds of extrasolar planets, and the eagerly anticipated detection of Earth-like planets orbiting other stars. These ongoing discoveries bring a sense of immediacy to the age old question of whether we are alone in the Universe or live in a cosmos that is teeming with life. The contemporaneous discovery of the extremophiles that populate such surprising environments on our planet strongly suggests that many other planets could, in principle, support life. But how likely is it that there is in fact life on these other planets? The answer to this question depends strongly on whether the emergence of simple biology from prebiotic planetary chemistry is easy or hard, common or rare. This question can be addressed by scientific studies of the origin of life on Earth, studies that encompass the full panoply of events ranging from the formation of the Earth, through its early geochemistry and later prebiotic organic chemistry, to the synthesis of the necessary biomolecular building blocks and finally the self-assembly of primitive replicating cell-like entities that could then use the power of Darwinian evolution to adapt to an ever wider range of environments. It is these studies that are the subject of this volume.

To get some idea of the scope of the question of life's origins, consider for a moment what the Earth's surface was like 4 billion years ago, before life began. There were no genes to tell a living organism what proteins to make. There were no enzymatic catalysts, no photosynthesis, and no metabolism. Instead, on Mars and the Earth, there were sterile mineral crusts, salty oceans containing a dilute solution of thousands of organic compounds, volcanic land masses rising from hot seas, and wet-dry cycles where seas met land. Water continuously evaporated from the interface between sea and atmosphere, condensed as rain and fell on the lava of volcanic islands where it formed small pools containing organic solutes, then evaporated again. From this unpromising chaos of land, sea, and atmosphere, the first life somehow emerged, certainly on the Earth, perhaps on Mars. (See Fig. 1.)

Because life is an emergent phenomenon of chemistry, it has been mostly chemists who are attracted to the question of how life began. When the first microorganisms began to grow and reproduce on the early Earth, chemical reactions associated with growth, metabolism, and replication were among the earliest adaptations to life in a harsh and increasingly competitive environment. But how could the chemistry begin? The answer to this question involves a much larger scope that views life not just from an Earth-centric perspective, but instead as part of a universal process involving the birth and death of stars, planet formation, interfaces between minerals, water, and the atmosphere, as well as the physics and chemistry of carbon compounds. The chemistry of life only becomes possible after physical processes permit life-specific chemical reactions to begin.

The main thrust of this book is to move away from classical scenarios that emphasize the unconstrained synthesis of organic molecules by the treatment of simpler molecules with intense sources of energy. Instead we treat the origin of life as the emergence of molecular systems contained in some form of semipermeable compartment. Containment will be

Figure 1. An artist's conception of the early Earth, sometime between 4.4 and 3.8 billion years ago. Oceans had condensed, volcanic land masses resembling Hawaii and Iceland rose out of the seas, and rainstorms produced fresh water ponds on the newly formed land masses. The physical and chemical processes leading to the first forms of life were taking place wherever there were interfaces between minerals, water and the atmosphere. (Figure by Don Dixon, FIAAA © 2005 and reprinted here with express permission from the artist.)

considered not just an afterthought, but instead as essential for life to begin and evolve. On the early Earth, over tens to hundreds of millions of years, vast numbers of microscopic compartments were produced at interfaces of minerals, water, and atmosphere. Within this multitude, some happened to contain genetic molecules that could guide the capture of chemical energy and smaller molecules from the surrounding environment and use this energy to facilitate their own replication. The emergence of compartmented sets of polymers capable of energy-dependent growth, reproduction, and evolution marked the beginning of life as we know it today.

A second theme will be to constrain our guesswork by a serious consideration of what the early Earth was really like, an effort that no matter how difficult is essential to direct realistic and informative laboratory simulations. Questions of impact history, the Hadean/Archaean environment, volcanism, and the establishment of plate tectonics are all critical if we are to understand the chemical and physical environment that prevailed at the time of life's origin.

The invited chapters in this book therefore represent an integrated set of primary concepts that we believe will foster new approaches to the origin of life. Each chapter presents aspects resembling pieces of a puzzle, and they are organized in such a way as to form an approximate chronological narrative. Because science necessarily progresses by filling in gaps with essential knowledge, we have invited authors to make explicit the challenges and future directions they perceive as scientists working on questions related to life's origin. The first chapter introduces the history of the primary ideas that guide research on life's beginning. This is followed by a section entitled Setting the Stage, which describes conditions on the early Earth that can be used to constrain laboratory simulations of the Hadean and Archaean environment. The next section, entitled Components of First Life covers the kinds of molecules that were likely to be available in the prebiotic environment. This is followed by a section called Primitive Systems that discusses the interactions between molecules that are fundamental to our understanding of the origin of living systems. The fourth section is dedicated to polymers that show how the requirements for life can be fulfilled if we treat them as the emergence of systems incorporating a flux of energy and nutrients that drives catalyzed polymerization and replication. The section entitled Transition to a Microbial World describes how the origin and evolution of ribosomes and a genetic code can be understood in terms of the RNA world that preceded it. A chapter on synthetic biology describes how we can use newly developed methods as tools to fabricate artificial cellular systems that show certain properties of the living state, including protein synthesis encoded by encapsulated genes. The book concludes with a final chapter that considers the possibility that life also arose on Mars.

Science is commonly viewed as primarily a body of knowledge, but for those engaged in research it is much more than that. Individual scientists are motivated by questions, not answers. The authors in this volume are each contributing to the progress being made toward answering one of the great questions of biology: How does life begin?

Historical Development of Origins Research

Antonio Lazcano

Facultad de Ciencias, Universidad Nacional Autónoma de México, Apdo. Postal 70-407, Cd. Universitaria, 04510 México D.F., Mexico

Correspondence: alar@correo.unam.mx

Following the publication of the Origin of Species in 1859, many naturalists adopted the idea that living organisms were the historical outcome of gradual transformation of lifeless matter. These views soon merged with the developments of biochemistry and cell biology and led to proposals in which the origin of protoplasm was equated with the origin of life. The heterotrophic origin of life proposed by Oparin and Haldane in the 1920s was part of this tradition, which Oparin enriched by transforming the discussion of the emergence of the first cells into a workable multidisciplinary research program.

On the other hand, the scientific trend toward understanding biological phenomena at the molecular level led authors like Troland, Muller, and others to propose that single molecules or viruses represented primordial living systems. The contrast between these opposing views on the origin of life represents not only contrasting views of the nature of life itself, but also major ideological discussions that reached a surprising intensity in the years following Stanley Miller's seminal result which showed the ease with which organic compounds of biochemical significance could be synthesized under putative primitive conditions. In fact, during the years following the Miller experiment, attempts to understand the origin of life were strongly influenced by research on DNA replication and protein biosynthesis, and, in socio-political terms, by the atmosphere created by Cold War tensions.

The catalytic versatility of RNA molecules clearly merits a critical reappraisal of Muller's viewpoint. However, the discovery of ribozymes does not imply that autocatalytic nucleic acid molecules ready to be used as primordial genes were floating in the primitive oceans, or that the RNA world emerged completely assembled from simple precursors present in the prebiotic soup. The evidence supporting the presence of a wide range of organic molecules on the primitive Earth, including membrane-forming compounds, suggests that the evolution of membrane-bounded molecular systems preceded cellular life on our planet, and that life is the evolutionary outcome of a process, not of a single, fortuitous event.

It is generally assumed that early philosophers and naturalists appealed to spontaneous generation to explain the origin of life, but in fact, the possibility of life emerging directly from nonliving matter was seen at first as a nonsexual reproductive mechanism. This changed with the transformist views developed by Erasmus Darwin, Georges Louis Leclerc de Buffon, and, most importantly, by Jean-Baptiste de Lamarck, all of whom invoked spontaneous generation as the mechanism that led to the emergence of life, and not just its reproduction. "Nature, by means of of heat, light, electricity and moisture", wrote Lamarck in 1809, "forms direct or spontaneous generation at that extremity of each kingdom of living bodies, where the simplest of these bodies are found".

Like his predecessors, Charles Darwin surmised that plants and animals arose naturally from some primordial nonliving matter. As early as 1837 he wrote in his *Second Notebook* that "the intimate relation of Life with laws of chemical combination, & the universality of latter render spontaneous generation not improbable." However, Darwin included few statements about the origin of life in his books. He avoided the issue in the *Origin of Species*, in which he only wrote "... I should infer from analogy that probably all organic beings which have ever lived on this Earth have descended from some one primordial form, into which life was first breathed" (Peretó et al. 2009).

Darwin added few remarks on the origin of life his book, and his reluctance surprised many of his friends and followers. In his monograph on the radiolaria, Haeckel wrote "The chief defect of the Darwinian theory is that it throws no light on the origin of the primitive organism—probably a simple cell—from which all the others have descended. When Darwin assumes a special creative act for this first species, he is not consistent, and, I think, not quite sincere ..." (Haeckel 1862).

Twelve years after the first publication of the *Origin of Species*, Darwin wrote the now famous letter to his friend Hooker in which the idea of a "warm little pond" was included. Mailed on February 1st, 1871, it stated that "It is often said that all the conditions for the first production of a living organism are now present, which could ever have been present. But if (and Oh! what a big if!) we could conceive in some warm little pond with all sorts of ammonia and phosphoric salts—light, heat, electricity &c. present, that a protein compound was chemically formed, ready to undergo still more complex changes, at the present day such matter w^d be instantly devoured, or absorbed, which would not have been the case before living creatures were formed." Although Darwin refrained from any further public statements on how life may have appeared, his views established the framework that would lead to a number of attempts to explain the origin of life by introducing principles of historical explanation (Peretó et al. 2009). Here I will describe this history, and how it is guiding current research into the question of life's origins.

BACKGROUND

The Search for the Physicochemical Basis of Life

In 1805 the German naturalist Lorenz Oken wrote a small booklet titled The Creation, in which stated that "all organic beings originate from and consist of vesicles of cells." Several decades later the jellylike, water-insoluble substance that was found inside all cells was termed "protoplasm" by the physician Johann E. Purkinje and the botanist Hugo von Mohl, who like others argued that it was the basic physicochemical component of life. This was followed by Thomas Graham's 1861 proposal that the protoplasm was a colloid formed by a homogenous, proteinaceous substance, which was understood by many as implying, as Thomas Henry Huxley would write a few years later, that the basic traits of life could be understood in terms of the chemical and physical properties of the molecules that made up protoplasm.

The birth and development of organic chemistry as a prominent scientific field very rapidly helped to bridge the gap separating organisms from the nonliving, paving the way to biochemistry. In 1827 Berzelius had written that "art cannot combine the elements of inorganic matter in the manner of living nature", but one year later his friend and former student Friedrich Wöhler showed that urea could be formed in high yield by heating ammonium cyanate "without the need of an animal kidney" (Leicester 1974).

Wöhler's work represented the first synthesis of an organic compound from inorganic starting materials and signals the tremendous advances in organic chemistry that would play a key role in our understanding of biology. Although it was not immediately recognized as such, a new era in chemical research had been begun: in 1850 Adolph Strecker achieved the laboratory synthesis of alanine from a mixture of acetaldehyde, ammonia and hydrogen cyanide. This was followed by the experiments of Alexander M. Butlerov showing that the treatment of formaldehyde with strong alkaline catalysts, such as calcium hydroxide, leads to the synthesis of sugars.

The laboratory synthesis of biochemical compounds was soon extended to include more complex experimental settings, some of which attempted to explain biological synthesis. For example, in 1877 Mendeleyeev, who was skeptical of the biological origin of oil, reported the synthesis of hydrocarbons from hot metallic carbides and water. By the end of the 19th century a large amount of research on organic synthesis had been performed, which showed the abiotic formation of fatty acids and sugars using electric discharges with various gas mixtures. This trend continued into the 20th century by Walther Löb, Oskar Baudish, and others, who reported the synthesis of amino acids by exposing wet formamide ($CHO-NH_2$) to a silent electrical discharge and to UV light (cf. Bada and Lazcano 2003).

In retrospect, these efforts to produce simple organic compounds heralded the dawn of what is termed today prebiotic chemistry. However, there are no indications that the researchers who performed these studies were interested in how life began on Earth, or in the synthesis of biochemical molecules under primitive conditions. It was generally assumed that that the first living beings had been autotrophic, plantlike organisms, so the abiotic synthesis of organic compounds did not appear to be a necessary prerequisite for the emergence of life. These organic syntheses were not conceived as laboratory simulations of Darwin's warm little pond, but rather as attempts to understand the autotrophic mechanisms of nitrogen assimilation and CO_2 fixation in green plants.

Spontaneous Generation or Cosmic Origins?

The Origin of Species was published in 1859, the very same year in which Pasteur began the experiments that would lead him to disprove spontaneous generation of living organisms. His results had implications that went well beyond the limits of academia. Since the times of Lamarck and Buffon, spontaneous generation had been associated in France not only with evolutionary theory but also with secular attitudes and radical political views. The publication in 1862 of the French translation The Origin of Species by the notorious atheist and republican Madame Clémence Royer rekindled the debate. Her version included a lengthy preface that was, in essence, a fierce attack against the Catholic Church, by then a powerful ally of Napoleon III. In such entangled atmosphere, spontaneous generation embodied not only support for evolution, but also a radical, anticlerical political stance (Farley 1977; Fry 2002; Strick 2009).

Pasteur was fully aware of the ideological implications of his discoveries. In a famous lecture delivered at La Sorbonne in 1864, he not only denied the possibility that inanimate matter could organize itself into living systems, but also stated that "what a victory for materialism if it could be affirmed that it rests on the established fact that matter organizes itself, takes on life itself; matter which has in it already all known forces. Ah! If we could add to it this other force which is called life ... what could be more natural than to deify such matter? Of what good would it be then to have recourse to the idea of a primordial creation? To what good the would be the idea of a Creator God? ... if we admit the idea of spontaneous generation, than it would not be surprising to assume that living beings "transformed themselves and climb from rank to rank, for example to insects after 10,000 years and no doubt to monkeys and man after 100,000 years" (cf. Farley 1977).

Regardless of their political ramifications, Pasteur's results made it difficult to advocate spontaneous generation as an explanation for the ultimate origin of life. As a result, a number of philosophers and naturalists promptly dismissed the study of the origins of life as senseless speculation, whereas the willful distortion of Pasteur's results by others raised vitalistic expectations once again. Several devoted materialists like Emil du Bois-Reymond, Karl von Nageli, and August Weismann continued to support the idea of spontaneous generation, but others, like Hermann von Helmholtz, felt that they could side-step the issue by assuming that viable microbes—"cosmozoa"—had been delivered to the primitive Earth by meteorites, thus maintaining the significance of evolution.

Cosmozoa became an alternative for those unwilling to accept the idea of a nonmaterial basis for life, but also for staunch opponents of evolution like Lord Kelvin, who since 1871 had argued for the extraterrestrial origin of life. Toward the end of the 19th century, the belief that life on Earth had evolved from extraterrestrial organisms elicited a number of proposed mechanisms that could have transported microbes between planets, but little attention was given to the central issue of the actual origin of the life forms (Kamminga 1982; Fry 2002). With formidable disregard for plausibility, the panspermia hypothesis has been repeatedly proposed in a variety of contexts, but of course does not solve the problem of the origin of life, instead merely transferring its origin to another habitable planet in our galaxy.

Life and the Single Molecule

Not surprisingly, the idea that living organisms were the historical outcome of gradual transformation of lifeless matter became widespread soon after the publication of *The Origin of Species*. Despite their diversity, most of these explanations went unnoticed, in part because they were incomplete, speculative schemes largely devoid of direct evidence and not subject to fruitful experimental testing. Although some of these hypotheses considered life as an emergent feature of nature and attempted to understand its origin by introducing principles of historical explanation, the dominant view was that the first forms of life were structureless droplets of protoplasm endowed with the ability fix atmospheric CO_2 and to use it with water to synthesize organic compounds.

The ideas of Jerome Alexander, Stephane Leduc, and Alfonso L. Herrera epitomize this trend. Like many of his contemporaries, the Mexican A.L. Herrera was convinced that life could be created in the laboratory, and proposed an autotrophic theory known as plasmogenesis. Herrera devoted more than 50 yr to experimenting with different kinds of substances, attempting to "illustrate the physico chemical concomitants of life" (Herrera 1902). At first he used mixtures of water and oil (or gasoline) to understand the shape, size and movement of cell-like structures. He would later refined his ideas and, despite the academic isolation in which he worked, developed his theory of "plasmogeny," which attempted to explain the origin of primitive photosynthetic protoplasm. This led him to experiment with formaldehyde and hydrogen cyanide derivatives like NH_4SCN (Herrera 1942), a combination that we now know produces sugars and highly colored polymers, which unfortunately he mistook for photosynthetic pigments (Perezgasga et al. 2003).

The rapid development of biochemistry and the characterization of an increasingly large number of proteins signaled the idea that life could be associated with specific enzymes and that submicroscopic colloidal aggregates or micelles could show the properties of life. Enzymes were seen as colloidal catalysts, which led to the hypothesis that entities smaller and simpler than protoplasm itself could be alive (cf. Fry 2006). In 1917 Felix D'Herelle discovered a self-propagating filterable "substance" that attacked and dissolved bacilli, which were later identified as bacteriophage viruses, and these supposedly simple submicroscopic particles were assumed to be primordial entities (D'Herelle 1926; Summers 1999).

Some investigators proposed even smaller structures as primordial. For instance, in a series of papers published between 1914 and 1917, the American physicist Leonard Troland suggested that the first living entity had been nothing more than a self-replicating enzymelike molecule that had suddenly appeared in early oceans. This primordial enzyme was assumed by Troland to be endowed with autocatalytic properties that allowed it to self-multiply, as well as having heterocatalytic abilities that could alter its surroundings, therefore giving rise to metabolism. Troland's hypothesis is hindered both by its highly reductionist nature and by the impossibility of empirical analysis of a proposal that depends on the chancelike association of a "living molecule." As summarized by Fry (2006), although Troland had originally proposed a primordial enzyme as the starting point of life, he modified his hypothesis to speak of a "genetic enzyme" which he eventually identified with nucleoproteins present in nuclei (Troland, 1914, 1916, 1917).

It did not take long for Hermann J. Muller, an American geneticist who would play an important role in the understanding of Mendelian heredity, to modify Troland's hypothesis and propose that the ancestral molecule had been, in fact, a gene. Even more explicitly than Troland, Muller argued that the first living material was formed abruptly and consisted of little more than a mutable gene, or set of genes, endowed with catalytic and autoreplicative properties, which, he hinted, were autotrophic (Muller 1922, 1926).

It is easy to understand Muller's proposal in terms of his commitment to Mendelian genetics. Muller was a founding member of Thomas Hunt Morgan's fly room in Columbia Uiversity, where he had spent several years working with Calvin Bridges, Alfred Sturtevant and Morgan himself, on the linear arrangement of genes in the Drosophila chromosomes (Carlson 1981). He had been pondering for some time on the autocatalytic properties of chromatin (Ravin 1977), and his appreciation of genetic mutation as the fundamental mechanism of evolutionary novelties developed at a time when the appeal of Darwin's ideas on the role of natural selection had diminished. Accordingly, given the appearance of a genetic material capable of replication, mutation and further replication of mutant forms, "evolution would automatically follow" (Pontecorvo 1982). Muller's explanation of the origin of life reveals a mutationist's attitude, not a Darwinian one.

Toward the Primitive Soup

In November 1923 a small book titled The Origin of Life was published in Moscow by the young Russian biochemist Alexander Ivanovich Oparin. Like many of his contemporaries, Oparin (1924) accepted the idea of a primordial protoplasm but proposed that life had been preceded by a lengthy period of abiotic syntheses and accumulation of organic compounds that had led to the accumulation of what we call today the primitive soup. Oparin's central thesis was that the first organisms to emerge in the anaerobic environment of the primitive Earth must have been heterotrophic bacteria.

As a young student at the University of Moscow, Oparin had joined the laboratory of Alexei N. Bakh, an eminent scientist and political figure at the Karpov Physicochemical Institute. There he worked on photosynthesis and, like most biochemists of his generation, quickly adopted the idea that metabolism was the outcome of oxidation and reduction reactions that were coupled inside cells. By then Oparin was also a convinced evolutionist. As an undergraduate he had attended the lectures given regularly by Kliment A. Tymiriazev, a renowed plant physiologist, agronomer, and the main advocate of Darwinism in Russia. Starting in 1865, Tymiriazev actively promoted Darwin's idea, an effort that would play a major role in the secularization of Russian society, and endeared him to both the liberal and the revolutionary intelligentsia (Vucinich 1988).

Tymiriazev had left the university and, because of his ill-health, did not teach but limited his meetings with students and colleagues to small gatherings in his Moscow flat. By the time Oparin graduated, he had an academic background that combined natural history, biochemistry, and plant physiology, a knowledge acquired within a research tradition strongly committed to integral approaches in the analysis of natural phenomena. He was not only familiar with nearly all the literature on evolution available in Russia but, perhaps even more important, with the Darwinian method of comparative analysis and historical interpretation of life features (Lazcano 1992).

Like many of his fellow students and colleagues, Oparin was well acquainted with Haeckel's work, in which the transition of the nonliving to the first organisms was discussed but always under the assumption that the first forms of life had been autotrophic microbes. Analysis of Oparin's writings shows that throughout his entire life he remained faithful to the Haeckelian division of life into plants, animals and protists. However, from the very beginning it was impossible for him to reconcile his biochemical understanding of the sophistication of photosynthesis and the Darwinian credence in a gradual, slow evolution from the simple to the complex, with the suggestion that life had emerged already endowed with an autotrophic metabolism that included enzymes, chlorophyll and the ability to synthesize organic compounds from CO_2 and water.

Because a heterotrophic anaerobe is metabolically simpler than an autotrophic one, Oparin argued, the former would necessarily have evolved first. Thus, based on the simplicity and ubiquity of fermentative reactions, he proposed that the first organisms must have been heterotrophic bacteria that could not make their own food but obtained organic material present in the primitive milieu. To buttress his intuition, Oparin needed to show that organic material could form in the absence of living beings. Two important pieces of evidence supported his claim that the first organisms were more likely to have been heterotrophic. First, hydrocarbons and other organic material were known to be present in meteorites, and perhaps even in comets. These facts had been known since the middle of the 19th century. Second, his proposal was sustained by the striking 19th century experimental synthesis of organic compounds discussed earlier, including the 1877 abiotic formation of long chain hydrocarbons reported by Mendeleyeev.

Careful reading of Oparin's 1924 book shows that, in contrast to common belief, at first he did not assume an anoxic primitive atmosphere. In his original scenario he argued that whereas some carbides, i.e., carbon-metal compounds, extruded from the young Earth's interior would react with water vapor leading to hydrocarbons, others would be oxidized to form aldehydes, alcohols, and ketones (such as acetone). These molecules would then react among themselves and with NH_3 originating from the hydrolysis of nitrides (nitrogen-metals),

$$Fe_mC_n + 4mH_2O \longrightarrow mFe_3O_4 + C_{3n}H_{8m}$$
$$FeN + 3H_2O \longrightarrow Fe(OH)_3 + NH_3$$

to form "very complicated compounds," as Oparin (1924) wrote, from which proteins and carbohydrates would form, that would rapidly form droplets of gel-like material ancestral to the first cells.

Similar proposals for a heterotrophic origin were also published in 1924 by the geochemist Charles Lipman and the microbiologist R.B. Harvey, although they were not as refined as Oparin's book. Quite significantly, Harvey argued that life had first evolved in a hot spring, where the high temperature would allow chemical reactions to proceed at a significant rate even if the first living beings were endowed with just a few enzymes. The most significant proposal, however, came from John B.S. Haldane, a versatile British biologist who became one of the founding fathers of neodarwinism. Haldane, like Oparin, argued that the origin of life had been preceded by the synthesis of organic compounds (Haldane, 1929). Based on experiments by the British chemist E.C.C. Baly, who claimed that he had synthesized amino acids (Baly 1924) and sugars by the UV irradiation of a solution of CO_2 in water (Baly et al. 1927), Haldane suggested that the absence of oxygen in a CO_2-rich primitive atmosphere had led to the synthesis of organic compounds and the formation of a "hot dilute soup." Haldane was also influenced by D'Herelle's discovery of phages, and suggested that viruses represented an intermediate step in the transition from the prebiotic soup to the first heterotrophic cells. Life may have remained, wrote Haldane (1929) "in the virus stage for many millions of years before a suitable assemblage of elementary units was brought together in the first cell."

The Painful Maturation of the Heterotrophic Theory

Oparin belonged to a generation that was experiencing the liberal, high-bourgeois cultural and scientific circles of Saint Petersburg and Moscow formed by broad-minded scholars like Pavlov and Vernandsky. He was also encouraged to develop and support materialistic ideas by the secular atmosphere that followed the 1917 Bolschevik revolution. His 1924 book can be read as the work of a young, bold, and talented researcher with abundant enthusiasm and free of intellectual prejudices, who was able to look beyond the boundaries separating different scientific fields. In retrospect, it can be also considered the harbinger of his major work, a 1936 volume in Russian also called *Origin of Life*, whose English translation became available 2 years later (Oparin 1938).

The new volume was far more mature and profound in its philosophical and evolutionary analysis, as argued forcefully by Graham (1972), reflecting the changes in a society that was attempting to develop science, art and culture within the framework of dialectical materialism. In his second book Oparin (1938) not only abandoned his naïve and crude materialism, but also provided a thorough presentation and extensive analysis of the literature on the abiotic synthesis of organic material. His original proposal was revised, leading to the assumption of a highly reducing primitive mileu in which iron carbides of geological origin would react with steam to form hydrocarbons. Their oxidation would yield alcohols, ketones, aldehydes, etc., that would then react with ammonia to form amines, amides and ammonium salts. The resulting proteinlike compounds and other molecules would form a dilute solution, where they would aggregate to form colloidal systems from which the first heteretrophic microbes evolved (Oparin 1938).

Oparin further argued that coacervate drops represented the optimal mechanism to concentrate organic material on the primitive Earth. Coacervates are charged, microscopic organic colloidal droplets that can concentrate organic materials existing in the medium. Because coacervates form spontaneously when two solutions of macromolecules with opposite charges are mixed, it is quite possible that they were present in the prebiotic milieu. However, they lack the lipid bilayers, present in all cells that retain organic matter in high concentrations inside a self-constructed boundary. Therefore, coacervates are no longer considered as potentially ancestral to life itself. Coacervates were the favorite model for a considerable time after Oparin's views became widely known, because they were perceived as mimicking the surmised properties of precellular systems, but the development of a more sophisticated understanding of cells led to their dismissal as constituting any step toward the origins of life (Deamer 1977).

A highly reducing atmosphere would be, for Oparin, a mixture of CH_4, NH_3, and H_2O with or without added H_2. The atmosphere of Jupiter contains these chemical species, with H_2 in large excess over CH_4. Oparin's proposal of a primordial reducing atmosphere was a brilliant inference from the then fledging knowledge of solar atomic abundances and planetary atmospheres, as well as from Vernadsky's idea that the early Earth would be anoxic in the absence of life because molecular oxygen is a product of photosynthesis. As summarized elsewhere (Miller et al. 1997) the benchmark contributions of Oparin's 1938 book include not only the hypothesis that heterotrophs and anaerobic fermentation were primordial and the proposal of a reducing atmosphere for the prebiotic synthesis and accumulation of organic compounds, but also the idea that the association of molecules in precellular systems was a necessary prerequisite for their evolution (Miller et al. 1997). Oparin was aware that the significance of coacervates as laboratory models of such polymolecular systems had been diminished by more recent developments. When he visited Mexico to receive an honorary degree that my university had granted him, Celia Ramírez and I gave him a copy of Light Transducing Membranes, which David W. Deamer (1977) had edited two years before. "If I had the chance to start all over again," he remarked, "I would work on liposomes rather than on coacervates as models of precellular systems."

As Farley (1977) wrote, Oparin's 1938 book may be the most significant work ever published on the origin of life. It is true that many of his original ideas have been superseded. However, over the years it has become clear that the open character of his theory of chemical evolution has allowed the incorporation of new discoveries and the development of more accurate descriptions of possible primitive scenarios without destroying its overall structure and premises. The heterotrophic theory has not been belittled, for instance, but magnified by the recognition of the key role that genetic material must have placed in the origin of life. Perhaps the most important scientific achievements of Oparin may be his insistence that life is the evolutionary outcome of a process and not of a single event, as well as the methodological breakthrough that transformed the study of the origin of life from a purely speculative problem into a workable multidisciplinary research program.

The Miller-Urey Experiment: The Birth of Prebiotic Chemistry

The English translation of Oparin's second book caught the attention of biologists such as Norman Horowitz and Cornelius van Niel, but during the next 10 years, while World War II raged, little progress was made in research on the origins of life. Driven by his interest in evolutionary biology, Melvin Calvin attempted to simulate the synthesis of organic compounds under primitive Earth conditions using the high-energy radiation sources available at the Lawrence Berkeley Laboratory. He and his group had limited success: The irradiation of CO_2 solutions with the Crocker Laboratory's 60-inch cyclotron led only to formic acid, albeit in fairly high yields (Garrison et al., 1951).

By the time Calvin and his colleagues published their results, Harold C. Urey had already moved from Columbia to the University of Chicago, where he started to work on cosmochemistry. He rapidly became convinced the Earth's earliest atmosphere was highly reducing, with CH_4 and NH_3 instead of CO_2 and N_2. In 1951 Urey gave a seminar dealing with the origin of the Solar System, and argued that the reducing conditions on the early Earth may have been important to the emergence of life. As he wrote in his 1952 book *The Planets: Their origin and development*, "if half the present surface carbon existed as soluble organic compounds and only 10 per cent of the water of the present oceans existed on the surface of the primitive earth, the primitive oceans would have been approximately a 10 per cent solution of organic compounds. This would provide a very favorable situation for the origin of life" (Urey 1952).

Stanley L. Miller, who had arrived to Chicago in the spring of 1951 after graduating from the University of California, Berkeley, attended Urey's lecture, who like Oparin suggested that it would be interesting to simulate the proposed reducing conditions of the primitive Earth to test the feasibility of organic compound synthesis. "Urey's point immediately seemed valid to me," wrote Miller many years afterward. "After this seminar someone pointed out to Urey that in his book Oparin had discussed the origin of life and the possibility of synthesis of organic compounds in a reducing atmosphere. Urey's discussion of the reducing atmosphere was more thorough and convincing than Oparin's; but it is still surprising that no one had by then performed an experiment based on Oparin's ideas" (Miller 1974).

Almost a year and a half after Urey's lecture, Miller approached Urey about the possibility of doing a prebiotic synthesis experiment using a reducing gas mixture. After overcoming Urey's initial resistance, he designed three apparatuses meant to simulate the ocean-atmosphere system on the primitive Earth by investigating the action of electric discharges acting for a week on a mixture of CH_4, NH_3, H_2, and H_2O; racemic mixtures of several protein amino acids were produced, as well as hydroxy acids, urea, and other organic molecules (Miller 1953, 1955; Johnson et al. 2008).

Miller achieved his results by means of an apparatus in which he could simulate the interaction between an atmosphere and an ocean. To activate the reaction, Miller used an electrical spark, which was considered to be a significant energy source on the early Earth in the form of lightning and coronal discharges. The apparatus was filled with various mixtures of methane, ammonia, and hydrogen as well as water, the latter being heated to boiling during the experiment. A spark discharge between the tungsten electrodes was produced by a high frequency Tesla coil with a voltage of 60,000 V. The reaction time was usually a week or so and the maximum pressure 1.5 bars. With this relatively simple experimental setup, Miller (1953) was able to transform almost 50% of the original carbon (in the form of methane) into organic compounds. Although most of the synthesized organic material was an insoluble tarlike solid, he was able to isolate amino acids and other simple organic compounds from the reaction mixture. Glycine, the simplest amino acid, was produced in 2% yield (based on the original amount of methane carbon), whereas alanine, the simplest amino acid with a chiral center, showed a yield of 1%. Miller was able to show that the alanine was a racemic mixture (equal amounts of D- and L-alanine). This provided convincing evidence that the amino acids were produced in the experiment and were not biological contaminants somehow introduced into the apparatus.

DNA versus Coacervates? The Reshaping of an Old Debate

The Miller paper (1953) was published only a few weeks after Watson and Crick's (1953) classic article revealed their double helix model for the structure of DNA. With few exceptions, like Sidney W. Fox's work on thermal polypeptides (cf. Fox and Dose 1977), modern attempts to understand the origin of life have been shaped by our burgeoning knowledge of DNA replication and protein biosynthesis. Prebiotic chemistry and molecular biology began to converge, albeit slowly, when Oró (1960) showed the remarkable ease with which adenine could be produced through the oligomerization of HCN.

Hermann J. Muller quickly used the developments in molecular genetics and the success in prebiotic syntheses to update his gene-first proposal by arguing that what had emerged in the primitive oceans had been, in fact, a primordial DNA molecule: "... it is to be expected that at last, just before the appearance of life, the very ocean had become, in Haldane's (1929, 1954) vivid phraseology, a gigantic bowl of soup," wrote Muller, and added "drop into this a nucleotide chain and it should eventually breed!" (Muller 1961). A few years later he would state that "...life as we know it, if stripped of all its superstructures, lies in the three faculties possessed by the gene material. These may be defined as, firstly, the self-specification, after its own pattern, of new material produced by it or under its guidance; secondly, of performing this operation even when it itself has undergone a great succession of permanent pattern changes which, taken in their totality, can be of a practically unlimited diversity; thirdly, of, through these changes, significantly and (for different cases) diversely affecting other materials and, therewith, its own success in genetic survival." Muller added that "the gene material alone, of all natural materials, possesses these faculties, and it is therefore legitimate to call it living material, the present-day representative of the first life" (Muller 1966). In other words, for Muller (and many others) the essence of life lies in the combination of autocatalysis, heterocatalysis, and mutability, i.e., evolvability.

Muller's proposal was brilliantly reductionist, and was soon contested by Oparin and others in a now largely forgotten debate. Although Muller (1947) had once expressed sympathy to Oparin's idea, they soon became engaged in an entangled debate in which science, philosophy, and politics mixed in an excruciating discussion that was shaped in part by the Cold War atmosphere (Lazcano 1992, 1995; Fry 2002). In sharp contrast with Muller's ideas, Oparin (1938) had argued that the essence of life was metabolic flow. For Oparin, life is "a special form of the motion of matter," always in flow, which included enzymatically based assimilation, growth, and reproduction, but not nucleic acids, whose genetic role was not even suspected during the 1930s. Biological inheritance was assumed by Oparin to be the outcome of growth and division of the coacervate drops he had suggested as models of precellular systems, a view that led Muller (1966) to state that "the Russian Oparin has since the early 1930s espoused this view and has followed the official Communist Party line by giving the specific genetic material a back seat."

Oparin and Muller came from different scientific backgrounds and almost opposite intellectual traditions. Their common interest in the origin of life did nothing to assuage their opposing views and their ideological clashes. Oparin was a convinced evolutionist, and, like many of his contemporaries, his original genetics were pre-Mendelian. Oparin's Darwinism had been nurtured by Tymiriazev, who had famously identified in 1912, many years before the Lysenko affair, Mendelians and mutationists as the opponents to be defeated in the war against anti-Darwinism (Vucinich 1988). For Muller, who remained bitterly disillusioned by Stalin's regime and Lysenko's tragic affair, life could be so well-defined that the exact point at which it started could be established with the sudden appearance of the first DNA molecule. Oparin, on the other hand, refused to admit that life could arise all at once by a spontaneous generation, and argued that it was the outcome of a slow, stepwise evolutionary developmental process.

Oparin's refusal to assume that nucleic acids had played a unique role in the origin of life resulted not only from his unwillingness to assume that life can be reduced to a single compound such as the "living DNA molecule" advocated by Muller and others, but also within the framework of Cold War politics, his complex relationship with Lysenko, and his long association with the Soviet establishment. As shown by his extensive work with RNA-containing coacervates (Oparin and Yevreinova, 1947; Oparin and Serebroskaya, 1963; Oparin et al. 1961, 1963, 1964) and his complete acceptance, based on the suggestions of Belozerskii (1959), Brachet (1959), and others, that RNA could have preceded DNA as genetic material, Oparin (1961) eventually acknowledged the role of nucleic acids in the origin of life and assumed, until the very end, that protein synthesis was the evolutionary outcome of the interaction of primordial polypeptides and polynucleotides within the boundaries of precellular systems (Oparin 1972).

Paving the Road to the RNA World

The launching of the Sputnik in 1957 signaled not only the start of space exploration but also a new epoch in the study of the origins of life, which acquired a novel perspective, as shown by the publication of *Life in the Universe*, a book by Oparin and Fesenkov (1961). The first chapter, written by Oparin, examined the conditions under which life was assumed to have originated on Earth, and the rest of the book by Vasily Fesenkov, an astronomer with considerable following in the USSR. The premise that life appeared throughout the Universe whenever the conditions for its appearance were present set the tone the book, and reflected optimistic views regarding the possibility of inhabited planets shared by many astronomers both in the Soviet Union and in Western countries. By then, the development of space programs and agencies had started to play a key role not only in transforming the issue of extraterrestrial life into a legitimate scientific question, but also to shape the study of origin and early evolution of life in new ways (Dick 1998; Wolfe 2002; Strick 2004).

If the onset of the so-called Space Age (which led to substantial funding from NASA to the origins-of-life community) set the emergence of living systems within a cosmic context, the work of Elso S. Barghoorn and his students and associates pushed the microbial fossil record back in time to the early Precambrian (Cloud 1983). Although it seems that life appeared as soon as environmental conditions permitted, identification of the oldest paleontological traces of life remains a contentious issue. Although it is not possible to assign a precise chronology to the origin and earliest evolution of cells, the recognition that life is a very ancient phenomenon runs parallel to the limits imposed by a geological record that becomes increasingly blurred as we go back in time (Schopf 1999; Knoll 2003).

Direct information is lacking not only on the composition of the terrestrial atmosphere during the period of the origin of life, but also on the temperature, ocean pH values, and other general and local environmental conditions which may or may not have been important for the emergence of living systems. However, the lack of detailed understanding of the conditions of the primitive environment did not stop prebiotic chemists from attempting the synthesis of a wide number of compounds of biochemical significance under highly reducing conditions but with few other environmental constraints. The robustness of this type of chemistry is supported by the occurrence of most of these biochemical molecules in the Murchison meteorite, reinforcing, but not proving, the idea that comparable compounds were present in the primitive Earth (Ehrenfreund et al. 2002).

From the late 1960s onward, however, it became clear that our understanding of the origin of life was troubled by two major issues: The possibility that the young Earth had been endowed with a highly reducing atmosphere was viewed with considerable skepticism by most planetary scientists, whose preference for a CO_2-rich atmosphere weakened the assumption that the primitive soup had formed by the accumulation of organic compounds synthesized under highly reducing conditions. Secondly, the emergence of nucleic acid-directed protein synthesis, which is recognized as a central feature of all extant life, appeared to be an insurmountable problem. At the time, few molecular biologists were inclined to evolutionary explanations and many, like Muller, relied on chance events to understand the basic molecular traits of cells. As the influential French biologist Jacques Monod wrote in his 1970 book Chance and Necessity, "... it might be thought that the discovery of the universal mechanisms basic to the essential properties of living beings would have helped solve the problem of life's origins. As it turns out, these discoveries, by almost entirely transforming the question, have shown it to be even more difficult than it formerly appeared" (Monod 1971). Monod's attitude had far reaching consequences; as summarized by Fry (2002), it would eventually lead the philosopher Karl Popper (1974) and his followers to argue that the emergence of life is "an impenetrable barrier to science and a residue to all attempts to reduce biology to chemistry and physics."

A possible solution to the problem posed by the lack of understanding of the relationship between nucleic acids and proteins was suggested by Carl Woese (1967), Leslie Orgel (1968), and Francis Crick (1968), who independently proposed the idea that the first living entities were based on RNA as both the genetic material and as catalyst. Surprisingly, these pioneering proposals of an RNA world received little attention. The relationship between evolutionary issues and molecular biology was slow to develop, and during several decades was embittered by frequent clashes during which evolutionary analysis was frequently dismissed as little more than useless speculation.

This skeptical attitude changed with the awareness that genes and proteins are rich historical documents from which a wealth of evolutionary information can be retrieved (Zuckerkandl and Pauling 1965). A major achievement of this approach was the use of small subunit ribosomal RNA as a phylogenetic marker, which led Carl Woese and his associates to the construction of a trifurcated, unrooted tree in which all known organisms can be grouped in one of three major cell lineages, i.e., eubacteria, archaeabacteria, and the eukaryotic nucleocytoplasm, all of which share a common ancestry (Woese and Fox 1977). The variations of traits common to extant species can be explained as the outcome of divergent processes from an ancestral life form that existed before their separation of the three major biological domains, i.e., the last common ancestor. Although no evolutionary intermediate stages or ancient simplified versions of the basic biological processes have been discovered in contemporary organisms, the differences in the structure and mechanisms of gene expression and replication among the three lineages have provided insights on the stepwise evolution of the replication and translational apparatus, including some late steps in the development of the genetic code. All of a sudden, it deemed possible to distinguish the origin of life problem from a whole series of other issues, often confused, that belong to the domain of the evolution of microbial life.

RECENT RESULTS

So Far from the Origin of Life, so Close to the RNA World

It is difficult to see how inferences based on universal phylogenies can be extended beyond a threshold that corresponds to a period of cellular evolution in which protein biosynthesis was already in operation, i.e., an RNA/protein world. Older stages are not yet amenable to molecular phylogenetic analysis (Becerra et al. 2007). A cladistic approach to the origin of life itself is not feasible, because all possible intermediates that may have once existed have long since vanished, and the temptation to do otherwise is best resisted. The most basic questions pertaining to the origin of life relate to much simpler entities predating by a long (but not necessarily slow) series of evolutionary events the oldest branches in universal phylogenetic trees.

Nevertheless, the examination of the prokaryotic branches of unrooted rRNA trees had already suggested that the ancestors of both Bacteria and Archaea were extreme thermophiles growing optimally at temperatures in the range of 90°C or above (Achenbach-Richter et al. 1987). Rooted universal phylogenies appeared to confirm this possibility, because heat-loving prokaryotes occupied short branches in the basal portion of molecular cladograms (Stetter 1994). Attempts to correlate the antiquity of hyperthermophiles with extreme environments such as those found today in deep-sea vents (Holm 1992) or in other sites in which mineral surfaces may have fueled the appearance of primordial chemoautolithotrophic life forms (Wächtershäuser 1988, 1992) became almost unavoidable.

Wächtershäuser's explicit adherence to Karl Popper's philosophical stand (Popper 1959) played a major role in his idea that life began with the appearance of an autocatalytic two-dimensional autochemolithotrophic metabolic system based on the formation of pyrite. His insightful prediction that ferrous sulfide in the presence of hydrogen sulfide (H_2S) is an efficient reducing agent should not be understated. Pyrite formation can produce molecular hydrogen, promote the formation of ammonia from nitrogen nitrogen, and can reduce a few organic molecules under mild conditions. However, compared with the surprising variety of biochemical compounds that are readily synthesized in one-pot Miller-Urey type simulations, the suite of molecules produced under the conditions suggested by Wächtershäuser is quite limited.

Based on the hypothesis that core metabolic processes have not changed since the emergence of life, Morowitz (1992) has argued that intermediary metabolism recapitulates prebiotic chemistry. He maintains that the basic traits of metabolism could only evolve after the closure of an amphiphilic bilayer membrane into a vesicle, that is, that the appearance of membranes represents the discrete transition from nonlife to life. According to his hypothesis, reverse Krebs cycle-dependent life appeared with "minimal protocells" formed by bilayer vesicles made up of small amphiphiles and endowed with pigments capable of absorbing radiant energy stored as a chemiosmotic proton gradient across the membrane.

These and other explanations of the origin of life are based on the idea that the emergence of autocatalytic "metabolic" cycles in the primitive Earth was an essential prerequisite for the appearance of genetic systems. According to this approach, life can be considered an emergent interactive system endowed with dynamic properties that exist in a state close to chaotic behavior. Some of these proposals reflect a (healthy) reaction against molecular biology reductionism, as well as the adherence to all-encompassing views based on complexity theories and self-assembly phenomena that are quite popular among physical scientists. In fact, the background of current metabolism-based explanations of the origin of life lies not in Oparin's proposals, but in the attempts to extrapolate to biology the deeply rooted tendency in physical sciences to search for all encompassing laws that can be part of grand theory which explain many, if not all, complex systems.

The many examples of self-organizing physical systems that lead to highly ordered structures show that, in addition to natural selection, other mechanisms of ordered complexity can come into play. Self-assembly is not unique to biology, and may indeed be found in a wide variety of systems, including cellular automata, the complex flow patterns of many different fluids, in cyclic chemical phenomena (such as the Belousov-Zhabotinsky reaction) and, quite significantly, in the self-assembly of amphiphilic lipid-like molecules in bilayers, micelles, and liposomes (cf. Lazcano 2009). There are indeed some common features among these systems, and it has been claimed that they follow general principles that are in fact equivalent to universal laws of nature (Kauffman 1993). Perhaps this is true. The problem is that such all-encompassing principles, if they exist at all, have so far remained undiscovered (Farmer 2005).

As discussed elsewhere, the experimental evidence that has been recently used to argue in favor of the metabolism-first theory is equally consistent with a genetic-first description of life (Lazcano 2009). What the metabolic-first approaches require is the confirmation that metabolic (or protometabolic) routes can replicate and evolve. So far, there are no indications that this is the case: As summarized by Leslie Orgel in a posthumous paper, theories that advocate the emergence of complex, self-organized biochemical cycles in the absence of genetic material are hindered not only by the lack of empirical evidence, but also by a number of unrealistic assumptions about the properties of minerals and other catalysts required to spontaneously organize such sets of autocatalytic chemical reactions (Orgel 2008).

CHALLENGES AND FUTURE DIRECTIONS

The remarkable coincidence between the monomeric constituents of living organisms and those synthesized in laboratory simulations of the prebiotic environment appears to be too striking to be fortuitous. Although we are far from understanding how life appeared, the available experimental evidence strongly suggest that the prebiotic environment was already endowed with a wide range of monomers of biochemical significance, many organic and inorganic catalysts, purines and pyrimidines, i.e., the potential for template-directed polymerization reaction, and membrane-forming compounds. Nevertheless, at present the hiatus between the primitive soup and the RNA world is discouragingly enormous.

There are many definitions of the RNA World. However, the discovery of ribozymes does not imply that wriggling autocatalytic nucleic acid molecules ready to be used as primordial genes were floating in the primitive oceans, or that the RNA world emerged completely assembled from simple precursors present in the prebiotic broth. Although it is true that genetic-first proposals do not require enclosure within compartments, the emergence of life may be best understood in terms of the dynamics and evolution of sets of chemical replicating entities. Whether such entities were enclosed within membranes is not yet clear, but given the prebiotic availability of amphiphilic compounds this may have well been the case (Deamer 2002). Indeed, the evidence supporting the presence of lipidic molecules in the prebiotic environment and their natural ability to self-organize into vesicular compartments underlines the significance of theoretical models of simple cells involving an evolving ribozymic RNA polymerase (Szostak et al. 2001; Deamer and Dworkin 2005) and increasingly sophisticated laboratory models of precellular systems (Mansy et al. 2008).

For obvious methodological reasons, experimental simulations of prebiotic events have concentrated on the empirical analysis of single variables. The study of more specific conditions, including the laboratory simulation of localized environments such as volcanic islands, tidal zones and microenviroments, including liposomes, clays and mineral surfaces, and volcanic ponds, which could have been prevalent in the primitive environment, are likely to yield promising results. It is reasonable to assume that the association and interplay of different biochemical monomers and oligomers in more complex experimental settings would lead to physicochemical properties not shown by their isolated components (Deamer et al. 2002). This is not purely speculative; that interactions between liposomes and different water-soluble polypeptides lead to major changes in the morphology and permeability of liposomes of phosphatidyl-L-serine, and to a transition of poly-L-lysine from a random coil into an α-helix that shows hydrophobic bonding with the lipidic phase, has been documented in the laboratory (Hammes and Schullery 1970).

Additional examples include experimental models of compartmentalized catalytic RNA (Mansy et al. 2008), which, although they do not necessarily correspond to particular stages in the origin of life, nonetheless illustrate how individual components of a system dynamically interact and lead to unexpected new properties. This approach, which falls within the venerable tradition of synthetic biology (Peretó and Catalá 2007), complements the attempts to work backward to reduce extant cell genomes and achieve the laboratory synthesis of minimal life forms.

There are inevitable gaps in the story, but reports on the death of the heterotrophic theory have been greatly exaggerated. The remarkable coincidence between the surprising variety of biochemical constituents that can be readily synthesized in experiments simulating the prebiotic environment and those found in some carbon-rich meteorites appears to be too striking to be fortuitous. The Earth's primitive atmosphere may have not been as strongly reducing as assumed by the early proponents of the prebiotic broth, but there is experimental evidence showing that amino acids can be synthesized in a CO_2-rich model atmosphere (Cleaves et al. 2008). It is true that the classical recipe for cooking a primitive soup needs to be updated to acknowledge, in an eclectic fashion, the contribution of extraterrestrial organic compounds, the role of catalytic minerals like pyrite, and the synthesis of organic molecules in hydrothermal vents, however limited it may have been, but this poses no threat to the idea of chemical evolution as a prerequisite to an heterotrophic origin of life.

We will never know how life first appeared. However, the study of the appearance of life is a mature, well-established field of scientific inquiry. As in other areas of evolutionary biology, answers to questions on the origin and nature of the first life forms can only be regarded as inquiring and explanatory rather than definitive and conclusive. This does not imply that all origin-of-life theories and explanations can be dismissed as pure speculation, but rather that the issue should be addressed conjecturally, in an attempt to construct not a mere chronology but a coherent historical narrative by weaving together a large number of miscellaneous observational findings and experimental results. It is probably useful to remember the line from Goethe's Faust that Oparin included in his 1924 book, "My worthy friend, gray is all theory, and green alone is life's golden tree."

ACKNOWLEDGMENTS

This article was completed during a sabbatical leave in which I enjoyed the hospitality of Professor Jeffrey L. Bada and his associates at the Scripps Institution of Oceanography, University of California San Diego. Support from a UC Mexus-CONACYT Fellowship is gratefully acknowledged.

REFERENCES

Achenbach-Richter L, Gupta R, Stetter KO, Woese CR. 1987. Were the original eubacteria thermophiles? *System Appl Microbiol* **9:** 34–39.

Bada JL, Lazcano A. 2003. Prebiotic soup: Revisiting the Miller experiment. *Science* **300:** 745–746.

Baly ECC. 1924. Photosynthesis. *Industrial Eng Chem* **16:** 1016–1018.

Baly ECC, Davies JB, Johnson MR, Shanassy H. 1927. The photosynthesis of naturally occurring compounds. 1. The action of ultraviolet light on carbonic acid. *Proc Roy Soc London A* **116:** 197–208.

Becerra A, Delaye L, Islas A, Lazcano A. 2007. Very early stages of biological evolution related to the nature of the last common ancestor of the three major cell domains. *Annu Rev Ecol Evol Syst* **38:** 361–379.

Belozerskii AN. 1959. On the species specificity of the nucleic acids of bacteria. In A.I. Oparin, A.G. Pasynskii, A.E. Braunshtein, T.E. Pavloskaya (Eds.), *The Origin of Life on Herat* Pergamon Press, New York, pp. 322–321.

Brachet J. 1959. Les acides nucléiques et l'origine des protéines. In A.I. Oparin, A.G. Pasynskii, A.E. Braunshtein,

T.E. Pavloskaya (Eds.), *The Origin of Life on Earth* New York: Pergamon Press. pp. 361–367.

Carlson EA. 1981. *Genes, radiation, and society: The life and work of H J Muller* Ithaca: Cornell University Press.

Cleaves JH, Chalmers JH, Lazcano A, Miller SL, Bada JL. 2008. Prebiotic organic synthesis in neutral planetary atmospheres. *Origins Life Evol Biosph* **38**: 105–155.

Cloud P. 1983. Early biogeologic history: The emergence of a paradigm. In J.W. Schopf (ed), *Earth's earliest biosphere: Its origin and evolution* Princeton: Princeton University Press pp. 14–31.

Crick FHC. 1968. The origin of the genetic code. *J Mol Biol* **39**: 367–380.

Deamer DW. 1977. (ed.) *Light transducing membranes* New York: Academic Press.

Deamer DW, Dworkin JP. 2005. Chemistry and physics of primtive membranes. In P. Walde (ed.), *Prebiotic chemistry: From simple amphiphiles to protocell models* pp. 1–27. Berlin: Springer.

Deamer DW, Dworkin JP, Sanford SA, Bernstein MP, Allamandola LJ. 2002. The first cell membranes. *Astrobiology* **2**: 371–382.

Dick SJ. 1998. *Life on other worlds* Cambridge: Cambridge University Press.

D'Herelle F. 1926. *The Bacteriophage and its behaviour* Baltimore: William and Wilkins.

Ehrenfreund P, Irvine W, Becker L, Blank J, Brucato J, Colangeli L, Derenne S, Despois D, Dutrey A, Fraaije H, Lazcano A, Owen T, Robert F. 2002. Astrophysical and astrochemical insights into the origin of life. *Reports Prog Phys* 65: 1427–1487.

Farley J. 1977. *The spontaneous generation controversy from Descartes to Oparin* Baltimore and London: Johns Hopkins University Press.

Farmer DJ. 2005. Cool is not enough. *Nature* **436**: 627–628.

Fox SW, Dose K. 1977. *Molecular evolution and the origin of life* New York: Marcel Dekker Inc.

Fry I. 2002. *The emergence of life on earth* New Brunswick: Rutgers University Press.

Fry I. 2006. The origins of research into the origins of life. *Endeavour* **30**: 24–28.

Garrison WM, Morrison DC, Hamilton JG, Benson A, Calvin M. 1951. Reduction of carbon dioxide in aqueous solutions by ionizing radiation. *Science* **114**: 416.

Graham LR. 1972. *Science and philosophy in the Soviet Union* New York :Alfred A. Knopf.

Haeckel E. 1862. *Die Radiolarien (Rhizopoda Radiaria)* Eine Monographie Berlin: Druck und Verlag Von Georg Reimer.

Haldane JBS. 1929. The origin of life. *Rationalist Annual* **148**: 3–10.

Haldane JBS. 1954. The origins of life. *New Biol* **16**: 12–27.

Hammes GG, Schullery SE. 1970. Structure of molecular aggregates. II. Construction of model membranes from phospholipids and polypeptides. *Biochemistry* **9**: 2555–2558.

Herrera AL. 1902. Note sur l'imitation du protoplasme. *Bull Soc Zoo France* **26**: 144.

Herrera AL. 1942. A new theory on the origin and nature of life. *Science* **96**: 14.

Holm N.G., ed. 1992. *Marine hydrothermal systems and the origin of life* Dordrecht: Klüwer Academic Publ.

Johnson AP, Cleaves HJ, Dworkin JP, Glavin DP, Lazcano A, Bada JL. 2008. The Miller volcanic spark discharge experiment. *Science* **322**: 404.

Kauffman SA. 1993. *The Origins of Order: Self organization and selection in evolution* New York: Oxford University Press.

Kamminga H. 1982. Life from space—a history of panspermia. *Vistas in Astronomy* **26**: 67–86

Knoll AH. 2003. *Life on a young planet: The first three billion years of evolution on Earth* Princeton: Princeton University Press.

Lazcano A. 1992. *La Chispa de la Vida: Alexander I. Oparin* Editorial Pangea/CONACULTA, México, D.F. (in Spanish).

Lazcano A. 1995. Aleksandr I. Oparin, the man and his theory. in B.F. Poglazov, B.I. Kurganov, M.S. Kritsky (eds), *Frontiers in Physicochemical Biology and Biochemical Evolution* Bach Institute of Biochemistry and ANKO, Moscow, p. 49–56.

Lazcano A. 2009. Which way to life? Origins Life *Evol Biosph* (in press).

Leicester HM. 1974. *Development of Biochemical Concepts from Ancient to Modern Times* Cambridge: Harvard University Press.

Mansy SS, Schrum JP, Krishnamurthy M, Tobé S, Treco DA, Szostak JJW. 2008. Template-directed synthesis of a genetic polymer in a model protocell. *Nature* **454**: 122–125.

Miller SL. 1953. A production of amino acids under possible primitive Earth conditions. *Science* **117**: 528.

Miller SL. 1955. Production of some organic compounds under possible primitive Earth conditions. *J Am Chem Soc* **77**: 2351–2361.

Miller SL. 1974. The first laboratory synthesis of organic compounds under primitive conditions. In: J. Neyman (ed) *The Heritage of Copernicus: Theories "more pleasing to the mind"* pp 228–24. Cambridge: MIT Press.

Miller SL, Schopf JW, Lazcano A. 1997. Oparin's "Origin of Life": Sixty years later. *J Mol Evol* **44**: 351–353.

Morowitz HJ. 1992. *Beginnings of Cellular Life: Metabolism Recapitulates Biogenesis* Binghamton: Yale University Press.

Monod J. 1971. *Chance and Necessity* Glasgow: Fontana Books.

Muller HJ. 1922. Variation due to change in the individual gene. *The American Naturalist*, **56**: 32–50.

Muller HJ. 1926. The gene as the basis of life. *Proceedings of the 1st International Congress of Plant Sciences*, Ithaca, p. 897–921.

Muller HJ. 1947. The gene. *Proc Royal Soc* **B134**: 1–37.

Muller HJ. 1961. Genetic nucleic acid: Key material in the origin of life. *Persp Biol Med* **5**: 1–23.

Muller HJ. 1966. The gene material as the initiator and the organizing basis of life. *Am Naturalist* **100**: 493–502.

Oparin AI. 1924. *Proiskhozhedenie Zhizni* Mosckovskii Rabochii, Moscow,. Reprinted and translated in J. D. Bernal (1967) The Origin of Life London: Weidenfeld and Nicolson.

Oparin AI. 1938. *The Origin of Life* New York: McMillan.

Oparin AI. 1961. *Life: Its nature, origin, and development* New York: Academic Press.

Oparin AI. 1972. The appearance of life in the Universe. In C. Ponnamperuma (Ed.), *Exobiology* pp 1–15 Ámsterdam: North-Holland.

Oparin AI, Fesenkov V. 1961. Life in the Universe. New York: Twayne Publishers.

Oparin AI, Serebroskaya KB. 1963. Formation of coacervates drops during the synthesis of polyadenylic acid by polynucleotide phosphorylase. *Dokl Akad Nauk SSSR* **148**: 943–947 (in Russian).

Oparin AI, Yevreinova TN. 1947. The effect of nucleic acid on thermostability of proteins. *Dokl Akad Nauk SSSR* **58**: 253–259 (in Russian).

Oparin AI, Serebroskaya KB, Auerman TL. 1961. Synthesizing effect of polynucleotide phosphorylase of *Micrococcus lyzodeikticus* in solution and in coacervate systems. *Dokl Akad Nauk SSSR* **126**: 499–504 (in Russian).

Oparin AI, Serebroskaya KB, Pantskhava SN, Vasil'yeva NV. 1963. Enzymatic synthesis of polyadenylic acid in coacervate drops. *Biokhimiya* **28**: 671–673 (in Russian).

Oparin AI, Serebroskaya KB, Vasil'yeva NV, Balayevskaya TO. 1964. The formation of coacervates from polypeptides and polynucleotides. *Dokl Akad Nauk SSSR* **154**: 407–412 (in Russian).

Orgel LE. 1968. Evolution of the genetic apparatus. *J Mol Biol* **38**: 381–392.

Orgel LE. 2008. The implausibility of metabolic cycles in the primitive Earth. *PLoS Biol* **6**: e18.

Oró J. 1960. Synthesis of adenine from ammonium cyanide. *Biochem Biophys Res Comm* **2**: 407–412.

Oró J. 1961. Comets and the formation of biochemical compounds on the primitive earth. *Nature* **190**, 442–443.

Peretó J, Catalá J. 2007. The renaissance of synthetic biology. *Biological Theory* **2**: 128–130.

Peretó J, Bada JE, Lazcano A. 2009. Charles Darwin and the origin of life. *Origins Life Evol Biosph* **39**: 395–406.

Perezgasga L, Silva E, Lazcano A, Negrón-Mendoza A. 2003. Herrera's sulfocyanic theory on the origin of life: A critical reappraisal. *Int Jour Astrobiol* **2**: 1–6.

Pontecorvo G. 1982. Who was H. J. Muller (1890–1967)? *Nature* **298**: 203–204.

Popper K. 1959. *The logic of scientific discovery*. New York: Basic Books.

Popper K. 1974. Reduction and the incompleteness of science. In F. Ayala, T. Dobzhanzky (eds), *Studies in the Philosophy of Biology* University of California Press, Berkeley, pp 251–267.

Ravin AW. 1977. The gene as catalysts; the gene as organism. W. Coleman, C. Limoges (eds) *Studies in the History of Biology* 1: 1–45.

Schopf JW. 1999. The Cradle of Life: The discovery of Earth's earliest fossils. Princeton: Princeton University Press.

Stetter KO. 1994. The lesson of archaebacteria. In S. Bengtson (ed), *Early Life on Earth: Nobel Symposium No 84* Columbia University Press, New York 114–122.

Strick JE. 2004. Creating a cosmic discipline: The crystallization and consolidation of exobiology, 1957–1973. *Jour History Biol* **37**: 131–180.

Strick JE. 2009. Darwin and the origin of life: Public versus private science. *Endeavour* (in press).

Summers WC. 1999. *Felix d'Herelle and the origins of molecular biology* New Haven: Yale University Press.

Szostak JW, Bartel DP, Luisi PL. 2001. Synthesizing life. *Nature* **409**:387–390.

Troland LT. 1914. The chemical origin and regulation of life. *The Monist* **22**: 92–133.

Troland LT. 1916. The enzyme theory of life. *Cleveland Me Jour* **15**: 377–385.

Troland LT. 1917. Biological enigmas and the theory of enzyme action. *Am Naturalist* **51**: 321–350.

Urey HC. 1951. On the early chemical history of the Earth and the origin of life. *Proc Natl Acad Sci USA* **38**: 351–363.

Urey HC. 1952. *The planets: their origin and development* Chicago: University of Chicago Press.

Vucinich A. 1988. *Darwin in Russian Thought* Berkeley: University of California Press.

Wächtershäuser G. 1988. Before enzymes and templates, a theory of surface metabolism. *Microbiol Rev* **52**: 452–84.

Wächtershäuser G. 1992. Groundwork for an evolutionary biochemistry: The iron-sulphur world. *Prog Biophys Molec Biol* **58**: 85–201.

Watson JD, Crick FHC. 1953. Molecular structure of nucleic acids. *Nature* **171**: 737.

Woese CR. 1967. The Genetic Code: the molecular basis for gene expression. New York: Harper and Row.

Woese CR, Fox GE. 1977. The concept of cellular evolution. *J Mol Evol* **10**:1–6.

Wolfe A. 2002. Germs in space: American life scientists, space policy, and the public imagination, 1958–1963. *Isis* 93: 183–205.

Zuckerkandl E, Pauling L. 1965. Molecules as documents of evolutionary history. *J Theor Biol* **8**: 357–366.

Cosmic Carbon Chemistry: From the Interstellar Medium to the Early Earth

Pascale Ehrenfreund[1,2] and Jan Cami[3,4]

[1]Leiden Institute of Chemistry, 2300 RA Leiden, The Netherlands
[2]Space Policy Institute, Washington DC
[3]Department of Physics and Astronomy, UWO, London, ON, Canada
[4]SETI Institute, Mountain View, California 94043

Correspondence: pascale@strw.leidenuniv.nl

Astronomical observations have shown that carbonaceous compounds in the gas and solid state, refractory and icy are ubiquitous in our and distant galaxies. Interstellar molecular clouds and circumstellar envelopes are factories of complex molecular synthesis. A surprisingly large number of molecules that are used in contemporary biochemistry on Earth are found in the interstellar medium, planetary atmospheres and surfaces, comets, asteroids and meteorites, and interplanetary dust particles. In this article we review the current knowledge of abundant organic material in different space environments and investigate the connection between presolar and solar system material, based on observations of interstellar dust and gas, cometary volatiles, simulation experiments, and the analysis of extraterrestrial matter. Current challenges in astrochemistry are discussed and future research directions are proposed.

Carbon is a key element in the evolution of prebiotic material (Henning and Salama 1998), and becomes biologically interesting in compounds with nitrogen, oxygen and hydrogen. Our understanding of the evolution of organic molecules—including such compounds—and their voyage from molecular clouds to the early solar system and Earth provides important constraints on the emergence of life on Earth and possibly elsewhere (Ehrenfreund and Charnley 2000). Figure 1 shows the cycle of organic molecules in the universe. Gas and solid-state chemical reactions form a variety of organic molecules in circumstellar and interstellar environments. During the formation of the solar system, this interstellar organic material was chemically processed and later integrated in the presolar nebula from which planets and small solar system bodies formed. The remnant planetesimals in the form of comets and asteroids impacted the young planets in the early history of the solar system (Gomes et al. 2005). The large quantities of extraterrestrial material delivered to young planetary surfaces during the heavy bombardment phase may have played a key role in life's origin (Chyba and Sagan 1992, Ehrenfreund et al. 2002). How elements are formed, how

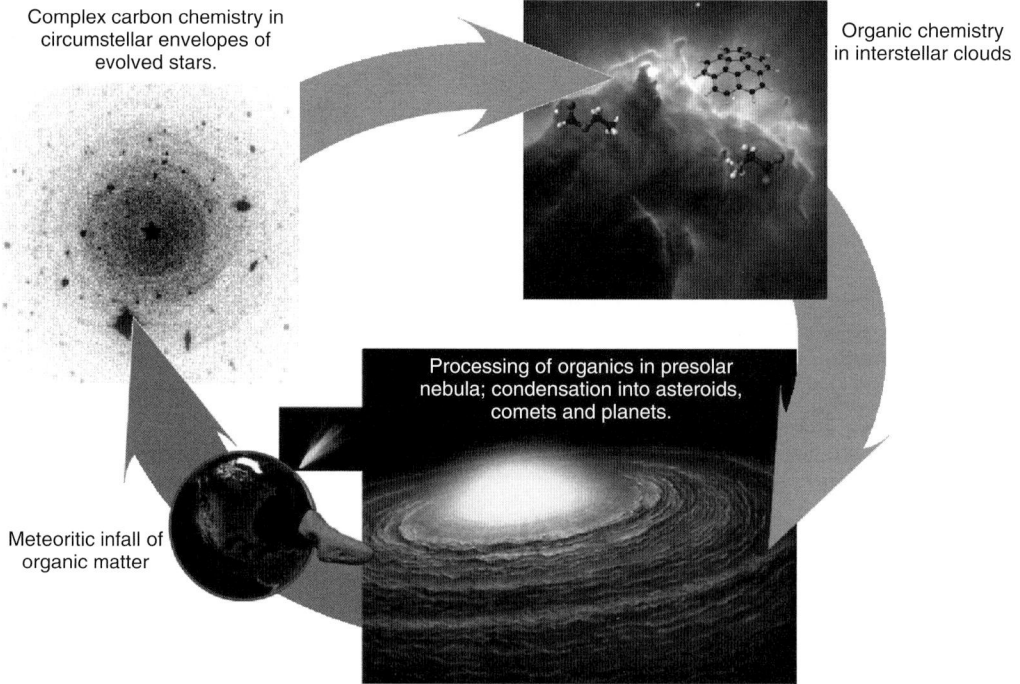

Figure 1. Carbon pathways between interstellar and circumstellar regions and the forming solar system.

complex carbonaceous molecules in space are, what their abundance is and on what timescales they form are crucial questions within cosmochemistry.

Inventory of Cosmic Carbon

Carbon is found in space in all its allotropic forms: diamond, graphite, and fullerene (Cataldo et al. 2004). Astronomical observations in the last decade have shown that carbonaceous compounds (gaseous molecules and solids) are ubiquitous in our own as well as in distant galaxies (Ehrenfreund et al. 2006a). The first chemical enrichment of the universe may likely be connected to the first generation of stars (Spaans 2004). Large carbon abundances are already extrapolated from observations of the strong C[II] and CO lines in the hosts of the most distant quasars (Bertoldi et al. 2003).

Carbon in space was first produced in stellar interiors in fusion reactions and was later ejected into interstellar and intergalactic space during stellar collapse and supernova explosions. In the denser regions of interstellar space, the so-called interstellar clouds, active chemical pathways form simple and complex carbon molecules from carbon atoms (van Dishoeck and Blake 1998). Circumstellar envelopes are regarded as the largest factories of carbon chemistry in space (Kwok 2004, 2009).

Organic molecules in the solar system are found in planetary atmospheres and on the surface of many outer solar system moons (e.g., Cruikshank et al. 2005; Raulin 2008; Lorenz et al. 2008). More than 50 molecules have been identified in cometary comae (Crovisier et al. 2009). Many small organic molecules observed in cometary comae probably originate wholly or partially from the decomposition of larger molecules or particles, indicating that large polymers such as polyoxymethylene and HCN-polymers are present in comets (Ehrenfreund et al. 2004; Cottin and Fray 2008). Carbonaceous chondrites (meteorites) and micrometeorites do contain a variety of organics (e.g., see

Alexander et al. 2007; Sephton 2002; Sephton and Botta 2005 for reviews). They are fragments of cometary and asteroidal bodies. Investigating their organic composition often indicates the nature of the parental body (Hiroi et al. 1993; Ehrenfreund et al. 2001; Nesvorny et al. 2009).

Cosmic Cycling of Organics

The interstellar medium, the space between the stars, is composed primarily of H and He and constitutes a few percent of the galactic mass. Interstellar material is dominated by interstellar gas (99%). The remaining 1% is composed of solid silicate and carbon-based μm-sized dust particles present throughout interstellar clouds that provide surfaces for accretion of gas phase species and subsequent grain surface chemistry (Ehrenfreund and Charnley 2000; Ehrenfreund and Fraser 2003). Fundamental physical parameters such as temperature and density vary strongly across the spectrum of interstellar regions. The diversity of interstellar clouds results from energy injected by supernova shockwaves and stellar outflows and radiative losses (Wolfire et al. 2003). The most recent classification by Wooden et al. (2004) describes in detail very low-density hot gas, environments with warm intercloud gas, and regions with denser and colder material, see Table 1. Cosmic abundances in the interstellar medium are derived by measuring elemental abundances in stellar atmospheres. These cosmic elemental abundances determine the amount of elements available for the formation of molecules and particles. Gas phase and solid-state reactions and gas-grain interactions lead to the formation of complex molecules. H_2 is by far the most abundant molecule in cold interstellar regions, followed by CO, the most abundant carbon-containing species, with $CO/H_2 \sim 10^{-4}$.

In cold dark clouds with a temperature of 3–10 K the sticking coefficient of most atoms and molecules is close to unity and all species (except H_2 and He) freeze out (Ehrenfreund and Charnley 2000). At 10 K only H, D, C, O, and N atoms have sufficient mobility to interact on the surfaces of grains. Dark clouds offer a protected environment for the formation of larger molecules. Those regions have a rather high density ($\sim 10^6/cm^3$) and experience a low radiation field of $\sim 10^3$ photons cm^2/s induced by cosmic rays (Prasad and Tarafdar 1983).

The diffuse interstellar medium is characterized by a low density ($\sim 10^3$ atoms/cm^3) and temperatures ≤ 100 K. Diffuse clouds are filamentary structures that surround cold dense interstellar regions. Ices are not present in those regions and a strong radiation field of $\sim 10^8$ photons/cm^2/s (Mathis et al. 1983) dominates the formation and evolution of molecules and larger structures. Small carbonaceous molecules in the gas phase are easily destroyed by radiation. Atoms with ionization potentials less than 13.6 eV are photoionized. The identification of many small molecules in dense clouds implies that their destruction is well balanced by active formation routes. Ion-molecule reactions, dissociative recombination with electrons, radiative association reactions and neutral-neutral reactions contribute to gas phase processes and influence interstellar chemistry (Snow and McCall 2006).

Table 1. Phases of the interstellar medium (adapted from Wooden et al. 2004).

ISM component	Designation	Temperature (K)	Density [cm^{-3}]
Hot ionized medium	coronal gas	10^6	0.003
Warm ionized medium	diffuse ionized gas	10^4	>10
Warm neutral medium	intercloud HI	10^4	0.1
Atomic cold neutral medium	diffuse clouds	100	10–100
Molecular cold neutral medium	dark clouds, molecular clouds, dense clouds	<50	10^3–10^5
Molecular hot cores	protostellar cores	100–300	>10^6

Adapted from Wooden et al. 2004.

"Stardust," in the form of dust and molecules, is injected by stellar sources in their late stage of evolution into interstellar clouds. Whereas the low temperature dust in dense interstellar clouds is covered by ice that experiences low UV radiation flux, dust in diffuse clouds is strongly processed by UV radiation and shocks. The distinct and surprising differences in the dust component of dense and diffuse interstellar clouds do not suggest a rapid cycling of cloud material (Chiar and Pendleton 2008). Previous ideas of periodical cycling of dust are currently revisited (e.g., Chiar et al. 2007). Understanding interstellar cloud evolution and dust cycling provides important insights into the nature of the material that is later incorporated into proto-planetary material (see *Solar System Formation* below).

The Conditions on the Young Terrestrial Planets

Planet Earth was formed through a hot accretion process that allowed only the rocky material from the inner solar system to survive. Ices were sublimed and existing carbon material (volatile and refractory) pyrolized. Therefore, organic molecules found on terrestrial planets must have formed after the planetary surface cooled, or were delivered via impacts by small bodies. All terrestrial planets have been seeded with organic compounds through the impact of small bodies during solar system formation. Part of this material may have been important starting material for life (Chyba et al. 1990; Ehrenfreund et al. 2002). At present, very little data are available regarding the atmospheric, oceanic, or geological conditions on the prebiological Earth. It is assumed, however, that conditions on the young Earth were very hostile due to volcanism, radiation, and bombardment by comets and asteroids. Primitive life, in the form of bacteria, emerged approximately 3.5 billion years ago (Derenne et al. 2008).

Earth provides an ideal environment for life to persist: Dynamic processes in the Earth's interior have established a magnetosphere that protects life from harmful cosmic ray particles impinging the Earth's atmosphere; in turn, the atmosphere shields life from radiation and allows for a stable climate and temperature cycle. A brief look at our planetary neighbors shows that Venus, with an average surface temperature of 500°C and Mars, with an average surface temperature of -60°C and a thin atmosphere, are both apparently unable to sustain life at their surfaces (Lammer et al. 2009). Oxidizing compounds are held responsible for the degradation of organics in the Martian soil (Quinn et al. 2005).

BACKGROUND

Formation of Carbon Compounds in Space Environments

Interstellar molecules can be identified through their electronic transition in the UV and optical part of the spectrum or in the infrared range through vibrational transitions. Molecules with a dipole moment display rotational lines that can be observed at radio wavelength. At the beginning of the 20th century, astronomers were skeptical about the presence of molecular species in the interstellar medium. However, the presence of simple molecules such as CN, CH, and CH^+ was confirmed in the 1940s by optical absorption spectroscopy. Telescopes capable of observing rotational molecular emissions at millimeter wavelengths in the 1970s allowed the discovery of many molecules.[1]

Carbon chemistry occurs efficiently in diffuse interstellar clouds. Simple molecules such as CO, CH, CN, OH, C_2, C_3, and others can be observed throughout the electromagnetic spectrum (Liszt and Lucas 2000). Cosmic dust models indicate that the majority of carbon in diffuse clouds (up to 80%) is incorporated into carbonaceous grains and gaseous polycyclic aromatic hydrocarbon molecules (PAHs) (Pendleton and Allamandola 2002; Snow and McCall 2006).

Amorphous carbon, hydrogenated amorphous carbon, diamond, refractory organics, and carbonaceous networks such as coal, soot, graphite, quenched-carbonaceous condensates, and others have been proposed as possible solid

---

[1] http://www.astrochymist.org/astrochymist_ism.html

carbon components of interstellar clouds (see Henning and Salama, 1998, Ehrenfreund and Charnley, 2000, Pendleton and Allamandola 2002, Tielens 2008). Recent spectroscopic evidence indicates that carbonaceous grains are predominantly made of amorphous carbon (Mennella et al. 1998; Henning and Mutschke 2004).

PAHs are observed widely distributed in galactic and extragalactic regions (Genzel et al. 1998; Peeters et al. 2004a,b; Smith et al. 2007; Tielens 2008). Their abundance is estimated to be 5×10^{-7} (with respect to H). Laboratory studies and theoretical calculations have provided important insights into their size and charge state distribution (e.g., Salama 1999; Allamandola et al. 1999; Ruiterkamp et al. 2005a; Bauschlicher et al. 2008, 2009).

Observations at infrared, radio, millimeter, and submillimeter frequencies show that a large variety of organic molecules are present in the dense interstellar gas (www.astrochemistry.net lists more than 150 molecules, Charnley et al. 2003). These include organics such as nitriles, aldehydes, alcohols, acids, ethers, ketones, amines, and amides, as well as long-chain hydrocarbons. Infrared observations of icy dust particles in the last decade with the Infrared Space Observatory, the Spitzer Space telescope and complementary ground based observations revealed a large variety of organic compounds. The nature of specific ice mixtures and the abundance of the individual compounds in grain mantles have been measured with unprecedented accuracy (Gibb et al. 2004; van Dishoeck 2004). In dense clouds, atoms accreted on dust particles can enter reaction pathways such as exothermic hydrogenation reactions, which result in the formation of the simplest mantle molecules (water, ammonia, methane etc.). The main species observed in interstellar ice mantles are H_2O, CO_2, CO, and CH_3OH, with smaller admixtures of CH_4, NH_3, H_2CO, and $HCOOH$ (Gibb et al. 2004).

The circumstellar envelopes of carbon-rich stars are the heart of the most complex carbon chemistry that is analogous to soot formation in candle flames or industrial smoke stacks (Henning and Mutschke 2004). Laboratory simulations of gas-phase condensation reactions such as laser pyrolysis and laser ablation showed that the temperature in the condensation zone determines the formation pathway of carbonaceous particles. At temperatures lower than 1700 K, the condensation by-products are mainly PAHs with three to five aromatic rings (Jäger et al. 2008). At condensation temperatures higher than 3500 K, fullerene-like carbon grains and fullerene compounds are formed. Molecular synthesis may occur in the circumstellar environment on timescales as short as several hundred years (Kwok 2004, 2009).

Solar System Formation

Interstellar matter provides the raw material for the formation of stars and planets. Approximately 4.6 billion years ago, the gravitational collapse of an interstellar cloud led to the formation of a protosolar disk (the solar nebula) with a central condensation developing into our Sun. Interstellar dust clumped together to form small particles in the solar nebula that grew bigger, accreted more and more material, eventually forming planets and small bodies such as comets and asteroids (Boss 2004; Blum 2004).

Volatile and robust carbon compounds residing in interstellar clouds were recycled during solar system formation (Ciesla 2008). The dynamic environment of the solar nebula with the simultaneous presence of gas, particles, and energetic processes, including shock waves, lightning, and radiation can trigger a rich organic chemistry leading to organic molecules (Chick and Cassen 1997, Gorti et al. 2009). Turbulent motion led to radial mixing of the products within the disk (Markwick and Charnley 2004; Visser et al. 2007; Dullemond et al. 2008).

The carbonaceous inventory of our solar system therefore contains highly processed material that was exposed to high temperatures and radiation, newly formed compounds, and some relatively pristine material with significant interstellar heritage. Organic compounds observed or sampled from our solar system, such as planetary surfaces/atmospheres, comets, and interplanetary dust, thus hold clues to processes that occurred during the origin of our solar system (Ehrenfreund and Charnley 2000; Cruikshank et al. 2005; Septhon and Botta 2005; Raulin 2008).

Extraterrestrial Delivery

The large quantities of extraterrestrial material delivered to young terrestrial planetary surfaces in the early history of our solar system may have provided the material necessary for the emergence of life (Chyba et al. 1992; Ehrenfreund et al. 2002). Comets are predominantly icy bodies containing some silicates and refractory organic material formed in the region beyond Jupiter (Greenberg 1998; Di Santi and Mumma 2009). The NASA Stardust comet sample-return mission captured cometary dust intact at a velocity of 6.1 km/s at a distance of about 300 km from the nucleus of comet Wild-2. Most of the >5 μm solid particles collected by the mission are mineral grains or assemblages of high-temperature minerals that condense at 1400 K or above. The data provided evidence for radial transport of large solid grains from the center of the solar nebula to the Kuiper belt (Brownlee et al. 2006). Comets probably contributed most of the carbonaceous compounds during the heavy bombardment phase 4.5–4 billion years ago (Ehrenfreund et al. 2002). Fragments of asteroids and comets such as interplanetary dust particles (IDPs) and carbonaceous meteorites were probably among the other major extraterrestrial contributors of carbon (Chyba et al. 1990).

Carbonaceous meteorites contain a substantial amount of carbon (up to 5% by weight) most of which is in organic compounds (Mullie and Reisse 1987); inorganic carbon is present as diamond, graphite and carbonate materials. They exhibit evidence of thermal and aqueous alteration believed to have occurred on their parent bodies (Sephton 2002; Botta and Sephton 2005; Martins et al. 2007). In the soluble fraction of the Murchison meteorite, more than 70 extraterrestrial amino acids have been identified in addition to many other organic compounds, including N-heterocycles, carboxylic acids, sulfonic and phosphonic acids, and aliphatic and aromatic hydrocarbons (Cronin et al. 1993; Martins et al. 2007, 2008). However, the major carbon component in meteorite samples is composed of a macromolecular organic fraction (Alexander et al. 2007). Although several classes of organic compounds important in contemporary biochemistry are found in the interstellar medium, comets and meteorites, the dominant form of carbonaceous material that will be delivered is likely aromatic in nature.

RECENT RESULTS OF SELECTED ABUNDANT ORGANICS IN SPACE FROM OBSERVATIONS, LABORATORY STUDIES AND MODELLING

Gas Phase Molecules

Polycyclic aromatic hydrocarbons (PAHs) are composed of aromatic rings and are characterized by a high stability against radiation, in particular when in pericondensed form. PAHs are observed widely distributed in galactic and extragalactic regions and have been identified by their characteristic emission features in the near and mid infrared at 3.3, 6.2, 7.7, 8.6, 11.2, and 12.7 μm, representative of vibrational modes of C–C and C–H bonds, see Figure 2. They are the most abundant organic molecules in the gas phase apart from CO. Figure 3 shows examples of several PAH structures. Their reasonably large surface area can facilitate molecular synthesis. In the diffuse interstellar medium PAHs are supposed to be present in ionized form (as cations). No PAH molecule has been unambiguously identified yet. However, the infrared emission bands observed ubiquitously in the interstellar medium may represent families of PAH molecules (Peeters et al. 2002) with a range of size distribution (Bauschlicher et al. 2009), a number of different structures (Hony et al. 2001) and ionization states (Bauschlicher et al. 2009). Variations in the relative strength of infrared emission bands and variations in the peak position and profiles have contributed many suggestions of possible PAHs species as well as large species that may include heteroatoms and form clusters up to nm sized small carbonaceous grains (Peeters et al. 2002; Hudgins et al. 2005; Bauschlicher et al. 2008, 2009;

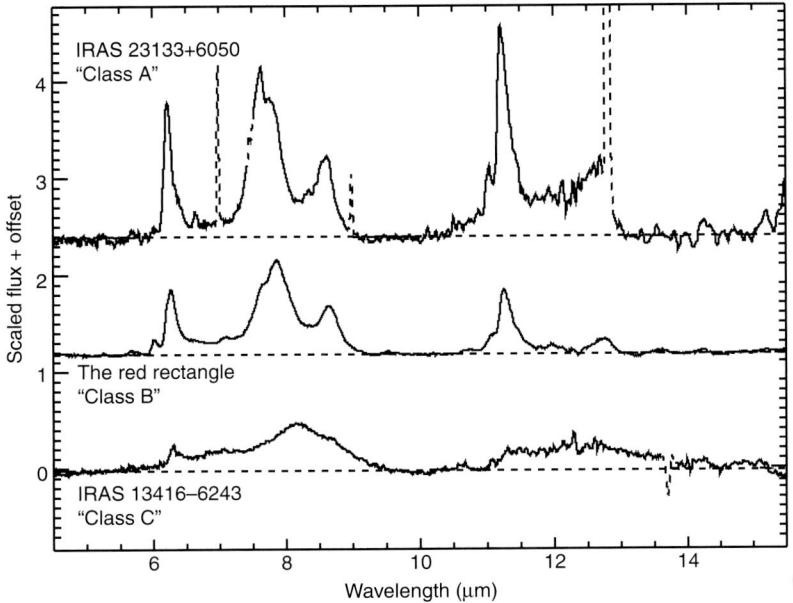

Figure 2. Astronomical spectra of the infrared vibrational modes (at 3.3, 6.2, 7.7, 8.6, 11.2, and 12.7 μm) of PAH molecules in three objects that represent the different peak positions and relative intensities observed in various galactic and extragalactic regions (see Peeters et al. 2002). Classes A, B and C represent the three different astronomical environments; (*A*) ISM, reflection nebulae, HII regions; (*B*) a few post-AGB and Herbig Ae/Be stars and most Planetary Nebulae; (*C*) a few peculiar post-AGB stars.

Tielens 2008). Only small species up to 50 carbon atoms can be measured in the laboratory. The spectrum of larger PAHs can only be derived from theoretical calculations, such as density functional theory.

Refractory Compounds

Carbon solids are ubiquitous material in interstellar space. However, the formation pathway of carbonaceous matter in astrophysical environments, as well as in terrestrial gas-phase condensation reactions, is not yet understood (Jaeger et al. 2008). Laboratory simulations in combination with interstellar observations support the idea that the predominant fraction of carbon in space is present as solid macromolecular carbon (e.g., Pendleton and Allamandola, 2002) or amorphous and hydrogenated amorphous carbon (Mennella et al. 1998; Duley

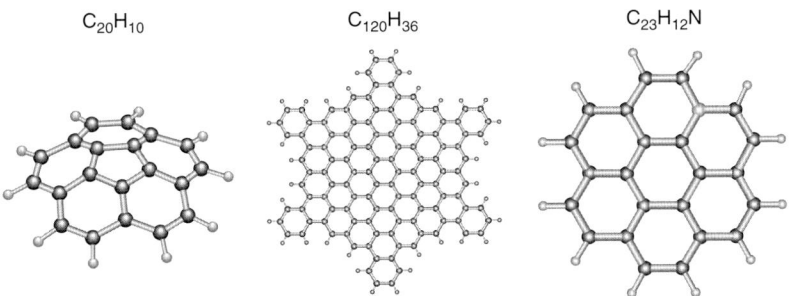

Figure 3. A few examples of PAH structures that might be present in the ISM.

and Lazarev 2004; Dartois et al. 2005). Fullerenes (Iglesias-Groth 2004) or defective carbon "onions" (Tomita et al. 2004) have also been proposed. Fullerenes of astronomical origin have been detected in meteorites and in and around an impact crater on the Long Duration Exposure Facility spacecraft; evidence for C_{60}^+ in the interstellar gas is provided by near infrared observations (Foing and Ehrenfreund 1996; see Ehrenfreund et al. 2006b for a review). Gas-phase condensation reactions in the laboratory demonstrate that the temperature in the condensation zone determines the formation pathway of carbonaceous particles. Condensation products in different astrophysical environments such as cool asymptotic giant branch stars or hot Wolf-Rayet stars should be different and should have distinct spectral properties (Jaeger et al. 2008).

Many of the described carbon compounds fail to match the observational constraints or account for the interstellar carbon budget (Snow and Witt 1995). The evolution of amorphous carbon measured by spectroscopy in the laboratory currently provides the best fit and satisfies the interstellar carbon balance of theoretical dust models (Menella et al. 1998). A combination of UV irradiated amorphous carbon (AC) and hydrogenated amorphous carbon (HAC) can explain the behavior of the bump structure (Mennella et al. 1998). Higher density regions display a larger width, compatible with hydrogenation of amorphous carbon. An alternative route to laboratory experiments at high temperature that simulate circumstellar environments is the production of hydrogenated carbon polymers through photolysis of a series of organics at low temperature (Dartois et al. 2005).

Interstellar Ices

Ground-based observations revealed already decades ago abundant water ice as well as CO (e.g. Whittet 1993). But other species remained undetected until infrared satellites opened up the 1–200 μm spectral window. Infrared Space Observatory (ISO) data were compiled to establish an inventory of interstellar ice species and measure their abundances in various interstellar environments (Gibb et al. 2004). After ISO, the Spitzer Space telescope (although operating at lower resolution) and ground-based observations have provided outstanding data on the main interstellar ice species. Recent c2d (core to disks) Spitzer surveys of ices investigated multiple sources, in particular low mass stars. Observations of the 6–8 μm region that displays the prominent bending mode of water ice shows five independent components that can be attributed to eight different carriers (Boogert et al. 2008). The spectrum is dominated by simple species formed by grain surface chemistry, which include CH_3OH (1%–30%), NH_3 (3%–8%), HCOOH (1%–5%), H_2CO (~6%), and $HCOO^-$ (0.3%) relative to water ice (Boogert et al. 2008). A comparison to high mass stars showed a rather similar ice distribution arguing against substantial UV radiation processing. CO_2 ice has been identified by ISO as a ubiquitous and abundant ice species (Gerakines et al. 1999). The characteristic band profile of the CO_2 bending mode at 15.2 μm has been used to identify the ice composition and characteristics in many lines of sight (Ehrenfreund et al. 1998, Boogert and Ehrenfreund 2004). Follow-up observations on a large sample of low mass stars during the c2d Spitzer survey showed higher abundances (average of 32% relative to water ice) (Pontoppidan et al. 2008). The survey also showed that CO_2 is mixed in different ice compositions within the line of sight, namely into a water rich component and in apolar ice mixtures (containing CO). Pure CO_2 ice layers were confirmed that indicate thermal processing actually described as "distillation" after CO has evaporated in cloud regions that exceed 20 K. A c2d Spitzer survey of CH_4 ice showed that 25 out of 52 targets displayed a feature at 7.7 μm attributed to CH_4. The abundances range from 2%–8% relative to water ice and reach 13% in a few sources (Oberg et al. 2008). The abundances are consistent with grain surface reaction formation of CH_4. Ice abundances of CH_4 seem to correlate with H_2O and CO_2 but not with CO and CH_3OH. Photodesorption of pure ices seem to be much more efficient according to laboratory results than previously assumed (Öberg et al. 2009).

CHALLENGES

Stability of Organic Molecules

To understand what compounds can survive solar system formation or extraterrestrial delivery the thermal and radiation stability of organic compounds and biomolecules has to be investigated in laboratory studies. During the formation of our solar system, interstellar gas and dust were mixed, processed and partly destroyed according to their distance from the forming star. Radiation chemistry involving X-rays and UV light acted on upper disk layers (Gorti et al. 2009). Results from the Sun in Time program suggests that the coronal X-ray EUV emissions of the young main sequence Sun were 100–1000 times stronger than those of the present Sun (Ribas et al. 2005). At the estimated time period of the origin of life on Earth $\sim$3.5 billion years ago, the solar high energy UV flux was 6 times the present value. The strong radiation emissions inferred, together with geological processes, may provide limits to the survival of organic compounds and biochemical pathways to create life.

Recent laboratory studies monitored the photostability of small N-heterocycles, nucleobases, benzene, and ethers using matrix-isolation spectroscopy at low temperature (Peeters et al. 2003, 2005, 2006). Amino acids have a limited life span in the interstellar medium (Ehrenfreund et al. 2001; Peeters et al. 2003). Amino acid photolysis under Martian conditions has been investigated by ten Kate et al. (2005) and showed a similar result, namely half lives of less than 22 h for glycine on the Martian surface. N-heterocycles are more easily destroyed than their carbonaceous cognate molecules such as benzene (Peeters et al. 2005). Furthermore, heterocycles containing several N-atoms in the ring, such as adenine have dramatically decreased half-lives when exposed to UV radiation.

In the gas phase, even benzene and small PAHs are not stable enough to survive the strong radiation fields in the diffuse interstellar medium (Ruiterkamp et al. 2005b, Peeters et al. 2005). Thin film radiation studies show that coronene and perylene, both pericondensed PAHs, are four to six times more stable than fluoranthene (which contains a pentagon ring) and 2,3-benzanthracene, respectively (Ehrenfreund et al. 2007). In comparison, the destruction of a thin film of the amino acid D-alanine measured under the same conditions proceeded $\sim$50 times faster than the PAH coronene. Detailed analyses of astronomical observations indicate sizes for PAHs of at least 50 carbon atoms and up to several hundred carbon atoms (Bauschlicher et al. 2009) (see Fig. 3).

Prebiotic Molecules in Space—Any Relevance for Early Earth

Aromatic molecules and biomarkers are widespread in our galaxy and beyond (Ehrenfreund et al. 2006a). The investigation of their life cycle is highly relevant for cosmochemistry. Molecules that are important in biochemical pathways such as nitriles, aldehydes, ethers, ketones, amines, and amides have been observed in space (www.astrochemistry.net). Even the amino acid glycine has been identified in hot molecular cores (Kuan et al. 2003).

However, as discussed earlier, most of those biomolecules are neither radiation nor thermal-resistant species, and are easily destroyed by UV radiation, shocks and thermal processing. Consequently they are unlikely to survive incorporation into solar system material without some degradation (see *Stability of Organic Molecules* above). A link between prebiotic compounds identified in short-lived hot core regions to the origin of life on Earth is therefore not realistic.

Prebiotically significant organic material identified in meteorites is formed within the solar system in radiation shielded environments and in the presence of liquid water. This is strongly supported by recent results that provide evidence that the organic composition in carbonaceous meteorites is dependent on parent body alteration processes, in particular aqueous alteration (Martins et al. 2007; Glavin and Dworkin 2009).

In addition to amino acids and sugar-related compounds, another important variety of precursor molecules, namely nucleobases, were recently detected in the Murchison meteorite (Martins et al. 2008). Compound-specific carbon isotope data of measured purine and pyrimidine compounds indicated a nonterrestrial origin for these compounds (see Fig. 4). Consequently, nucleobases delivered to these worlds together with sugar- related species and amino acids might have been beneficial to the origin of life on Earth, Mars, or elsewhere. The high abundance of aromatic material in meteorites led to the hypothesis of the "aromatic world" that describes PAH-based transitions from non-living to living matter as an interesting alternative to traditional origin of life models (Ehrenfreund et al. 2006a). Observations of cometary comae indicate that comets are rich in organic compounds and will have effectively delivered carbonaceous solids and volatiles to the surface of young planets. In contrast, the endogenous synthesis of prebiotic organic compounds may have been constrained by the conditions on the young Earth. Whatever the inventory of endogenous organic compounds on the ancient Earth, it would have been augmented by extraterrestrial material. It is estimated that these sources delivered $\sim 10^9$kg of carbon per year to the Earth during the heavy bombardment phase 4.5–3.9 billion years ago (Chyba and Sagan 1992).

RESEARCH DIRECTIONS AND CONCLUDING REMARKS

Several compounds are observed ubiquitously in our and external galaxies. Among them are PAHs, aliphatic hydrocarbons and ices (see Pendleton and Allamandola 2002; Ehrenfreund et al. 2006a; Tielens 2008 for reviews). The so-called "ultraviolet extinction bump" (at 2175Å) in the interstellar extinction curve is attributed to solid carbon-bearing material. The position of this feature attributed to amorphous carbon compounds is practically invariant in galactic and extragalactic regions, which argues for an abundant and stable carrier. To respond to current challenges described in Section 4, the following future research directions are proposed: (1) Investigation of the link between presolar and solar system material, (2) Revealing formation mechanisms/sites for macromolecular carbon networks, (3) Identification of specific PAHs, (4) Defining the role of small bodies in extraterrestrial delivery processes.

Understanding the implications of extraterrestrial delivery requires substantial knowledge

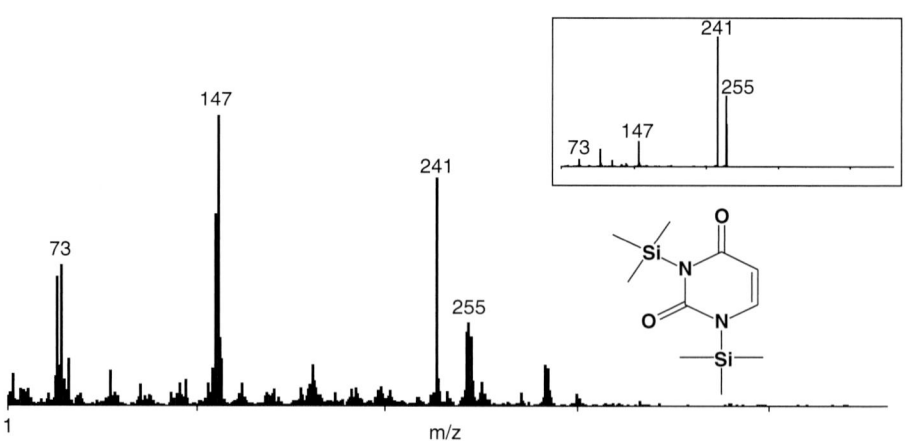

Figure 4. The gas chromatography/mass spectrometer spectrum for the peak assigned to BSTFA (N,o-Bis Trimethylsilyl (trifluoroacetamide) derivatization reagent) derivatized uracil and its structure. The inset shows the mass spectrum of a BSTFA-derivatized uracil standard (Martins et al. 2008).

of planetary dynamics, disk and solar nebula chemistry. There is a huge difference between making or discovering amino acids in space and creating life and the *origin* of life and the *survival* of existing life through adaptation in apparently impossible environments. It is questionable if the most abundant organic material, namely aromatic hydrocarbons and carbonaceous networks are formed only in circumstellar environments. Exploring mechanisms to form aromatics in the low temperature diffuse interstellar medium represent an important research avenue. The infrared emission bands that have identified aromatic compounds in galactic and extragalactic regions sample multi-cloud lines of sight and consequently a size and shape and likely ionization state distribution of PAHs. The identification of individual PAH molecules through their vibrational transitions will be very difficult. Only a combination of observations, laboratory data and theoretical calculations will provide constraints for the identification of a specific PAH class or family. Small bodies, such as comets, asteroids and their fragments, meteorites and interplanetary dust particles (IDPs) bear witness of processes occurring at the time of solar system formation. Astronomical observations, improved technology for extraterrestrial sample analysis, in-situ surface measurements of small solar system bodies and/or returned samples will strongly enhance the knowledge of the role of small bodies in impact and extraterrestrial delivery processes to young planets. These research avenues will also guide us to extend our knowledge to other habitable worlds.

ACKNOWLEDGMENTS

PE is supported by NASA grant NNX08AG78G and the NASA Astrobiology Institute NAI.

REFERENCES

Alexander CMO'D, Fogel M, Yabuta H, Cody GD. 2007. The origin and evolution of chondrites recorded in the elemental and isotopic compositions of their macromolecular organic matter. *Geochim Cosmochim Acta* **71:** 4380–4403.

Allamandola LJ, Hudgins DM, Sandford SA. 1999. Modeling the unidentified infrared emission with combinations of polycyclic aromatic hydrocarbons. *ApJ* **511:** 115–119.

Bauschlicher CW, Peeters E, Allamandola LJ. 2008. The infrared spectra of very large, compact, highly symmetric, polycyclic aromatic hydrocarbons (PAHs). *ApJ* **678:** 316–327.

Bauschlicher CW, Peeters E, Allamandola LJ. 2009. the infrared spectra of very large irregular polycyclic aromatic hydrocarbons (PAHs): Observational probes of astronomical PAH geometry, size, and charge. *ApJ* **697:** 311–327.

Bertoldi F, Carilli CL, Cox P, Fan X, Strauss MA, Beelen A, Omont A, Zylka R. 2003. Dust emission from the most distant quasars. *Astronomy and Astrophysics* **406:** L55–L58.

Blum J. 2004. Grain growth and coagulation. In *Astrophysics of Dust* (eds A.N. Witt, G.C. Clayton, and B.T. Draine) Vol. 309, pp. 369–392. ASP Conference Series.

Boogert ACA, Pontoppidan KM, Knez C, Lahuis F, Kessler-Silacci J, van Dishoeck EF, Blake GA, Augereau J-C, Bisschop SE, Bottinelli S, et al. 2008. The c2d Spitzer Spectroscopic Survey of Ices around Low-Mass Young Stellar Objects. I. H$_2$O and the 5–8 μm Bands. *ApJ* **678:** 985–1004.

Boogert ACA, Ehrenfreund P. 2004. Interstellar Ices. In *Astrophysics of Dust* (eds A.N. Witt, G.C. Clayton, and B.T. Draine) Vol. 309, pp. 547. ASP Conference Series.

Boss AP. 2004. From molecular clouds to circumstellar disks. In: *COMETS II*, (eds M.C. Festou, H.U. Keller, and H.A. Weaver) Univ. Arizona Press, 67–80.

Brownlee D, Tsou P, Aléon J, Alexander CMOD, Araki T, Bajt S, Baratta GA, Bastien R, Bland P, Bleuet P, et al. 2006. Comet 81P/Wild 2 Under a Microscope. *Science* **314:** 1711–1716.

Cataldo F. 2004. From Elemental Carbon to Complex Macromolecular Networks in Space. In *Astrobiology: Future Perspectives* (eds P. Ehrenfreund, L. Becker, and J. Blank) Astrophysics and Space Science Library, Vol. 305, pp. 97–126. Kluwer Academic Publishers: Dordrecht, The Netherlands.

Charnley SB, Ehrenfreund P, Kuan Y. 2003. Molecules in Space, *Physics World*, Institute of Physics Publishing Ltd, October 2003, 35–38.

Charnley SB, Kuan Y, Huang H, Botta O, Butner HM, Cox N, Despois D, Ehrenfreund P, Kisiel Z, Lee Y, et al. 2005. Astronomical searches for nitrogen heterocycles. *Adv Space Res* **36:** 137–145.

Chiar JE, Pendleton Y. 2008. The origin and evolution of interstellar organics. *Organic Matter in Space*. Proceedings of the International Astronomical Union, IAU Symposium, Volume 251: 35–44.

Chiar JE, Ennico K, Pendleton YJ, Boogert ACA, Greene T, Knez C, Lada C, Roellig T, Tielens AGGM, Werner M, et al. 2007. The relationship between the optical depth of the 9.7 μm silicate absorption feature and infrared differential extinction in dense clouds. *ApJ* **666:** L73–L76.

Chick K, Cassen P. 1997. Thermal Processing of Interstellar Dust Grains in the Primitive Solar Environment. *ApJ* **477:** 398–409.

Chyba C, Thomas P, Brookshaw L, Sagan C. 1990. cometary delivery of organic molecules to the early Earth. *Science* **249:** 366–373.

Chyba C, Sagan C. 1992. Endogenous production, exogenous delivery and impact-shock synthesis of organic molecules: an inventory for the origins of life. *Nature* **355:** 125–132.

Ciesla F. 2009. Observing our origins. *Science* **319:** 1488–1489.

Cottin H, Fray N. 2008. Distributed Sources in Comets. *Space Sci Rev* **138:** 179–197.

Cronin JR, Chang S. 1993. Organic matter in meteorites: Molecular and isotopic analyses of the Murchison meteorite. In *The chemistry of life's origin* (eds J.M. Greenberg, C.X. Mendoza-Gomez, V. Pirronello) pp. 209–258. Kluwer Academic Publishing, Dordrecht, The Netherlands.

Crovisier J, Biver N, Bockelée-Morvan D, Boissier J, Colom P, Dariusz C. 2009. The chemical diversity of comets: synergies between space exploration and ground-based radio observations. *Earth, Moon, and Planets* **105:** 267–272.

Cruikshank D, Imanaka H, Dalle O, Cristina M. 2005. Tholins as coloring agents on outer Solar System bodies. *Adv Space Res* **36:** 178–183.

Dartois E, Muñoz Caro GM, Deboffle D, Montagnac G, D'Hendecourt L. 2005. Ultraviolet photoproduction of ISM dust. Laboratory characterisation and astrophysical relevance. *Astronomy and Astrophysics* **432:** 895–908.

Derenne S, Robert F, Skrzypczak-Bonduelle A, Gourier A, Binet L, Rouzaud JN. 2008. Molecular evidence for life in the 3.5 billion year old Warrawoona Chert. *Earth and Planetary Sci Lett* **272:** 476–480.

Di Santi M, Mumma M. 2009. Reservoirs for comets: compositional differences based on infrared observations, origin and early evolution of comet nuclei. Space Sciences Series **28:** 127.

Duley WW, Lazarev S. 2004. Ultraviolet absorption in amorphous carbons: Polycyclic aromatic hydrocarbons and the 2175 Å extinction feature. *ApJ* **612:** L33–L35.

Dullemond C, Pavlyuchenkov Y, Apai D, Pontoppidan K. 2008. Structure and evolution of protoplanetary disks. In *Structure and evolution of protoplanetary disks*, *Phys Conf Ser* **131:** 012018.

Ehrenfreund P, Dartois E, Demyk K, d'Hendecourt L. 1998. Ice segregation toward massive protostars. *AstronomyAstrophysics Lett* **339:** L17–L21.

Ehrenfreund P, Charnley SB. 2000. Organic molecules in the interstellar medium, comets, and meteorites: A voyage from dark clouds to the early Earth. *Ann Rev Astron Astrophys* **38:** 427–483.

Ehrenfreund P, Glavin D, Botta O, Cooper G, Bada J. 2001. Extraterrestrial amino acids in Orgueil and Ivuna: Tracing the parent body of CI type carbonaceous chondrites. *PNAS Special Issue on Astrobiology* **98:** 2138.

Ehrenfreund P, Irvine W, Becker L, Blank J, Brucato JR, Colangeli L, Derenne S, Despois D, Dutrey A, Fraaije H, et al. 2002. Astrophysical and astrochemical insights into the origin of life. *Reports on Progress in Physics* **65:** 1427–1487.

Ehrenfreund P, Fraser H. 2003. Ice chemistry in space. In: *Solid state Astrochemistry*, NATO ASI Series, (eds V. Pirronello, K. Krelowski, and G. Manicò) pp. 317–356. Kluwer Academic Publishers.

Ehrenfreund P, Charnley SB, Wooden DH. 2004. From ISM material to comet particles and molecules. In *COMETS II*, (eds M.C. Festou, H.U. Keller, and H.A. Weaver), pp 115–133. Univ. Arizona Press.

Ehrenfreund P, Rasmussen S, Cleaves JH, Chen L. 2006a Experimentally tracing the key steps in the origin of life: The aromatic world. *Astrobiology* **6/3:** 490–520.

Ehrenfreund P, Cox N, Foing BH. 2006b. Fullerenes and related carbon structures in stellar atmospheres and the interstellar medium. In: *Natural Fullerenes and related structures of elemental carbon.* (ed. F. Rietmeijer) Series Developments in Fullerene Science, Vol. 6, 57–63.

Ehrenfreund P, Ruiterkamp R, Peeters Z, Foing B, Salama F, Martins Z. 2007. The ORGANICS experiments on BIOPAN V: UV and space exposure of aromatic compounds. *Planetary Space Sci* **55:** 383–400.

Foing BH, Ehrenfreund P. 1994. Detection of two interstellar absorption bands coincident with spectral features of C_{60}^+. *Nature* **369:** 296–298.

Genzel R, Lutz D, Sturm E, Egami E, Kunze D, Moorwood AFM, Rigopoulou D, Spoon HWW, Sternberg A, Tacconi-Garman LE, et al. 1998. What powers ultraluminous IRAS Galaxies? *Astrophysical J* **498:** 579.

Gerakines PA, Whittet DCB, Ehrenfreund P, Boogert ACA, Tielens AGGM, Schutte WA, Chiar JE, van Dishoeck EF, Prusti T, Helmich FP, et al. 1999. ISO-SWS observations of solid carbon dioxide in molecular clouds. *Astrophysical J* **522:** 357.

Gibb E, Whittet D, Boogert A, Tielens AGGM. 2004. Interstellar ice: The infrared space observatory legacy. *Astrophysical J Supp* **151:** 35–73.

Glavin DP, Dworkin JP. 2009. Enrichment in L-Isovaline by Aqueous Alteration on CI and CM Meteorite Parent Bodies. *PNAS* **106/14:** 5487–5492.

Gomes R, Levison HF, Tsiganis K, Morbidelli A. 2005. Origin of the cataclysmic Late Heavy Bombardment period of the terrestrial planets. *Nature* **435:** 466–469.

Gorti U, Dullemond CP, Hollenbach D. 2009. Time Evolution of Viscous Circumstellar Disks due to Photoevaporation by Far-Ultraviolet, Extreme-Ultraviolet, and X-ray Radiation from the Central Star. *Astrophysical J* **705:** 1237–1251.

Greenberg J. 1998 Making a comet nucleus. *Astronomy and Astrophysics* **330:** 375–380.

Henning T, Mutschke H. 2004. In *Astrophysics of Dust* (eds A.N. Witt, G.C. Clayton, and B.T. Draine) Vol. 309, p. 603, ASP Conference Series.

Henning T, Salama F. 1998. Carbon in the Universe. *Science* **282:** 2204–2210.

Hiroi T, Pieters C, Zolensky ME, Lipschutz ME. 1993. Evidence of thermal metamorphism on the C, G, B, and F asteroids. *Science* **261:** 1016–1018.

Hony S, Van Kerckhoven C, Peeters E, Tielens AGGM, Hudgins DM, Allamandola LJ. 2001. The CH out-of-plane bending modes of PAH molecules in astrophysical environments. *Astronomy and Astrophysics* **370:** 1030–1043.

Hudgins DM, Bauschlicher CW, Allamandola LJ. 2005 Variations in the peak position of the 6.2 μm interstellar emission feature: A tracer of n in the interstellar polycyclic aromatic hydrocarbon population. *Astrophysical J* **632**: 316–332.

Iglesias-Groth S. 2004. Fullerenes and buckyonions in the interstellar medium. *Astrophysical J* **608**: L37–L40.

Jäger C, Mutschke H, Henning Th, Huisken F. 2008. Spectral properties of gas-phase condensed fullerene-like carbon nanoparticles from far-ultraviolet to infrared wavelengths. *ApJ* **689**: 249–259.

Kuan Y, Charnley S, Huang H, Tseng W, Kisiel Z. 2003. Interstellar glycine. *Astrophysical J* **593**: 848–867.

Kwok S. 2004. The synthesis of organic and inorganic compounds in evolved stars. *Nature* **430**: 985–991.

Kwok S. 2009. Delivery of complex organic compounds from planetary nebulae to the solar system. *Intern. J Astrobiol* **8/3**: 161–167.

Lammer H, Bredehöft JH, Coustenis A, Khodachenko ML, Kaltenegger L, Grasset O, Prieur D, Raulin F, Ehrenfreund P, Yamauchi M, et al. 2009. What makes a planet habitable? *Astron Astrophys Rev* **17**: 181–249.

Liszt H, Lucas R. 2000. The structure and stability of interstellar molecular absorption line profiles at radio frequencies. *Astronomy and Astrophysics* **355**: 333–346.

Lorenz RD, Mitchell KL, Kirk RL, Hayes AG, Aharonson O, Zebker HA, Paillou P, Radebaugh J, Lunine JI, Janssen MA, et al. 2008. Titan's inventory of organic surface materials. *Geophys Res Lett* **35**: L02206.

Markwick A, Charnley SB. 2004. Chemistry of Protoplanetary Disks. In *Astrobiology: Future Perspectives* (eds P. Ehrenfreund, L. Becker, and J. Blank) Astrophysics and Space Science Library, Vol. 305, pp. 33–66. Kluwer Academic Publishers: Dordrecht, The Netherlands.

Martins Z, Alexander OD, Orzechowska G, Fogel M, Ehrenfreund P. 2007. Indigenous amino acids and chiral excess identified in CR primitive meteorites. *Meteoritics and Planetary Sci* **42/12**: 2125–2136.

Martins Z, Botta O, Fogel ML, Sephton MA, Glavin DP, Watson JS, Dworkin JP, Schwartz AW, Ehrenfreund P. 2008. Extraterrestrial nucleobases in the Murchison meteorite. *Earth and Planetary Sci Lett* **270**: 130–136.

Mathis J, Mezger P, Panagia N. 1983. Interstellar radiation field and dust temperatures in the diffuse interstellar matter and in giant molecular clouds. *Astronomy and Astrophysics* **128**: 212–229.

Mennella V, Colangeli L, Bussoletti E, Palumbo P, Rotundi A. 1998. A new approach to the puzzle of the ultraviolet interstellar extinction bump. *Astrophysical Jl* **507**: 177–180.

Mullie F, Reisse J. 1987. Organic Matter in carbonaceous chondrites. *Topics in Current Chemistry* **139**: 83.

Nesvorny D, Jenniskens P, Levison HF, Bottke WF, Vokrouhlicky D. 2009. Cometary origin of the zodiacal cloud and carbonaceous micrometeorites. *Astrophysical J*, in press.

Öberg KI, Boogert ACA, Pontoppidan KM, Blake GA, Evans NJ, Lahuis F, van Dishoeck EF. 2008. The c2d spitzer spectroscopic survey of ices around low-mass young stellar objects. III. CH_4. *Astrophysical J* **678**: 1032–1041.

Öberg K, Bottinelli S, van Dishoeck EF. 2009. Cold gas as an ice diagnostic toward low mass protostars. *Astronomy and Astrophysics* **494**: L13–L16.

Peeters E, Hony S, van Kerckhoven C, Tielens AGGM, Allamandola L, Hudgins DM, Bauschlicher CW. 2002. The rich 6 to 9 μm spectrum of interstellar PAHs. *Astronomy and Astrophysics* **390**: 1089–1113.

Peeters E, Allamandola LJ, Hudgins DM, Hony S, Tielens AGGM. 2004a. The unidentified infrared features after ISO. In *Astrophysics of Dust* (eds A.N. Witt, G.C. Clayton, and B.T. Draine) **309**: 141. ASP Conference Series.

Peeters E, Spoon HWW, Tielens AGGM. 2004b. polycyclic aromatic hydrocarbons as a tracer of star formation? *Astrophysical J* **613**: 986–1003.

Peeters Z, Botta O, Ruiterkamp R, Charnley SB, Ehrenfreund P. 2003. The astrobiology of nucleobases. *Astrophysical J Lett* **593**: L129–L132.

Peeters Z, Botta O, Charnley SB, Kuan YL, Kisiel Z, Ehrenfreund P. 2005. Formation and photostability of N-heterocycles in space: The effect of nitrogen on the photostability of small aromatic molecules. *Astronomy and Astrophysics* **433**: 583–590.

Peeters Z, Rodgers S, Charnley S, Schriver A, Schriver-Mazzuoli L, Keane J, Ehrenfreund P. 2006. Astrochemistry of dimethyl ether. *Astronomy and Astrophysics* **445**: 197.

Pendleton YJ, Allamandola LJ. 2002. The organic refractory material in the diffuse interstellar medium: mid-infrared spectroscopic constraints. *Astrophysical J Supplement* **138**: 75–98.

Pontoppidan KM, Boogert ACA, Fraser HJ, van Dishoeck EF, Blake GA, Lahuis F, Öberg KI, Evans J, Salyk C. 2008. The c2d spitzer spectroscopic survey of ices around low-mass young stellar objects. II. CO_2. *Astrophysical J* **678**: 1005–1031.

Prasad S, Tarafdar SP. 1983. UV radiation field inside dense clouds - Its possible existence and chemical implications. *Astrophysical J* **267**: 603–609.

Quinn R, Zent A, Grunthaner F, Ehrenfreund P, Taylor C, Garry J. 2005. Detection and characterization of oxidizing acids in the Atacama desert using the Mars Oxidation Instrument. *Planetary Space Sci* **53**: 1376–1388.

Raulin F. 2008. Astrobiology and habitability of Titan. *Space Sci Rev* **135**: 37–48.

Ribas I, Guinan E, Güdel M, Audard M. 2005. Evolution of the solar activity over time and effects on planetary atmospheres. I. High-Energy irradiances (1–1700 Å). *Astrophysical J* **622**: 680–694.

Ruiterkamp R, Cox N, Spaans M, Kaper L, Salama F, Foing B, Ehrenfreund P. 2005a. The PAH charge state distribution in diffuse and translucent clouds: Implications for DIB carriers. *Astronomy and Astrophysics* **432**: 515.

Ruiterkamp R, Peeters Z, Moore M, Hudson R, Ehrenfreund P. 2005b. A quantitative study of proton irradiation and UV photolysis of benzene in interstellar environments. *Astronomy and Astrophysics* **440**: 391.

Salama F. 1999. Polycyclic aromatic hydrocarbons in the interstellar medium. In *Solid Interstellar Matter - The ISO Revolution* (eds L. d'Hendecourt, A. Jones, and C. Joblin) p. 65, EDP Sciences and Springer-Verlag.

Sephton MA. 2002. Organic compounds in carbonaceous meteorites. *Nat Prod Rep* **19**: 292–311.

Sephton MA, Botta O. 2005. Recognizing life in the solar system: guidance from meteoritic organic matter. *Int J Astrobiol* **4:** 269–276.

Smith JDT, Draine BT, Dale DA, Moustakas J, Kennicutt RC Jr, Helou G, Armus L, Roussel H, Sheth K, Bendo GJ. 2007. The mid-infrared spectrum of star-forming galaxies: global properties of polycyclic aromatic hydrocarbon emission. *Astrophysical J* **656:** 770–791.

Snow T, Witt A. 1995. The interstellar carbon budget and the role of carbon in dust and large molecules. *Science* **270:** 1455–1460.

Snow TP, McCall BJ. 2006. Diffuse atomic and molecular clouds. *Ann Rev Astronomy Astrophys* **44:** 367–414.

Spaans M. 2004. The synthesis of the elements and the formation of stars. In *Astrobiology: Future Perspectives* (eds P. Ehrenfreund, L. Becker, and J. Blank) *Astrophysics and Space Science Library*, Vol. 305, pp. 1–16. Kluwer Academic Publishers: Dordrecht, The Netherlands.

ten Kate I, Garry J, Peeters Z, Quinn R, Foing BH, Ehrenfreund P. 2005. Amino acid photostability on the Martian surface. *Meteoritics and Planetary Science* **40:** 1185–1193.

Tielens AGGM. 2008. Interstellar polycyclic aromatic hydrocarbon molecules. *Ann Rev of Astronomy and Astrophysics* **46:** 289–337.

Tomita S, Fujii M, Hayashi S. 2004. Defective carbon onions in interstellar space as the origin of the optical extinction bump at 217.5 nanometers. *ApJ* **609:** 220–224.

Van Dishoeck EF, Blake G. 1998. Chemical evolution of star-forming regions. *Ann Rev of Astronomy and Astrophysics* **36:** 317.

Van Dishoeck EF. 2004. ISO Spectroscopy of gas and dust: from molecular clouds to protoplanetary disks. *Ann Review of Astronomy and Astrophysics* **42:** 119–167.

Visser R, Geers VC, Dullemond CP, Augereau J, Pontoppidan K, van Dishoeck EF. 2007. PAH chemistry and IR emission from circumstellar disks. *Astronomy and Astrophysics* **466:** 229–241.

Whittet D. 1993. Observations of Molecular Ices. Dust and Chemistry in Astronomy. *The Graduate Series in Astronomy Dust* (eds T.J. Millar and D.A. Williams) Institute of Physics Publishing, PA, pp. 9. Philadelphia.

Wolfire MG, McKee CF, Hollenbach D, Tielens AGGM. 2003. Neutral atomic phases of the interstellar medium in the galaxy. *ApJ* **587:** 278–311.

Wooden DH, Charnley SB, Ehrenfreund P. 2004. Composition and evolution of molecular clouds. In: *COMETS II* (eds M.C. Festou, H.U. Keller, and H.A. Weaver) Univ. Arizona Press, 33–66.

Yan L, Chary R, Armus L, Teplitz H, Helou G, Frayer D, Fadda D, Surace J, Choi P. 2005. Spitzer detection of polycyclic aromatic hydrocarbon and silicate dust features in the mid-infrared spectra of z ~2 ultraluminous infrared galaxies. *ApJ* **628:** 604–610.

The Hadean-Archaean Environment

Norman H. Sleep

Department of Geophysics, Stanford University, Stanford, California 94305

Correspondence: norm@stanford.edu

A sparse geological record combined with physics and molecular phylogeny constrains the environmental conditions on the early Earth. The Earth began hot after the moon-forming impact and cooled to the point where liquid water was present in ~10 million years Subsequently, a few asteroid impacts may have briefly heated surface environments, leaving only thermophile survivors in kilometer-deep rocks. A warm 500 K, 100 bar CO_2 greenhouse persisted until subducted oceanic crust sequestered CO_2 into the mantle. It is not known whether the Earth's surface lingered in a ~70°C thermophile environment well into the Archaean or cooled to clement or freezing conditions in the Hadean. Recently discovered ~4.3 Ga rocks near Hudson Bay may have formed during the warm greenhouse. Alkalic rocks in India indicate carbonate subduction by 4.26 Ga. The presence of 3.8 Ga black shales in Greenland indicates that S-based photosynthesis had evolved in the oceans and likely Fe-based photosynthesis and efficient chemical weathering on land. Overall, mantle derived rocks, especially kimberlites and similar CO_2-rich magmas, preserve evidence of subducted upper oceanic crust, ancient surface environments, and biosignatures of photosynthesis.

Life possibly originated and definitely evolved on the early Earth. Several other articles in this collection pose specific questions for the Earth scientist. Benner et al. (2010) discuss the origin of ribose in boron containing fluids in serpentinite. One would like to know when, where, and if this was a likely rock environment. Deamer and Weber (2010) discuss energy sources and Hazen (2010) discusses mineral surfaces. The mineralogical composition of likely rocks and its tendency to provide chemical equilibrium directly involve geology. Gaucher et al. (2010) use the molecular biology of extant life to infer conserved traits that involved on the early Earth. Zahnle et al. (2007) and Nisbet et al. (2007) review conditions on the early Earth. I focus on illustrative specific issues that relate to molecular phylogeny and/or to use of the geological record.

Earth scientists ask whether inferences from molecular phylogeny make geological and ecological sense. They also use the Earth's geological record to tie phylogenic events to the Earth's absolute time scale. The lack of ancient preserved rocks hinders all these efforts to the point of grasping at straws. One may use the pittance of ancient rock samples working back from younger better recorded times. A parallel approach uses physics to infer conditions are the earliest Earth and then works forward. Overall, the geologist provides a shopping list of likely conditions to biologists and draws biological inferences from the Earth's meager record.

As a caveat, it is conceivable that terrestrial life did not originate on our planet. Panspermia from outside the solar system is highly unlikely and replaces well-posed questions with speculations about an unknowable world. Yet the transfer of life between terrestrial planets in our solar system seems feasible. In fact, the origin of terrestrial life on Mars (McCay 2010) and/or the asteroid Ceres are more testable hypotheses than its origin on the Earth. These inactive bodies preserve very ancient geological records. Logistics preclude immediate appraisal. This subject is mostly beyond the scope of this review.

Finally, some definition of the terms "Hadean" and "Archaean" is necessary. Defining Hadean as predating the oldest rock is awkward in that a Hadean rock can then never be found. For this review, Hadean includes the late heavy bombardment before $\sim$3.8 Ga. It follows convention of considering the oldest well-preserved sedimentary rocks in Isua Greenland at $\sim$3.8 Ga as Archaean (e.g., Rosing and Frei 2004). The Archaean ends at $\sim$2.5 Ga. The advent of O_2 in the atmosphere at $\sim$2.45 Ga is a practical boundary involving recognizable global event in the sedimentary record.

BACKGROUND

Some definition of life is necessary to focus discussion. Here, a self-organized and reproducing consortium of complicated chemical compounds that includes catalysts that extract energy from the environment suffices. The complexity of terrestrial life has a high degree of order, using a limited number of organic compounds as building blocks. It resembles writing in an alphabet, much more than the scribbling of a three year old.

I limit the scope of the article to life as we find on Earth. Today we have chemoautotrophic life that extracts energy ultimately from rocks and from photolysis in the air and water, photosynthetic life, and heterotrophs that eat products of other life. This partition creates semantic problems in discussing the earliest Earth. For example, organisms similar to modern heterotrophs could have eaten carbon compounds brought in by meteorites by reacting them with ferric iron from rocks. This resource persists today but is extremely meager compared with the bountiful products of photosynthesis. This potential prebiotic and early biotic substrate is beyond the scope of this article mainly because my knowledge of the geological record offers little insight; see Ehrenfreud (2010) and Zahnle (2010).

It is unproductive to precisely define when autocatalysis became life. It is also unnecessary for the Earth. Nascent life competes with nonlife (Nowak and Ohtsuki 2008). There is selection both for efficient gathering of resources and for faithful reproduction. Once the fidelity of reproduction crosses a threshold, life wins. The population explosion colonizes all connected environments on a time scale of years to thousands of years. It is highly unlikely that the Earth's meager geological record preserves this event.

The Earth scientist thus asks whether the Earth was habitable at a given time and whether it was in fact inhabited. A minimum requirement for habitability is temperature. There is a real upper limit for mundane life; the current record is 122°C (Takai et al. 2008). Gradual gradients to higher temperatures at depth in the rocks and sharp gradients around hydrothermal events have highly existed throughout geological time and with them the opportunity for evolution of more extreme thermophiles. Life cannot function at temperatures where water is solid. However, life likely arose in or around rocks rather than photosynthetically at the surface (Gaucher et al. 2010). Liquid water existed in the deep oceans and rocks on the early Earth even if the surface froze over. The second requirement is energy. As discussed later, the Earth began hot. There have always been hydrothermal systems to provide chemical disequilibria.

A rock record is necessary to tell whether the Earth was in fact inhabited. It is unlikely that morphological fossils will be found in rocks older than 3.8 Ga. Neither will biomarkers, that is, molecular fossils than can be associated with specific groups of organisms. The earth scientist is thus left with "biosignatures," indications of life that can range from weak to strong. This situation implies a functional discussion of life forms. For example, metamorphosed black shales in Isua Greenland contain abundant reduced carbon and some pyrite (Rosing and Frei 2004). This association is a biosignature for anoxygenic photosynthesis that oxidized sulfide to sulfate and the heterotrophic reaction of sulfate and organic matter in the sediments to produce pyrite. It is not *a priori* evident that the ancient organisms are the ancestors of modern organisms that fill these niches.

The section begins with the earliest Earth and considers the aftermath of its formation and the effects of large asteroid impacts, using physics, as the relevant terrestrial rock record is sparse. It then reviews the advent of continents where a meager but relevant rock record exists.

Approach to Clement Conditions on the Earliest Earth

The current paradigm is that the current Earth-Moon system formed when a Venus-sized ($\sim$0.9 Earth mass) "target" planet collided with a Mars-sized ($\sim$0.1 Earth mass) "projectile" planet. Zahnle et al. (2007) and Nisbet et al. (2007) reviewed its aftermath and the early Earth in general. This subsection concentrates on features than might be preserved in the rock record and aspects directly related to life.

The Moon condensed from vaporized ejecta from the mantle of the projectile. It is sterile because it lost its volatile components including water in this process, rather than because its surface is airless. The energy of the impact melted the mantle of the Earth and vaporized significant volumes of rock. The top of the vapor atmosphere (that is the photosphere) radiated its heat to space at an effective temperature of $\sim$2300 K. The entire mass of the mantle needed to pass repeatedly through the photosphere to vent its heat to space. The end of the process after $\sim$1000 years left an atmosphere of traditional volatiles, a few hundred bars of water and 100–200 bars of CO_2.

Water clouds soon condensed at the top of the atmosphere limiting the escape of heat to the runaway greenhouse threshold. The heat flow from the Earth's interior was $\sim$140 W/m^2. The surface remained hot 1800–2000 K, partially molten with some solid scum. Tidal heating from the Moon prolonged the episode. In $\sim$20 million years, the surface and mantle of the Earth were solid rock and the heat flow waned to $\sim$0.5 W/m^2, similar to 1 million-year-old modern oceanic crust.

Considerable CO_2 $\sim$100 bars likely remained in the atmosphere at this stage as this compound is nearly insoluble in magma at this pressure and carbonates are unstable at the temperatures of molten rock, here $\sim$1800 K. This amount of CO_2 was insufficient to trigger a runaway greenhouse on the early Earth, but enough to maintain a surface temperature of $\sim$500 K above a liquid water ocean (Kasting and Ackerman 1986; Sleep et al. 2001). (As in a pressure cooker, liquid water is stable at the high pressures, here a dense CO_2 atmosphere and hydrothermal systems on the modern seafloor. The steam saturation pressure at 500 K is 26.5 bars, compared to the pressure of $\sim$250 bars for a uniform layer of water with the mass of the present oceans.)

Calcium and magnesium carbonates were stable at the surface in equilibrium with basaltic rocks. However, carbonate minerals were stable only in the uppermost relatively cool region ($\sim$500 m) of the oceanic crust. The limited mass of CaO and MgO (each $\sim$10% by weight) could take up worldwide only $\sim$10 bars of CO_2 at any one time in carbonates. Repeated carbonatization of the oceanic crust and it subsequent subduction of some kind was necessary to sequester all the CO_2 in the Earth's deep interior. This process became more efficient as the interior cooled. Continental weathering and continental formation of carbonate need not have been involved. Note that subduction implies island arcs in the tectonic sense, but not necessarily islands above sea level.

It is unclear how long a warm 500 K CO_2 greenhouse persisted. Zahnle et al. (2007) prefer a range between the minimum time of $\sim$10 million years to subduct the available CO_2 to

100 million years Sleep et al. (2001) and Zahnle et al. (2007) with caveats could not identify any mechanism that would prolong a thermophile environment at 70–100°C. Once the CO_2 was subducted, a dynamic balance between rock sinks including weathering at the surface and reaction of CO_2 with the oceanic crust to form carbonates maintained a modest concentration of CO_2 in the air and the ocean. I return to this topic in the next subsection on asteroid impacts.

With regard to the implications of early life, mantle-derived igneous rocks (once the mantle became essentially solid) were similar to modern ones, that is, basalts and more Mg-rich and higher temperature komatiites. The precise composition of komatiite depends on its depth of origin and hence of the temperature in the upwelling mantle (Arndt 2003). It is also conceivable that a ~150-km-thick "magma ocean" of basaltic mush analogous to the mush chamber at fast ridge axis (Fig. 1) capped the mantle at some times (Zahnle et al. 2007; Sleep 2007). Igneous activity and hydrothermal circulation have been present throughout Earth history. The Earth has always recycled buried volatiles including carbon and water back to the surface.

With regard to the composition of the ocean, the behavior of modern hydrothermal systems provides analogy (Sleep et al. 2001). The ancient ocean as a whole cooled over millions of years toward clemency, passing through the P-T conditions beneath modern ridge axes. In particular, Na is much more abundant than Cl on the Earth. A dense liquid NaCl brine formed by the time the temperature reached 500°C if not before.

Basalt likely buffered the pH and the water chemistry; thermodynamic calculations indicate that carbonates are stable phases in water that has reacted with basalt (Sleep et al. 2001). The pH like the modern ocean was near neutral by the time it became habitable. A komatiite or peridotite/serpentinite buffered ocean would be more alkaline like modern vents in serpentinite. There was never a $HCl–H_2SO_4$ ocean nor a $NaHCO_3–Na_2CO_3$ one, as they would have been strongly out of equilibrium with basalt and also with komatiite or peridotite/seprentinite.

The precise amounts of NaCl and water in the ancient surface ocean, however, are not evident. Knauth (2005) points out the higher salinity 1.5–2.0 times present level is likely in the Hadean and Archaean because there

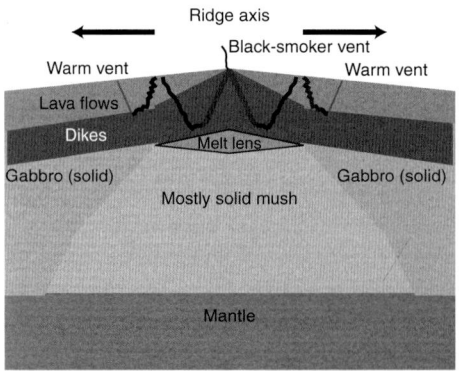
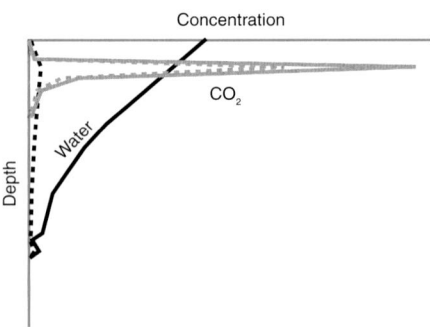

Figure 1. Schematic diagram shows cross section of ridge axis (*left*). Hot black smoker vents cool the magma lens and underlying mush chamber. Warm 20–60°C vents occur off axis. Carbonate minerals remove CO_2 from seawater. CO_2 is highly concentrated in a thin layer. The concentration before subduction (solid line) is somewhat greater than the concentration after subduction (dashed line). The concentration of water decreases slowly with depth (solid line). The subduction process removes most of the water into arc volcanics and returns it to the surface (dashed line). Mantle CO_2 hence resides at high concentrations in rocks formed from the upper oceanic crust. These domains are the source for kimberlites and other CO_2-rich magmas. Not to scale.

was insufficient continental area to form closed seas where halite (rock salt, solid NaCl) could accumulate by local evaporation of seawater. The direct geological record is not particularly helpful as halite is soluble in water and hence has a poor preservation potential compared with other rock types.

Ground Zero on the Early Earth

Zahnle et al. (2007) reviewed knowledge of the asteroid impacts on the early Earth. Little rock record exists, but the Moon provides a nearby well-preserved surface. Mars provides a most distant one. The impacting objects were mostly asteroids in solar orbit. It is straightforward to include the effect of the Earth's gravity in extrapolating its impact rate from the lunar one. Comets appear to have been a small fraction of the projectiles. The approach velocity of most comets is high enough compared to the escape velocities of the Moon and the Earth, that their gravity has little effect on the impact probability and impact energy.

The largest bodies are of interest here as they may sterilize an otherwise habitable planet. More interesting, they may have decimated life, leaving only predictable thermophile survivors. The killing mechanism of large projectiles is heat and the conservation of energy simplifies calculations. Basically an object larger that $\sim$300 km boils part of the ocean and leaves the rest of it uninhabitably hot. A 400–500 km diameter will boil the ocean. Thereafter water rains out (initially at $\sim$350°C) of the atmosphere at $\sim$1 m/yr. The surface returns to clement conditions in a few thousand years.

A refuge exists in the surface below $\sim$1 km depth if the geothermal gradient is low enough that this "Goldilocks" region is cool enough to be habitable. Shallower regions are normally habitable but are sterilized. Deeper regions are too hot for life all the time. Some Goldilocks regions existed in oceanic crust by 4.37 Ga in the thermal history calculation of Sleep (2007). As discussed in the next subsection, low-heat flow continental regions may have existed as early as 4.5 Ga.

The lunar record provides some calibration of the rate of large impacts on the Earth. It is obvious that the highland primordial crust still covers much of the lunar surface. That is, the Moon was hit by only a modest number of $>$100 km diameter objects that would have exhumed deeper rocks. In comparison, the resurfacing rate of the Moon by large impacts between 4.5 and 3.8 Ga was less than the rate that plate tectonics renew the oceanic crust over $\sim$100 million years on the Earth.

Quantitatively, one object hits the Moon compared to $\sim$20 hitting the much larger target Earth. That is, $\sim$14 objects larger than the largest lunar impact are expected for the Earth. The largest lunar projectile was $\sim$200 km diameter. So a few objects greater than $\sim$300 km diameter are expected to have hit the Earth. Prediction is thus complicated by the statistics of small numbers. Zahnle and Sleep (2006) preferred zero to four impacts large enough to leave only thermophile survivors. Ryder (2002) preferred zero but acknowledged the possibility of a sterilizing event using essentially the same data, but a different scaling relationship between basin size and projectile size on the Moon.

The timing of the impacts is of interest. One view is that the projectiles were gradually used up as a tail to accretion. Zahnle et al. (2007) review that this hypothesis is currently not favored by orbital dynamicists, nor by what is know about absolute ages on the moon. Rather, changes in the orbits of Jupiter and Saturn perturbed asteroids out of previously stable orbits at $\sim$3.9 Ga. This hypothesis is obviously testable by renewed sampling of the Moon. It implies that the Earth's interior had cooled significantly by the time of serious impacts and the Goldilocks zones existed in its subsurface. Hopkins et al. (2008) infer the pressure and temperature of formation of zircon crystals of this age. They conclude that regions of modest heat flow and modest thermal gradient $\sim$30 K/km existed then within continents.

Three additional environmental effects involve impacts. (1) Impacts of Fe-metal-bearing asteroids release Fe vapor into the air and Fe-metal into the ocean and ejecta (Kasting 1990). The metal reacts with water to form ferrous oxide and hydrogen gas. Methane forms from these reactants and CO_2. Ammonia forms from hydrogen and nitrogen gas. This may have led to transient reducing conditions in atmosphere, as in the experiments of Miller and Urey (1959). (2) Large impacts may exhume the mantle, providing serpentine at the surface and an environment where ribose may form as in Benner et al. (2010). In this vein, metamorphosed rock that was likely once serpentine (e.g., Friend et al. 2002) exists in the 3.8 Ga Issua sequence in Greenland. Studies of other altered rocks indicate that circulating seawater contained significant boron (Chaussidon and Appel 1997). (3) Ejecta are highly reactive pulverized and partly glassy rocks. Basaltic, komatiitic, and mantle ejecta are potent CO_2 sinks. They may have well drawn down atmospheric CO_2 to the point that the Hadean oceans froze over (Sleep and Zahnle 2001).

Zircons and the Earliest Continents

Hadean and Archaean detrital zircons from the 3.3 Ga Jack Hills formation in Australia have been extensively studied (see Cavosie et al. 2007). The mere survival of these samples over time indicates that regions akin to continental crust existed by 4.4 Ga and possibly as early as 4.5 Ga. For present purposes, zircon ($ZrSiO_4$) is very stable from weathering temperatures to low igneous temperatures $<700°C$. It contains U, Th, Hf, and Li as trace components. It sequesters radiogenic lead that facilities absolute age determination. The chemical diffusion of its trace elements slow facilitating studies of its source rock.

Overall these zircons indicate that their granitic host rocks formed partly at the expense of clay-rich sediments, hence indicate the existence of subaerial land masses. Oxygen isotopic variations (Cavosie et al. 2005, 2006, 2007) and lithium abundance and isotopic variations (Ushikubo et al. 2008) are the best evidence. These data do not bear on whether weathering occurred in a CO_2-rich atmosphere at ~ 500 K or at clement conditions.

Ushikubo et al. (2008) prefer the former as an explanation for the scarcity of Hadean rocks. This group correctly states that high temperature and high CO_2 facilitate chemical weathering. However, the exhumation of granitic rocks ceases once the surface has been beveled to sea level. Tectonics in addition of climate are thus necessary to recycle crustal hard rocks.

Continuing with tectonics, the Jack Hills zircons provide a potential constraint on the rate of tectonic activity. Cavosie et al. (2004) report that peaks and gaps exist in their zircon age suite. The gaps are 50–100 million years compared with 500–1000 million years for modern suites. Zahnle et al. (2007) and Sleep (2007) made the straightforward assumption that the duration of the gaps indicates the typical rate of tectonics. That is, tectonic processes recycled oceanic and continental crust at ~ 10 times the present rate. With this reasoning, Hadean heat flow was approximately three times its present value or ~ 0.2 W/m^2, compatible with their thermal modeling. Older lithosphere was ~ 40 km thick, enough to behave like modern plates and to transport continents.

RECENT RESULTS

I focus on recent work that relates to recognition of specific environmental conditions or specific functional biology on the early Earth. The subfield is new because evidence comes partly from "hard" igneous and metamorphic rocks, rather than "soft" sedimentary and low-grade metamorphic rocks. Hard rock petrologists have not been traditionally trained to seek biosignatures and paleontologists have traditionally focused on soft rocks.

World's Oldest Rocks Locality near Hudson Bay

O'Neil et al. (2009) described a locality of Hadean rocks near Hudson Bay. Their methodology used a short-lived radioactive isotopic system where isotope ^{146}Sm decayed to the stable isotope ^{142}Nd with a half-life of ~ 103 million years O'Neil et al. (2009) obtained an isochron age of 4.28 Ga from their suite of samples. At a minimum, their data point to the early formation and persistence of crust. The Hudson

Bay suite has $^{142}Nd/^{144}Nd$ that is less than the bulk mantle value; the daughter element Nd preferentially enters mantle derived melts and continental crust relative to Sm. Significant changes in the daughter/stable isotope ratio $^{142}Nd/^{144}Nd$ could have occurred only within a few half-lives after the Earth's formation and the rocks must have remained as a chemically isolated domain since then.

It is unclear whether ~4.3 Ga is actually the eruption age, rather than the age of the source region of the melt. If eruption occurred at ~4.3 Ga, the rocks provide the first direct evidence of the Hadean surface environment. Chemically these "faux amphibolites" appear to be metamorphosed basaltic rocks, possibly pyroclastic flows. They are greatly depleted in calcium relative to the composition of pristine basalts, but otherwise resemble mantle-derived igneous rocks. Such depletion does not occur in modern weathering.

Two candidate Hadean environments that might selectively deplete Ca are obvious from the previous section: (1) Hot 200°C rain falls during the warm CO_2 greenhouse, indicating an uninhabitable abiotic environment until at least ~4.3 Ga. (2) Hot 350°C low CO_2 rain falls in the few 1000 year aftermath of an ocean boiling impact. Geochemical studies as to what alteration occurs in these environments are not available. Note that precipitation-fed freshwater hydrothermal systems exist in Iceland. Some of these systems are likely to have the temperature, pressure, and CO_2 concentrations discussed earlier and thereby provide some analogy.

Advent of Photosynthesis

I take the availability of essentially modern sunlight as a given in discussing the advent of photosynthesis. The Sun had approached the main sequence (hydrogen burning to helium) and was ~70% as bright as present by the time the Earth had cooled to habitable conditions (e.g., Zahnle et al. 2007). Dust from impacts and volcanoes transiently clouded the sky, but there is no evident mechanism that would keep light from usually reaching the surface. Environments where any viable photosynthetic organisms were able to endure transient darkness certainly existed, including polar regions and lakes and land prone to burial by snow, mud, and dust. Modern sulfur-based photosynthetic microbes (see the following discussion) are able to act facultatively as heterotrophs and survive transient darkness. (Canfield and Raiswell 1999; Nisbet and Fowler 1996; Xiong et al. 2000).

Still the first terrestrial life forms were probably chemoautotrophic rather than photoautotrophic (Gaucher et al. 2010). It is unclear whether the first cellular life form lived at high or low temperatures. The Earth's climate may have lingered at a thermophile, but habitable condition ~70°C, well into the Archaean. The geological record is not definitive at present (e.g., Zahnle et al. 2007). The weathering rate would obviously be higher at high CO_2 atmospheric concentrations and high surface temperatures. The Archaean rock record, however, has been used to support both modern (Holland 1984; Condie et al. 2001; Sleep and Hessler 2006) and rapid rates (Schwartzman 2002). The straightforward interpretation of oxygen (e.g., Knauth and Lowe 2003) and silicon (Robert and Chaussidon 2006) isotopic ratios in ancient marine cherts indicates high temperatures.

Gaucher et al. (2008, 2010) query molecular phylogeny in this regard. The last universal common ancestor (LUCA) is often but not universally regarded to be thermophile. The maximum temperature of stability of reconstructed nodal proteins (from the last common ancestors of major clades in the tree of life) systematically increases from young to old nodes.

The discussion of the early Earth in this article introduces three well-known end-member possibilities with these results. (1) The Hadean climate after the early warm greenhouse ceased was clement or icy. Life originated and colonized the planet, some species adapting to thermophile lifestyles at hydrothermal events and in the kilometer-deep surface. A large asteroid impact boiled much of the ocean leaving only thermophile survivors. The descendants of the survivors adaptively radiated colonize low-temperature environments. High-temperature proteins were lost over time in the low-temperature branches. (2) The same phylogeny occurred except that the thermophile organisms outcompeted their low-temperature relatives, leaving an apparent LUCA bottleneck without a sudden mass extinction (Miller and Lazcano 1995). (3) The climate cooled slowly after the end of the CO_2 greenhouse and only thermophile organisms existed at the end of the Hadean and the temperature was still 50–70°C at ~3.3 Ga (Gaucher et al. 2008, 2010).

The first two possibilities have similar molecular genetic implications and geological implications until one finds evidence of a Hadean massive impact. Mat et al. (2008) call such phylogenies "hot cross." The accessible Archaean geological record bears on third possibility, and by potential elimination the other two.

It is relevant to attempt to tie phylogeny to absolute geological time. Sleep and Bird (2007, 2008) noted a difference between ecologies based on photosynthesis with those based on chemoautotrophy that this allows correlation in a functional sense. Briefly, a chemoautotrophic niche depends on a flux of chemical disequilibrium from rocks in the earth or photolysis in the air or water. For example, consider methanogenesis

$$CO_2 + 4H_2 \Rightarrow CH_4 + 2H_2O \quad (1)$$

The flux of H_2 mainly from serpentinite limits the productivity of this niche. Acetogenesis is a related niche

$$2CO_2 + 4H_2 \Rightarrow 2CH_2O + 2H_2O \quad (2)$$

which provides complex organic matter (idealized formula CH_2O) for the cell as well as energy if the reactants CO_2 and H_2 are abundant (e.g., Hoehler et al. 1998). As the forward reactions provide energy, there are no heterotrophic niches for the back reaction.

Overall, such niches are dependable, but involve very modest primary productivities. Organic carbon forms in small amounts and there is little tendency for it to accumulate in shale, a weathered product that is already near chemical equilibrium with its environment. The microbes have little available energy to accelerate weathering of the rocks. Hence, recognition of pre-photosynthetic biology in the rock record is quite difficult.

In contrast, photosynthesis has produced an obvious ubiquitous rock record on the Earth. For example, the crust of the Earth is grossly modified by photosynthesis. It is far more oxidized than mantle-derived igneous rocks (Lecuyer and Ricard 1999). The excess oxygen exceeds that to balance buried reservoirs of organic carbon (see Sleep and Bird 2008). It is likely that biological methane from photosynthesis allowed H_2 to escape to space before the oxidation of the Earth's atmosphere at ~2.45 Ga (Catling et al. 2001), leaving the oxygen from water behind. Subduction of organic carbon-rich sediments into the mantle is less likely but possible. These processes need not have involved oxygenic photosynthesis.

Photosynthesis may have evolved from hydrogen-based photocatalysis with the net effect of reaction (2). Sulfur and iron have multiple oxygen states suitable for photosynthetic ecology with heterotrophy. The idealized reactions are

$$2CO_2 + S^{-2} + 2H_2O + h\nu$$
$$\Rightarrow 2CH_2O + SO_4^{-2} \quad (3)$$

(e.g., Grassineau et al. 2001) and

$$CO_2 + 4FeO + H_2O + h\nu$$
$$\Rightarrow CH_2O + 2Fe_2O_3 \quad (4)$$

(e.g., Ehrenrich and Widdel 1994; Kappler and Newman 2004). Modern heterotrophic microbes obtain usable energy from the back reaction (3) with sulphate (Nisbet and Fowler 1996; Canfield and Raiswell 1999; Xiong et al. 2000), and back reaction (4) with ferric iron (Schröder et al. 2003; Luu and Ramsey 2003).

The presence of heterotrophy causes each C atom liberated by erosion, metamorphism, or volcanism to cycle many times between organic carbon and CO_2 before it is buried again. The organic carbon thus accumulates in preferred environments like black shale where quick burial sequesters it from oxidants. Metamorphosed black shale occurs in ~3.8 Ga rocks in Isua Greenland with pyrite (Rosing and Frei 2004). This indicates that S-based photosynthesis and heterotrophy existed by that time.

As with the origin of life itself (Nowak and Ohtsuki 2008), an unstable threshold exists. The nascent anoxygenic photosynthetic microbes likely used hydrogen as their main substrate (reaction 2). They facultatively used FeO and/or sulfide to increase their productivity. There was thus selection toward using less scarce H_2 in the mix. The first microbe that could dispense altogether with H_2 proliferated with its productivity increasing by a factor of thousands above the previous total primary productivity of the Earth. It occupied all suitable environments on the scale of years to thousand of years. Its descendents held most of the tickets in the subsequent evolutionary lottery.

Available molecular phylogeny is compatible with this inference. Bacterial photosynthesis evolved just once. Photosystems I and II are related (e.g., Sadakar et al. 2006; Barber 2008). Battistuzzi et al. (2004) and Battistuzzi and Hedges (2009) proposed that most bacteria clades descend from the original photosynthetic ancestor. Sleep and Bird (2008) called the clade descending from the original photosynthesizer with the obvious informal name Photobacteria. Battistuzzi et al. (2004) and Battistuzzi and Hedges (2009) further divided Photobacteria into Hydrobacteria that remained in the ocean and Terrabacteria that colonized land.

With regard to the rock record, Sleep and Bird (2008) noted that the photosynthetic Terrabacteria colonists faced a dearth of FeO and sulfide on land. Sulfur (0.1% in basalt [e.g., and Bird 1995]) is much less common than FeO (~10% in basalt [e.g., Krauskopf and Bird 1995]) in most common rocks. Note that casual observations of volcanoes over estimate the abundance of sulfur. Elemental sulfur "brimstone" is obviously common at volcanic vents. Physically, sulfur species (H_2S, S_8, and SO_2) exsolve from magma at low pressures. The FeO component of the magma stays in the solid. Local sulfur-rich oases existed on land;

I do not intend to contend that S-rich land volcanic vents should be excluded as potential prebiotic and early biotic environments.

Returning to land away from volcanoes, sulfate from reaction (3) is quickly leached from soils. Moreover, the FeO was locked up in minerals and volcanic glass. There was selection for soil organisms that could efficiently weather Fe-rich rocks. Notably Terrabacteria include Actinobacteria that live in soil and photosynthetic Cyanobacteria. Blank and Sánchez-Baracaldo (2010) traces the molecular phylogeny of cyanobacteria to fresh water organisms. Extraction of FeO from rocks by an anoxygenic photosynthetic community may well produce distinctive geochemical signals.

Returning the protein phylogeny (Gaucher et al. 2008, 2010), the occurrence of 3.8 Ga black shale (Rosing and Frei 2004) ties their nodes to geological time. The implications do not depend on the precise phylogenic tree of Battistuzzi et al. (2004) and Battistuzzi and Hedges (2009), only that efficient photosynthesis and the colonization of land predate 3.8 Ga. If this is correct, it is reasonable that the high-temperature proteins at the nodes of Gaucher et al. (2008) occurred in the Hadean following a high temperature bottleneck from an asteroid impact.

Paleontology of the Earth's Mantle

Subduction has recently been found to sequester material from early in the Earth's history. The occurrence of diamonds ultimately derived from subducted organic carbon in kimberlites (used loosely throughout this article) has been known from some time (Nisbet et al. 1994). The subsection reviews other examples that can be tied to environmental and biological processes.

Kimberlites form by partial melting of carbonate-rich domains in the Earth's mantle (Fig. 1). They react extensively with the Earth's lithosphere on the way up (Francis and Patterson 2009). Still, they retain isotope and trace element signatures of their pre-subduction history and pre-eruption source region. The carbonate-rich sources of kimberlites come significantly from subducted oceanic crust that reacted with warm 20–60°C circulating seawater near ridge axes. As already noted, this process is the major mantle sink of CO_2; all the Earth's CO_2 was sequestered in this way at the end of the warm CO_2 greenhouse. The process involves chemical reaction with the surface environment and the shallow (less than 1 km) oceanic crust started subduction near sediments.

Upadhyay et al. (2009) reported low values of $^{142}Nd/^{144}Nd$ in a 1.48 Ga alkalic complex in India intruded into 3.6 Ga crust in India. These rocks contain calcite as a trace phase and other trace elements indicating that a small fraction of partial melting occurred in a CO_2-rich source region. They authors preferred a sequence of events where the anomalous Nd domain formed in the Hadean, rocks derived from it were emplaced in the deep lithosphere at ~3.6 Ga and then remobilized at 1.48 Ga. In general, kimberlites liberated from the mantle often freeze in the lithosphere and form secondary melts when reheated or assimilated by subsequent magma (Tappe et al. 2008). With regard to habitability, these data indicate the subduction of carbonatized oceanic crust at ~4.26 Ga.

Data on kimberlites indicate that these rocks in fact come from subducted carbonates and retain a strong biosignature. For example, Zartman and Richardson (2005) found that that U/Th doubled from its bulk Earth value from 2.7 Ga to the recent. Chemically, U and Th behave as immobile elements during anoxic weathering. However, U is oxidized to the soluble oxidation state +6 if oxygen is present and is carried by rivers into the ocean and eventually into the oceanic crust, increasing the U/Th ratio above the bulk Earth value.

Thallium isotopes are another mantle biosignature that can be associated with a specific biological process and subduction. Circulating oxic seawater leaches thallium from the shallow oceanic crust (layer 2A). This thallium accumulates in manganese nodules on the seafloor, and the process fractionates thallium isotopes between the two reservoirs (Nielsen et al. 2006a). Nielsen et al. (2006b) detected thallium isotope variations with the expected coupled chemical variations in recent Hawaiian basalts. The deep ocean became suboxic at ~1.8 Ga (Slack et al. 2007, 2009).

CHALLENGES AND FUTURE RESEARCH DIRECTIONS

There is enough information from physics and the rock record to provide biologists with candidate environmental conditions on the early Earth. The Earth began at rock-vapor temperatures and gradually cooled to habitability over 20–120 million years By the time it was habitable, its mantle was mostly solid, but a few 100 K hotter than present. Volcanic rocks erupted to the surface and hydrothermal water circulated through hot rocks. At least some continental mass was exposed above sea level. The modern Earth provides a qualitative analog with basalts, continental granites, and weathering. I did not attempt to deduce the relatively and absolute land areas of various rock types, as applicable data on ancient sediments is not available. Geochemists are able to predict differences in composition between Hadean and modern, for example, with hot mantle-derived komatiitic rocks (Arndt 2003).

River run-off and reaction with seafloor basalt and seafloor sediment dynamically buffer the chemistry of seawater. Overall, evaporation does not affect ocean chemistry on the sense that the water eventually returns in rain and run-off. Continents with local evaporation provide a much wider repertoire of fluid and rock chemistries. Lakes may be essentially fresh water. Other lakes in closed basins accumulate alkalis as inflow continually evaporates. Lakes near volcanoes may become acid. Such environments are far from equilibrium from basalt. These types of Darwin's warm pond are potential venues for the origin of life.

Asteroid impacts punctuated an otherwise habitable environment. Until direct record is found, the number of impacts that left only thermophile survivors is unknown because of the statistics of small numbers. Renewed lunar exploration, however, will give much better constraints on the flux and timing of these objects.

Finally, photosynthesis appeared before the good rock record at 3.8 Ga. Its effects are pervasive in both the crust and mantle. A biological process in one place thus may have profound effects elsewhere. Work on recognizing biosignatures in mantle derived igneous rocks is just beginning. Kimberlites that sequester subducted carbonate in the oceanic crust (Fig. 1) are especially promising.

REFERENCES

Arndt N. 2003. Komatiites, kimberlites, and boninites. *J Geophys Res* **108**: 2293. doi: 10.1029/2002JB002157.

Barber J. 2008. Photosynthetic generation of oxygen. *Phil Trans R Soc* **B363**: 2665–2674.

Battistuzzi FU, Hedges SB. 2009. A major clade of prokaryotes with ancient adaptations to life on land. *Mol Biol Evol* **26**: 335–343.

Battistuzzi FU, Feijao A, Hedges SB. 2004 A genomic timescale of prokaryote evolution: Insights into the origin of methanogenesis, phototrophy, and the colonization of land. *Evolutionary Biol* **4**: 44. doi: 10.1186/1471-2148-4-44.

Benner SA, Kim H.-Y, Kim M.-J, Ricardo A. 2010. Planetary organic chemistry and the origins of biomolecules. *Cold Spring Harb Perspect Biol* **2**: a003467.

Blank CE, Sánchez-Baracaldo P. 2010. Timing of morphological and ecological innovations in the cyanobacteria—A key to understanding the rise in atmospheric oxygen. *Geobiology* **10**: (in press).

Canfield DE, Raiswell R. 1999. The evolution of the sulfur cycle. *Am J Sci* **299**: 697–723.

Catling DC, Zahnle KJ, McKay CP. 2001. Biogenic methane, hydrogen escape, and the irreversible oxidation of early Earth. *Science* **293**: 839–843.

Cavosie AJ, Valley JW, Wilde SA. 2007. The oldest terrestrial mineral record: A review of 4400 to 3900 Ma detrital zircons from Jack Hills, Western Australia. In *Earth's oldest rocks, developments in precambrian geology* (ed. Van Kranendonk M.J., Smithies R.H., Bennett V.), pp. 91–111. Elsevier, Amsterdam.

Cavosie AJ, Valley JW, Wilde SA, E.I.M.F. 2005. Magmatic $\delta^{18}O$ in 4400–3900 Ma detrital zircons: A record of the alteration and recycling of crust in the Early Archean. *Earth Planet Sci Lett* **235**: 663–681.

Cavosie AJ, Valley JW, Wilde SA, E.I.M.F. 2006. Correlated microanalysis of zircon: Trace element, $\delta 18O$, and U–Th–Pb isotopic constraints on the igneous origin of complex N3900 Ma detrital grains. *Geochim Cosmochim Acta* **70**: 5601–5616.

Cavosie AJ, Wilde SA, Liu D, Weiblen PW, Valley JW. 2004. Internal zoning and U-Th-Pb chemistry of Jack Hills detrital zircons: A mineral record of early Archean to Mesoproterozoic (4348–1576 Ma) magmatism. *Precambrian Res* **135**: 251–279.

Chaussidon M, Appel PWU. 1997. Boron isotopic composition of tourmalines from the 3.8-Ga-old Isua supracrustals, West Greenland: Implications on the delta B-11 value of early Archean seawater. *Chem Geol* **136**: 171–180.

Condie KC, DesMarais DJ, Abbott D. 2001. Precambrian superplumes and supercontinents: A record in black shales, carbon isotopes, and paleoclimates? *Precambrian Res* **106**: 239–260.

Deamer D, Weber A. 2010. The cell. *Cold Spring Harb Perspect Biol* **2**: a004929.

Ehrenfreund P. 2010. Intersteller organics & delivery. *Cold Spring Harb Perspect Biol* **2**: a002097.

Ehrenreich A, Widdel F. 1994 Anaerobic oxidation of ferrous iron by purple bacteria, a new type of phototrophic metabolism. *Appl Environ Microbiol* **60**: 4517–4526.

Francis D, Patterson M. 2009. Kimberlites and aillikites as probes of the continental lithospheric mantle. *Lithos* **109**: 72–80.

Friend CRL, Bennett VC, Nutman AP. 2002. Abyssal peridotites >3,800 Ma from southern West Greenland: Field relationships, petrography, geochronology, whole-rock and mineral chemistry of dunite and harzburgite inclusions in the Itsaq Gneiss Complex. *Contrib Mineral Petrol* **143**: 71–92.

Gaucher EA, Govindarajan S, Omjoy K, Ganesh OK. 2008. Palaeotemperature trend for Precambrian life inferred from resurrected proteins. *Nature* **451**: 704–708.

Gaucher EA, Kratzer JT, Randall RN. 2010. Deep phylogeny-how a tree can help characterize early life on earth. *Cold Spring Harb Perspect Biol* **2**: a002238.

Grassineau NV, Nisbet EG, Bickle MJ, Fowler CMR, Lowry D, Mattey DP, Abell P, Martin A. 2001. Antiquity of the biological sulphur cycle: Evidence from sulphur and carbon isotopes in 2700 million-year-old rocks of the Belingwe Belt, Zimbabwe, *Proc Roy Soc Biol Sci Series B* **268**: 113–119.

Hazen W. 2010. Mineral surfaces, geochemical complexities, and the origins of life. *Cold Spring Harb Perspect Biol* **2**: a002162.

Hoehler TM, Alperin MJ, Albert DB, Martens CS. 1998. Thermodynamic control on H_2 concentrations in an anoxic marine sediment. *Geochimica Cosmochimica Acta* **62**: 1745–1756.

Holland HD. 1984. *The chemical evolution of the atmosphere and oceans*. Princeton University Press, Princeton.

Hopkins M, Harrsion MT, Manning CE. 2008. Low heat flow inferred from >4 Gyr zircons suggests Hadean plate boundary interactions. *Nature* **456**: 493–496.

Kappler A, Newman DK. 2004 Formation of Fe(III) minerals by Fe(II)-oxidizing photoautotrophic bacteria. *Geochim Cosmochim Acta* **68**: 1217–1226.

Kasting JF. 1990. Bolide impacts and the oxidation-state of carbon in the Earth's early atmosphere. *Origins Life Evolution Biosphere* **20**: 199–231.

Kasting JF, Ackerman TP. 1986. Climatic consequences of very high-carbon dioxide levels in the Earth's early atmosphere. *Science* **234**: 1383–1385.

Knauth LP. 2005. Temperature and salinity history of the Precambrian ocean: Implications for the course of microbial evolution. *Palaeogeo Palaeoclimatol Palaeoecol* **219**: 53–69.

Knauth LP, Lowe DR. 2003. High Archean climatic temperature inferred from oxygen isotope geochemistry of cherts in the 3.5 Ga Swaziland Supergroup, South Africa. *Geo Soc Am Bull* **115**: 566–580.

Krauskopf KB, Bird DK. 1995. Introductiion to geochemistry. 3rd edn. p. 647, McGraw-Hill, New York, NY.

Lecuyer C, Ricard Y. 1999. Long-term fluxes and budget of ferric iron: Implication for the redox states of the Earth's mantle and atmosphere. *Earth Planetary Sci Lett* **165**: 197–211.

Luu Y-S, Ramsey JA. 2003. Review: Microbial mechanisms of accessing insoluble Fe(III) as an energy source. *World J Microbiol Biotechnol* **19**: 215–225.

Mat W-K, Xue H, Wong T-F. 2008. The genomics of LUCA. *Frontiers Bioscience* **13**: 5605–5613.

McCay CP. 2010. An origin of life on Mars. *Cold Spring Harb Perspect Biol* **2**: a003509.

Miller SL, Lazcano A. 1995. The origin of life–Did it occur at high temperatures, *J Mol Evol* **41**: 689–692.

Miller SL, Urey HC. 1959. Organic compound synthesis on the primitive Earth. *Science* **130**: 245–251.

Nielsen SG, Rehkamper M, Norman MD, Halliday AN, Harrison D. 2006b. Thallium isotopic evidence for ferromanganese sediments in the mantle source of Hawaiian basalts. *Nature* **439**: 314–317.

Nielsen SG, Rehkamper M, Teagle DAH, Butterfield DA, Alt JC, Halliday AN. 2006a. Hydrothermal fluid fluxes calculated from the isotopic mass balance of thallium in the ocean crust. *Earth Planet Sci Lett* **251**: 120–133.

Nisbet EG, Fowler CMR. 1996. The hydrothermal imprint on life; did heat-shock proteins, metalloproteins and photosynthesis begin around hydrothermal vents? In *Tectonic and biological segmentation of mid-ocean ridges* (ed. MacLeod CJ, Tyler PA, Walker CL), pp. 239–251. Geological Society of London Special Publication.

Nisbet EG, Mattey DP, Lowry D. 1994. Can diamonds be dead bacteria? *Nature* **367**: 694–694.

Nisbet E, Zahnle K, Gerasimov MV, Helbert J, Jaumann R, Hofmann BA, Benzerara K, Westall F. 2007. Creating habitable zones, at all scales, from planets to mud microhabitats, on Earth and on Mars. *Space Sci Rev* **129**: 79–121.

Nowak MA, Ohtsuki H. 2008. Prevolutionary dynamics and the origin of evolution. *Proc Natl Acad Sci* **105**: 14924–14927.

O'Neil J, Carlson RW, Francis D, Stevenson RK. 2009. Neodymium-142 evidence for Hadean mafic crust. *Science* **321**: 1828–1831.

Robert F, Chaussidon M. 2006. A palaeotemperature curve for the Precambrian oceans based on silicon isotopes in cherts. *Nature* **443**: 969–972.

Rosing MT, Frei R. 2004. U-rich Archaean sea-floor sediments from Greenland - indications of >3700 Ma oxygenic photosynthesis. *Earth Planet Sci Lett* **217**: 237–244.

Ryder G. 2002. Mass flux in the ancient Earth-Moon system and benign implications for the origin of life on Earth. *J Geophys Res* **107**: 10.1029/2001JE001583

Sadekar S, Raymond J, Blankenship RE. 2006. Conservation of distantly related membrane proteins: Photosynthetic reaction centers share a common structural core. *Mol Biol Evol* **23**: 2001–2007.

Schröder I, Johnson E, de Vries S. 2003. Microbial ferric iron reductases. *FEMS Microbiology Reviews* **27**: 427–447.

Schwartzman D. 2002. *Life, temperature, and the Earth*, p. 272. Columbia University Press, NY.

Slack JF, Grenne T, Bekker A. 2009. Seafloor-hydrothermal Si-Fe-Mn exhalites in the Pecos greenstone belt, New Mexico, and the redox state of ca. 1720 Ma deep seawater. *Geosphere* **5**: 302–314.

Slack JF, Grenne T, Bekker A, Rouxel OJ, Lindberg PA. 2007. Suboxic deep seawater in the late Paleoproterozoic: Evidence from hematitic chert and iron formation related to seafloor-hydrothermal sulfide deposits, central Arizona, USA. *Earth Planetary Sci Lett* **255**: 243–256.

Sleep NH. 2007. *Plate tectonics through time, Treatise on Geophysics* Volume 9, (ed. Schubert G.), pp. 101–117. Oxford, Elsevier.

Sleep NH, Bird DK. 2007 Niches of the pre-photosynthetic biosphere and geologic preservation of Earth's earliest ecology. *Geobiology* **5**: 101–117.

Sleep NH, Bird DK. 2008. Evolutionary ecology during the rise of dioxygen in the Earth's atmosphere. *Philos Trans R Soc B* **363**: 2651–2664.

Sleep NH, Hessler AM. 2006. Weathering of quartz as an Archean climatic indicator. *Earth Planet Sci Lett* **241**: 594–602.

Sleep NH, Zahnle K. 2001. Carbon dioxide cycling and implications for climate on ancient Earth. *J Geophys Res* **106**: 1373–1399.

Sleep NH, Zahnle K, Neuhoff PS. 2001. Initiation of clement surface conditions on the early Earth. *Proc Natl Acad Sci* **98**: 3666–3672.

Takai K, Nakamura K, Toki T, Tsunogai U, Miyazki M, Miyazaki J, Hirayama H, Nakagawa S, Nunoura T, Horikoshi K. 2008. Cell proliferation at 122°C and isotopically heavy CH_4 production by a hyperthermophilic methanogen under high-pressure cultivation. *Proc Natl Acad Sci* **105**: 10949–10954.

Tappe S, Foley SF, Kjarsgaard BA, Romer RL, Heaman LM, Stracke A, Jenner GA. 2008. Between carbonatite and lamproite - Diamondiferous Torngat ultramafic lamprophyres formed by carbonate-fluxed melting of cratonic MARID-type metasomes. *Geochim Cosmochim Acta* **72**: 3258–3286.

Upadhyay D, Scherer EE, Mezger K. 2009. ^{142}Nd evidence for an enriched Hadean reservoir in cratonic roots. *Nature* **459**: 1118–1121.

Ushikubo T, Kita NT, Cavosie AJ, Simon A, Wilde SA, Rudnick RL, Valley JW. 2008. Lithium in Jack Hills zircons: Evidence for extensive weathering of Earth's earliest crust. *Earth Planet Sci Lett* **272**: 666–676.

Xiong J, Fischer WM, Inoue K, Nakahara M, Bauer CE. 2000. Molecular evidence for the early evolution of photosynthesis. *Science* **289**: 1724–1730.

Zahnle K. 2010. Atmospheric composition. *Cold Spring Harb Perspect Biol* **2**: a004895.

Zahnle KJ, Sleep NH. 2006. Impacts and the early evolution of life. In *Comets and the origin and evolution of life*, 2nd ed. (ed. Thomas P.J., Hicks R.D., Chyba C.F., McKay C.P.), pp. 207–251. Springer, New York.

Zahnle K, Arndt N, Cockell C, Halliday A, Nisbet E, Selsis F, Sleep NH. 2007. Emergence of a Habitable Planet. *Space Sci Rev* **129**: 35–78.

Zartman RE, Richardson SH. 2005. Evidence from kimberlitic zircon for a decreasing mantle Th/U since the Archean. *Chem Geol* **220**: 263–283.

GLOSSARY

Basalt: The black volcanic rock that erupts for example in Hawaii and Iceland. It forms by moderate <25% partial melting of the earth's mantle. It is the most common volcanic rock on the Earth, the Moon and Mars.

Fast ridge axis: Originally, a ridge where the spreading rate is rapid enough that a central peak rather than a central valley exists at the axes. The morphology occurs when the full spreading rate is greater than ∼60 km/million years Fast ridges have steady state magma lens underlain by mush as shown in Figure 1.

Faux amphibolite: (transferred with partial translation from Québécois French): Rocks formed by the metamorphism of basalt containing the mineral amphibole. Ordinary amphibole in common amphibolite that formed from basalt is rich is calcium, idealized formula $Ca_2(Mg,Fe,Al)_5(Al,Si)_8O_{22}(OH)_2$; The amphibole in faux amphibolite has a major cummingtonite $Fe_2Mg_5Si_8O_{22}(OH)_2$ component. The process by which it formed is poorly understood.

Gabbro: A course-grained igneous rock formed by slow cooling of a molten rock (magma) with the composition of basalt. This rock usually forms the lower part of the oceanic crust (Fig. 1).

Goldilocks situation: From the children's story where Goldilocks found Papa Bear's oatmeal too hot, Mama Bear's too cold, and Baby Bear's just right. We find many such situations on the Earth that favor habitability in general and the origin of our own species, both locally and globally. For example, Venus is too hot and Mars is too cold. The ubiquity of Goldilocks situations on our planet is to a major extent the result of sampling bias: we have to survive to observe so we can expect no events like recent extirpation of terrestrial life that would preclude our existence. Philosophers refer to this concept as the weak antropic principle.

Isochron: A line obtained from a graph of isotopic concentration ratios from a suite of rock or mineral samples. Each point is the ratio daughter isotope from radioactive decay divided by stable isotope of same element versus the radioactive parent isotope divided by same stable isotope. These points fall on a line (the isochron) within analytical error if the samples formed at the same time from starting material that had the same initial ratio of daughter to stable isotope but variable parent to daughter ratios. The method requires that the samples behaved thereafter as closed systems. It often works in practice because the time for an initially homogeneous magma to differentiate and crystallize into samples of different parent/daughter isotopic ratios (days to thousand of years) is much less than the half-lifes (billions of years) of parent isotopes in standard use. The differentiation and crystallization times are frequently much less than the geological times of interest, many millions of years.

Isua: A region in southwestern Greenland that has well studied exposures of old Archaean rocks.

Kimberlite (loose usage): A MgO, water, and CO_2-rich magma formed melting of CO_2-rich rocks at ∼200 km depth in the Earth's mantle. Some of these rocks are commercial sources of diamond. The magma ascends violently from depth to the surface in hours to days.

Komatiite: A black volcanic rock that is significantly more MgO-rich than basalt. It formed in the past when the mantle was hotter by a high >25% fraction of partial melting. A continuum exists between basalt and komatiite. In addition komatiite may partly crystallize at shallow depths leaving basalt. See Arndt (2003) for more on the chemistry and physics of these processes.

Lithosphere: The cool upper part of the Earth, composed of crust and mantle, that moves in rigid plates. The base of ocean lithosphere is a thermal boundary that deepens to ∼100 km for old oceanic crust as it spreads away from the ridge axis. The base of continental lithosphere may be either compositional or thermal. Continental lithosphere within stable regions is typically ~200 km thick.

Mantle: The part of the Earth beneath the crust and above the core (2900 km deep). The depth to the top of the mantle (the Moho) is ~6 km beneath ocean basins and ~40 km beneath continents. The mantle is made of an igneous rock called peridotite with olivine (Mg_2SiO_4-Fe_2SiO_4) as its major mineral at low pressures above ~400 km depth. This rock also contains silicates of calcium, aluminum, and some sodium. It forms basaltic magma with a moderate fraction of melting and komatiite with a large fraction.

Mush: A partially molten rock with a small (less than few percent) fraction of partial melt. The rock flows slowly over geological time but behaves as solid on the time scale of seismic waves, seconds. Mush of basaltic composition forms gabbro when it freezes at ridge axes (Fig. 1).

Oxic and **suboxic** (adjectives): Oxic refers to atmosphere or water in equilibrium with atmosphere with significant O_2. Crudely, enough free oxygen to support many species of animals. Suboxic refers to lower concentrations of O_2, air with tenths of percent O_2 and water in equilibrium with it. It supports oxygen-based heterothrophy but not most animals.

Pyroclastic (advective): Refers to the violent eruption of a gas-rich magma as with Pompeii and Mount St. Helens and the deposits of rock fragments formed when the material comes to rest.

Serpentinite: A rock composed mainly of the mineral serpentine, idealized formula $(Mg,Fe)_6 Si_4O_{10}(OH)_8$ and some brucite $(Mg,Fe)(OH)_2$. The rock forms from the reaction of peridotite and water at temperatures below 350°C. The ferrous iron in the rock reacts with water to form hydrogen gas and magnetite Fe_3O_4. The hydrogen with CO_2 provides a substrate for modern methanogens at serpentine-fed vents. Serpentine also forms by reaction of water with olivine in basalt and komatiite.

Shale: A rock formed form mud deposited on the seafloor (or sometimes lake or river floor). It is fine grained and rich in clays. Reduced organic carbon accumulates in black shales when it burial is rapid. Shales have very low permeability to water circulation so that carbon once buried is not readily oxidized.

Subduction: The process that returns oceanic lithosphere into the deep mantle. Geologists call the returning lithosphere the slab. The top of the slab (upper oceanic crust and sediments) becomes hotter with increasing depth. At ~100 km depth, the silicates in the uppermost slab dehydrate, producing a hydrous fluid that ascends into the overlying hot mantle. The hydrous fluid mixes with the hot mantle, which partially melts to form hydrous basalt. The basalt ascends toward the surface feeding volcanoes in island arcs. Fujijama, the Aleutians, and Mount St. Helens are examples. Most of the water in the slab returns to the surface in this way, but most of the carbonate in the slab continues to depth.

Earth's Earliest Atmospheres

Kevin Zahnle[1], Laura Schaefer[2], and Bruce Fegley[2]

[1]Space Science Division, NASA Ames Research Center, MS 245-3, Moffett Field, California 94035
[2]Planetary Chemistry Laboratory, Dept of Earth & Planetary Sciences & McDonnell Center for the Space Sciences, Washington University, St Louis, Missouri 63130

Correspondence: kevin.j.zahnle@nasa.gov

Earth is the one known example of an inhabited planet and to current knowledge the likeliest site of the one known origin of life. Here we discuss the origin of Earth's atmosphere and ocean and some of the environmental conditions of the early Earth as they may relate to the origin of life. A key punctuating event in the narrative is the Moon-forming impact, partly because it made Earth for a short time absolutely uninhabitable, and partly because it sets the boundary conditions for Earth's subsequent evolution. If life began on Earth, as opposed to having migrated here, it would have done so after the Moon-forming impact. What took place before the Moon formed determined the bulk properties of the Earth and probably determined the overall compositions and sizes of its atmospheres and oceans. What took place afterward animated these materials. One interesting consequence of the Moon-forming impact is that the mantle is devolatized, so that the volatiles subsequently fell out in a kind of condensation sequence. This ensures that the volatiles were concentrated toward the surface so that, for example, the oceans were likely salty from the start. We also point out that an atmosphere generated by impact degassing would tend to have a composition reflective of the impacting bodies (rather than the mantle), and these are almost without exception strongly reducing and volatile-rich. A consequence is that, although CO- or methane-rich atmospheres are not necessarily stable as steady states, they are quite likely to have existed as long-lived transients, many times. With CO comes abundant chemical energy in a metastable package, and with methane comes hydrogen cyanide and ammonia as important albeit less abundant gases.

The origin of life has not been seen in Earth's rock record, and poor preservation of Earth's oldest rocks suggests that it will not be. The oldest bits of rock that might plausibly retain geochemical hints of habitable conditions on Earth's surface—a handful of zircons—have been subject to fierce debates regarding whether they contain such evidence. However, the weight of evidence does suggest that Earth has supported microbial life since the Hadean ("Hadean" refers to the geologic Eon preceding the Achaean. It can be regarded as Earth before the appearance of a true-rock record 3.9 Ga—the precise definition remains to be agreed on by the proper authorities). This would make Hadean Earth the one known planet where life has begun.

The modern focus on the atmosphere as the source of prebiotic chemistry dates to the famous Miller-Urey experiments of the 1950s (Miller 1953, 1955; Miller and Urey 1959; Oró and Kamel 1961; Johnson et al. 2008). These experiments were intended to model the kinds of disequilibrium chemistry that would have resulted from electrical discharges in, or ultraviolet radiation being absorbed in, highly reduced atmospheres in which methane, ammonia, and water were all major constituents. These experiments were driven by Harold C. Urey's theory that Earth accreted as a cool body and that its atmosphere was dominated by hydrogen and the hydrides of common volatiles (Urey won the Nobel Prize for Chemistry in 1934 for the discovery of deuterium). Miller and his colleagues, and many other investigators who have since performed similar experiments, have consistently found that a wide range of amino acids and other prebiotically interesting molecules form readily in such environments (e.g., see the review by Oró et al. 1990). These experiments were very influential in directing the attention of prebiotic chemists to a highly reduced primordial atmosphere.

However, photochemical studies showed that any methane (Lasaga et al. 1971) or ammonia (Kuhn and Atreya 1979; Kasting 1982) in the atmosphere would quickly be destroyed. Meanwhile geologically based arguments, which treat the atmosphere as outgassed from the solid Earth, were taken as strongly suggesting that Earth's original atmosphere was composed mostly of H_2O, CO_2, and N_2, with only small amounts of CO and H_2, and essentially no CH_4 or NH_3 (Poole 1951; Holland 1962; Abelson 1966; Holland 1984). This composition of volcanic gases is determined by temperature and the QFM (quartz-fayalite-magnetite) mineral buffer pertinent to the modern mantle. Nor is there evidence of a time on Earth when things were clearly different. Geochemical evidence in Earth's oldest igneous rocks indicates that the redox state of the Earth's mantle has not changed over the past 3.8 Gyr (Delano 2001; Canil 2002). Miller-Urey-type experiments performed in the more oxidized mixtures of modern volcanic gases generate relatively little of prebiotic interest, especially when CO_2 is abundant (Miller and Urey 1959; Schlessinger and Miller 1983; Stribling and Miller 1987). New work suggests that spark yields of ammonia, HCN, and amino acids in CO_2-N2-water mixtures can be less disappointing if the water is allowed to become acidic (Cleaves et al. 2008). Nevertheless, the contrast between methane and ammonia on the one hand and carbon dioxide and dinitrogen on the other led many prebiotic chemists, Miller and Urey prominent among them, to regard the presence of life on Earth as providing a strong boundary condition on the nature of Earth's early atmosphere.

The sense of poor prospects led some to abandon the atmosphere in favor of the hydrosphere. At the low temperatures and high water activities of hydrothermal systems, it is in theory possible to get non-negligable amounts of methane and ammonia at the QFM buffer (French 1966; Shock and Schulte 1990, 1998). Shock and Schulte (1990) used this approach to explain the abundances of organic molecules in asteroids (as sampled by carbonaceous meteorites) and suggested that such a model might have application to the origin of life on Earth (Shock 1990; Shock et al. 1995; Shock and Schulte 1998). The issue of a submarine (as opposed to a subaerial) origin of life became contentious (Miller and Bada 1988, 1993; Shock and Schulte 1993). Significant features of the hydrothermal hypothesis are that (1) it ties the origin of life to the process of making organic molecules, and (2) it implies that life is widespread in the Solar System, because hydrothermal systems may exist in many moons (Shock and McKinnon 1993).

Another workaround is to abandon the idea that organic molecules were generated in situ here on Earth. Instead the organic molecules would be delivered by comets and asteroids and interplanetary dust particles (IDPs, Anders 1989; Chyba and Sagan 1992; Whittet 1997). The basis of this proposal is that organic molecules are abundant in the Solar System. Many meteorites and dust grains are rich in complex organic molecules, and there is little doubt that comets are at least as rich. The chief difficulty is that, apart from special cases, only a small fraction of the more interesting and more delicate organic materials in comets and asteroids would survive impact (Clark 1988; Anders 1989; Chyba et al. 1990; Chyba and Sagan 1992; Whittet 1997; Pierazzo and Chyba 1999; Pasek and Lauretta 2008). The importance of an exogenic source of organics to the origin of life has probably been overstated. In the median case the quantities aren't large and the biological potential of a modest cosmic windfall of IDPs is unclear, although a slow soft collision by a big organic-rich comet—possible but by construction unlikely—could have a huge unique effect (Clark 1988). An alternative lesson to be taken from abundant organics in the Solar System is that organic molecules are not hard to make, and so were probably also made here.

A third perspective to the origin of the atmosphere—that the earliest atmosphere was degassed from impacting material as it arrived rather than outgassed from the solid Earth into a primordial vacuum (Arrhenius et al. 1974; Lange et al. 1985; Tyburczy et al. 1986; Abe and Matsui 1986; Zahnle et al. 1988; Ahrens et al. 1989)—has gotten comparatively little traction. Recently three new theoretical studies (Schaefer and Fegley 2007, 2010; Hashimoto et al. 2007) show that atmospheres dominated by impact degassing would be much more reduced than atmospheres dominated by Earth's mantle. A fourth recent study (Sugita and Schultz 2009) addresses impact degassing and impact synthesis in possible cometary matter experimentally. The latter can be regarded as parallel to the experiments performed by nature when the pieces of comet Shoemaker-Levy 9 struck Jupiter in July 1994. The SL9 impacts generated vast quantities of small molecules, especially CO, but the list of apparently synthetic products also included HCN, C_2H_2, C_2H_4, S_2, CS, CS_2, OCS, and CO_2 (Zahnle et al. 1995; Harrington et al. 2004).

Here we address the state and properties of Earth's primordial atmosphere. Our review is presented in three parts: the origin of the atmosphere, the Moon-forming impact, and events taking place after the Moon-forming impact. Placing the origin of the atmosphere before the Moon-forming impact is a choice that is founded on the high volatile contents of all known chondritic meteorites: To build Earth without volatiles is difficult if all known examples of possible source materials are more volatile-rich than Earth. Nonetheless we will also consider the alternative—volatile delivery after the Moon-forming impact—because it is one of the concepts in debate (cf. Albarède 2009) and because the hypothesis of transient, strongly reduced impact-degassed atmospheres applies obviously and directly to it.

BACKGROUND

Before the Moon-forming Impact: Origin of Planets and Atmospheres

The origin of Earth's atmosphere is a profound question of comparative planetology. The basic alternatives are a primary atmosphere captured from the solar nebula or a secondary atmosphere degassed from condensed materials accreted by Earth. Although the debate has long appeared settled in favor of secondary atmospheres, primary atmospheres probably did exist and may have had a minor role to play.

Primary atmospheres are by definition composed of gases captured from the solar nebula (Hayashi et al. 1979; Lewis and Prinn 1984; Ikoma and Genda 2006). In this model *planetesimals* (the first generation of solid worlds to form in the nebula) and proto-planets (the stage of growth between planetesimals and planets) capture nebular gases by gravitational attraction. The process is therefore much more efficient if the protoplanet is massive and the nebula cool. The captured gas is mostly H_2 — by far the most abundant element in the Universe—and interesting volatiles would tend to have been present as simple hydrides (H_2O, CH_4, NH_3). Such atmospheres are common. Jupiter, Saturn, Uranus, and Neptune are examples of big ones.

To the extent that the planetesimals and proto-planets that accumulated to make the Earth did so inside the nebula, they too would have been immersed in primary H_2-rich atmospheres. There is very little evidence left of Earth's primary atmosphere. It is possible that

^{3}He in the mantle is a fossil of the primary atmosphere, but the very low abundances of the noble gases relative to chemically reactive volatile elements such as carbon, hydrogen, nitrogen, and sulfur indicates that they have had profoundly different histories. The difference is nicely illustrated by nitrogen and neon. Nitrogen is the most volatile element other than hydrogen and the noble gases. Nitrogen and neon have very similar solar abundances and therefore would have been about equally abundant in the solar nebula, and they have similar atomic weights and so are comparably subject to escape, but in Earth's atmosphere the Ne/N ratio is just 10^{-5}. Likewise, argon and sulfur have similar solar abundances and capture of solar nebula gas would give a terrestrial Ar/S abundance ratio of ~ 0.2 instead of 0.00003 as observed.

When the solar nebula disappears the primary atmosphere escapes. Although escape of a primary atmosphere from a planet the size of Earth is difficult (Sekiya et al. 1980, 1981), escape is easy if the planet or planetesimal is the size of Mars or smaller: the first collision between two such planetesimals will raise the thermal velocities of a hydrogen-rich atmosphere above the escape velocity (Zahnle et al. 2006), and thus the primary atmosphere as a whole would be lost over the course of hours. Before it was lost the primary atmosphere may have influenced the redox state of the planetesimals directly, or indirectly by being a source of reduced solids that subsequently were folded into the planet. For example, nitrogenous organics or highly reduced nitrides may have been synthesized in the primary atmosphere and precipitated as solids to the surface, and thus remained behind when the atmosphere escaped.

Secondary Atmospheres

The disparity between the abundances of the noble gases and other volatile elements was first realized 85 years ago by F. W. Aston, who received the Nobel Prize in Chemistry for his pioneering work in mass spectroscopy. Aston (1924) concluded that the noble gases are abnormally scarce on the Earth in comparison to other elements. Aston also argued that the most plausible explanation for the rarity of the noble gases is that Earth is depleted in these elements. His conclusion was reinforced by Russell and Menzel (1933). They noted that astronomical spectra showed neon was abundant in the cosmos yet geochemists found it was scarce on Earth. This prior work and new data on the solar system abundances of the elements were incorporated into the secondary atmosphere model in the late 1940s by Brown (1949).

In the secondary model the atmosphere and oceans are produced by chemical reactions that released water and gases from condensed volatile-bearing materials during and/or after Earth's accretion. Brown showed that all volatile elements are depleted on Earth relative to their solar abundances. However, as noted earlier, the noble gases are much more depleted on Earth than the chemically reactive volatiles (H, C, N, O gases). This is true even if one considers the probable noble gas inventories left in the undegassed portion of Earth's mantle and possible trapping of xenon in Xe clathrate hydrates $Xe \cdot 6H_2O$, which may be found as pure phases or a $Xe-CH_4$ clathrate hydrate solid solution $(CH_4, Xe) \cdot 6H_2O$. The reason for this is simple—chemically reactive volatiles were incorporated into the solid grains accreted by the Earth during its formation but the noble gases were not.

The types of grains accreted by the Earth during its formation are exemplifed by chondritic meteorites (chondrites), which are primitive material from the solar nebula and are generally believed to be the building blocks of the Earth and other rocky asteroids, planets and satellites. Chondrites are undifferentiated (i.e., unmelted) stony meteorites containing metal+sulfide+silicate. The ordinary chondrites constitute about 97% of all chondrites. The rest of the chondrites fall mainly into two major groups, which are the carbonaceous and enstatite chondrites. Many geochemists think the Earth accreted from a mixture of chondritic materials (Wänke 1981; Lodders and Fegley 1997). Achondritic (i.e., melted and differentiated) material may also have been accreted by the Earth during its formation, but the volatile content of achondrites is generally much smaller than that of chondrites. This is possibly due to volatile loss from achondritic material during the heating that led to its melting and differentiation. Thus achondritic material probably made only minor contributions to the volatiles in Earth's early atmosphere and oceans.

Volcanic Atmospheres

The secondary origin of Earth's atmosphere led many to argue that the primordial atmosphere had the rather oxidized composition of volcanic gases (Poole 1951; Holland 1962; Abelson 1966). In modern volcanic gases the source is usually hot ($\sim$1500 K) and only weakly reduced because the mantle is metal-free and the redox state is set by the QFM (Quartz-Fayalite-Magnetite) buffer. The resulting atmosphere would be mostly H_2O, CO_2, and N_2, with small amounts of CO and H_2 (Poole 1951; Holland 1962; Abelson 1966; Holland 1984). Geochemical proxies for the mantle redox state (e.g., the vanadium and chromium contents) in Earth's oldest igneous rocks indicate that the redox state of the Earth's mantle has not changed over the past 3.8 Gyr (Delano 2001; Canil 2002). It is understood that Earth's mantle cannot always have been metal-free, because Earth was made from matter in which rocky materials and metals were mixed, but U-Pb and Hf-W radiometric ages make clear that core formation in Earth was complete within a hundred million years of the origin of the solar system.

Urey (1951, 1952a, 1952b) argued instead that the secondary atmosphere of an accreting Earth would be cool and strongly reducing. He made the case that a reducing atmosphere would be better for the origin of life. Urey's model was inspired. He was the first to emphasize the importance of impact degassing and of the chemical reactions between the atmosphere and hot reduced impact ejecta (including iron rain), and he emphasized that the relatively oxidized state of modern volcanic gases is an evolutionary feature of Earth and not indicative of the primordial gases. But Urey left out two key processes: (1) He neglected the deep burial of impact energy, because he did not imagine collisions between planet-sized bodies—in Urey's model, Earth grew by accumulating asteroids. (2) He neglected the infrared thermal blanketing effect of the atmosphere; i.e., he ignored the greenhouse effect. His calculations explicitly assume blackbody cooling at the surface temperature. The result is that Earth accumulates as a cold body, an intimate mixture of metallic iron and silicates similar to that seen in ordinary chondrites, and that as a consequence the atmosphere is profoundly reducing.

Impact Degassing

Elemental analyses show that chondritic material is generally more volatile-rich than the Earth, but until recently the outgassing of volatiles from chondritic material during and/or after accretion was a largely unexplored topic. In his unpublished masters thesis, Bukvic (1979) performed chemical equilibrium calculations for gas—solid equilibria in the upper layers of an Earth-like planet (suggested by his thesis advisor J. S. Lewis). He modeled the planet's composition using H-chondritic or a mixture of 90% H-and 10% C-chondritic material. Bukvic used a volcanic outgassing scenario, assuming that the gas moved from depth to the surface without interacting or equilibrating with the intervening layers, to predict the composition of the initial atmosphere. He found that the major outgassing products were N_2 at low temperatures and CH_4 and H_2 at higher temperatures. Bukvic never pursued this work further and it did not attract much attention for nearly 30 years, despite being discussed by Lewis and Prinn (1984).

Schaefer and Fegley (2007) reviewed the volatile contents of average ordinary chondrites. The data show that ordinary chondrites always contain substantial amounts of C and H: about 0.1% C and 0.3% H_2O by weight. These fractions are both about ten times what is now present on Earth. It would be impossible to build a volatile-poor planet from these sorts of building blocks without granting a central role to volatile escape (Cameron 1983). Ordinary chondrites are volatile-poor compared to other primitive meteorites. The CI carbonaceous chondrites—which are admittedly volatile-rich—have reduced carbon inventories of 3.5% and water up to 10% by weight (Hashimoto et al. 2007). This means that a planetary atmosphere outgassed from materials like those in primitive meteorites would be substantial, with scope for thousands or tens of thousands of bars of steam and other gases.

The implication of the above is that Earth accreted as a volatile-rich body. Lange et al. (1985) and Tyburczy et al. (1986) addressed the experimental conditions under which water is degassed on impact. They found that this occurs from carbonaceous chondrites when impact velocities exceed 5 km/s. Thus, when planetesimals had become large enough (about the size of Mars) for impact velocities to routinely exceed 5 km/s, most of the volatiles in the impactors were delivered directly into the growing atmosphere. The first generation of impact degassing models emphasized water and steam (Abe and Matsui 1986; Matsui and Abe 1986a,b; Zahnle et al. 1988; Abe et al. 2000) because water vapor is an exceptionally potent greenhouse gas. These workers found that Earth's atmosphere, at least during the interval between the dissipation of the primary atmosphere and the Moon-forming impact, was for most of the time in a runaway greenhouse state and the surface would have been molten. The water was kept in the vapor state in part by sunlight but also in part by the energy released by Earth's accretion. Escape was undoubtedly important at the time, but it probably did not greatly influence the chemical composition of the atmosphere, given that Earth would also have been accreting vast amounts of metallic iron at the same time. Indeed the steam atmosphere was likely to have been strongly or mildly reduced.

Three recent studies address the chemical composition of the impact-degassed steam atmosphere. Ordinary chondrites are usually regarded as a useful albeit imperfect guide to what Earth was made of (Drake and Righter 2002). Schaefer and Fegley (2007) used chemical equilibrium calculations to model gas–solid equilibria during outgassing of ordinary (H, L, LL) chondritic material over wide pressure and temperature ranges. They are strongly reducing because they contain metallic iron and iron sulfides. The three major types of ordinary chondrites—H, L, and LL— reflect high, low, and very low metal abundances. Schaefer and Fegley's results turn out to be not very sensitive to type.

Because conventional wisdom said otherwise, Schaefer and Fegley were surprised to find that CH_4, H_2, H_2O, N_2, and NH_3 are the major gases produced by heating ordinary chondritic material up to the solidus temperature ($\sim$1225 K) where Fe-Ni sulfide begins to melt. Thus, their results confirmed those obtained by Bukvic (1979) for outgassing of ordinary (H) chondritic material. Furthermore, Schaefer and Fegley (2007) found that their conclusions were remarkably insensitive to variations in the assumed temperature, total pressure and volatile element abundances over wide ranges. Carbon monoxide replaces CH_4 as the major carbon gas at higher temperatures, with the exact temperature depending on the total pressure. For example, 1:1 ratios of CO to CH_4 occur at $\sim$1000 K at one bar total pressure and $\sim$1200 K at 100 bar total pressure. To first approximation their model is dominated by the quartz-fayalite-iron (QFI) buffer, which is significantly more reduced than the better known iron-wustite (IW) buffer. Experimental work confrms their result. Although the focus of their study was asteroid-sized bodies, the prominence of methane in particular led them to extend the work to Earth. They concluded that Earth's early atmosphere contained CH_4, H_2, H_2O, N_2, and NH_3, similar to the gas mixtures used in the Miller–Urey synthesis of organic compounds. Illustrative results from calculations of this type are shown in Figure 1.

Hashimoto et al. (2007) also addressed the composition of gases generated in equilibrium with the projectile in the context of an impact degassed primordial atmosphere, and they too concluded that primordial methane-ammonia atmospheres were plausible. In their model, gas compositions at 1 and 100 bar fixed pressures are computed as a function of temperature assuming volatile abundances and oxygen

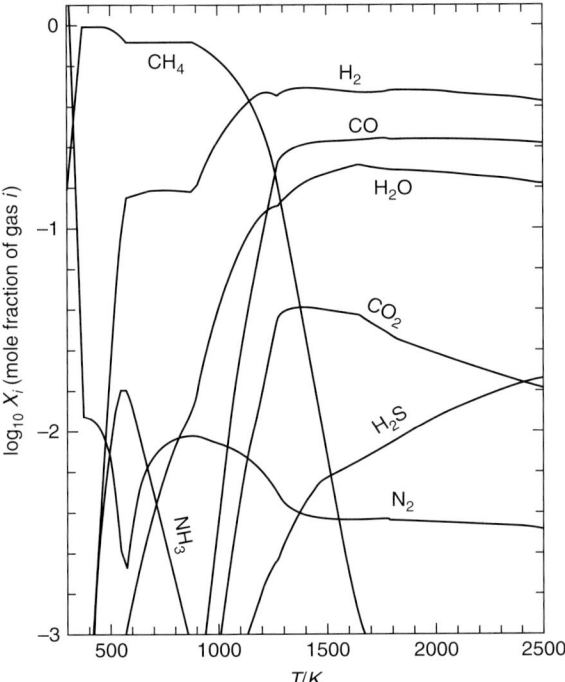

Figure 1. Gas compositions in equilibrium with ordinary H-type (high iron) chondrites at 100 bars (Schaefer and Fegley 2010). Abundances and oxygen fugacities are consistent with what is measured in ordinary H-type chondrites. Ordinary chondrites are often regarded as indicative of the bulk material of the Earth, although in detail the match is imperfect (Drake and Righter 2002). The gases are very reduced and at 100 bars pressure methane is strongly favored. The disappearance of CH_4 at very low temperatures indicates that graphite has become the stable form of carbon. The temperature axis can also be interpreted as quench temperatures in the cooling impact ejecta plume or after some other cause of flash heating, for example by lightning or by impact. Generally similar results are obtained for other ordinary chondrites and enstatite chondrites (both EL and EH), because all contain significant amounts of metallic iron and iron sulfides (Schaefer and Fegley 2010).

fugacities appropriate to CI carbonaceous chondritic materials. CI chondrites do not contain metallic iron, but they do contain a lot of reduced carbon, which is the source of their reducing power. These are the most highly oxidized meteorites and should generate the mostly highly oxidized gases that could be derived from a meteoritic source. Their results are indeed more oxidized than the gases from ordinary chondrites, but they are still notably reduced so that both CH_4 and NH_3 are produced. However, Hashimoto et al. do not show H_2O, which makes it impossible to determine what the gas composition actually is, because the major gas is usually water vapor. Figure 2, adapted from Schaefer and Fegley (2010), provides a clearer picture of the scenario addressed by Hashimoto et al.

Schaefer and Fegley (2010) extended their earlier work to consider a wider range of meteorite types, including carbonaceous and enstatite chondrites and mixtures of carbonaceous, ordinary, and enstatite chondritic material, which span the range of plausible planetary materials. Schaefer and Fegley again used chemical equilibrium calculations over a wide pressure and temperature range. They also considered the chemical kinetics of key reactions involved in the formation of CH_4. Schaefer and Fegley (2010) found that CO_2 is generally the major carbon-containing gas produced by heating CI and CM carbonaceous material.

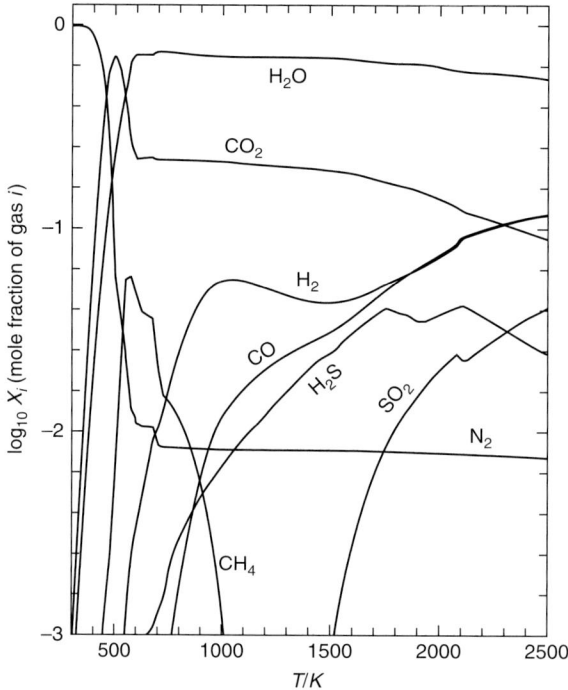

Figure 2. Gas compositions in equilibrium with CI chondrites at 100 bars (Schaefer and Fegley 2010). Abundances and oxygen fugacities are consistent with what is measured in carbonaceous chondrites. Carbonaceous chondrites are the most highly oxidized meteorites, and thus the gases that result are relatively highly oxidized; the chief source of reducing power is the reduced carbon itself. The temperature axis can also be interpreted as quench temperatures. Hashimoto et al. (2007) obtained generally similar results for similar assumptions.

In contrast, CI and CM carbonaceous material does not produce CH_4 in any appreciable amounts on heating. Carbon monoxide or CH_4 are the major carbon gases released by outgassing CV carbonaceous material, whereas methane (at low temperatures) or CO (at high temperatures) are the major carbon gases released by heating enstatite chondritic material. The line along which CH_4 and CO have 1:1 ratios is very similar to that for ordinary chondritic material. Finally, Schaefer and Fegley (2010) found that mixtures of chondritic material also give CH_4 or CO-bearing gases unless CI or CM carbonaceous chondritic material makes up most of the mixture. Figures 2 and 1 are modified from this work.

To a first approximation the three new studies reach essentially the same conclusion from the same premise: If the gases of Earth's earliest atmosphere had equilibrated with material like that in primitive meteorites, the atmosphere that results would be much more reduced than modern volcanic gases. These results apply to impact-degassing both before and after the Moon-forming impact.

RECENT RESULTS

The Moon-Forming Impact

It is now widely accepted that the Moon formed by the collision of two planets, one roughly the size of Mars and the other roughly the size of Earth (Benz et al. 1986; Cameron and Benz 1991; Canup and Asphaug 2001; Canup 2004). The collision took place 30–100 Myr after the formation of the Solar system—there is some disagreement over how the key Hf-W and U-Pb chronometers are to be interpreted. The impact was a major punctuating event in Earth's history. Zahnle et al. 2007 recently developed a speculative narrative history of the Hadean Earth beginning with the Moon-forming impact and ending with the rock record in the Archean. Figure 3 reduces the story to a single picture in four stages.

The first stage is a hot silicate vapor atmosphere that would have cooled and condensed over ~1000 years. As the gas cooled the silicates would have rained out according to a condensation sequence, with the relatively volatile rock-forming elements such as Na remaining in the atmosphere for a relatively long time. At first the metallic iron from the core of the impacting body will not have fully migrated to the core, as some of the iron will have been emulsified by its high speed interaction with Earth's mantle, which suggests that the rock vapor might have been rather reduced. In hot rock vapor two of the most abundant gases are SiO and O_2, and as the vapor cools several geochemical volatiles—sodium in particular—remain abundant in the gas phase (Schaefer and Fegley 2009). It is reasonable to expect that the most volatile elements would have partitioned between the atmosphere and the mantle according to their solubility in silicate melt, with much of the H_2O and perhaps the sulfur entering the mantle, whereas CO, CO_2 and most other gases likely going into the atmosphere. Whether the atmosphere is reduced (mostly H_2 and CO, as in Fig. 1) or oxidized (mostly H_2O or CO_2, as

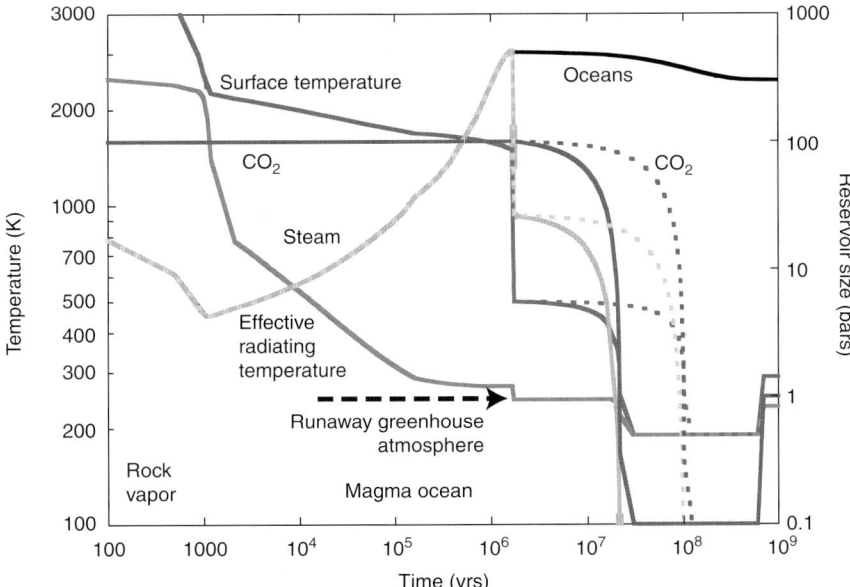

Figure 3. A schematic but energetically self-consistent history of temperature, water, and CO_2 during the Hadean (after Zahnle et al. 2006). Steam and CO_2 are in bars and read off the right hand axis; the pressure equivalent of the modern ocean is 270 bars. The Hadean began with a bang. (i) For 100–1000 yr after the Moon-forming impact Earth was enveloped in rock vapor. (ii) This was followed by a deep magma ocean that lasted for some 2 million years. The cooling rate was controlled by the thermal blanketing effect of water vapor and other greenhouse gases. (iii) Once the mantle solidified the steam in the atmosphere condensed to form a warm (~500 K) liquid water ocean under ~100 bars of CO_2. This particular model assumes that CO_2 was the major atmospheric gas apart from water. The warm early Earth would have lasted while Earth's CO_2 remained in the atmosphere. Here we assume that CO_2 reacts with the seafloor and is subducted into the mantle on either a 20 Myr (solid curves) or a 100 Myr (dotted curves) time scale. (iv) The combination of the high chemical reactivity of ultramafc impact ejecta with CO_2 and the faint early Sun suggests that, in the absence of another abundant greenhouse gas, the Earth should have become ice covered and very cold (Koster van Groos 1988; Sleep and Zahnle 2001). Occasional big impacts would have brought brief thaws (not shown).

in Fig. 2) probably depends on how quickly the stranded metallic iron from the impactor regathers itself into drops and rains to the core. With the metals gone, the mantle is likely to support a relatively oxidized atmosphere dominated by H_2O and CO_2.

Genda and Abe (2003, 2005) showed that Earth is unlikely to lose much of its water as a result of the Moon-forming impact, although loss of a significant amount of atmosphere is possible. Hence Earth's cooling after the silicates had rained out would have been controlled by the steam in the atmosphere, with the asymptotic cooling rate set by the $\sim$300 K radiating temperature of a water vapor runaway greenhouse. Because the liquid mantle convects easily, it cools as soon as the atmosphere permits. During this time the geothermal heat flow rivals sunlight in magnitude. It is probably also during this time that any iron from the intruder's core that was stranded in the mantle descended as a molten rain to the core and the mantle became metal-free. The mantle would have frozen from the bottom up, because the melting point of silicates is a strong function of pressure. With the cooling rate controlled by the atmosphere it takes $\sim$2 Myrs for magma ocean to freeze to the surface.

Most of the water that had been dissolved in the mantle degassed as the volume of magma shrank. Two million years is not enough time for selective hydrogen escape to oxidize the planet significantly, so the relevant buffer is likely to have remained reduced. Although the radiating temperature is cool the surface temperature is >1500 K. Kinetic inhibitions against forming CH_4 and NH_3 in the gas phase suggest that the gas composition was mostly H_2, CO, H_2O, CO_2, and N_2, probably in this order.

Once the mantle has mostly solidified, the mantle no longer convects easily and the geothermal heat flow drops below the point where it can support a runaway greenhouse atmosphere. At this point the steam rains out to make hot salty oceans, salty because NaCl had been an abundant gas in the atmosphere. Other water soluble volatiles would also likely enter the ocean; these may have included various forms of sulfur compounds. Gases more volatile than H_2O and only sparingly soluble in hot water would stay in the atmosphere. Based on the modest solubility of CO_2 in magma we estimated that the atmosphere would have contained 100–200 bars of CO_2 and the surface temperature would have been $\sim$500 K. The preference for CO_2 over CO or CH_4 assumes a QFM-like (or CI- or CV-like) redox buffer. However, if a more reduced QFI-like buffer (EH-, EL-, or H-like) still applied, the 500 K atmosphere would be mostly methane with very little CO or CO_2. This seems unlikely, because the ocean and atmosphere were in close contact with a balsaltic seafloor crust that is not very reducing.

The fourth stage begins when CO_2 is removed from the atmosphere. As Urey (1952b) pointed out, a thick primordial CO_2 atmosphere, a liquid water ocean, and a basalt crust do not make a stable tripod. The CO_2 will react to make carbonates and thus be removed from the atmosphere (Sleep et al. 2001). Vigorous hydrothermal circulation through the oceanic crust and rapid mantle turnover could have moved 100 bars of CO_2 from the atmosphere into the mantle in less than 10 million years (Sleep 2010). However, the rest of the CO_2 cycle remains obscure, which makes it difficult to predict how quickly this would have happened or what the asymptotic atmospheric CO_2 level would have been (Sleep et al. 2001).

After the Moon-forming Impact

Without potent greenhouse gases, the surface of the Hadean Earth might have been extremely cold because the young Sun was faint. This would be the case if most of Earth's CO_2 had reacted with rock to form carbonates (Sleep and Zahnle 2001). The surface temperature could then drop as low as $\sim$220 K and the oceans would be ice-covered, somewhat resembling the oceans of Europa. But unlike Europa, Earth's atmosphere was thick and the heat flow would have been high enough to ensure that the ice cover would have been thin (<100 m) in many places. Such thin ice would break and form leads. Emergent active volcanoes would each have skirts of liquid water. Volcanic

CO_2 would only build up to a point, because it would enter the ocean through the holes in the ice. This describes a negative feedback, in which CO_2 is regulated so that its greenhouse effect is just big enough to ensure some open water.

Another factor affecting the climate is that hundreds or thousands of substantial asteroid impacts would have melted the ice, each impact triggering a brief impact summer. Iceball Earth is an interesting possibility that has at times been regarded as being favorable to the origin of life because freezing pools concentrate HCN and H_2CO, which combine in interesting ways (Bada et al. 1994). This path to the origin of life was featured by Leslie Orgel and Stanley Miller in Bronowski's The Ascent of Man (Bronowski 1973).

Several workers have shown that lightning (Chameides and Walker 1981) or bolide impacts (Fegley et al. 1986; Chyba and Sagan 1992) can produce modest amounts of HCN from strong shock heating in CO-N_2-H_2O atmospheres. Yields are about three orders of magnitude larger in methane-ammonia atmospheres, and three orders smaller if the atmosphere is mostly CO_2 rather than CO. Somewhat surprisingly, given the null results obtained in similar experiments by Schlessinger and Miller (1983), Cleaves et al. (2008) report modest yields of ammonia, HCN, and amino acids in spark discharge experiments in CO_2-N_2-water systems, especially if they allow the pH of the water to become acidic. Recently Sugita and Schultz (2009) report that experiments simulating strong shocks in possible cometary materials destroy complicated organic matter but also generate HCN.

Very large impacts can affect atmospheric chemistry in another way by wholly resetting the chemistry of the atmosphere, in manner akin to the Moon-forming impact but on a much smaller scale. South Pole-Aitken is the biggest impact crater on the Moon (it is on the far side). Orientale is representative of the great craters of the lunar late bombardment: there are about a dozen of these on the Earth-facing side of the Moon; others include Crisium, Serenitatis, Nectaris, and Imbrium. Earth would have experienced hundreds of comparable impacts ca. 3.9 Ga during the late lunar bombardment. Figures 4 and 5 illustrate the effects of South Pole-Aitken and Orientale-scale impacts on the Earth.

Solar ultraviolet light (UV) is another source of disequilibrium chemistry. For the Miller-Urey methane-ammonia atmosphere UV is both good and bad: it drives the chemistry that makes interesting molecules, but that chemistry can also destroy methane and ammonia. These two consequences are inseparable.

It is sometimes suggested that high altitude organic hazes can have a shielding effect that protects the methane and ammonia from rapid destruction by UV (Sagan and Chyba 1997). Titan's atmosphere is often regarded as chemically analogous to early Earth but the comparison is strained because its atmosphere is too cold for oxygen (the element) to have a significant role. But Titan is enough of an analog to suggest that such hazes will be ineffective. Titan's organic hazes are produced by UV and solar wind irradiation of CH_4 and N_2. The rate that CH_4 is destroyed is equal to the rate at which hydrogen escapes to space. The hydrogen escape rate was measured by the Cassini spacecraft, which few through the expanding cloud of hydrogen, and was found to be comparable to the incident UV photon flux (Cui et al. 2008). The conclusion is that the organic haze offers very little protection against the destructive or constructive effects of incident radiation.

Pinto et al. (1980) and Wen et al. (1989) showed theoretically that formaldehyde can be produced photochemically in small amounts in CO_2-dominated atmospheres. Bar-Nun and Chang (1983) obtained small amounts of interesting organic matter by irradiating CO-H_2O mixtures at 185 nm. Zahnle (1986) showed that HCN can be produced in high yields photochemically from EUV photolysis of N_2 in CO_2-N_2 atmospheres provided that methane is present in significant amounts. The process exploits the high EUV radiation emitted by the young active Sun (the young Sun was only 70% as bright as it is now, but it was a much larger source of nonthermal radiation, including vacuum UV and X-rays, by factors of

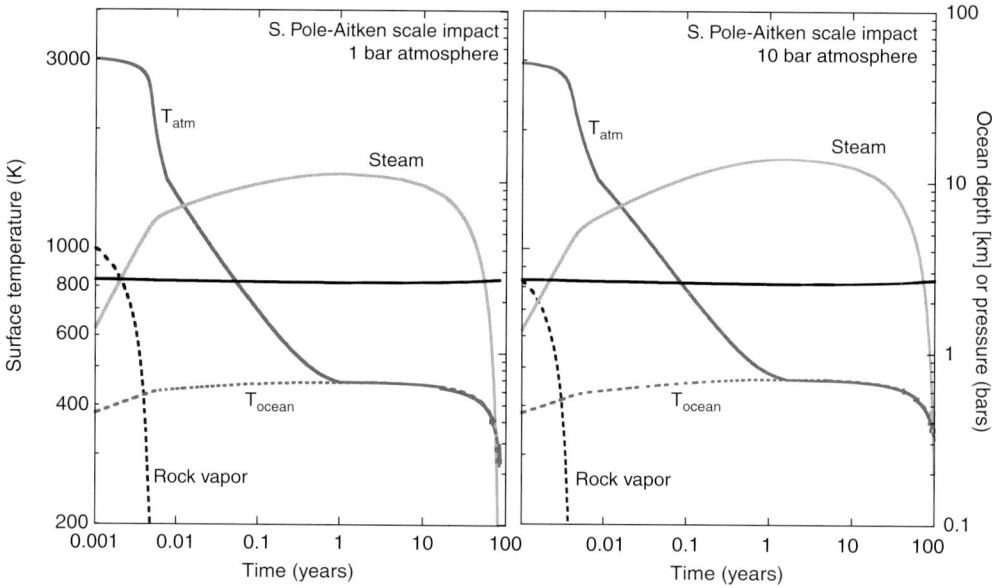

Figure 4. Thermal response of Earth to an impact on the scale of the impact that created the lunar S. Pole-Aitken basin. Tens of these occurred on Earth during the Hadean. The total energy released is 10^{34} ergs; we assume that half promptly escapes to space or is deeply buried. The calculations assume an ocean of water (or ice) and either 1 or 10 bars of preexisting atmosphere. The method of calculation is described in Nisbet et al. (2006). The plot shows the temperatures of the atmosphere and ocean, the effect of evaporation on the depth of the ocean, and the pressures of steam and rock vapor. These atmospheres are thin enough to be heated to the temperature of rock vapor, which resets the chemical state of the entire atmosphere. The atmosphere will likely equilibrate with the rock vapor, which in many cases will be dominated by materials from the impacting body. An upper limit on the consequences for an ordinary chondritic impactor can be inferred from Figure 1. If the effective quench temperature is ~ 1500 K, the result would be a CO-rich atmosphere. If the effective quench temperature were as low as ~ 1000 K, the result would be methane. In practice these are upper limits because the atmosphere itself before the impact will have been more oxidized (either by hydrogen escape or by interaction with the mantle), and these pre-existing oxidants would need to be accounted for.

10–100). Such an atmosphere would require a significant methane source. One possibility is low temperature equilibria pertinent to hydrothermal circulation through new seafoor, another would be impact degassing of chondritic impactors as discussed in detail earlier.

CHALLENGES AND FUTURE RESEARCH DIRECTIONS

Delivering all Earth's volatiles after the Moon-forming impact is quantitatively challenging. Extreme siderophile elements—elements such as osmium and iridium that are strongly concentrated in the metallic core—are found in Earth's mantle in the relative proportions that they are found in chondritic meteorites, but at much smaller concentrations. The extreme siderophiles imply that Earth accreted the equivalent of a 20 km thick blanket of chondritic materials after iron had stopped migrating to the core (Anders 1989). Some of this may represent material from the core of the Moon-forming impactor. In either case, osmium isotopes show that this material resembled ordinary chondrites rather than the more volatile-rich carbonaceous chondrites (Meisel et al. 2001; Drake and Righter 2002). The total volatile load delivered by 20 km of ordinary chondrites that are 0.3% water by

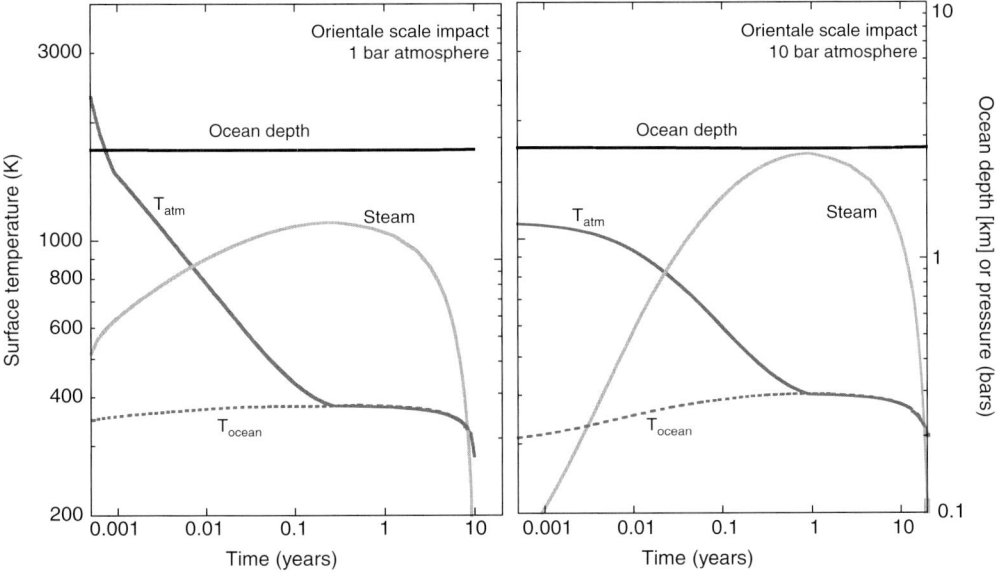

Figure 5. Thermal response of Earth to an impact on the scale of the impact that created the lunar Orientale basin. Hundreds of these occurred on Earth during the Hadean. The total energy is 10^{33} ergs. The plot shows the temperatures of the atmosphere and ocean, the effect of evaporation on the depth of the ocean, and the steam pressure. The 1 bar atmosphere is thin enough that rock vapor should briefly raise the temperature of the atmosphere to $\sim$2500 K, but the event is brief and the system may not time enough to equilibrate thermally or chemically before the rock vapor precipitates. The 10-bar atmosphere is thick enough to effectively quench the impact, and a higher fraction of the impact's energy goes into evaporating rather than into thermal radiation to space. The chemical consequences of such impacts might be considerable, especially in the latter case where methane would seem strongly favored.

weight (Schaefer and Fegley 2007) is equivalent to the delivery of only 200 m of water, well short of an ocean. Only the wettest vectors, e.g., comets, could deliver an ocean of water in such a small total mass.

Application of Schaefer and Fegley (2007, 2010) and Hashimoto et al. (2007) scenarios to the late volatile hypothesis is straightforward. Earth is fully formed and impact velocities are high. Hence impacts are extremely energetic and to first approximation we can expect all the volatiles (and many of the refractory elements as well) carried by the impactors to be vaporized by impact, to equilibrate chemically with the other materials of the impactor, and to enter the atmosphere. This situation was addressed by Hashimoto et al. (2007) and Schaefer and Fegley (2010) using carbonaceous chondrites as the starting material. Both find that the gases that result are rather strongly reducing despite the minerals being rather oxidized. This occurs because these meteorites are carbon rich, and the bulk of the carbon is present in reduced form (Fig. 2). Thus the impacts of the late bombardment represent a substantial stochastic source of reduced gases to the atmosphere.

Kasting (1990) used an atmospheric photochemical model with redox tracking to show that meteoritic material falling on young Earth can produce a CO-rich atmosphere. In these models the atmospheric chemistry is computed assuming a steady state redox balance between reduced sources (volcanic gases and exogenic input of reduced matter, principally meteoritic iron and iron sulfides) and hydrogen escape. The latter is presumed to take place at the physical upper limit imposed by diffusion of hydrogen through the background gases. The presumption is that escape is easy once the hydrogen reaches the thermosphere. Kasting did not predict the synthesis of CH_4 or NH_3 because he knew no effective photochemical pathways for converting CO to CH_4 or N_2 to NH_3, and he assumed the geologist's CO_2-N_2-H_2O atmosphere as a base state. Kasting found that the incoming flux of meteoritic iron could be big enough to flip the redox state of the atmosphere from $CO_2 \gg CO$ to $CO \gg CO_2$, although to do so the mean impact flux had to be comparable to the upper bound inferred from the lunar impact record. On the other hand, Kasting treated the input of meteoritic iron as a continuous function, which is not a good approximation for a distribution of impacts, which tend to be dominated by the few largest objects. This means that the delivery of new iron and iron sulfide would be stochastic, with very large excursions from the mean. If the transient effect of impacts is considered, it seems very likely that each major impact would convert a photochemically sensitive atmosphere from CO_2 to CO. It has also been suggested that iron grains can catalyze the conversion of CO to CH_4 (Kress and McKay 2004). The central role of iron in an impact-driven atmospheric chemistry was anticipated by Urey (1952b).

A recent model suggesting that hydrogen escape may not have been as rapid as most photochemical calculations have taken it to be (Tian et al. 2005) is directly relevant to Kasting's model. If hydrogen escape were truly inefficient, the lifetime of transient reduced atmospheres would be greatly extended, and the balance between CO and CO_2 would tip strongly in favor of the former. A consequence is that the atmosphere would be CO-dominated through most of the Hadean. Tian et al.'s argument that hydrogen escape was inefficient on early Earth assumes that the upper atmosphere would be cold as on Venus, because the atmosphere is composed mostly of CO_2. Their claim is controversial because their model depends on gases other than hydrogen to provide the radiative cooling, but their model does not actually include gases other than hydrogen. In particular, the key assumption that the upper atmosphere was cold is not obvious and needs to be addressed quantitatively. Visconti (1975) computed the temperature of Earth's upper atmosphere if the lower atmosphere were anoxic. He obtained thermospheric temperatures well over 1000 K for current solar max EUV fluxes. EUV fluxes from the young Sun would have much bigger (Zahnle and Walker 1982), so that hotter thermospheres would be expected. Visconti's calculations contrast markedly with the cold thermospheres in Tian et al.'s pure hydrogen escape models.

Carbon monoxide is probably the easiest prebiotically interesting gas to generate in the post-Moon-forming-impact Hadean atmosphere. CO is relatively easy to generate abiotically in a wide range of plausible atmospheres, and it is packed with energy (with it organisms can eat water). CO could be formed by lightning, by impact shocks, or photochemically if there is a significant source of meteoritic of volcanic reducing power. The cold dry atmosphere of an iceball Earth is especially favorable to CO (Zahnle et al. 2008). The possible origin-of-life aspects of CO and its derivatives formamide (for pyrimidine synthesis, Powner et al. 2009) and OCS (condensing agent for forming peptide bonds, Leman et al. 2004) are catching attention, especially among prebiotic chemists who are attempting to synthesize an RNA world (Ricardo et al. 2004; Powner et al. 2009). There is also evidence that CO metabolism is very ancient (Ragsdale 2004). Modern methanogens first convert CO_2 to CO with one enzyme (CO dehydrogenase, or CODH), and then apparently send the CO as a gas down a sealed tube to a second enzyme complex where it is used for energy or for cell material (Ragsdale 2004). This is known as the Wood-Ljundahl pathway of carbon assimilation; it can be regarded as parallel to the more famous Calvin cycle. The enzymes are both based on NiFeS cubes, fitting well with a separate speculation that the first metabolism made use of natural iron sulfides as catalysts (Wächtershäuser 1992).

ACKNOWLEDGMENTS

The authors would like to thank NASA's Exobiology and Astrobiology Program for support.

REFERENCES

Abe Y. 1993. Physical state of very early Earth. *Lithos* **30**: 223–235

Abe Y. 1997. Thermal and chemical evolution of the terrestrial magma ocean. *Phys Earth Planet Int* **100**: 27–39.

Abe Y, Matsui T. 1986. Early Evolution of the Earth: Accretion, Atmosphere Formation, and Thermal History. *J Geophys Res* **91**: E291–E302.

Abe Y, Ohtani E, Okuchi T, Righter K, Drake M. 2000. Water in the Early Earth. In *Origin of the Earth and Moon*. R.M. Canup, K. Righter (eds.), University of Arizona Press, pp. 413–433.

Abelson PH. 1966. Chemical Events on the Primitive Earth. *Proc Nat Acad Sci* **55**: 1365–1372.

Ahrens TJ, O'Keefe JD, Lange MA. 1989. Formation of atmospheres during accretion of the terrestrial planets. In *Origin and Evolution of Planetary and Satellite Atmospheres*. S.K. Atreya, J.B. Pollack, M.S. Matthews (eds.), University of Arizona Press, pp. 328–385.

Albarède F. 2009. Volatile accretion history of the terrestrial planets and dynamic implications. *Nature*. **461**: 1227–1233.

Anders E. 1989. Pre-biotic organic matter from comets and asteroids. *Nature* **342**: 255–257.

Arrhenius G, De BR, Alfvén H. 1974. Origin of the ocean. In *The Sea*, Vol. 5: E.D. Goldberg (Ed.) Wiley-Interscience, New York, pp. 839–861.

Aston FW. 1924. The rarity of the inert gases on Earth. *Nature* **114**: 786.

Bada JL, Bigham C, Miller SL. 1994. Impact melting of frozen oceans on the early Earth: Implications for the origin of life. *Proc Natl Acad Sci USA* **91**: 1248–1250.

Bar-Nun A, Chang S. 1983. Photochemical reactions of water and carbon monoxide in Earth's primitive atmosphere. *J Geophys Res* **88**: 6662–6672.

Benz W, Slattery WL, Cameron AGW. 1986. The origin of the moon and the single-impact hypothesis. *Icarus* **66**: 515–535.

Bronowski J. 1973. The Ascent of Man. Little Brown & Co. 448 pp.

Brown H. 1949. Rare gases and the formation of the Earth's atmosphere. In *The Atmosphere of the Earth and Planets*. G. Kuiper (Ed.). Univ. Chicago Press, Chicago, pp. 258–266.

Cameron AGW. 1983. Origin of the atmospheres of the terrestrial planets. *Icarus* **56**: 195–201.

Cameron AGW, Benz W. 1991. The origin of the moon and the single impact hypothesis. *Icarus* **92**: 204–216.

Canil D. 2002. Vanadium in peridotites, mantle redox and tectonic environments: Archean to present. *Earth Planetary Science Letters* **195**: 75–90.

Canup RM. 2004. Simulations of a late lunar-forming impact. *Icarus* **168**: 433–456.

Canup RM, Asphaug E. 2001. Origin of the Moon and the single impact hypothesis. *Nature* **421**: 708–712.

Chameides WL, Walker JCG. 1981. Rates of fxation by lightning of carbon and nitrogen in possible primitive atmospheres. *Origins of Life* **11**: 291–302.

Chyba CF, Sagan C. 1992. Endogenous production, exogenous delivery and impact-shock synthesis of organic molecules: An inventory for the origins of life. *Nature* **355**: 125–132.

Chyba CF, Thomas PJ, Brookshaw L, Sagan C. 1990. Cometary Delivery of Organic Molecules to the Early Earth. *Science* **249**: 366–373.

Clark BC. 1988. Primeval procreative comet pond. *Orig Life Evol Biosph* **18**: 209–238.

Clark PD, Dowling NI, Huang M. 1998. Comments on the role of H_2S in the chemistry of Earth's early atmosphere and in prebiotic synthesis. *J Mol Evol* **47**: 127–32.

Cleaves HJ, Chalmers JH, Lazcano A, Miller SL, Bada JL. 2008. A reassessment of prebiotic organic synthesis in neutral planetary atmospheres. *Orig Life Evol Biosph* **38**: 105–1156.

Cui J, Yelle RV, Volk K. 2008. Distribution and escape of molecular hydrogen in Titan's thermosphere and exosphere. *J Geophys Res* **113**: E10004.

Delano JW. 2001. Redox history of the Earth's interior since approximately 3900 Ma: Implications for prebiotic molecules. *Orig Life Evol Biosph* **31**: 311–341.

Drake M, Righter K. 2002. Determining the composition of the Earth. *Nature* **416**: 39–44.

Fegley B, Prinn RG, Hartman H, Watkins GH. 1986. Chemical effects of large impacts on the Earth's primitive atmosphere. *Nature* **319**: 305–308.

Ferris JF, Joshi PC, Edelson EH, Lawless JG. 1978. HCN: A plausible source of purines, pyrimidines and amino acids on the primitive Earth. *J Mol Evol* **11**: 293–311

Frei R, Gaucher C, Poulton SW, Canfield DE. 2009. Fluctuations in Precambrian atmospheric oxygenation recorded by chromium isotopes. *Nature* **461**: 250–253.

French BM. 1966. Some Geological Implications of Equilibrium between Graphite and a C-H-O Gas Phase at High Temperatures and Pressures. *Rev Geophys Space Phys* **4**: 223–253.

Genda H, Abe Y. 2003. Survival of a proto-atmosphere through the stage of giant impacts: The mechanical aspects. *Icarus* **164**: 149–162.

Genda H, Abe Y. 2005. Enhanced atmospheric loss on protoplanets at the giant impact phase in the presence of oceans. *Nature* **433**: 842–844.

Harrington J, de Pater I, Brecht SH, Deming D, Meadows V, Zahnle K, Nicholson P. 2004. Lessons from Shoemaker-Levy 9 about Jupiter and Planetary Impacts. In *Jupiter: The Planet, Satellites and Magnetosphere*. F. Bagenol, T. Dowling, W. McKinnon, Eds. Cambridge Univ. Press. pp 158–184.

Hashimoto GL, Abe Y, Sugita S. 2007. The chemical composition of the early terrestrial atmosphere: Formation of a reducing atmosphere from CI-like material. *J Geophys Res* **112**: E05010.

Hayashi C, Nakazawa K, Mizuno H. 1979. Earth's melting due to the blanketing effect of the primordial dense atmosphere. *Earth Planet Sci Lett* **43**: 22–28.

Holland HD. 1962. Model for the evolution of the earths atmosphere. In *Petrologic studies: a volume in honor of A.G. Buddington*, E.J. Engle, H.L. James, B.F. Leonard (eds.). Geological Society of America, Boulder, pp 447–477.

Holland HD. 1984. The Chemical Evolution of the Atmosphere and Oceans. Princeton University Press, Princeton, pp. 582.

Holland HD. 2002. Volcanic gases, black smokers, and the Great Oxidation Event. *Geochimica et Cosmochimica Acta* **66:** 3811–3826.

Ikoma M, Genda H. 2006. Constraints on the Mass of a Habitable Planet with Water of Nebular Origin. *Astrophys J* **648:** 696–706.

Irvine WM. 1998. Extraterrestrial organic matter: A review. *Orig Life Evol Biosph* **28:** 365–83.

Johnson AP, Cleaves HJ, Dworkin JP, Glavin DP, Lazcano A, Bada JL. 2008. The Miller Volcanic Spark Discharge Experiment. *Science* **322:** 404.

Kasting JF. 1982. J Geophys Res **87:** Pages 3091–3098.

Kasting JF. 1990. Bolide impacts and the oxidation state of carbon in the Earth's early atmosphere. *Orig Life Evol Biosph* **20:** 199–231.

Kasting JF. 1991. CO_2 Condensation and the Climate of Early Mars. *Icarus* **94:** 1–13.

Kasting JF. 1993. Earth's early atmosphere. *Science* **259:** 920–926.

Kasting JF, Catling D. 2003. Evolution of a habitable planet. *Ann Rev Astron Astrophys* **41:** 429–463.

Koster van Groos AF. 1988. Weathering, the carbon cycle, and the differentiation of the continental crust and mantle. *J Geophys Res* **93:** 8952–8958.

Kress ME, McKay CP. 2004. Formation of methane in comet impacts: Implications for Earth, Mars, and Titan. *Icarus* **168:** 475–483.

Kuhn WR, Atreya SK. 1979. Ammonia photolysis and the greenhouse effect in the primordial atmosphere of the earth. *Icarus* **37:** 207–213.

Lange MA, Lambert P, Ahrens TJ. 1985. Shock effects on hydrous minerals and implications for carbonaceous meteorites. *Geochim Cosmochim Acta* **49:** 1715–1726.

Lasaga AC, Holland HD, Dwyer MJ. 1971. Primordial Oil Slick. *Science* **174:** 53–55.

Leman L, Orgel L, Ghadiri MR. 2004. Carbonyl Sulfide Mediated Prebiotic Formation of Peptides. *Science* **306:** 283–286.

Lewis JS, Prinn RG. 1984. Planets and their atmospheres—Origin and evolution. Academic Press, Orlando, 480 pp.

Lodders K, Fegley B. 1997. An oxygen isotope model for the composition of Mars. *Icarus* **126:** 373–394.

Matsui T, Abe Y. 1986a. Impact-induced atmospheres and oceans on earth and Venus. *Nature* **322:** 526–528.

Matsui T, Abe Y. 1986b. Evolution of an impact-induced atmosphere and magma ocean on the accreting earth. *Nature.* **319:** 303–305.

Maurette M. 1998. Carbonaceous micrometeorites and the origin of life. *Orig Life Evol Biosph* **28:** 385–412.

Meisel T, Walker RJ, Irving AJ, Lorand J-P. 2001. Osmium isotopic compositions of mantle xenoliths: A global perspective. *Geochim Cosmochim Acta* **65:** 1311–1323.

Miller SL. 1953. A Production of Amino Acids under Possible Primitive Earth Conditions. *Science* **117:** 528–529.

Miller SL. 1955. Production of some organic compounds under possible primitive Earth conditions. *J Am Chem Soc* **77:** 2351–2361.

Miller SL. 1986. Current status of the prebiotic synthesis of small molecules. *Chem Scr* **26B:** 5–11.

Miller SL, Bada JL. 1988. Submarine hot springs and the origin of life. *Nature* **334:** 609–611.

Miller SL, Bada JL. 1993. Comment on "Summary and implications of reported amino acid concentrations in the Murchison meteorite" by EL Shock and MD Schulte. *Geochim Cosmochim Acta* **57:** 3473–3474.

Miller SL, Urey HC. 1959. Organic compound synthesis on the primitive Earth. *Science* **130:** 245–251.

Nisbet EG, Zahnle KJ, Gerasimov MV, Helbert J, Jaumann R, Hofmann BA, Benzerara K, Westall F. 2006. Creating habitable zones, at all scales, from planets to mud microhabitats, on Earth and on Mars. *Space Sci Rev* **129:** 79–121.

Oró J, Kamat S. 1961. Amino-acid synthesis from hydrogen cyanide under possible primitive Earth conditions. *Nature* **190:** 442–443.

Oró J, Miller SL, Lazcano A. 1990. The origin and early evolution of life on Earth. *Ann Rev Earth Planet Sci* **18:** 317–56.

Pasek M, Lauretta D. 2008. Extraterrestrial fux of potentially prebiotic C, N, and P to the early Earth. *Orig Life Evol Biosph* **38:** 5–21.

Pierazzo E, Chyba CF. 1999. Amino acid survival in large cometary impacts. *Meteoritics Planet Sci* **34:** 909–918.

Pinto JP, Gladstone GR, Yung YL. 1980. Photochemical Production of Formaldehyde in Earth's Primitive Atmosphere. *Science* **210:** 183–185

Poole JHJ. 1951. The evolution of the earth's atmosphere. *Sci Proc Roy Dublin Acad* **25:** 201–224.

Powner MW, Gerland B, John D, Sutherland JD. 2009. Synthesis of activated pyrimidine ribonucleotides in prebiotically plausible conditions. *Nature* **459:** 239–242.

Ragsdale SW. 2004. Life with carbon monoxide. *Crit Rev Biochem Mol Biol* **39:** 165–195.

Ricardo A, Carrigan MA, Olcott AN, Benner SA. 2004. Borate Minerals Stabilize Ribose. *Science* **303:** 196.

Russell HN, Menzel DH. 1933. The Terrestrial Abundance of the Permanent Gases. *Proc Nat Acad Sci* **19:** 997–1001.

Sagan C, Chyba CF. 1997. The early faint sun paradox: Organic shielding of ultraviolet-labile greenhouse gases. *Science* **276:** 1217–1221.

Schaefer L, Fegley B. 2007. Outgassing of ordinary chondritic material and some of its implications for the chemistry of asteroids, planets, and satellites. *Icarus* **186:** 462483.

Schaefer L, Fegley B. 2009. Chemistry of Silicate Atmospheres of Evaporating Super-Earths Submitted to Astrophys. *J Lett* [in press].

Schaefer L, Fegley B. 2010. Chemistry of Atmospheres Formed during Accretion of the Earth and Other Terrestrial Planets, Submitted to Icarus 17 Sept. 2009.

Schlesinger G, Miller SL. 1983. Prebiotic synthesis in atmospheres containing CH_4, CO and CO_2. I Amino acids *J Mol Evol* **19:** 376–382.

Sekiya M, Hayashi C, Kanazawa K. 1981. Dissipation of the primordial terrestrial atmosphere due to irradiation of the solar far-UV during T tauri stage. *Prog Theor Phys* **66:** 1301–1316.

Sekiya M, Nakazawa K, Hayashi C. 1980. Dissipation of the primordial terrestrial atmosphere due to irradiation of the solar EUV. *Prog Theor Phys* **64:** 1968–1985.

Shock EL. 1990. Geochemical constraints on the origin of organic compounds in hydrothermal systems. *Orig Life Evol Biosph* **20:** 331–367.

Shock EL, McCollom T, Schulte M. 1995. Geochemical constraints onbchemolithoautotrophic reactions in hydrothermal systems. *Orig Life Evol Biosph* **25:** 141–159.

Shock EL, McKinnon WB. 1993. Hydrothermal processing of cometary volatiles—Applications to Triton. *Icarus* **106:** 464–477.

Shock EL, Schulte MD. 1990. Amino-acid synthesis in carbonaceous meteorites by aqueous alteration of polycyclic aromatic hydrocarbons. *Nature* **343:** 728–731.

Shock EL, Schulte M. 1993. Reply to the comment by S. L. Miller and J. L. Bada on "Summary and implications of reported amino acid concentrations in the Murchison meteorite." *Geochim Cosmochim Acta* **57:** 3475–3477.

Shock EL, Schulte M. 1998. Organic synthesis during fluid mixing in hydrothermal systems. *J Geophys Res* **103:** 28513–28528.

Sleep NH. 2010. The Hadean-Achaean environment. *Cold Spring Harb Perspect Biol* **2:** a002527.

Sleep NH, Zahnle KJ. 2001. Carbon dioxide cycling and implications for climate on ancient Earth. *J Geophys Res* **106:** 1373–1399.

Sleep NH, Zahnle KJ, Neuhoff PS. 2001. Initiation of clement surface conditions on the earliest Earth. *Proc Nat Acad Sci* **98:** 3666–3672.

Stribling R, Miller SL. 1987. Energy yields for hydrogen cyanide and formaldehyde syntheses: The HCN and amino acid concentrations in the primitive ocean. *Origins Life* **17:** 261–273.

Sugita S, Schultz P. 2009. Efficient cyanide formation due to impacts of carbonaceous bodies on a planet with a nitrogen-rich atmosphere. *Geophys Res Lett* **36:** L20204.

Summons RE, Bradley AS, Jahnke LL, Waldbauer JR. 2006. Steroids, triterpenoids and molecular oxygen Philosophical Transactions. *Royal Society B-Biological Sciences* **361:** 951–968.

Tian F, Toon O, Pavlov A, Sterck De H. 2005. A hydrogen-rich early Earth atmosphere. *Science* **308:** 1014–1017.

Tyburczy JA, Frisch B, Ahrens TJ. 1986. Shock-induced volatile loss from a carbonaceous chondrite Implications for planetary accretion. *Earth Planet Sci Lett* **80:** 201–207.

Urey HC. 1951. The origin and development of the earth and other terrestrial planets. *Geochim Cosmochim Acta* **1:** 209–277.

Urey HC. 1952a. The origin and development of the earth and other terrestrial planets: A correction. *Geochim Cosmochim Acta* **2:** 263–268.

Urey HC. 1952b. On the early chemical history of the earth and the origin of life. *Proc Nat Acad Sci* **38:** 351–363.

Visconti G. 1975. The exospheric temperature of a primitive terrestrial atmosphere with evolving oxygen content. *J Atmos Sci* **32:** 1631–1637.

Wächtershäuser G. 1992. Groundworks for an evolutionary biochemistry: The iron-sulfur world. *Prog Biophys Molec Biol* **58:** 85–201.

Wänke H. 1981. Constitution of the terrestrial planets. *Phil Trans R Soc London* **A303:** 287–302.

Wen JS, Pinto JP, Yung YL. 1989. Photochemistry of CO and H_2O: Analysis of laboratory experiments and applications to prebiotic Earth's atmosphere. *J Geophys Res* **94:** 14957–70.

Whittet DC. 1997. Is extraterrestrial organic matter relevant to the origin of life on Earth? *Orig Life Evol Biosph* **27:** 249–62.

Zahnle KJ. 1986. Photochemistry of methane and the formation of hydrocyanic acid (HCN) in the Earth's early atmosphere. *J Geophys Res* **91:** 2819–2834.

Zahnle KJ. 1990. Atmospheric chemistry by large impacts. In *Global Catastrophes in Earth History*. V. Sharpton, P. Ward, Eds., GSA Special Paper. 247: 271–288.

Zahnle KJN, Arndt N, Cockell C, Halliday A, Nisbet E, Selsis F, Sleep NH. 2006. Emergence of a habitable planet. *Space Sci Rev* **129:** 35–78.

Zahnle KJ, Kasting JF, Pollack JB. 1988. Evolution of a steam atmosphere during Earth's accretion. *Icarus* **74:** 62–97.

Zahnle K, Mac Low M-M, Lodders K, Fegley B. 1995. Sulfur chemistry in the wake of Comet Shoemaker-Levy 9. *Geophys Res Lett* **22:** 1593–1596.

Zahnle KJ, Walker JGG. 1982. Evolution of solar ultraviolet luminosity. *Rev Geophys Space Phys* **20:** 280–292.

Planetary Organic Chemistry and the Origins of Biomolecules

Steven A. Benner, Hyo-Joong Kim, Myung-Jung Kim, and Alonso Ricardo

Foundation for Applied Molecular Evolution and The Westheimer Institute for Science and Technology, Gainesville, Florida 32601

Correspondence: sbenner@ffame.org

Organic chemistry on a planetary scale is likely to have transformed carbon dioxide and reduced carbon species delivered to an accreting Earth. According to various models for the origin of life on Earth, biological molecules that jump-started Darwinian evolution arose via this planetary chemistry. The grandest of these models assumes that ribonucleic acid (RNA) arose prebiotically, together with components for compartments that held it and a primitive metabolism that nourished it. Unfortunately, it has been challenging to identify possible prebiotic chemistry that might have created RNA. Organic molecules, given energy, have a well-known propensity to form multiple products, sometimes referred to collectively as "tar" or "tholin." These mixtures appear to be unsuited to support Darwinian processes, and certainly have never been observed to spontaneously yield a homochiral genetic polymer. To date, proposed solutions to this challenge either involve too much direct human intervention to satisfy many in the community, or generate molecules that are unreactive "dead ends" under standard conditions of temperature and pressure. Carbohydrates, organic species having carbon, hydrogen, and oxygen atoms in a ratio of 1:2:1 and an aldehyde or ketone group, conspicuously embody this challenge. They are components of RNA and their reactivity can support both interesting spontaneous chemistry as part of a "carbohydrate world," but they also easily form mixtures, polymers and tars. We describe here the latest thoughts on how on this challenge, focusing on how it might be resolved using minerals containing borate, silicate, and molybdate, inter alia.

Interesting organic chemistry occurs throughout the cosmos, including in presolar nebulae (see the article in this collection by Pascale Ehrenfreund), asteroidal bodies (see the article in this collection by Sandra Pizzarello) and icy bodies near the outer boundary of our solar system (Bernstein et al. 2002). Although organic molecules made in off-Earth locales almost certainly contributed to the reduced carbon inventory on Earth before life emerged, planetary processing on Earth undoubtedly also contributed to the inventory of prebiotic molecules that were available to life as it originated (assuming that Earth was the site of life's origin). Indeed, in the RNA first model for the origin of life on Earth (Joyce and Orgel 1999)(Benner 2009), it is often proposed that terran-based chemistry produced RNA in oligomeric form to initiate Darwinian evolution.

How are we to constrain models for planetary processing to converge on a model for what actually happened on Earth four billion years ago? Today, atmospheric dioxygen (O_2) readily converts organic materials to carbon dioxide, making it essentially impossible to observe such processing on the surface of Earth. Furthermore, the ubiquity of life on modern Earth means that any organic processing is more likely to reflect biology than prebiology. The closest we may come today to observe organic transformations absent biology on a planetary scale might be on Titan, a moon of Saturn whose atmosphere and surface is rich in reduced carbon.

Nevertheless, it is possible to apply a general understanding of organic chemical reactivity to suggest chemical reactions that might have occurred on early Earth and the products that they might have produced. These suggestions are constrained by models for the atmosphere and mineralogy of early Earth, although these constraints might change as models improve.

In this article, we assume that the atmosphere of early Earth was less oxidizing than today's atmosphere, although not as rich in methane as the simulated atmosphere used in the classic experiments of Stanley Miller (Miller 1955). Further, we assume that the atmosphere on early Earth had access to many sources of energy. These include electrical discharge, ultraviolet and visible light (although the Sun was almost certainly dimmer then than now, a Titan-like haze may have prevented high energy photons from reacting the Earth's surface), volcanism (providing not only heat but also reactive species and mixtures not at thermodynamic equilibrium), ionizing radiation, and impacts. (See Pizzarello and Shock 2010 for a discussion of such energy sources.)

We also assume that life emerged after the planet underwent a geological fractionation in which heavier minerals and elemental iron sank towards the core, leaving lighter rocks to form the crust. Open questions concern the inventory of water relative to the surface of early Earth, an inventory that determined whether planetary organic transformations might have occurred on dry land or below water on a planetary surface that was totally submerged.

BACKGROUND

Organic Molecules with Energy Spontaneously Yield Polymers and Complex Mixtures

Any model for planetary organic chemistry must recognize that very few organic molecules are thermodynamically stable in water, either with respect to conversion to their fully hydrated state or decomposition upon heating to elemental carbon (charcoal), carbon dioxide, or other thermodynamic "end points." Accordingly, any model for planetary prebiotic chemistry must address the *metastability* of organic species. This word captures the concept that organic molecules that have appeared through the interaction of precursors with energy, water, and other organics can then disappear upon further interaction. As many authors have noted (Cairns-Smith 1982; Shapiro 1987; Shapiro 2007), any prebiotic reaction scheme that requires two or more organic species must be concerned about the metastability of two or more species; any scheme that does not produce both components in useful concentrations *at the same time* will not meet a standard of proof that the community need to accept a solution to the problem.

Here, too, the mineral inventory of early Earth cannot be ignored. Minerals of many kinds may have guided the reactivity of organic species that emerged on early Earth, altered their metastability, and influenced the time when specific organic species were available to emerging life.

Carbohydrates Embody this Natural Propensity to Polymerize

To explore these points and develop the scientific methods that enable this exploration, we will use carbohydrates as a focus. Carbohydrates are organic species having carbon, hydrogen and oxygen atoms in a ratio of 1:2:1. They are therefore at the same oxidation level of elemental carbon. Furthermore the simplest carbohydrate, formaldehyde (HCHO, $H_2C=O$, or $C_1H_2O_1$) is easily generated by electrical discharge or ultraviolet radiation impinging on moist atmospheres that are rich in carbon dioxide; nearly every contemporary model for early Earth permits such atmospheres (Pinto et al. 1980; Cleaves 2008). Because they contain ketone and aldehyde groups (see later), carbohydrates have interesting reactivity that includes the ability to form new carbon-carbon bonds under "standard conditions," defined as those where water is liquid under contemporary terran atmospheric pressure. By comparison, the compounds that often concern prebiotic chemists (carboxylic acids, fatty acids, and amino acids, for example) have essentially no such reactivity. Indeed, carbohydrates are "high energy" because they can rearrange their constituent atoms to give carboxylic acid derivatives and other more stable end point "sinks."

Recognizing this, some authors, most notably Arthur Weber (Weber 2001a; Weber 2001b; Weber 2007) (see Blackmond 2010), have exploited the reactivity, energy, and prebiotic accessibility of carbohydrates to suggest entire "carbohydrate world" metabolisms at or near life's origins. This exploits the energy of formaldehyde, hydroxyaldehydes and hydroxyketones relative to isomeric forms that have carboxylate groups, to do chemistry, much of it reminiscent of modern metabolism.

Unfortunately, others use the very same reactivity and energy to argue that carbohydrates could not have been present on early Earth (Shapiro 1988). For example, based on the short survival time of ribose, especially at high temperature and high pH, Stanley Miller and his coworkers (Larralde et al. 1995) concluded that "ribose and other sugars were not components of the first genetic material" and *precluded* their presence in prebiotic scenarios. This is despite the fact that simple carbohydrates such as glycolaldehyde are well known in the cosmos (Hollis et al. 2000).

ORGANIC CHEMISTRY SKILLS: A REMINDER

This apparent paradox, where experts within the same community simultaneously exploit and preclude certain compound classes based on the same data, characterizes the "origins" field in general (Benner 2009). To evaluate such contradicting views, participants in this field must understand relevant features of bonding and reactivity in organic molecules. This understanding requires in turn that they understand the formalisms used by organic chemists to describe bonding and reactivity in organic molecules. Although the features and formalisms are taught in introductory organic chemistry courses, it is worth summarizing them here a prelude to discussing constraints on reactions that might be expected to occur in prebiotic environments.

First, a single bond joining two carbon atoms is strong, on the order of 400 kJ ($\sim$100 kcal) per mole. This means that a pair of carbon atoms joined by a typical single bond will remain joined for many millions of years at temperatures when water is a liquid at sea level on Earth near neutral pH ("standard conditions"). This is also true for single bonds between carbon and hydrogen.

In contrast, single bonds between carbon and oxygen carbon and nitrogen, although similarly strong, but tend to be less persistent over time because they confer reactivity upon organic species that possess them. *Heteroatoms* (meaning neither carbon nor hydrogen) create centers of reactivity that are "weak" spots in organic compounds. In particular, they provide paths under standard conditions where bonds between carbon atoms or between carbon and hydrogen atoms can be broken *at the same time* as a bond to *another* atom is formed. This compensates the energy lost in the breaking bond with energy gained with a simultaneously forming bond, allowing the reaction to occur under relatively mild conditions.

Pairs of Electrons form Bonds between Atoms

Understanding ways that new bonds can form as old bonds are breaking is a key to understanding what reactions actually will occur under standard conditions. Recalling general chemistry, covalent bonds between two atoms are formed by the sharing of pairs of electrons (Fig. 1). Thus, in water (H-O-H), the lines between the hydrogen atoms and the central oxygen atom each represent a pair of electrons that form a single bond holding the two hydrogen atoms to the oxygen. In formaldehyde ($H_2C=O$), the double line between the carbon

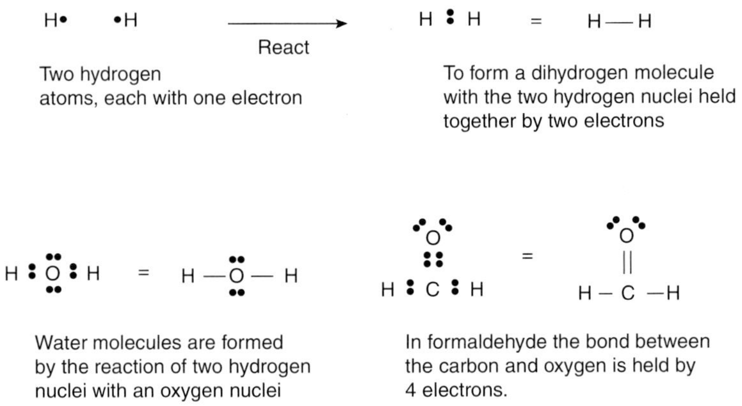

Figure 1. Lewis structures of chemical bonds for dihydrogen (H_2), water (H_2O) and formaldehyde ($H_2C=O$, or CH_2O, or HCHO).

and the oxygen represents two pairs of electrons; four electrons in total bind the C atom to the O atom.

Chemists frequently do not write symbols representing all of the atoms and electrons. Therefore, the first skill required in analyzing reactivity in organic chemistry requires that one put back into a structure the symbols that practiced chemists do not write (but understand are there), completing the structure of the organic molecule. This ensures that the analysis reflects all of the atoms and electrons in a molecule.

A structure completed in this way is known as a *Lewis structure*. A Lewis structure explicitly indicates electrons that are *not* involved in bonding. For example, in the Lewis structure of water, oxygen carries two pairs of unshared electrons from the outer valence shell. We represent each of these valence electrons not involved in a bond by a dot. Hence, the oxygen in the H-O-H structure has four dots, representing electrons on the oxygen but not involved in bonding. Likewise, the oxygen in formaldehyde carries two pairs of unshared electrons, represented again by four dots on the oxygen in the Lewis structure.

A Nucleophilic Center Brings a Pair of Electrons to Form a New Bond

As a chemical bond is a pair of electrons between two atoms, any unshared pair of electrons is available in principle to form a new bond. Further, if a bond breaks, the electrons that *were* in that bond are available to form a new bond. Atoms that contain pairs of electrons available to form a new bond (or can get them by breaking a bond) are called *nucleophilic centers* (Fig. 2).

To form a bond, the electron pair from the nucleophilic center must find an atom that has a vacant orbital or can get one by losing a bond through breakage. This atom is called an *electrophilic center*. An archetypal electrophile is a proton (H^+). A proton is not bonded to anything, has a vacant 1s bonding orbital, so it can form a bond with a single partner (Fig. 3). An archetypal nucleophilic center is the oxygen atom on water. It carries an unshared pair of electrons that can form a new bond to H^+. The product of the reaction between water and H^+ is H_3O^+ (the hydronium ion).

Curved Arrows Describe the Movement of Pairs of Electrons

Organic chemists use *curved arrows* to describe reactions between nucleophilic and electrophilic centers that produce a new bond. The curved arrow begins with an unshared pair of electrons on the nucleophile, the pair that will form the new bond in the product. The arrow is drawn to end at a position (on the structures of the reactants) where the electron pair *will be* after the bond is formed. Figure 3 shows the reaction of the unshared pair of

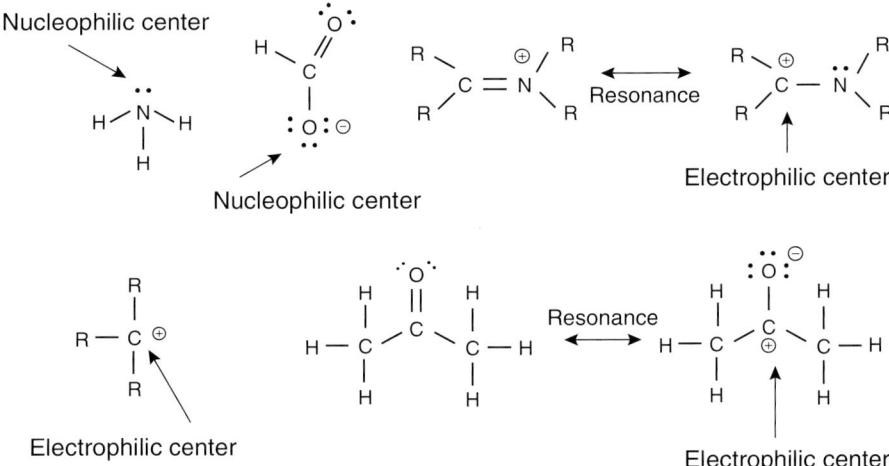

Figure 2. Nucleophilic centers have an unshared pair of electrons that can form a new bond, or can get one (via resonance, for example). Electrophilic centers have a vacant orbital (or can get one via resonance, for example) that can accept an unshared pair of electrons from a nucleophilic center to form a new bond. In a resonance form a pair of electrons moves between adjacent atoms (electrons are dynamic entities), creating a new representation for the same molecule. The resulting resonance forms are joined by a double headed arrow to indicate equivalence.

electrons on the oxygen of water (the nucleophilic center) with H^+ (the electrophilic center) to give H_3O^+.

When nucleophilic centers bear an electron pair prominently placed in a correctly drawn Lewis structure, they are easy to spot. They are less easy to spot when they get the electron pair needed to form a future bond by breaking an existing bond. The same is true for electrophilic centers that obtain vacant orbitals available for forming new bonds only after an existing bond is broken.

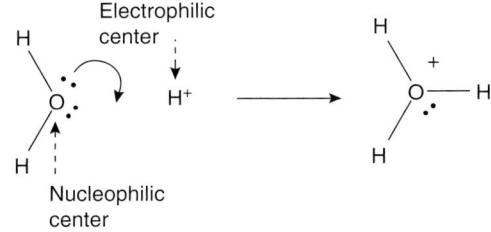

Fig. 3. Reaction of the nucleophilic center on the oxygen of water with an electrophilic center, H^+. The movement of a pair of electrons in the reaction is illustrated using a curved arrow. The result is H_3O^+, the hydronium ion.

For example, the carbon of formaldehyde ($H_2C=O$) has all of its four valences occupied. That carbon does not seem to have a valence available to form a new bond with anything. If, however, one of the two bonds between carbon and oxygen breaks, with the electron pair moving from a position between the carbon and the oxygen to a new position on the oxygen, then the carbon center has a valence free and can undergo nucleophilic attack by an oxygen atom, for instance, from a nearby H_2O molecule.

This process is shown using curved arrows in Figure 4. Here, a bond between carbon and oxygen is broken at the same time as the carbon forms a new bond to an incoming oxygen atom. At the same time as the energy in the second C-O bond is lost through breakage, the energy of a new C-O bond is gained. The resulting product is the hydrate of formaldehyde (H_4CO_2).

CURVED ARROW MECHANISMS FOR THE PREBIOTIC SYNTHESIS OF BIOLOGICAL MOLECULES

The curved arrow tool can be used to describe most reactions of organic molecules under standard conditions. This includes transformations

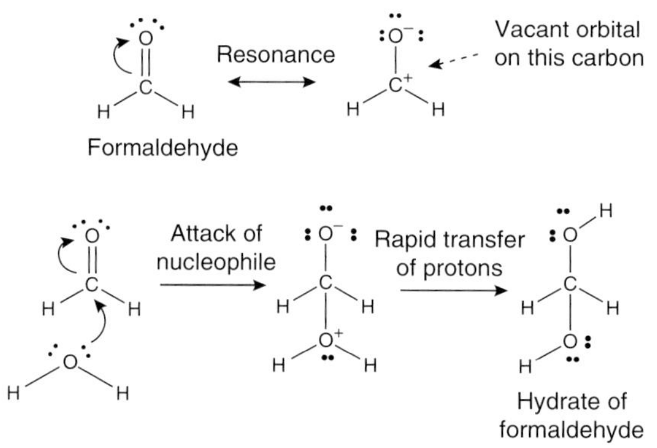

Figure 4. Reaction of the nucleophilic center on the oxygen of water with an electrophilic center, the carbon atom of formaldehyde forms the hydrate of formaldehyde. The movement of a pair of electrons in the reaction is illustrated using a curved arrow.

that might have converted formaldehyde in a prebiotic world into molecules that are characteristic of contemporary terran life.

Curved Arrow Mechanisms for the Formation of Amino Acids

The curved arrow formalism can be used to describe the synthesis from formaldehyde, ammonia, cyanide and water of a simple amino acid, glycine (NH_2CH_2COOH), one of the basic building blocks of proteins. The steps are shown in Figure 5 with species both named and numbered for future reference.

Ammonia (**2**) has three hydrogen atoms and one nitrogen atom. A Lewis structure shows that the nitrogen in ammonia also carries an unshared pair of electrons. The nitrogen atom is therefore a nucleophilic center. Ammonia should therefore react with formaldehyde (**1**) for the same reason that water does. In this reaction, the unshared pair of electrons on ammonia forms a new bond between its nitrogen and the carbon of formaldehyde, just as the pair of electrons forming the second carbon-oxygen bond leaves to form a new bond between the oxygen and H^+. This generates an "amino alcohol" (**3**).

After the transfer of some of the hydrogen atoms, the nitrogen of the amino alcohol again has an unshared pair of electrons, and is able to form a second bond with the carbon atom. The resulting compound is known as an imine (**4**), which contains a C=N unit having a carbon atom bonded twice to a nitrogen atom.

The carbon of the C=N unit is also an electrophilic center. This sets the stage for the next reaction, where the carbon of the cyanide anion (**5**) attacks the imine (**4**) carbon to form an aminonitrile (**6**). The nitrile has a C≡N unit, where the nitrogen is bonded three times to the carbon. This carbon is again an electrophilic center. If a pair of electrons forming one of the bonds between carbon and nitrogen leaves to form a bond with H^+, then the carbon has a free valence. It is therefore available to form a bond with a nucleophilic oxygen atom from water.

The product, again after H^+ atoms are transferred, has another C=O group in a unit known as an *amide* (**7**). The carbon atom of the amide is again an electrophilic center; it can be attacked by the nucleophilic oxygen of another water molecule. This leads to the *hydrolysis* (taking on a water molecule) of the amide and the formation of the amino acid glycine, together with an ammonia molecule.

The net process is the reaction of one molecule of formaldehyde, one molecule of hydrogen cyanide, and one molecule of water to give one molecule of glycine (**8**). In terms of chemical formulas, the reaction is $HCHO + HCN + H_2O = C_2H_5NO_2$; this equation "balances." Ammonia is used in the first step, and is released in the last step. Therefore, ammonia is a

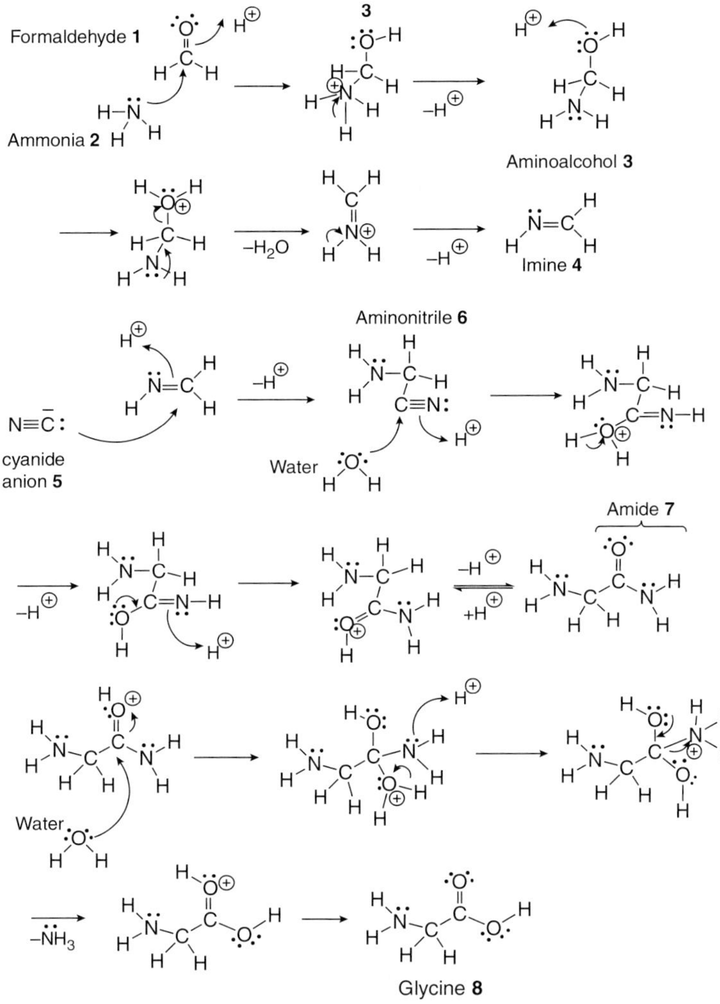

Figure 5. The Strecker synthesis of glycine, an amino acid. Reaction of the nucleophilic centers on the nitrogen of ammonia, the carbon of the cyanide anion and the oxygen of water with electrophilic centers on formaldehyde and key intermediates. The movements of pairs of electrons in the reactions are illustrated using curved arrows.

catalyst for the reaction, being consumed and formed in equal amounts in the reaction cycle.

This sequence of reactions is known as the *Strecker synthesis* of amino acids, named after the chemist who developed it in the 1860s. The Strecker synthesis is driven by the innate reactivity of nucleophiles and electrophiles, and proceeds spontaneously and in reasonable yield. Further, the Strecker synthesis is quite general. It can be used to prepare any amino acid for which the corresponding aldehyde is available, not just formaldehyde. For example, if we start with acetaldehyde (CH_3CHO) rather than formaldehyde, the amino acid alanine ($CH_3CH(NH_2)COOH$) is formed (Fig. 6). Analogous processes can be drawn for many of the amino acids found commonly in proteins from terran organisms.

The Strecker pathway accounts for some products found in Stanley Miller's experiments attempting to reproduce early Earth's atmosphere (Miller 1955). Miller found that amino acids were generated after electrical discharges from electrodes were passed through an

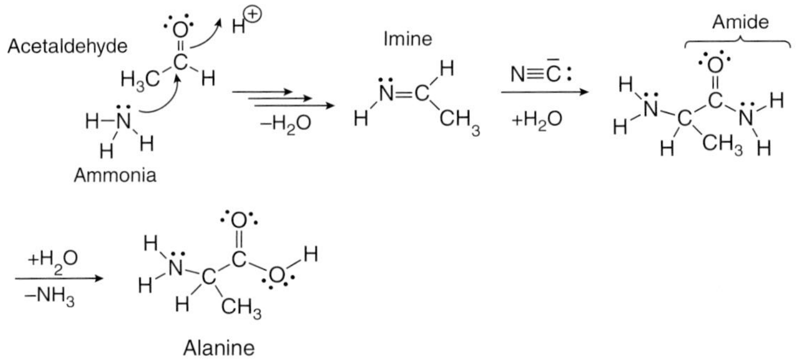

Figure 6. Strecker synthesis of alanine starting from acetaldehyde.

atmosphere of hydrogen, methane, and ammonia over water. The electrical discharge appears to have generated the formaldehyde and cyanide needed as precursors for Strecker syntheses. Once these were formed, the synthesis of amino acids occurred spontaneously.

The simulated atmosphere chosen by Miller for his laboratory experiments was considered at the time to approximate the atmosphere of early Earth. Today, many models hold that the amount of methane on early Earth was much smaller than that used in the Miller experiments. Instead, the carbon inventory of the early Earth is today modeled as being present largely as carbon dioxide (Kasting 1993).

Accordingly, nonbiological syntheses of biomolecules under these conditions have been sought. Again, organic molecules are easy to get when carbon dioxide and water is subjected to electrical discharge, ionizing radiation, and ultraviolet light. The results obtained under a variety of conditions are complex red-brown mixtures of organic molecules, often called "tholins" (without making a distinction between what those mixtures contain) (Sagan et al. 1978) (Sagan and Khare 1979). A partial inventory of molecules comprising certain tholins is shown in Table 1. As the atmosphere of Titan is also red-brown complex mixture of organics, it too may be said to contain "tholins," although by ignoring the details of the mixture, this name is not particularly useful. However, there is little doubt that both formaldehyde and hydrogen cyanide are key intermediates being generated from these atmospheres as well.

Curved Arrow Mechanisms for Forming Nucleobases for RNA and DNA

Curved arrow mechanisms can be used to generate nonbiological routes for the synthesis of many molecules in biology. For example, the

Table 1. Some organic compounds identified in tholin mixtures. (Sagan et al., 1978; Sagan & Khare, 1979; Pietrogrande et al., 2001).

Hydrogen sulfide	Hexene	Formamide*
Hydrogen cyanide	Heptene	Pyridine*
Ammonia	Butadiene	Styrene*
Ethane	Benzene	2,3 Pentadiene*
Propane	Toluene	2-Methylpyrimidine*
Butane	Thiophene	4-Methylpyrmidine*
Ethene	2-Methyltiophene	3- Butenenitrile*
Propene	Methylmercaptan	Butyne*
Butene	Ethylmercaptan	Acetonitrile*
Pentene	Propylmercaptan	Carbon dioxide*
Carbon disulfide	Methylisocyanate	Acetamide*

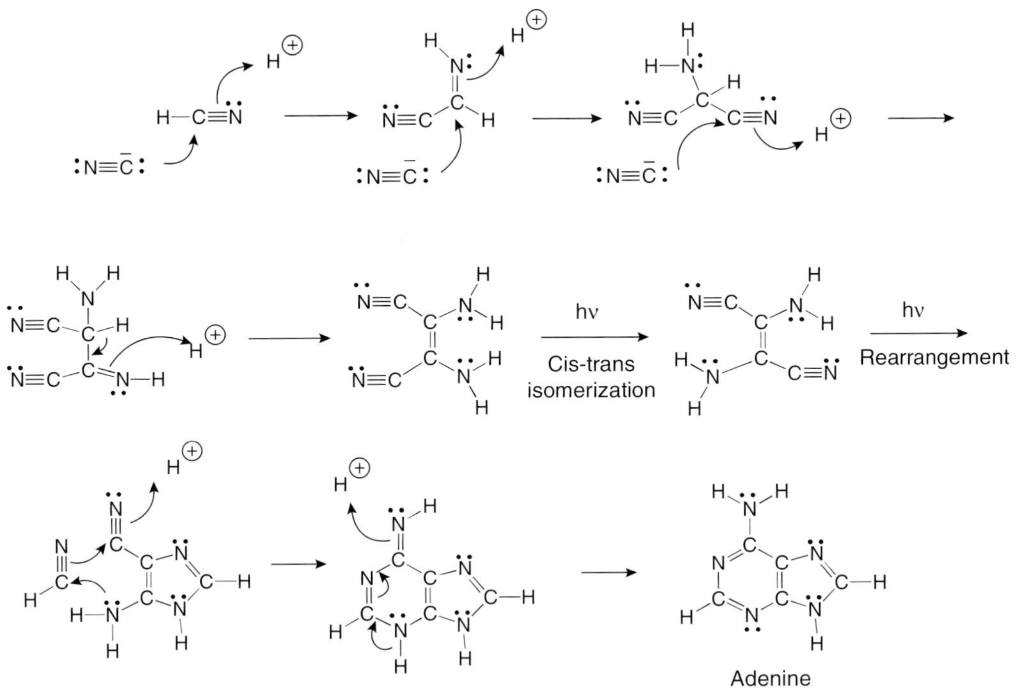

Figure 7. HCN yields adenine via the Oró-Orgel synthesis.

Oró-Orgel synthesis (Fig. 7) exploits the reactivity of HCN to make adenine ($C_5H_5N_5$), one of the five *nucleobases* (symbolized as A, C, T, G, and U) used to store information in DNA and RNA. The cyanide anion again reacts as a nucleophile, this time with a molecule of HCN serving as an electrophile. The combination of reactions was studied in the laboratory Sanchez, Ferris, and Orgel, and provides one of the canonical examples of nucleophile-electrophile chemistry (supplemented with some photochemistry) in the prebiotic chemistry literature (Oró 1960; Sanchez et al. 1967).

Curved Arrow Mechanisms for Forming Carbohydrates

Amino acids are "easy" components of modern terran biochemistry from the perspective of metastability. The carboxyl group of an amino acid has a C=O unit. Although this carbon is potentially an electrophilic center just as the C=O unit of formaldehyde is, the C=O unit of a carboxylic acid is also attached to an −OH unit. This, at neutral pH, loses an H^+ and becomes a COO^- carboxylate anion, a species having a negative charge. This negative charge discourages any electron rich nucleophile from attacking the C=O unit of a carboxylic acid, making it rather stable under standard conditions. The other bonds in an amino acid are single bonds, including C-C, C-H, and C-N bonds. These are, as noted earlier, also rather stable. This means that amino acids are quite metastable, a feature of their reactivity that undoubtedly accounts for their presence in meteorites.

Metastability is a more serious issue with respect to carbohydrates. Many reaction schemes might give carbohydrates under prebiotic conditions. However, carbohydrates are often unstable under the conditions where they are formed. Let us review just two examples where synthesis of carbohydrates is possible, at least the laboratory, before turning to metastability as an issue.

For example, Eschenmoser exploited the reactivity C=O aldehyde groups to generate ribose, the carbohydrate found in RNA. His

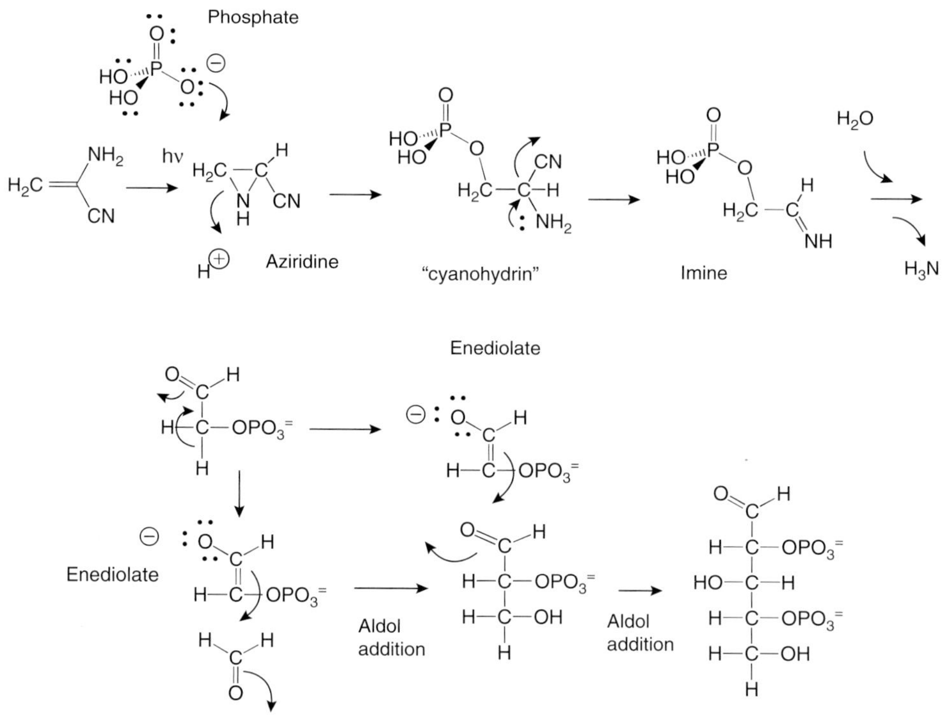

Figure 8. The Eschenmoser synthesis of ribose-2,4-diphosphate from a proposed starting material derived from HCN.

proposed reaction sequence begins with an interesting starting material derived from HCN having two C-N single bonds in a three-membered ring (Fig. 8). Key in this process is the use of phosphate as a nucleophile to break one of these C-N single bonds, replacing a C-N bond by a C-O bond. The breakability of the C-N bond is enhanced because it is strained in a three-membered ring.

The resulting intermediate is called a *cyanohydrin*. It can fall apart to give an imine (a structure with a C=N unit that we saw in the Strecker synthesis). The imine can be hydrolyzed by water to forms an aldehyde having an -O-phosphate unit next to the C=O unit; this is called *glycolaldehyde phosphate*. This species can *enolize* and add formaldehyde to give glyceraldehyde (we will discuss *enolization* and *addition* reactions more in a bit), which can serve as an electrophile to form a new bond to the enol of a second glycolaldehyde, to give ribose-2,4-diphosphate as a major product.

Not quite ribose, but a very interesting product nevertheless. This process occurs in the laboratory when the starting materials are mixed with formaldehyde (Mueller et al. 1990). The starting material has not been detected naturally in the cosmos by microwave spectroscopy, but this is hardly dispositive; the material would be difficult to detect.

Another route for the prebiotic synthesis of carbohydrates is based on the *formose process*, described more than a century ago by the Russian chemist Aleksandr Butlerov (Butlerov 1861). If one incubates formaldehyde (HCHO) in a hot solution of calcium hydroxide $(Ca(OH)_2$, pH 12.5), nothing happens at first. However, after some time, the formaldehyde is rapidly consumed. The mixture turns brown and acquires a sweet taste; the product *formose* smells like toasted marshmallows. Further, if the reaction is stopped at the right time, some five-carbon carbohydrates can be isolated from it, perhaps even ribose.

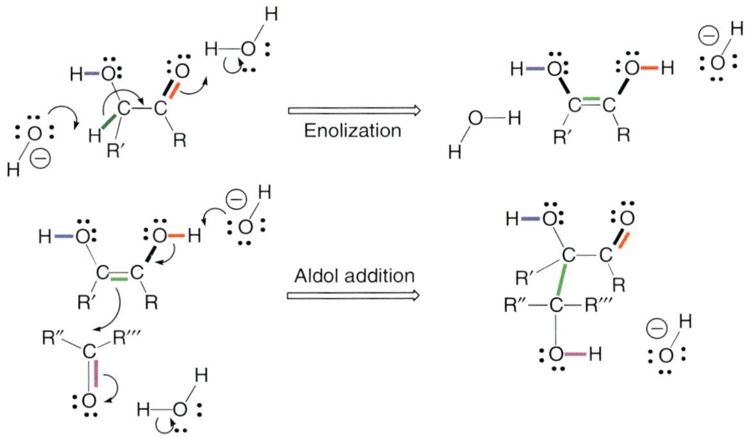

Planetary Organic Chemistry and the Origins of Biomolecules

Figure 9. Two reaction types create formose. *Top.* In an enolization reaction, a base (here, hydroxide) removes a proton (H+) from a carbon adjacent to a C=O unit to give an enediol. The enediol can then react as a nucleophile in an aldol addition reaction. *Bottom.* In an aldol addition reaction, an enediol reacts as a nucleophile to form a new bond to the C of a C=O unit (the electrophilic center). If the carbonyl species is formaldehyde, R"=R"'=H.

A half century ago, modern chemical analysis was brought to bear on the process by Ronald Breslow (Breslow 1959). It involves the repetition of two reactions shown in Figure 9:

1. removal of a proton (H^+) from a carbon next to a C=O (carbonyl) group, known as *enolization*, and

2. attack of the resulting enediolate (a nucleophile) on HCHO (an electrophile) to form a new carbon-carbon bond, known as an *aldol addition*.

By repeating these reactions again and again, complex mixtures of organic species, all having a ratio of carbon:hydrogen:oxygen of 1:2:1 (just like the starting formaldehyde) can be generated. A subset of these reactions is shown in Figure 10.

Many popular books on the origin of life regard the formose process as a solution to the problem of prebiotic carbohydrate formation (Dyson 1985); some see it as a satisfactory approach for the prebiotic synthesis of ribose. This is not the current consensus, however (Shapiro 2007). Both the complexity of the product mixture (as shown in Fig. 10) and the metastability of its components are at issue. Many of those components themselves have C=O units (and therefore can react as electrophiles). Many of these compounds also have a C-H unit adjacent to the C=O unit. From these, H^+ can be lost in an enolization reaction, allowing them to react further as nucleophiles. Eventually, the products in the mixture undergo still more reactions that are too many to capture in a single figure (and this is *before* considering chirality). Unfortunately, five carbon carbohydrates such as ribose, of particular interest for the RNA-first model for the origin of life, are also metastable. They react further, especially at high pH.

SOLVING THE METASTABILITY PROBLEM IN CARBOHYDRATE CHEMISTRY

Appendages to Carbohydrates to Improve their Metastability

Two general classes of solution have been proposed to address the complexity and metastability of carbohydrate formed by prebiotic processes. One proposes that *free* carbohydrates were never formed on early Earth as prebiotically relevant species. In this class of solution, carbohydrates are proposed to have entered early metabolism joined to another chemical moiety that improves their stability. One proposal is described by John Sutherland, who appends

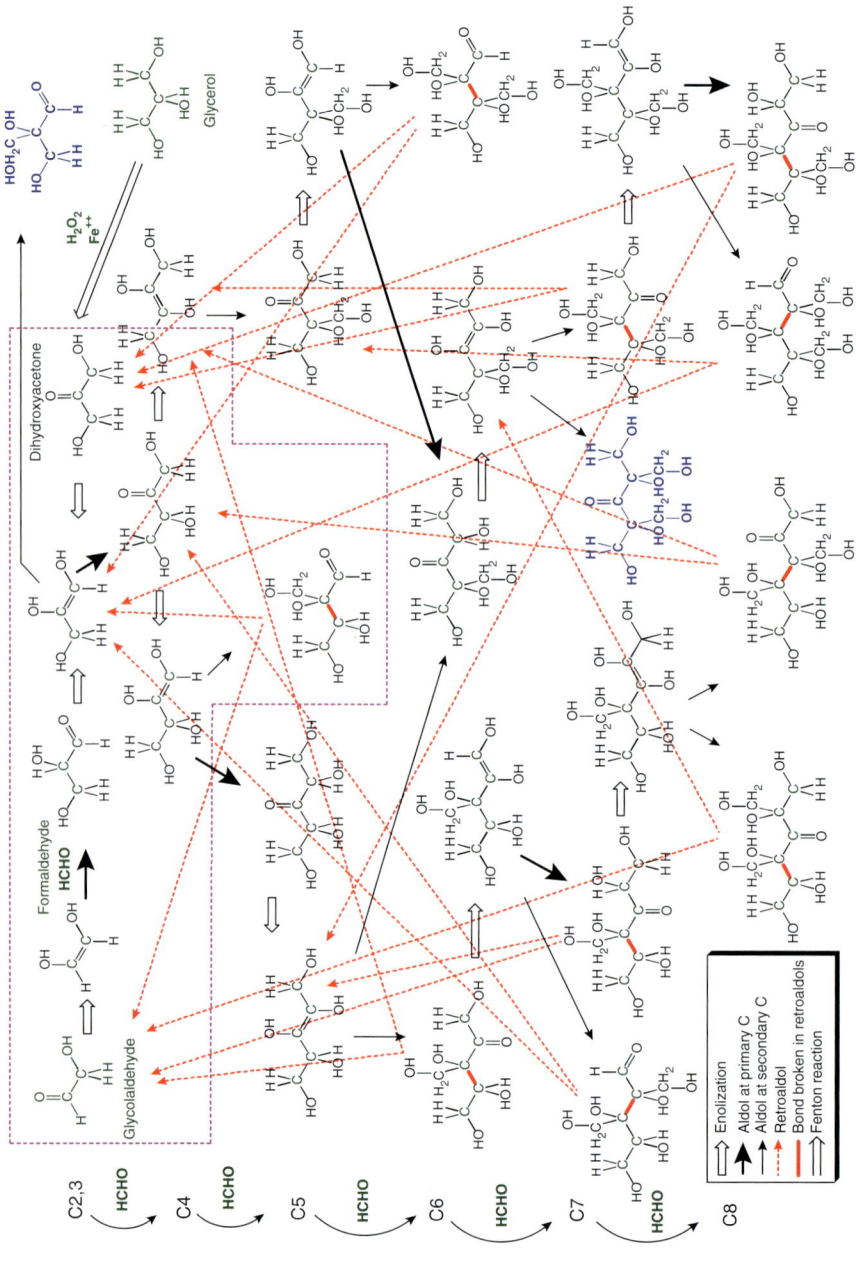

Figure 10. This figure shows the complexity that is possible simply by repeating the two reactions shown in Figure 9. It shows the structures of organic molecules made of only carbon atoms (C), hydrogen atoms (H), and oxygen atoms (O) in a ratio of 1:2:1. The compounds are ordered by size, with compounds containing two and three carbon atoms (C2 and C3, respectively) at the top, and compounds containing four, five, six, seven and eight carbon atoms (C4, C5, C6, C7, and C8) ordered in rows below. The arrows show reactions that interconvert these compounds. The heavy black arrows show the addition of formaldehyde (HCHO) to a species in the row above; this converts that species to a new species with one more carbon atom. The open arrows show reactions that interconvert species having the same number of carbon atoms. Red arrows show reactions that fragment a larger molecule to give two smaller molecules, where the bond that is broken in the fragmentation is red. Blue compounds are dead-end compounds that accumulate in the reaction. A chemist must intervene to prevent this mixture from evolving further to give still more complexity.

the emerging carbohydrate to an emerging nucleobase (Powner et al. 2009).

Eschenmoser's proposal (Fig. 8) embodies a similar idea. By introducing phosphate early in the sequence, the intermediates are not glycolaldehyde, glyceraldehyde, and ribose, but rather glycolaldehyde phosphate, glyceraldehyde phosphate, and ribose-2,4-diphosphate. The negative charges on the phosphate groups diminish the rate at which H^+ is removed, therefore decreasing the rate of enolization. The phosphates attached to these carbohydrates therefore decreases the reactivity of these species as nucleophiles, allowing them to survive longer under conditions where they are formed.

It remains to be seen whether attaching carbohydrate fragments to these or other groups increases metastability sufficiently to allow the carbohydrate derivatives to be plausible prebiotic participants in subsequent reactions proposed in an RNA-first model for the origin of life. It is clear, however, that the universe of chemical structures is far from explored for possible attachments that might confer metastability with respect to decomposition and (still better) activation for the next steps in a useful prebiotic sequence that does not require continued human intervention. Much more work in this direction is needed.

Minerals to Improve the Metastability of Carbohydrates

A second line of thinking looks to the mineral world to preserve carbohydrates long enough to allow them to accumulate to an extent useful for them to participate in subsequent steps in a prebiotic synthesis of Darwinian chemical systems (Prieur 2001; Ricardo et al. 2004). The role of mineral elements to *create* new reactivity has been reviewed by Robert Hazen. Here, we consider the opposite, the use of mineral elements to *remove* reactivity.

Several elements in mineral form are known to bind to carbohydrates in a way that diminishes their reactivity. The best known of these are borate (Chapelle and Verchere 1988), silicate (Lambert et al. 2004), vanadate, molybdate, germanate, and (to a lesser extent) aluminate (Schilde et al. 1994). Purely formally, these are derived from the acids H_3BO_3, H_4SiO_4, HVO_3, H_2MoO_4, H_4GeO_4, and H_3AlO_3, although many of these acids are not observed, preferring to precipitate or form complex ions. Silicic acid, for example, precipitates as SiO_2, the mineral known as *quartz*, in many forms, including gem opal.

Consider, for example, boron. Standing right before carbon in the Periodic Table, boron has only three electrons in its outer shell. This means that boron must pick up five more electrons to get the eight that it wants in its outer shell.

Oxygen atoms in water or the -OH hydroxyl groups of organic molecules offer those electrons. Each of these oxygens has two unshared pairs of electrons. Therefore, boron (III) atoms bind to water oxygens. In geology, borate is an anion whose central boron atom is bonded to four oxygen atoms.

Boron bonds especially well to organic molecules that have two adjacent hydroxyl groups, molecules known as 1,2-diols. Pentoses like ribose in their cyclic forms (Fig. 11) are 1,2-diols. Their adjacent -OH groups bind borate tightly (Fig. 12). Still better, their cyclic borate complexes lack C=O groups necessary for carbohydrates to react to form tar.

As many have noted, the borate complex of ribose is stable, even at high pH. Indeed, for over 40 years, this stability has been used as part of synthetic procedures in the laboratory that interconvert ribose and other five-carbon sugars, including ribulose and arabinose (Angyal 2001; Mendicino 1960).

To convert this common knowledge into a useful prebiotic hypothesis, two questions are relevant. The first asks whether minerals containing borate are likely to be available on prebiotic Earth. As an element, boron is scarce in the cosmos (Zhai and Shaw 1994), especially relative to silicon, carbon, and other familiar elements. Balancing this is the fact that boron is not a good mineral-forming species. It therefore tends to be concentrated in residual igneous melts, where it appears on the surface. The predominant mineral species containing boron are tourmalines. These are easily weathered in acid to form borate salts, which are quite water-soluble.

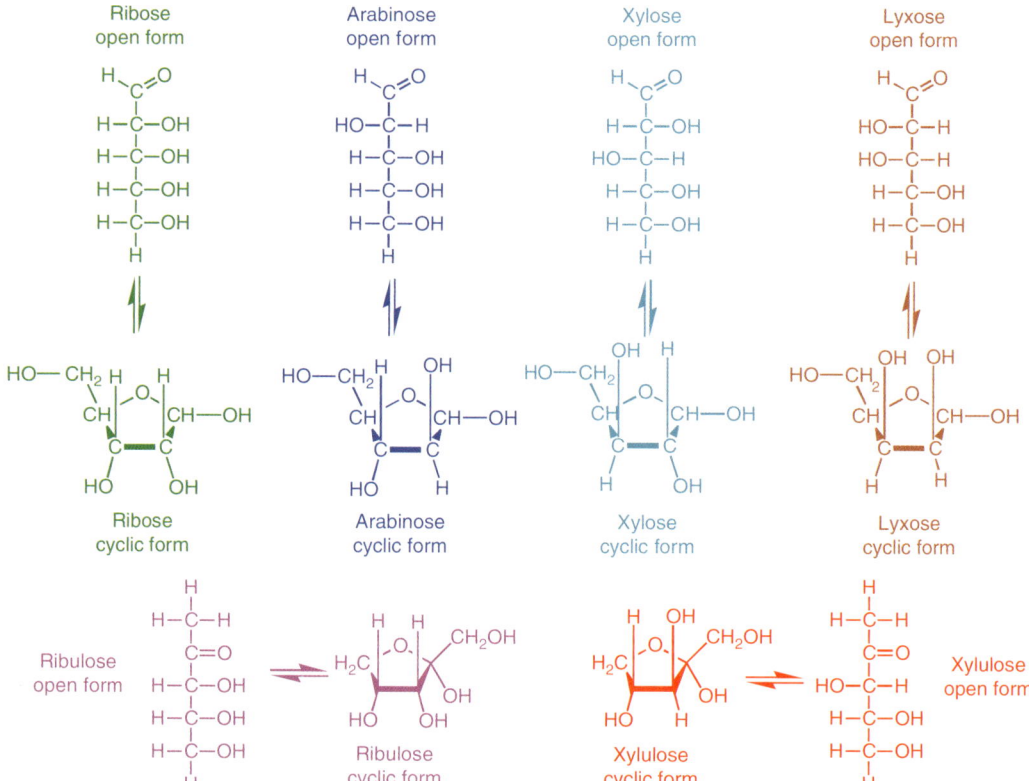

Figure 11. Carbohydrates with 5 carbons (pentoses or pentuloses, penta=5). Pentoses have a HC=O group in their open form; pentuloses have a C=O unit flanked by two carbons All have the formula $C_5H_{10}O_5$. All can exist in both open and cyclic forms, as the C of the C=O unit is an electrophile and can react with the O of an –OH group as a nucleophile to form a ring. In many cases, more than one cyclic form is possible. Different pentoses differ in how their atoms are arranged in space. We represent these 3D orientations on a 2D sheet of paper by drawing thicker bonds (which are forwards) and placing the H and OH groups up or down. Cyclic pyranose forms (with six atoms in the ring) are not shown.

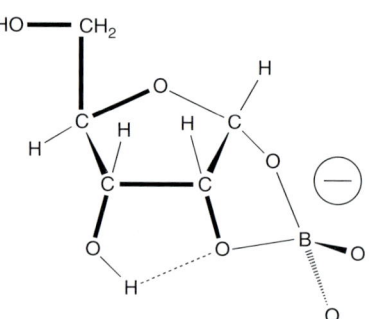

Figure 12. The NMR structure of the borate-ribose complex, with boron complexing adjacent 1,2-hydroxyl groups.

The fate of water-soluble borate salts depends on whether run-off streams deliver them. If they run into the ocean, borate is diluted. However, if they end up in a dry basin such as Death Valley or the Dead Sea, borates end up in minerals evaporated from water (*evaporites*). Borax, a sodium borate salt, is one of these minerals, and is mined in Death Valley. Other boron minerals found in Death Valley include ulexite, a sodium calcium borate mineral, and colemanite, which is a calcium borate.

Alkali is also easily generated from igneous rocks. One source for alkalinity in geology is the mineral peridotite (its gem form is peridot). This green mineral is a magnesium iron silicate, and is widespread in rocks called *serpentines*. When serpentines erode in water, *serpentinization* occurs. Serpentinization is a process that creates dihydrogen gas, the related reducing power, reduced organic molecules, and magnesium hydroxide, a base. For example, Lake Mono, just to the east of the Sierra Nevada mountains in California, has a pH as high as 12, similar to the pH in the laboratory formose reaction, all arising via serpentinization.

The second question is more problematic. Simply adding borate does not solve problems with the prebiotic formose synthesis of carbohydrates from formaldehyde. The synthesis begins with an exceptionally slow reaction between two HCHO molecules, a reaction that creates just single molecules of glycolaldehyde. The formose process therefore relies on the complex reaction cycles that use glycolaldehyde molecules (Hollis 2000) to "fix" more HCHO to give higher carbohydrates (Fig. 10).

Unfortunately, intermediates in these cycles *also* have 1,2-dihydroxyl units that bind to borate. Accordingly borate (at 100 mM) prevents formose cycling, with HCHO disproportionating instead to give formate ($HCOO^-$) and methanol (CH_3OH) as the dead-end products. Even at 6 mM, borate significantly slows formose cycling.

Obviously, the ability of borate to stabilize pentoses (such as ribose and arabinose) and pentuloses (such as ribulose and xylulose) is useless if borate prevents their formation. Thus, the second question asks whether processes exist that might form carbohydrates from HCHO in the presence of borate minerals *without* the slow reaction between two HCHO molecules, and therefore without the need to have all of the cycles shown in Figure 10.

One particularly promising cycle that is guided by borate was hypothesized (Fig. 13). It starts with dihydroxyacetone, which lacks a

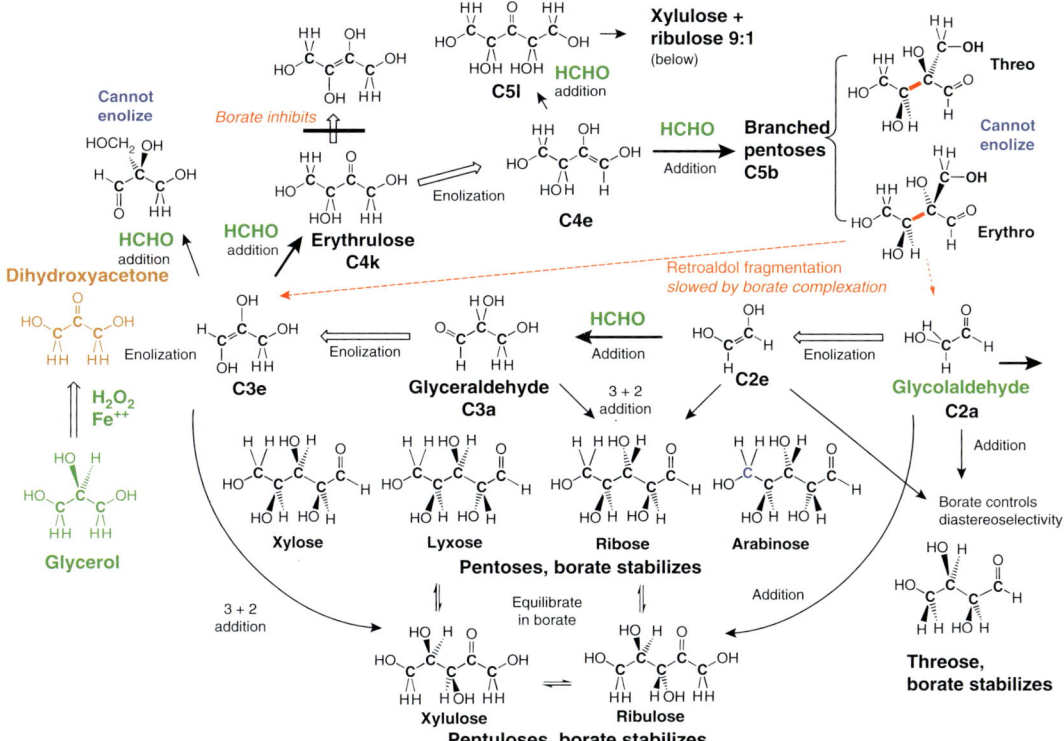

Figure 13. Borate-guided prebiotic path to form carbohydrates from formaldehyde. Compounds in green are known to be prebiotic. Compounds in blue cannot enolize. Dihydroxyacetone is shown in brown.

1,2-dihydroxyl unit. This should, according to the hypothesis, enolize in the presence of borate and initiate a simple catalytic cycle to fix HCHO (reactions within the red dotted box in Fig. 10). Proceeding clockwise in Figure 13, the enediol of dihydroxyacetone should react as a nucleophile with HCHO (**C1**) to give erythrulose (**C4 k**, 3 + 1 = 4).

Erythrulose cannot form a cyclic hemiacetal as do pentoses (Fig. 11), and therefore does not bind borate tightly. It can, however bind borate less tightly through its 3,4-diol unit (Fig. 14). As the boron atom in the complex carries a negative charge, this binding should direct the enolization of erythrulose *away* from the borate to give the 1,2-enediol (**C4e**), which should fix a

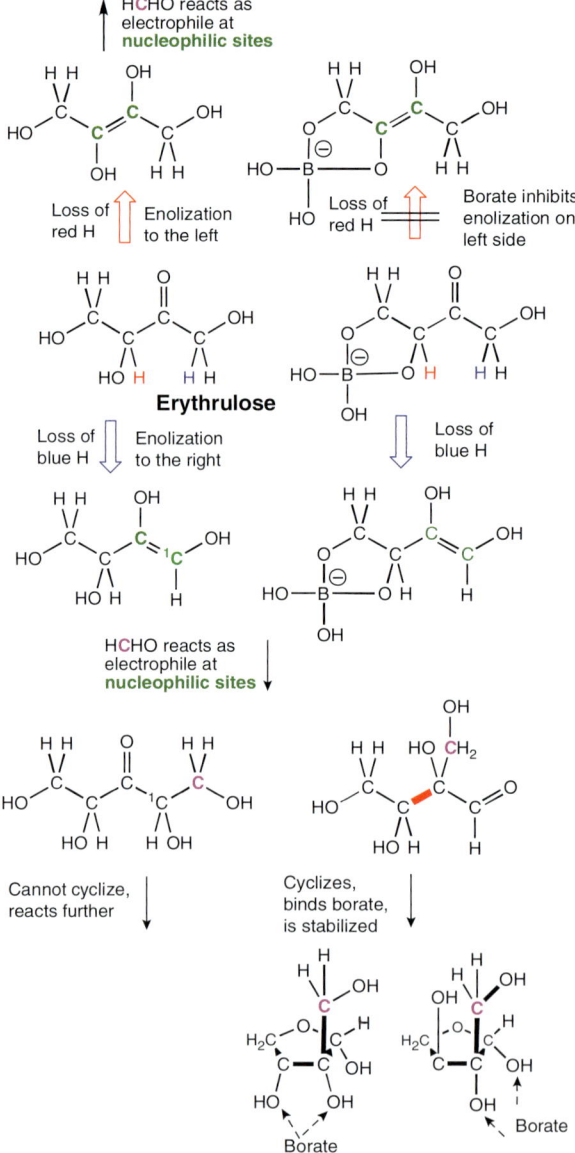

Figure 14. Binding of borate to the 3,4-diol unit of erythrulose should direct enolization to give the 1,2-enediolate, as this removes the H^+ from the carbon the farthest from the negative charge on boron.

third HCHO to give either the linear or branched C5 species at the top (**C5l**) and top-right (**C5b**) of Figure 13 by aldol reaction at the less or more hindered enediol carbon of **C4e** (these are both $4 + 1 = 5$ reactions).

The branched pentoses **C5b** do not have enolizable hydrogens, and therefore cannot react as nucleophiles. They can, however, cyclize to give species that have adjacent 1,2-diol units that can bind borate. Therefore, their complexes with borate should accumulate. The uncomplexed branched pentoses **C5b** species can, however, undergo retroaldol fragmentation to generate **C2a** and **C3e** (Fig. 13, $5 = 2 + 3$). The **C3e** is, of course the very same enediol that was generated from dihydroxyacetone, and can start the cycle anew. The enediol of newly formed glycolaldehyde (**C2e**) should react with HCHO (**C1**) to form **C3a** (a $2 + 1 = 3$ reaction), which is hypothesized to enolize to form enediol **C3e**, again the same enediol that is formed by enolization of dihydroxyacetone. Thus, we can go around the cycle again, fixing more HCHO.

A rich collection of labeling experiments with H^{13}CHO and 1-^{13}C-glycolaldehyde show that each of these steps is possible under conditions that might have occurred on early Earth (Benner, Kim, and Kim unpublished). These studies often use alkaline borate solutions buffered to pH 9 to 11 by carbonate, which is available from atmospheric carbon dioxide. Formation of 2'-^{13}C and 4-^{13}C 2-hydroxymethylerythrose (**C5b**), 2'-^{13}C-2-hydroxymethylthreose (**C5b**), and linear 1-^{13}C -1,2,4,5-tetrahydroxypentan-3-one (**C5l**) from H^{13}CHO proceeds easily without human intervention in these buffers, a fact that was established by comparing carbon-13 nmr signals in the products with the signals of labeled authentic material obtained by direct synthesis.

These steps in the presence of borate can be examined individually to determine how robust they are with respect to changes in conditions. For example, if the concentration of glycolaldehyde is present in two-fold excess over HCHO, addition of H^{13}CHO to glycolaldehyde ($1 + 2 = 3$, at 65°C) followed by reaction with a second molecule of glycolaldehyde (in a $3 + 2 = 5$ reaction) gave 5-^{13}C-ribose, 5-^{13}C-arabinose, and 1-^{13}C-xylulose as major metastable products.

Reaction of glycolaldehyde (**C2e**) in the absence of CH$_2$O gave four-carbon sugars, threose, and erythrose in $2 + 2 = 4$ aldol addition reactions. The formation of threose is significant because this tetrose can replace ribose in the backbone of RNA-like genetic molecules (Schöning et al. 2000; Ebert et al. 2008; Horhota et al. 2005). Consistent with their ability to form a cyclic hemiacetals that coordinate borate, threose, and erythrose are also quite metastable in borate buffers at pH 10.4.

The same process can be observed with silicate. Silicate dissolves as its sodium salt only at high pH; at lower pH, silica precipitates. Nevertheless, a solution of concentrated sodium silicate (about 20%, "water glass") has a pH of 11.5 to 12.0. Even at 100:1 fold dilution, however, glycolaldehyde and glyceraldehyde react to form arabinose and other pentoses, which are also stabilized by complexation, this time to silicate (Lambert et al. 2004).

To complete the metabolic cycle, we begin by recognizing that the branched pentoses **C5b** are stabilized by borate, even though they suffer retroaldol fragmentations rapidly in the absence of borate (half life of ca. 30 minutes at 65 °C). The half-life for the retroaldol fragmentation is ca. 24 hour at 40 mM borate, and becomes still longer at still higher borate concentrations. Thus, on a prebiotic Earth in the presence of ∼200 mM borate, branched pentoses formed from HCHO and glycolaldehyde and/or glycerol (Cooper, 2001) should have formed stable borate minerals. These, in turn, could have been reservoirs for subsequent steps in the prebiotic synthesis of RNA.

Of course, they could also support the cycle by slow fragmentation, or by fragmentation after they are stripped of borate by another carbohydrate (e.g. ribose) that binds borate more tightly. To demonstrate the cycle, **C5b** was incubated with H^{13}CHO at 20 mM borate. The mixture gives rise to labeled **C5b** as well as other products that can arise only if **C5b** had suffered a retroaldol fragmentation to give glycolaldehyde and glyceraldehyde. This establishes the complete cycle.

These results show the possibility of a formaldehyde fixation cycle, where **C2** and **C3** add

HCHO to give branched species that either accumulate or suffer retroaldol fragmentation to give glycolaldehyde and glyceraldehyde. These can either capture HCHO to go around the cycle again, or can combine to give xylulose and ribulose as stable borate complexes.

Under kinetic control, the ratio of C5 species formed depends on the relative rates of aldol additions and fragmentations, the concentrations of various species, pH and borate concentration. However, under prolonged time on a primitive Earth, thermodynamics might well have determined the final products rather than kinetics. Equilibration within two sets of stereo-similar five-carbon species (ribulose, ribose and arabinose, and xylulose, xylose, and lyxose) is well known under a variety of mechanisms (Angyal 2001). Thus, a route that gives ribulose in the presence of borate also gives ribose, whereas a route that gives xylulose in the presence of borate also gives xylose. More slowly, the two pentose-pentulose sets themselves interconvert (Fedoronki and Linek 1967). Absent borate, the equilibrium ratios can be represented as a series (xylose 1:1 xylulose 85:15 ribulose 1:9 arabinose 4:1 ribose) (Sultana et al. 2003), with the ratio of any two C5 carbohydrates calculable by multiplying the appropriate ratios (thus the xylulose:ribose equilibrium is ~5:2). Borate favors the -uloses in each set over the -oses; among the pentoses, borate complexation favors ribose (Mendicino 1960; Li et al. 2005).

Challenges of Prebiotic Metabolic Cycling

Interestingly, borate-moderated cycles such as those in Figure 13 share some features proposed for cycles hypothesized for "metabolism first" models (Shapiro 2007). Although the "genetics first" versus "metabolism first" models for the origin of life on Earth are currently being presented as adversaries (Orgel 2008), no logic compels them to be. It is nearly certain that chemical processes that might be likened to metabolism occurred on Earth before genetics, providing components of whatever genetic system did first emerge.

However, a cycle will not operate if more than one equivalent of material leaks from the cycle before completion of the cycle produces a second equivalent of cyclable material. Prebiotic metabolic cycles can easily be defeated by leakage that allows enough material to escape from the cycle to prevent product accumulation.

To understand leakage in mineral-controlled cycles, we might focus on leakage products that may have emerged from the hypothetical cycle in Figure 13. One leakage product is 1,2,4,5-tetrahydroxypentan-3-one, shown the top of Figure 13 (**C5l**). It is a linear species formed by the addition of HCHO to the erythrulose enediol (**C4e**) at its less hindered center (4 + 1 = 5). This species *cannot* form a cyclic hemiacetal to bind borate; it therefore reacts further in carbonate-borate buffer.

Remarkably, the leakage is "productive." With a half-life of 4-5 days, synthetic $1-^{13}C$-**C5l** in carbonate-borate buffer gives 1- and $5-^{13}C$-xylulose (90%) and 1- and $5-^{13}C$-ribulose (10%). Ribulose is known to be isomerized in the presence of borate to ribose. Thus, leakage by this route forms the products that one wants, but in a different way from the aldol addition of a glycolaldehyde and glyceraldehyde.

More Mineralogy to Manage Carbohydrate Metastability

As before, borate's ability to stabilize carbohydrates offers both good news and bad news. In particular, retroaldol fragmentation of branched pentoses **C5b** is slowed dramatically by borate. Prebiotic chemical scenarios often assume millions of years of time to manage slow reactions (a thermodynamic control scenario), and we can easily imagine scenarios where the accumulation of branched carbohydrate outruns the concentration of borate, leaving some unbound complexes to suffer retroaldol reaction to give ribose, which extracts a borate from a branched pentose to permit the transformative process to go to completion.

Further, the relative scarcity of boron has become an issue. For example, Hazen has suggested that borate minerals are "exotic," in part because they are concentrated in pegmatites whose contents may have required too much time took to accumulate (Hazen et al 2009). It is difficult to know in this context what "too much time" might be, as weatherable igneous borates should have appeared on the surface as soon as the planetary material separated and concentrated as long as there was aqueous rain, dry land, and dry valleys.

Accordingly, some geologists are more comfortable with silicate control than they are with borate control, even considering the fact that silicates, although more abundant, are less soluble in water than borate. Aluminates are also candidate coordinating species, also abundant but also rather poorly soluble (aluminate and silicate together form clay).

Nevertheless, other mineral forming species are known to coordinate carbohydrates, including vanadate, germanate, and molybdate (Schilde et al. 1994). These are also "exotic," and are more oxidized with respect to the mantle than borate, silicate, and aluminate. Thus, they are not likely to have been present on a reducing early Earth in abundance. They may, however, have been present in significant amounts at specific locales.

Balancing their exotic nature is some very interesting chemistry produced by these mineral components, especially at moderate pH. For example, molybdate (Mo^{6+}) minerals at neutral (or slightly acidic pH) catalyze a *Bilik reaction*, a rearrangement of a hydroxyl-carbonyl compound to give an isomeric carbonyl-hydroxyl compound (Petrus et al. 2001). This can convert branched tetroses such as **C4b** into erythrulose and branched pentoses such as **C5b** to linear pentoses (Fig. 15). In both cases, "dead end" products might be returned to the catalytic cycle or to biologically interesting species.

Indeed, incubating branched pentose (**C5b**) in the presence of molybdate (65 °C, 24 hour) leads to an equilibrium mixture of **C5b** starting material and linear xylulose, with some linear pentulose 1,2,4,5-tetrahydroxypentan-3-one (Kim and Benner, unpublished). The rearrangement is stereospecific; *threo* branched pentose gives ribulose. As noted above, xylulose and ribulose equilibrate slowly with Mo^{6+} to give xylose and ribose. A similar reaction was observed with Ca^{++} as catalysts, but at high pH.

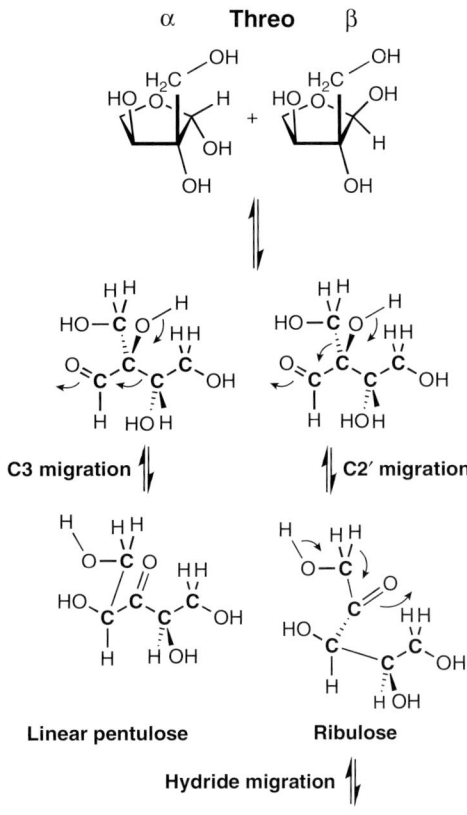

Figure 15. The Bilik reaction uses molybdate minerals to catalyze the stereospecific isomerization of carbohydrates such as the branched pentoses to give linear pentuloses such as ribulose at neutral pH under mild conditions.

The Need for Dry Land, Incompletely Reducing Conditions, and Incompatible Species

To exploit molybdate, the pH must fall. This could happen under a carbon dioxide atmosphere through dissolution of atmospheric CO_2 in an alkaline solution. Thus, coupled with the borate-guided (or silicate-guided or aluminate-guided) prebiotic fixation of HCHO and fluctuating pH, Mo^{6+} can generate free ribose and xylose from **C5b** branched pentoses. This model envisions a borate-rich (or silicate-rich) alkaline stream emerging from serpentinizing igneous rocks above the Earth's surface, with formaldehyde arriving from the atmosphere above as the product of electrical discharge through a moist atmosphere dominated by carbon dioxide. Under these conditions, complexes of borate or silicate are formed with branched pentoses, ribulose and xylulose, and their corresponding pentoses, arabinose, lyxose, ribose and xylose.

The alkalinity of such a stream would be buffered over time by the dissolution of atmospheric carbon dioxide. As the pH drops, the rate of formation of carbohydrates via enolization reactions slows, but so does their rate of decomposition. At the same time, borate complexes weaken and silicate precipitates, but molybdenum species capable of supporting the Bilik reaction become more abundant. According to this model, these conditions transform the branched carbohydrates to linear carbohydrates, which themselves form pentoses like ribose (Mendicino 1960). If the pH rose, the ribose would again become bound to and stabilized by borate. Such pH fluctuations would help equilibrium to be reached.

Certain models for early Earth are not compatible with this scenario, of course. For example, this scenario could not easily operate on a planet that was entirely flooded by water, as borate and alkali would be diluted into the planetary ocean whose oxidation state, pH, and mineral composition would be that of the ocean as a whole. Dry land is needed to permit the concentration of borate and borate-organic evaporite minerals for this model; alternatively, compartmentalization is needed within the ocean, such as near subsurface vents, to have such cycles. Thus, if the "super-Earths" being observed as extrasolar planets are indeed "water worlds," this cycle would be more difficult to operate, requiring sequestration in specific suboceanic environments.

Given land surfaces, however, the model makes sense. Fluctuating pH is not implausible, as the high pH emerging from serpentinizing rocks is buffered beneath a CO_2 atmosphere. Borate is readily delivered to alkaline washes through the erosion of tourmalines which were almost certainly present in igneous rock on early Earth, as they are today. Mo^{6+} is an oxidized form of the element, but not a strongly oxidizing form ($MoO_2 + H_2O + 2\,Fe^{3+} = MoO_3 + 2\,H^+ + 2\,Fe^{2+}$ at $+236$ mV; $2\,MoO_2 + SiO_2 = Si + 2\,MoO_3$ at -0.145 mV) Thus, that the formation of MoO_3, which is hydrated to give molybdate, would have been favored at neutral when the ferric:ferrous ratio is one part in ca 10^{10}.

Of course, other elements may be involved in such processes. Sulfur is an interesting example. Species containing sulfur interact with carbohydrates and other compounds containing C=O bonds. Hydrogen sulfide attacks C=O groups, just like water. Further, the bisulfite (HSO_3^-) anion reacts with C=O units to form reasonably stable bisulfite addition products. Bisulfite is formed from sulfur dioxide, which was undoubtedly produced on early volcanoes on Earth.

The reaction of carbohydrates under plausibly prebiotic conditions in the presence of these other species has only begun (Weber, 2001). Much more research is needed before a model can capture the entire space of variable contents in a way that suggests what mixtures are productive and what mixtures are nonproductive, for prebiotic synthesis on a complex planet. Ultimately, compatibility issues need to be captured in a table that includes both organic and inorganic species, including cations and anions that are incompatible with solubility, redox potential, and reactivity.

REFERENCES

Angyal SJ. 2001. The Lobry de Bruyn-Alberda van Ekenstein transformation and related reactions. *Glycoscience* **215**:1–14.

Benner SA. 2009. *Life, the Universe and the Scientific Method.* Gainesville FL, Foundation Press. 320 pp.

Bernstein MP, Dworkin J, Sandford SA, Cooper GW, Allamandola LJ. 2002. Racemic amino acids from the ultraviolet photolysis of interstellar ice analogues. *Nature* **416**:401–403.

Blackmond DG. 2010. The origin of biological homochirality. *Cold Spring Harb Perspect Biol* **2**: a002147.

Breslow R. 1959. On the mechanism of the formose reaction. *Tetrahedron Lett* **21**:22–26.

Butlerov A. 1861. Bildung einer zuckerartigen Substanz durch Synthese. *Annalen Chemie* **120**:295–298.

Cairns-Smith Alexander. 1982. *Genetic Takeover and the Mineral Origins of Life.* Cambridge UK: Cambridge University Press. 488 pages.

Chapelle S, Verchere J-F. 1988. A ^{11}B and ^{13}C NMR determination of the structures of borate complexes of pentoses and related species. *Tetrahedron* **44**:4469–4482.

Cleaves HJ. 2008. The prebiotic geochemistry of formaldehyde. *Precambrian Res* **164**: 111–118.

Cooper G, Novelle K, Belisle W, Sarinana J, Brabham K, Garrel L. 2001. Carbonaceous meteorites as a source of sugar-related organic compounds for the early earth. *Nature* **414**: 879–884.

Dyson F. 1985. *Origin of life*. Cambridge University Press. 94 pp.

Ebert MO, Mang C, Krishnamurthy R, Eschenmoser A, Jaun B. 2008. The structure of a TNA-TNA complex in solution: NMR study of the octamer duplex derived from α-(L-threofuranosyl-(3'-2'-CGAATTCG). *J Am Chem Soc* **130**: 15105–15115.

Fedoronki M, Linek K. 1967. Transformation of pentoses in pyridine. *Coll Czech Chem Comm* **32**: 2177.

Hazen RM, Ewing RJ, Sverjensky DA. 2009. The evolution of uranium and thorium minerals. *Am Mineral* **94**: 1293–1311.

Hollis JM, Lovas FJ, Jewell PR. 2000. Interstellar glycolaldehyde: The first sugar. *Astrophys J* **540**: L107–L110 Part 2.

Horhota A, Zou K, Ichida JK, Yu B, McLaughlin LW, Szostak JW, Chaput JC. 2005. Kinetic analysis of an efficient DNA-dependent TNA polymerase. *J Am Chem Soc* **127**: 7427.

Joyce GF, Orgel LE. 1999. Prospects for understanding the origin of the RNA world. In *The RNA world* 2nd edn. (eds Gestland R. et al.). Cold Spring Harbor Press Cold Spring Harbor NY.

Kasting JF. 1993. Earth's early atmosphere. *Science* **259**:920–925.

Lambert JB, Lu G, Singer SR, Kolb VM. 2004. Silicate complexes of sugars in aqueous solution. *J Am Chem Soc* **126**:9611–9625.

Larralde R, Robertson MP, Miller SL. 1995. Rates of decomposition of ribose and other sugars, implications for chemical evolution. *Proc Natl Acad Sci* **92**:8158–8160.

Li Q, Ricardo A, Benner SA, Winefordner JD, Powell DH. Desorption/ionization on porous silicon mass spectrometry studies on pentose-borate complexes. *Anal Chem* **77**: 4503. 2005.

Mendicino JF. 1960. Effect of borate on the alkali-catalyzed isomerization of sugars. *J Am Chem Soc* **82**:4975.

Miller SL. 1955. Production of some organic compounds under possible primitive earth conditions. *J Am Chem Soc* **77**:2351–2361.

Mueller D, Pitsch S, Kittaka A, Wagner E, Wintner CE, Eschenmoser A. 1990. Chemistry of alpha-aminonitriles Aldomerization of glycolaldehyde phosphate to rac-hexose 2 4 6-triphosphates and. in presence of formaldehyde. rac-pentose 2 4-diphosphates: rac-allose 2 4 6-triphosphate and rac-ribose 2 4-diphosphate are the main reaction products. *Helv Chim Acta* **73**:1410–1468.

Orgel LE. 2008. The implausibility of metabolic cycles on the prebiotic Earth. *PLOS Biol* **6**: e18.

Oró J. 1960. Synthesis of adenine from ammonium cyanide. *Biochem Biophys Res Commun* **2**:407–412.

Petrus L, Petrusová M, Hricovíniová Z. 2001. The Bilik reaction. *Glycoscience* **215**:15–41.

Pinto JP, Gladstone GR, Yung YL. 1980, Photochemical production of formaldehyde in Earth's primitive atmosphere. *Science* **210**:183–184.

Pizzarello S, Shock E. 2010. The organic composition of carbonaceous meteorites: The evolutionary story ahead of biochemistry. *Cold Spring Harb Perspect Biol* **2**: a002105.

Powner MW, Gerland B, Sutherland JD. 2009. Synthesis of activated pyrimidine ribonucleotides in prebiotically plausible conditions. *Nature* **459**: 239–242.

Prieur BE. Étude de l'activité prébiotique potentielle de l'acide borique. *C R Acad Sci Paris Chimie* **4**:667–670.

Sagan C, Khare NB. 1979. Tholins: Organic chemistry of interstellar grains and gas. *Nature* **277**:102–107.

Sagan C, Khare NB, Bandurski LE, Batholomew N. 1978. Ultraviolet-photoproduced organic solids synthesized under simulated jovian conditions: Molecular analysis. *Science* **199**: 1199–1201.

Sanchez RA, Ferris JP, Orgel LE. 1967. Studies in prebiotic synthesis II. Synthesis of purine precursors and amino acids from aqueous hydrogen cyanide *J Mol Biol* **30**:223–253.

Schilde U, Kraudelt H, Uhlemann E. 1994. Separation of the oxoanions of germanium, tin, arsenic, antimony, tellurium, molybdenum and tungsten with a special chelating resin containing methylaminoglucitol groups. *Reactive Polymers* **22**:101–106.

Schöning KU, Scholz P, Guntha S, Wu X, Krishnamurthy R, Eschenmoser A. 2000. Chemical Etiology of nucleic acid structure. The ⟨-threofuranosyl-(3'→2') oligonucleotide system. *Science* **290**:1347–1351.

Shapiro R. 1987. *Origins: A skeptic's guide to the creation of life on earth*. Bantam Books, New York.

Shapiro R. 1988. Prebiotic ribose synthesis. A critical analysis. *Origins Life Evol Biosphere* **18**:71–85.

Shapiro R. June 2007. A simpler origin for life. *Scientific American* **296**:46.

Sultana I, Mizanur RMD, Takeshita K, Takada G, Izumori K. 2003. Direct production of D-arabinose from D-xylose by a coupling reaction using D-xylose isomerase D-tagatose 3-epimerase and D-arabinose isomerase. *J Biosci Bioeng* **95**:342–347.

Weber AL. 2001a. The sugar model. Catalytic flow reactor dynamics of pyruvaldehyde synthesis from triose catalyzed by poly-D-lysine contained in a dialyzer. *Orig Life Evol Biosph* **31**:231–240.

Weber AL. 2001b. The sugar model. Catalysis by amines and amino acid products. *Orig Life Evol Biosph* **31**: 71–86.

Weber AL. 2007. The sugar model: Autocatalytic activity of the triose-ammonia reaction. *Orig Life Evol Biosphere* **37**: 105–111.

Zhai M, Shaw DM. 1994. Boron cosmochemistry Part I: Boron in meteorites. *Meteoritics* **29**:607–615.

The Organic Composition of Carbonaceous Meteorites: The Evolutionary Story Ahead of Biochemistry

Sandra Pizzarello[1] and Everett Shock[1,2]

[1]Department of Chemistry and Biochemistry, Arizona State University, Tempe, Arizona 85287-1604
[2]School of Earth and Space Exploration, Arizona State University, Tempe, Arizona 85287-1404
Correspondence: pizzar@asu.edu

Carbon-containing meteorites provide a natural sample of the extraterrestrial organic chemistry that occurred in the solar system ahead of life's origin on the Earth. Analyses of 40 years have shown the organic content of these meteorites to be materials as diverse as kerogen-like macromolecules and simpler soluble compounds such as amino acids and polyols. Many meteoritic molecules have identical counterpart in the biosphere and, in a primitive group of meteorites, represent the majority of their carbon. Most of the compounds in meteorites have isotopic compositions that date their formation to presolar environments and reveal a long and active cosmochemical evolution of the biogenic elements. Whether this evolution resumed on the Earth to foster biogenesis after exogenous delivery of meteoritic and cometary materials is not known, yet, the selective abundance of biomolecule precursors evident in some cosmic environments and the unique L-asymmetry of some meteoritic amino acids are suggestive of their possible contribution to terrestrial molecular evolution.

INTRODUCTION

Why Meteorites are Part of the Discourse about the Origin of Life

The studies of meteorites have long been part of investigations and discussions about the origin of life for the reason that some of these extraterrestrial bodies have reached the Earth containing abundant carbon since its accretion, provide a natural sample of abiotic organic chemistry, and may offer insights on the possible environments and physico-chemical processes that fostered biogenesis. These conditions are entirely unknown because geological and biological processes of over four billion years have long eradicated any traces of early Earth's chemistry. On the other hand, we know that life has embarked in a long evolutionary path all through its recorded history and it seems reasonable to extend to its unknown beginning the same evolutionary nature. Albeit *a posteriori* and without knowledge of the actual chemical steps that carried this evolution, therefore, the single assessment one can safely make about life's origin on the Earth is that it must have been an emergent process, through which biogenic atoms and molecules gained the complex associative and interactive states we observe in even the simplest forms of extant life. It is then easy to see why the discourse about the origins of life has been multidisciplinary, broad based, and fostered many theories, all of which, with the notable exception of the *panspermia* hypothesis (e.g., Crick and Orgel 1973), accept the fundamental emergent nature of life from simple molecules.

In exobiological (as well as astrobiological) terms, it has been proposed that life's fundamental evolutionary nature might have extended beyond its origin and might be rooted in the abiotic cosmochemical evolution of the biogenic elements. C, H, N, O, P, and S are known to be present as diverse and often complex organic molecules in a variety of extraterrestrial environments (Lazcano 2010) and their long cosmic history has supported the idea of a possible exobiology. However, its analytical basis comes from the study of carbon-containing meteorites that have provided the only natural sample of chemical evolution large enough for direct laboratory analyses. Uniquely, therefore, carbon containing meteorites record for us the abiotic organic chemistry that preceded life's origin and may as yet reveal whether it is realistic to assume that these or similar materials, i.e., either by direct delivery or analogy of formation, might have fostered or even inducted molecular evolution toward biogenesis.

The Early Solar System, Meteorites, and the Possible Survival of Cosmochemical Evolution

The meteorites that reach the Earth are for the most part fragments of asteroids, i.e., of those small planetesimals that orbit the Sun in great number between Mars and Jupiter. By the Titius-Bode law of a regular spacing of planets from the Sun, their orbit should be occupied by a planet; it is believed, however, that the small chunks of early solar materials reaching this area fell under the strong gravity of the already formed giant planets and were either scattered throughout or left unable to coalesce. That is how we still find them today, joined by icy objects from more distant locations of the solar system that were brought in by further dynamical evolution of giant planets' orbits (Levison et al. 2009). With their crowding, hazardous orbits, and constant collisions, all of these bodies put fragments on route to the Earth and have done so through the ages. The importance of meteorites for the study of prebiotic chemistry is a result of this failed planet formation and not just for their obvious delivery but also because many of the asteroid belt objects never had their composition drastically transformed by gravitational high temperatures and pressures as larger bodies did. Their meteoritic fragments, therefore, may carry unaltered a pristine record of early solar system chemistry as well as allow the deciphering of its cosmic history.

The meteorites that best fit this description are the carbonaceous chondrites (CCs), a primitive subgroup of stony meteorites having an elemental composition that is very similar to that of the Sun and the universe overall. As their name indicates, CCs have the distinction of containing several percent amounts ($\sim$1.5%–4%) of carbon, which is for the most part present as organic materials. These meteorites are aggregate rocks, i.e., consist mainly of a matrix made of packed together hydrous and anhydrous silicates that do not show signs of metamorphism or alteration by high heat. However, as part of their small planet parent bodies, CC mineralogy also shows that these rocks had experienced a liquid water phase as well as the effects of impact shocks. For example, a recent measurement of the optical activity of three CC surfaces (Arteaga et al. 2010) showed a circular birefringence bias to negative values that the authors attribute to chiral fractures and distortions in the clays following mechanical forces. The meteorites' aggregation also captured various inclusions; the chondrules, to whose name CCs owe their classification, are round beads of glassy appearance that have re-crystallized from a melt, i.e., high heat, and bring witnesses to the variety of materials and processes that must have contributed to CC parent bodies' formation (Fig. 1).

Overall, these meteorites do not seem to differ much from terrestrial rocks, a similarity that has not helped their collection or preservation because, if not seen to fall and promptly collected, they easily disappear in the environment. The Murchison meteorite was exceptional in

Figure 1. A CR2 meteorite stone found in the Antarctica Graves mountains (GRA 95229). The open faces show the large chondrules that characterize this family of meteorites. Chondrule and other inclusion abundance reduce the amount of matrix where organics are found to about 30% of the geology.

this respect because it fell at the very eve of lunar samples' return in 1969 and was analyzed directly by NASA laboratories as a possible analog of those samples. One hundred kilograms of this meteorite were recovered and have been used in 40 years of analyses for probably the most comprehensive study of any extraterrestrial organic material to date. As a result of this focus, the Murchison meteorite composition has long been considered representative not only of meteorites of the same type (Pizzarello et al. 2006) but, often (e.g., Luisi 2007), also of the capabilities of abiotic organic syntheses in general. Given our yet tentative knowledge of cosmochemical environments, it is not surprising that the latter assumption turned out to be premature and a new group of pristine meteorites found in the ice fields of Antarctica, the Renazzo-family of chondrites or CR, have been offering a novel view of the possible synthetic outcomes of abiotic chemical evolution as well as of its prebiotic relevance.

BACKGROUND

The Abiotic Organic Composition of Meteorites: Prebiotic Traits and Biochemical Counterparts

In spite of exhaustive chemical analyses, we still have a very vague idea of where Murchison organic materials are actually located vis-a-vis the inorganic components in the meteorite. The only successful description so far was obtained by X-ray microscopy of the meteorite surfaces after their exposure to selective staining with OsO_4 vapors (Pearson et al. 2007). From these analyses, they appear to be broadly distributed within the matrix, intermixed with hydrous silicate components. As in other CCs, Murchison organic materials can be broadly described in terms of their solubility in aqueous and organic solvent systems, a practical characterization that nevertheless leaves room for missed analytical targets and the possibility of unknowns (e.g., Deamer 1985). Insoluble and soluble components represent respectively 70% and 30% of total carbon and, within their molecular range, are both very complex and fundamentally heterogeneous.

Murchison Insoluble Organic Material (IOM)

The larger portion of Murchison organic carbon is often referred to as kerogen-like because, like terrestrial kerogens, it is an insoluble macromolecular material of complex composition that is not known in much molecular detail; its average elemental abundances are $C_{100}H_{46}N_{10}O_{15}S_{4.5}$. The bulk of the IOM can be inferred only indirectly from spectroscopy (e.g., nuclear magnetic resonance and infrared) and by decomposition studies, where it is pyrolyzed by heat or oxidized into its fragments. These analyses suggest a general structure composed of aromatic ring clusters, bridged by aliphatic chains containing S, N, and O, with peripheral branching and functional groups. By transmission electron microscopy, most of the IOM appears dispersed and amorphous but ~10% of it is found as self-contained nanostructures (Fig. 2), spheres as well as tubes, of diverse elemental composition that varies from close to pure graphitic C ($>$ 99%) to containing several percent amounts of O, N, and S. The IOM also contain minute amounts of "exotic" carbon, so called because it was likely formed in the envelopes of stars prior to the formation of the solar system.

On the whole, the large compositional heterogeneity of the IOM as well as the diversity of its phases strongly suggest that this material is the complex end product of cosmochemical regimes and environments that varied greatly. On the other hand, in spite of being insoluble

Figure 2. Scanning electron microscope image of the GRA 95229 acid residue showing an abundance of submicron-sized spherical carbonaceous particles. The particles are solid, single, and agglomerated with the largest close to 500 nm in diameter. The residue is deposited onto a carbon planchette and imaged with 5 kV electrons and current of 98 pA (reprinted with express permission from Laurence Garvie).

in acids and solvents, the IOM can free several individual compounds under conditions of high temperature and pressure similar to those of terrestrial hydrothermal vents (300°C, 100 MPa) (Yabuta et al. 2007). These are mainly a variety of aromatic and heteroaromatic hydrocarbons but also smaller noncondensed molecular species and a suite of alkyl dicarboxylic acids up to C_{18} chain length. In addition, the hydrothermolysis changed the IOM's chiral response to the Soai autocatalytic reaction in that it displayed a statistical R-chiral bias prior to the treatment but not afterwards (Kawasaki et al. 2006), suggesting that some chiral species are present in Murchison IOM but cannot be detected at the molecular level. These experiments show that portions of the IOM macromolecular structure can be modified at the molecular level, exchange species with the soluble organic pool, and possibly represent materials caught in flux between aggregation states. We may also assume that, were meteoritic materials exposed to hydrothermal conditions or prolonged exposure to water upon their fall, IOM release might have made an important contribution to the organic pool of the early Earth when CCs delivered an estimated 1%–3% of their weight in carbon during the early impact period (Mautner et al. 1995).

Murchison Soluble Organic Compounds

The soluble organic compounds of the Murchison meteorite make up an abundant and diverse group of well over a thousand molecular species that vary from smaller water-soluble compounds such as amino acids and polyols up to 30-carbon-long nonpolar hydrocarbons extracted only with solvents (Table 1). As their large number indicates, they are also present in multiple isomeric forms up to the limit of their solubility. This diversity is observed throughout most of the various compound types and is often a sign of their indigeneity because it contrasts starkly with the structural and functional selectivity displayed in biochemistry. It has been analyzed in particular detail for Murchison amino acids. For example, all the possible α-amino alkylamino acids up to seven-carbon were identified in Murchison extracts based on the reference of synthesized standards and several eight- and nine-carbon homologous species could also be easily recognized by chromatography-mass spectroscopy on the basis of their spectra even if their standards were not available. Similar large abundances of N-substituted, cyclic, β-, γ-, δ-, and ε-amino acids were also found and the total number of meteoritic amino acids can be placed at over one hundred. In contrast, the whole of terrestrial protein is made up of just 20 amino acids.

Within this overall diversity, several components of Murchison's organic suite have identical counterparts in the biosphere. Eight of the meteorite amino acids are also found in proteins (glycine, alanine, proline, valine, leucine, isoleucine, aspartic acid, and glutamic acid) and numerous other compounds are encountered in terrestrial metabolisms, as shown in Table 1. A very interesting similarity with biochemical traits was found in a group of chiral amino acids not present in terrestrial proteins, the 2-methyl 2-amino acids, which display in Murchison L-enantiomeric excesses (*ee*) that, if not as large, have the same configuration (L-) of terrestrial protein. The *ee* were first discovered in the diastereomers of the seven-carbon 2-amino 2,3-dimethylbutaoic acid (Fig. 3) (Cronin and Pizzarello 1997) and were

Table 1. Classes of organic compounds in the Murchison meteorite.

Compound Class	Structure	Example Molecule
Carboxylic acids	$H_3C-COOH$	Acetic acid
Amino acids	$H_3C-CH(NH_2)-COOH$	Alanine
Hydroxy acids	$H_3C-CH(OH)-COOH$	Lactic acid
Ketoacids	$H_3C-C(=O)-H$	Pyruvic acid
Dicarboxylic acids	$HOOC-CH_2-COOH$	Succinic acid
Sugar alcohols & acids	$H_2C(OH)-CH(OH)-CHO$	Glyceric acid
Aldehydes & Ketones	$H_3C-C(=O)-H$	Acetaldehyde
Amines & Amides	$H_3C\cdot CH_2NH_2$	Ethyl amine
Pyridine carb. acids	(pyridine-COOH)	Nicotinic acid
Purines & Pyrimidines	(adenine structure)	Adenine
Hydrocarbons: Alyphatic	$H_3C-CH_2-CH_3$	Propane
Aromatic	(naphthalene)	Naphthalene
Polar	(isoquinoline)	Isoquinoline
(complex macromolecular structure)		Insoluble Material (estimated)

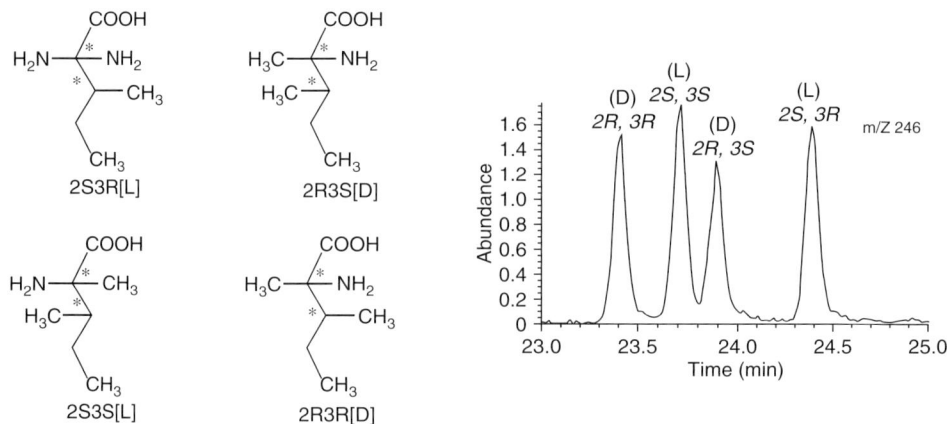

Figure 3. Chemical structure and chromatographic elution of the Murchison 2-amino 2,3-dimethylbutyric acid diastereomers.

later established for the whole homologous series of these chiral compounds up to eight-carbon long; their magnitude varies within the meteorite and is largest, up to 18%, for isovaline (2-methyl-2-aminobutyric acid). As the biochemical structures and functions of all life today are dependent upon the exclusive chiral homogeneity of their polymers, it appears reasonable to assume that a homochirality, albeit of unknown origin, was also essential to the origin and/or evolution of life. The ee of meteorites represent the only case so far of molecular asymmetry ever measured outside the biosphere and their indigeneity is supported by compound specific carbon-, and hydrogen isotopic data obtained for D-, and L-isovaline enantiomers (Pizzarello et al. 2003; Pizzarello and Huang 2005).

Overall, the study of Murchison has disclosed detailed insights on the capabilities and possible range of abiotic syntheses in cosmochemical environments. We have learned that this abiotic chemistry can form organic materials of considerable complexity and include compounds similar or identical to biomolecules. Particularly captivating is the finding of chiral asymmetry in abiotic amino acids and, although less defined at the molecular level, the fact that some macromolecular and inorganic phases of the meteorite show signs of optical activity is intriguing as well. Considered as a whole, these data support the conclusion that molecular chiral asymmetry preceded biochemistry.

Nevertheless, these studies also leave many questions unanswered as to the prebiotic potential of an organic suite of Murchison-like composition. In fact, the large heterogeneity of Murchison organic inventories and the apparent randomness involved in their formation led to question the means and opportunities by which such a diverse mixture of molecules, a majority of which are thermodynamically stable end products (e.g., carboxylic acids and hydrocarbons), could find an evolutionary path toward the selectivity and functional specificity displayed by even the simplest biochemistry.

RECENT RESULTS

The CR Antarctica Finds

Recorded falls of carbonaceous chondrites have been few (37 to date, since the first registered in 1806) and this record is needed, because these meteorites resemble terrestrial rocks, are porous in nature, and quickly disappear into the environments if not spotted soon. For the same reason, their organic analyses have also become increasingly limited in scope with their years of terrestrial residence, due to the ease with which CCs acquire biochemical contaminants. Fortunately, several of the meteorites recovered in

Antarctica are found unspoiled because of the unique shelter of the glaciers, where falling meteorites are quickly covered by snow, remain buried within the ice, and resurface only when the ice sheets, flowing toward the sea, encounter the obstacle of a mountain.

Renazzo family of meteorites make up a recent classification (CR) of several Antarctic "finds" that have petrology closely similar to that of the Renazzo meteorite, a CC that fell in 1864 and long remained unclassified. Two CR2[1] meteorites (the GRA95229 and LAP02342[2]) were analyzed recently for the major groups of organic compounds known to be present in the Murchison meteorite (Pizzarello et al. 2008; Pizzarello and Holmes 2009) and have shown an organic composition that differs dramatically from that of Murchison and, in fact, from any seen before in carbonaceous meteorites (Fig. 4). Their organic suite is composed mainly of water-soluble compounds, between which N-containing amino acids and amines are predominant. Ammonia is the single largest component of the suite, whereas hydrocarbons and carboxylic acids are only minor components. Novel were also the abundant distributions found within CR2 amino acids, where the shorter chain-length molecules of a homologous series, e.g., glycine, alanine, and α-amino isobutyric acid, are overabundant compared to longer chain species and, in effect, account for most of these compounds' abundance. Several reactive compounds are found in these meteorites as well, such as aldehydes, tertiary amines, and the hydroxy amino acids serine, threonine, *allo*threonine, and tyrosine (Pizzarello and Holmes 2009); the latter two groups of compounds were never detected in Murchison.

Another difference between CR2 and CM meteorites is found in their respective content of enantiomerically enriched chiral molecules. In CR2s, the same amino acid species having

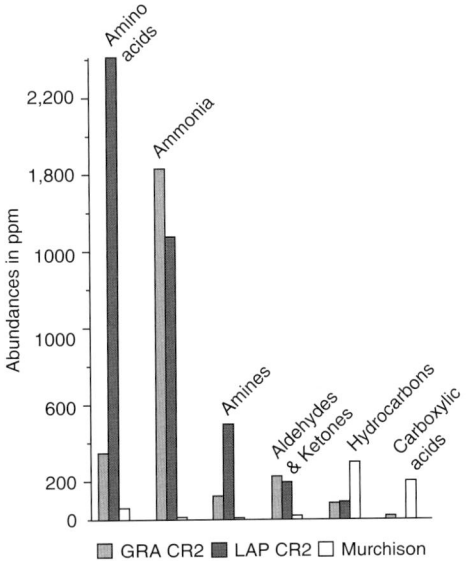

Figure 4. Comparative plot of major soluble organic compound abundances in the Murchison and CR2 meteorites (GRA 95229 and LAP 02342 shown).

ee in CMs display this trait to less extent (GRA95229) or not at all (LAP02342), whereas the abundance distribution of some of the meteorites' diastereomer amino acids allowed the inference of an original asymmetry of their precursor aldehydes (Pizzarello et al. 2008). This somewhat indirect reasoning concerns the diastereomers of the amino acid isoleucine and can be explained as follows. The molecule [2-amino3-methylpentanoic acid, CH_3CH_2-$C^*H(CH_3)$-$C^*H(NH_2)$-COOH] contains two chiral centers (C^*) and can be present as two different compounds (depending on the possible distribution of the methyl branching along the alkyl chain), each with two enantiomers, i.e., the pairs of D-, L-isoleucine (ile) and D-, L-*allo*isoleucine (*allo*) (called diastereomers and shown schematically in Fig. 5B). Only L-ile is present in terrestrial proteins, whereas all four diastereomers are found in meteorites.

A possible reaction for the formation of amino acids in meteorites is the addition of HCN to ketones and aldehydes in the presence of water and ammonia (Fig. 5A) (e.g., Peltzer and Bada 1978). Although producing an asymmetric carbon in most cases (and therefore

---

[1]The number represents a classification of petrographic type and estimates asteroidal secondary processes (were 2<1).

[2]The acronyms stand for the names of the Antarctica locations where the meteorites were found: Graves mountain and LaPaz ice fields, respectively.

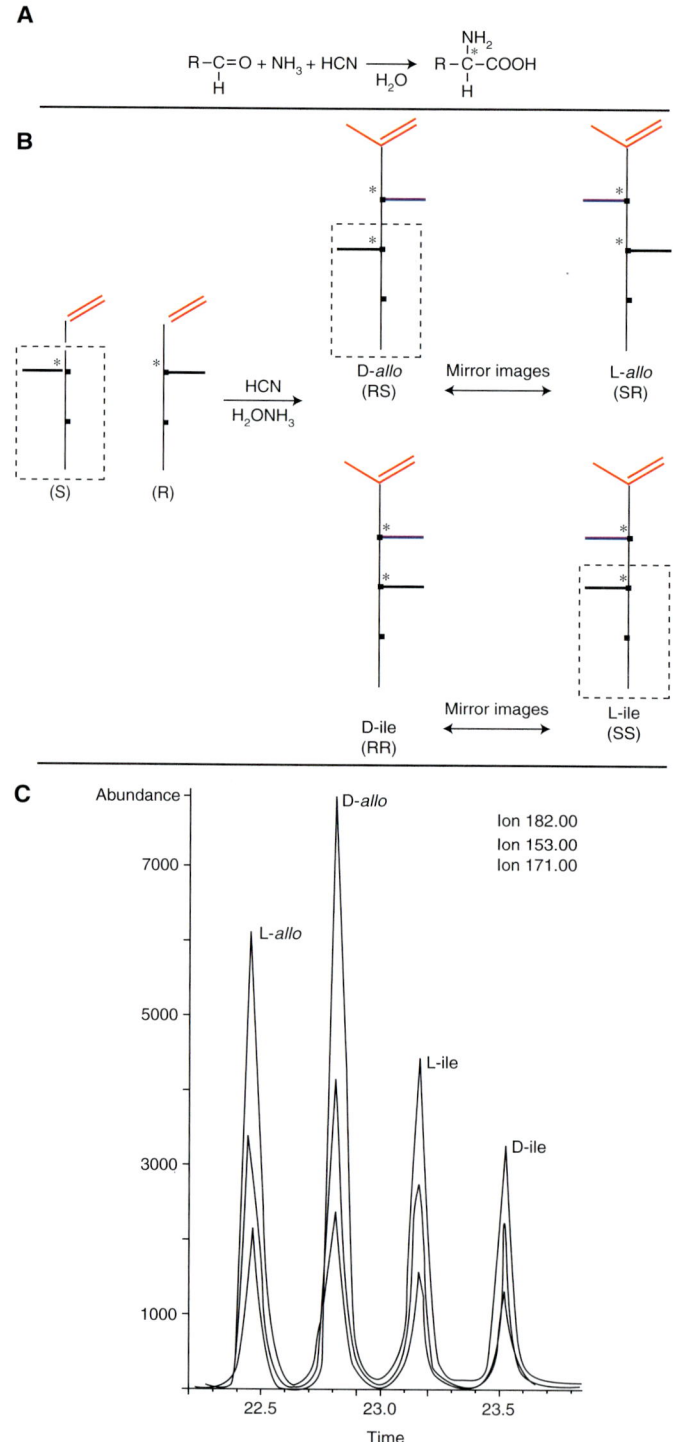

Figure 5. Possible formative pathway of the isoleucine (ile)-*allo*isoleucine (*allo*) diastereomers in meteorites. (*A*) The cyanohydrine reaction. (*B*) Schematic of the distribution of ile and *allo* following the same reaction with a 2-methylbutyraldehyde precursor. (*C*) A chromatogram of the ile-*allo* diatereomers in the GRA 95229 meteorite.

a chiral molecule), this type of synthesis is non-stereospecific because the HCN addition would be random and give equal amounts of D- and L-enantiomers. However, the reaction results become more complex for longer aldehydes that already contain an asymmetric carbon, for example, in the synthesis of the four ile and *allo* diastereomers from DL 2-methyl butanal. In this case, were an *ee* present in the aldehyde, e.g., of the (S) configuration, those amino acids that carried the S-portion of the molecule through their synthesis (shown between dotted squares in Fig. 2B) will be more abundant than their respective enantiomers. In the above example, this would be the (RS) *allo* and (SS) ile compounds or, in the formalism used for amino acids, D-*allo* and L-ile. Such was the distribution of isoleucine diastereomers found in the CR2 extracts (Fig. 5C) that, on the basis of the above formative premise, was interpreted to signify that their precursor aldehyde enantiomers carried an original S-bias to the meteorite's parent body.

Overall, the compositional differences between CR2- and Murchison-type meteorites make stark contrasts. Where the heterogeneity of Murchison compounds easily points to the difficulties that a primordial "soup" would encounter in molecular evolution, CR2 organic distributions and abundances have the unquestionable prebiotic appeal of being over abundantly water-soluble, N-containing, and of low molecular weight. Regardless of how CR2 organic material came to be, it is also clear that an unknown combination of elemental composition, energetic availabilities, and cosmic contingencies made CR2 precursor environments capable of a *de facto* selectivity of such "prebiotically desirable" molecular species.

Whether CR2 meteorite parent bodies would fit the new category (Levison et al. 2009) of "trans-Neptunian" objects or not, certainly the formative environments and histories of their organic materials must have differed from those of CMs. That the known *ee*-carrying amino acids as well as their *ee* are in lower abundance in CR2 than CM meteorites, whereas *ee* appear larger in a precursor aldehydes, would seem to further allow the general inference that abiotic organic pools in chemical evolution were diverse and differed in both their composition and exposure to asymmetric effects.

The Long Cosmic History of Meteorites' Organic Materials

The formation of meteoritic organic compounds was actively debated after Murchison's fall and the revelation that a large variety of extraterrestrial organic molecules with counterparts in the biosphere could be made abiotically. Clearly, to know the syntheses and locals responsible for their formation may have profound significance for the origin of extant life and even a broader exobiology. The earlier hypotheses all focused on solar system processes and, of these, the more influential were the suggestion of possible Miller-Urey type (Miller et al. 1976) syntheses in small planets, following production and recombination of radicals, and of catalytic, FisherTropsch-type, processes in the early stages of the solar nebula, where carbon monoxide could have undergone hydrogenation to hydrocarbons and other compounds (Lancet and Anders 1970). Eventually, the history of the organic compounds in carbonaceous meteorites was elucidated, at least in general terms, by the stable isotope analyses of several compounds and compound classes in the Murchison meteorite.

The isotopic composition is a good indicator of any molecule's synthetic history because the mass differences between isotopomers result in energy differences in their bond formation and may lead to a mass dependent fractionation, which becomes diagnostic of the physicochemical conditions affecting those reactions. Ultimately, isotopic fractionation is a function of zero point energy difference between isotopomers (ΔE) and local temperature, in the form: $\exp(-\Delta E/T)$. In other words, the larger the energy difference between isotope bonds and the lower the temperature, the greater the potential for heavy isotope enrichment. Between the biogenic elements, therefore, hydrogen has the potential for the most enrichment in ^2H (deuterium, D) at low T, because of the high relative mass difference of D/H isotope pair (2/1 u, e.g., compared to $^{13}C/^{12}C = 1.084$). The most dramatic demonstration of these capabilities is given by the spectroscopic observations of the D/H ratios of molecules formed in the dense clouds of the interstellar medium where temperatures are in the 10–30K range. Over a hundred such molecules have been described (e.g., Roueff and Gerin 2002), many of which show extremely high D/H ratios. For example, the average D/H ratio in terrestrial organic compounds is approximately 1.5×10^{-4}, whereas a D/H ratio as high as 0.33 has been observed in the interstellar molecule $D_2CO/DHCO$ (Loinard et al. 2000).

Most Murchison compounds were found enriched in D and ^{13}C to varying degrees and these data, which alone suggest a relation between such molecules or their direct precursors and cold synthetic environments, led to a general theory of formation of meteorite organics that involved interstellar as well as parent body processes. By this hypothesis, icy asteroidal bodies accreted with abundant volatiles, including water and deuterium-rich interstellar organics that, upon warming and a subsequent period of aqueous phase chemistry, yielded the various soluble organic compounds of meteorites.

The possibility of parent body aqueous syntheses seems confirmed by the likelihood that at least some of Murchison amino acids were formed via a Strecker-like reaction of precursor aldehydes and ketones, ammonia and HCN (Fig. 5A). The evidence supporting this hypothesis is the finding in the Murchison meteorite of comparable suites of α-amino and α-hydroxy linear acids (although this correspondence is not valid for the α-methyl compounds) and of imino acids (e.g., Pizzarello and Cooper 2001). These are compounds in which two carboxyl-containing alkyl chains are bonded at the same amino group and would likely result from a Strecker synthesis, e.g., when an amino acid product becomes the reactant in place of ammonia (Fig. 5).

However, there are isotopic as well as molecular trends within the Murchison organic suite that reveal significant formative distinctions between individual compounds and cannot be accounted for by any simplified model. For example, not all of Murchison amino acids fall in the same range of deuterium enrichment, and asymmetry-carrying 2-methyl amino acids display far larger δD values than the 2-H isomers (Fig. 6). Because a similar branched versus linear difference in D-enrichment was also observed between 3- and 4-amino isomers, it seems reasonable to assume that branched molecular species were processed in cold environments to a different degree than the linear ones. On the other hand, 2-, 3- and 4-amino acids also show different trends of ^{13}C abundance with increasing chain lengths, which decreases in the case of the 2-amino acids while remains level, or even slightly increases, in the case of the others. That is, within each level of D-enrichment, various processes of chain elongation seem to have been possible. The obvious conclusion from these Murchison detailed analyses is that diverse cosmic regimes and synthetic processes might have participated in producing the organic composition of this type of meteorites.

The isotopic analyses of CR meteorites added to the above scenario. The δD differences between 2-amino acid types are still present and further magnified, with the two GRA95229 2-methyl amino acids analyzed showing the highest δD values (+7200‰) ever measured for an extraterrestrial molecule by direct analyses. However, $\delta^{15}N$ values determined for CR2 amino acids have a distribution between molecular subgroups that is opposite to the one of their δD values, with 2-H amino acids having higher $\delta^{15}N$ than 2-methyl amino acids (Pizzarello and Holmes 2008).

Because of the near absence of molecular ^{15}N values for cosmic environments[3], only theoretical considerations can be offered for the CR2 findings. The ones offered by Charnley and Rodgers (2002, 2004, 2008) describe a mechanism for higher nitrogen fractionations in regions of the ISM, where the enhanced density and pressure that precede star formation

---

[3] The possibility of different stellar nucleosynthetic pathways for the element of nitrogen (e.g., Wannier et al. 1981) would also further complicate their interpretation.

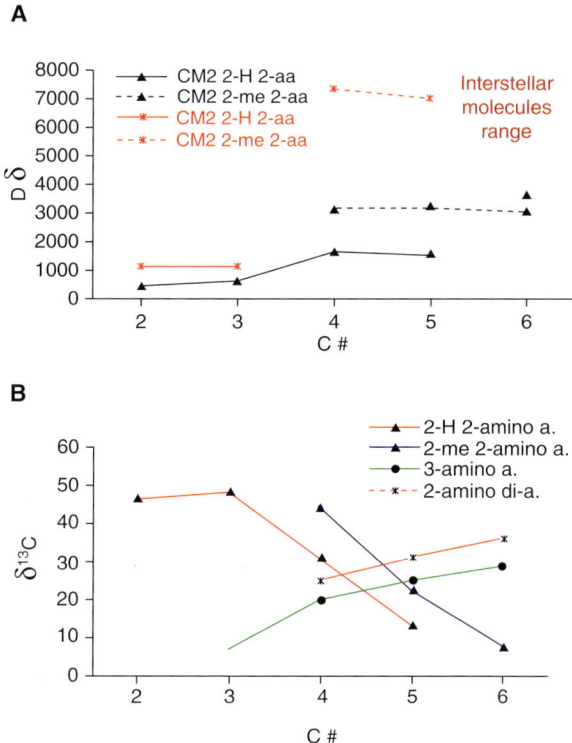

Figure 6. The hydrogen (*A*) and carbon (*B*) isotopic distributions of Murchison and CR2 amino acids.

would cause the freeze-out of most carbon- and oxygen containing molecules; with their disappearance, the disruption of N_2 formation pathways in clouds of lesser density would result in a prevalence of gas-phase atomic nitrogen. In turn, this would lead to the efficient production of ammonia and $^{15}NH_3/^{14}NH_3$ ratios higher than the cosmic $^{15}N/^{14}N$ ratios (to as much as by 80%).

These predictions are interesting in that they appear to match, albeit in broad terms, the findings in meteorites and the current interpretation of meteoritic amino acid formation. In fact, if the distinctly higher δD values of 2-methyl amino acids seem to point to their syntheses in cold ISM environments and the lower values of 2-H amino acids to suggest that their syntheses took place at a later stage in the presence of liquid water, their $δ^{15}N$ opposite trends would also fit with earlier (ISM) and later (prestellar cores) cosmochemical processes, albeit removed from a parent body environment.

Very little is known of the molecular sequence of events that would have taken place in a prestellar core; however, we can expect that several stages of temperature, pressure, and ensuing chemical regimes followed the initial collapse of the presolar portion of the ISM (e.g., Ceccarelli et al. 2007). We could hypothesize, therefore, that some of the warmer stages of star formation might have allowed selected environments, where the desorption, mixing, and reactions of radical, precursor molecules, water, and ammonia led to the syntheses of higher ^{15}N amino acids and favored shorter molecular species formation. It also appears that such locals and the kinetic processes they allow to envision could, rather than parent body reactions, explain some of the molecular distributions seen in the CR meteorites, such as: the far from unity diastereomer ratios seen for the thermodynamically similar amino acids *allo*ile and ile (Chaban and Pizzarello 2007), their erratic levels of enantiomeric excesses as well as the preponderance of lower chain length species and the abundance of unreacted carbonyl containing molecules (Pizzarello and Holmes 2009).

Abiotic Pathways to Biomolecules

Meteorites probably present just a minuscule sample of the prebiotic potential of cosmic synthetic processes but, through their studies, we may be able to infer how common or widespread they may be. Transformations of organic compounds, or their synthesis from inorganic compounds, occurs in response to thermodynamic drives, modulated by the kinetic properties of individual reactions. Setting aside the mechanistic details for a moment, it is useful to examine how reactions may or may not be favored by the thermodynamic properties of the system. Reactions involving organic compounds and occurring in aqueous solution may have occurred on meteorite parent bodies, smaller icy aggregates on their way to form asteroids or comets, and in selected prestellar environments; therefore, investigating relative stabilities of aqueous organic compounds may yield clues to these processes. This approach can help to answer specific questions about relative abundances of organic compounds found in carbonaceous meteorites. The following discussion illustrates this approach, with the specific goal of understanding the relative abundances of ammonia, amino acids, and aldehydes.

Stabilities of amino acids relative to other organic compounds during aqueous alteration can be assessed by considering a set of hypothetical overall reactions involving amino acids and other aqueous organic compounds. As an example, the stability of alanine relative to the aldehyde propanal can be assessed by considering a reaction in which carbon is conserved in the two aqueous organic compounds, given by

$$CH_3CH_2CHO(aq) + H_2O + NH_3(aq)$$
propanal
$$= CH_3CHNH_2COOH(aq) + 2\,H_2(aq),$$
alanine (1)

where the (aq) indicates that the compounds of interest are all dissolved in H_2O. This reaction is not meant to depict a specific synthetic process, but instead delineates relative stabilities. It is evident from reaction (1) that there could be abundances of $NH_3(aq)$ that would favor the stability of alanine relative to propanal. Likewise, at strongly reduced conditions, where there may be considerable $H_2(aq)$ present, alanine would become unstable relative to propanal and $NH_3(aq)$. Quantifying the activities (and concentrations) of $NH_3(aq)$ and $H_2(aq)$, where such transformations become possible, can be accomplished by considering the equilibrium constant for reaction (1), and manipulating its law of mass action expression. That expression, in its logarithmic form, is given by

$$\begin{aligned}\log K = {} & \log a_{CH_3CHNH_2COOH(aq)} \\ & + 2 \log a_{H_2(aq)} \\ & - \log a_{CH_3CH_2CHO(aq)} \\ & - \log a_{NH_3(aq)} - \log a_{H_2O}.\end{aligned} \quad (2)$$

In most dilute aqueous solutions (salinity < seawater, as a rule of thumb), it can be safely assumed that the activity of H_2O is so close to 1 that setting it equal to 1 introduces only trivial uncertainty. With this assumption, Equation (2) can be rearranged to give

$$\log a_{NH_3(aq)} = 2\log a_{H_2(aq)} - \log K \\ + \log\left(\frac{a_{CH_3CH_2CHO(aq)}}{a_{CH_3CHNH_2COOH(aq)}}\right), \quad (3)$$

which represents the equation of a line on a plot of log $a_{NH_3(aq)}$ vs log $a_{H_2(aq)}$ with a slope of 2 and an intercept equal to

$$-\log K + \log\left(\frac{a_{CH_3CH_2CHO(aq)}}{a_{CH_3CHNH_2COOH(aq)}}\right).$$

At constant temperature and pressure, log K is a constant, which means that various lines can be determined based on the activity ratio of alanine to propanal.

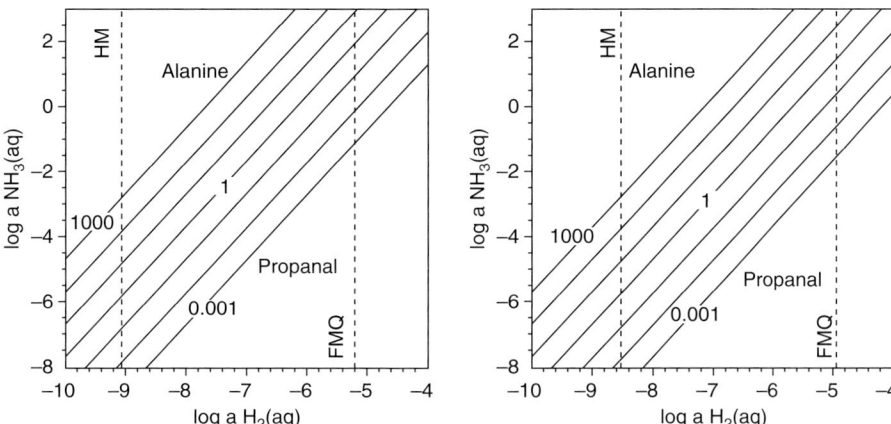

Figure 7. Equilibrium activity diagrams showing the relative stabilities of aqueous alanine and propanal in terms of the activities of $NH_3(aq)$ and $H_2(aq)$ at (*left*) 0°C and 1 bar and (*right*) 25°C and 1 bar. Selected contours of the equilibrium ratio of activities of alanine to propanal from 1000 to 0.001 are indicated. Equilibrium constants for reaction (1) were calculated with the revised Helgeson-Kirkham-Flowers equation of state (Shock et al. 1992) using data and parameters from Shock et al. (1989); Shock and Helgeson (1990) and Shulte and Shock (1993). Also shown are activities of $H_2(aq)$ corresponding to equilibrium between hematite and magnetite (HM, reaction 4), as well as magnetite, quartz, and fayalite (FMQ, reaction 5). Thermodynamic data for minerals come from Helgeson et al. (1978). All calculations were conducted with the software package SUPCRT92 (Johnson et al. 1992).

Plots of this type are shown in Figure 7 for 0°C and 25°C both at 1 bar, with contours of the activity ratio ranging from 0.001 to 1000. The bold contour labeled 1 in each plot shows the position of equal activities of the two organic solutes at equilibrium. Ranges of relative predominance of propanal and alanine are indicated, with that of alanine in each plot falling at higher activities of $NH_3(aq)$ and lower activities of $H_2(aq)$, consistent with Le Chatlier's principle applied to reaction (1). Also shown in these diagrams are the values of log a $H_2(aq)$, at which hematite (Fe_2O_3) would be reduced to magnetite (Fe_3O_4), consistent with

$$3\,Fe_2O_3 + H_2(aq) = 2\,Fe_3O_4 + H_2O, \quad (4)$$

and where magnetite would be reduced to the ferrous silicate fayalite (Fe_2SiO_4) in the presence of quartz (SiO_2) according to

$$Fe_3O_4 + H_2(aq) + 3/2\,SiO_2 = 3/2\,Fe_2SiO_4 + H_2O. \quad (5)$$

Magnetite, which is one of the aqueous alteration products identified in CI, CM, CO, CR, CV meteorites and some LL3 chondrites (Zolensky et al. 2008) would be stable between the two vertical dashed lines on each plot.

The presence of magnetite brackets the equilibrium activities of $H_2(aq)$ that could have attained during at least a portion of the aqueous alteration processes occurring on the Murchison parent body. If this alteration occurred at 0°C, then the equilibrium activity of $H_2(aq)$ fell between about $10^{-5.2}$ and $10^{-9.1}$. If, on the other hand, temperatures were warmer, these activities would change. As an example, at 25°C, the equilibrium activities of $H_2(aq)$ fall between $10^{-4.9}$ and $10^{-8.5}$. In dilute solutions, activities of neutral solutes correspond closely to concentrations (Amend and Shock 2001).

These plots reveal the ranges of $H_2(aq)$ concentrations that are consistent with the occurrence of magnetite, and the $NH_3(aq)$ concentrations that would provide a thermodynamic drive for the formation of an amino acid rather than an aldehyde, and *vice versa*.

At, for example, log a $H_2(aq) = -7$ (equal to about 100 nanomolar dissolved H_2), conditions in the middle of the range of magnetite stability, activities of $NH_3(aq) > 10^{-2}$ at 0°C, and $> 10^{-3}$ at 25°C, would favor the formation of alanine at abundances greater than those of propanal. Whether or not equilibrium is actually attained among organic compounds during aqueous alteration events on meteorite parent bodies, the persistent thermodynamic drive to form amino acids or aldehydes depends on the chemical composition of the system. The plots in Figure 7 reveal the quantitative nature of those thermodynamic drives. They also make it possible to begin to understand the amounts of $NH_3(aq)$ that would be required if amino acid concentrations were similar to aldehyde concentrations or vastly different.

Comparison of the data from the CR2 meteorites and Murchison shown in Figure 4 indicates that the ratio of total amino acids to aldehydes+ketones is on the order of 12 for LAP and about two for GRA and Murchison. It can also be seen that ammonia abundances are greatest in GRA, similarly large in LAP, and very low in Murchison. These data can be combined with the thermodynamic analysis depicted in Figure 7 in an attempt to evaluate what conditions were like during aqueous alteration, if the relative abundances of aldehydes and amino acids were influenced by that stage of meteorite history. It should be kept in mind that the data that exist are for what is present in the meteorites and not what may have been present in aqueous solution during the alteration process. Adsorption equilibria among solutions and various mineral phases may differ for these two classes of organic compounds, and much could have happened to alter ratios inherited from such an early stage in the history of the solar system. Let us assume that the relative abundances of organic compounds in meteorite extracts reflects conditions on the parent bodies at the time the compounds formed and that they were not radically reset by subsequent history.

Starting with LAP, conditions consistent with the overall amino acid to aldehyde ratio would fall just above and to the left of the 10 contour, which is the first above the equal activity (=1) contour in a plot like those shown in Figure 7 for the presently unknown temperature of aqueous alteration. If the amino acid to aldehyde ratio is a result of aqueous alteration, then it provides us with this locus of possibilities in log a $NH_3(aq)$ versus log a $H_2(aq)$ space. Likewise, ratios from GRA and Murchison indicate that conditions during aqueous alteration may have generated conditions near or slightly above the equal activity contour. If there were estimates of the activity of either $H_2(aq)$ or $NH_3(aq)$ that prevailed during aqueous alteration, then the equilibrium value of the other would be uniquely defined by the ratio of organic compounds.

Assuming that the relative abundances of ammonia in the extracts are analogous to the relative abundances during aqueous alteration leads to the following assessment of relative oxidation-reduction (redox) states during aqueous alteration events. The amino acid to aldehyde ratios in GRA and Murchison are about equal, but the abundance of ammonia that can be extracted from GRA is much greater. Therefore, it seems likely that conditions during aqueous alteration of the Murchison parent body would plot at a lower activity of $NH_3(aq)$ than those that attained during alteration of the GRA parent body. If so, then the fact that both meteorites fall on about the same contour means that the activity of $H_2(aq)$ was much greater during alteration of GRA than during alteration of Murchison if alteration processes happened at similar temperatures on both parent bodies. The abundance of ammonia in the LAP extracts is nearly as great as the GRA extracts, but the amino acid to aldehyde ratio is also greater. If the temperature of alteration of LAP was similar to that of GRA, then conditions during the alteration of LAP would fall somewhat lower in log a $NH_3(aq)$, but also considerably lower in the activity of $H_2(aq)$ to maintain the higher amino acid to aldehyde ratio. All else being equal, indications are that conditions were most oxidized during alteration of the Murchison parent body, most reduced during alteration of GRA, and intermediate during the alteration of LAP. Corroborating evidence may be found in the relative abundances of carboxylic acids, which are more oxidized than either amino acids or aldehydes. Redox conditions during alteration directly affect the potential for abiotic organic synthesis (Shock 1990; 1992a; Shock and Schulte 1990; 1998; Amend and Shock 1998; Shock and Canovas 2010). If the analysis outlined above survives deeper scrutiny, the *overall* potential for abiotic organic synthesis from inorganic starting compounds may have been greatest on the GRA parent body, despite its lower abundances of amino acids.

There are several ways that these predictions of relative redox states can be tested. One would be to examine the mineralogy of the alteration products in all three meteorites for evidence of mineral assemblages that could indicate redox conditions that prevailed during alteration. Another would be to seek evidence from mineral assemblages, organic compound associations, and isotopes (oxygen in pairs or suites of minerals that formed together, for example) that could bracket the temperatures of the alteration events on each parent body so that quantitatively appropriate versions of the plots in Figure 7 could be built. Also, experimental studies of the adsorption of ammonia, amino acids, aldehydes, and other organic compounds commonly extracted from meteorites on minerals found in meteorite alteration assemblages would enable estimation of aqueous concentrations or activities from the abundances of these compounds in the meteorites.

Exogenous Delivery and Molecular Evolution

The Monomers and Their Potential

If we trust the record of impact craters observed in most of solar planets and satellites, meteorites have showered the Earth throughout geological ages and certainly did so soon after its accretion (e.g., Chyba and Sagan 1992). Abundant organic materials were just as certainly delivered to the early Earth and, it is reasonable to assume, a good portion of them survived the process. We have learned from the analyses of two largely different types of meteorites that this exogenous input delivered both complex macromolecules of uncertain composition and free soluble compounds. Various molecular species must have interacted in the meteorites already prior to their fall, to a certain degree, because some derivative compounds such as the carboxamides (Cooper and Cronin 1995) are released from their extracts upon hydrolysis; however, peptides have been carefully searched for in the Murchison meteorite and not found, with the exception of diglycine (Shimoyama and Ogasawara 2002). If we are trying to estimate the potential of this delivery for prebiotic evolution and we believe that such evolution had to gain some polymeric complexity for life to ensue, then, we have to conclude that the bulk of meteoritic compounds could have provided, at best, monomeric constituents. In general, however, any evolutionary path has to rely on monomeric material as well and, just comparing with other early planetary processes that could have led to organic compounds such as atmosphere-mediated Miller-Urey-type syntheses or the environment of hydrothermal vents, the molecular species ready-made in meteorites would not appear as too bad of a start. Of these, meteoritic amino acids appear as likely candidates for further molecular evolution, particularly considering their selective and abundant suites found in CR2 chondrites.

Amino acids, the components of extant proteins, are able to polymerize under a variety of laboratory conditions and could have done so in early Earth environments. For example, Oro' and Guidry (1961) first showed that glycine readily polymerizes in the presence of ammonia and little water at temperatures of about $140°C$. Also Leman, Orgel, and Ghadiri (2004) showed that the presence of carbonyl sulfide, such as it is found around volcanoes, could lead to easy formation of peptides. When of the type found nonracemic in Murchison, amino acids readily form an activated carboxyl, e.g., as an oxazolone by intramolecular dehydration, and polymerize conforming into helixes at lengths as short as three-amino acid units (Crisma et al. 2004).

These findings suggest that it is plausible that exogenous amino acids acquired at least some polymeric complexity during early terrestrial evolution; it is as likely that their overall molecular properties might have been evolutionary factors as well. For example, all *ee* found so far for meteoritic amino acids have just one configuration, L-, whereas those obtained for chiral molecules in natural processes, designed experiments, or via theoretical schemes are all subjected to chance outcome in the absence of asymmetric influences. Similarly, several terrestrial crystals such as quartz are chiral but their world-wide production is about equal in *d*- and *l*-forms. Also, so far, *ee* have been found in amino acids that do not racemize[4], meaning that their *ee* could have been preserved in prebiotic aqueous environments.

Most importantly, amino acids as well as peptides are molecules with diverse catalytic properties that are readily displayed experimentally and in biochemistry. They can also be asymmetric catalysts, a fact suggesting that the unique molecular asymmetry of meteoritic amino acids might have been a particularly useful evolutionary tool. A set of experiments have been conducted with this theme, to assess the possibility that the nonracemic amino acids of meteorites could have acted as catalysts during early Earth molecular evolution and transferred their asymmetry to other prebiotic building blocks such as sugars. It was found that both amino acids and dipeptides can catalyze the asymmetric aldol condensation of glycolaldehyde, or glycolaldehyde and glyceraldehyde, to produce tetrose (Pizzarello and Weber 2004; Weber and Pizzarello 2006) and pentose (Pizzarello and Weber 2010) sugars with significant *ee*. It is interesting that these syntheses singled out D-erythrose and D-ribose in forming *ee* with LL dipeptides catalysts, whereas all other sugars acquired either *ee* of the same configuration as the catalyst or, in some cases, no *ee* at all. These reactions were conducted in buffered water solution, made use of simple reactant realistically available to the early Earth, and implied likely catalytic pathways under mild conditions. They have, therefore, some prebiotic credibility and support the conclusion that, whereas the extent to which meteoritic catalysts might have been effective in a mixture is entirely unknown, their possible inductive effect toward chiral asymmetry in the monomeric interactions of molecular evolution cannot be disregarded.

Energetic Contingencies

Knowing that life ensued rather quickly in early Earth history, it seems also realistic to assume that the planet environments might have been part of the unknown contingencies that fostered the transition from abiotic chemistry to the molecular evolution that preceded the emergence of life. If so, these environments would have combined available organic compounds with favorable catalysts, which might have been organic as well as inorganic. Just as all known life forms have habitats, the emergence of life may have had a habitat as well (Shock et al. 1998, 2000).

Because the record of Earth's first geological era was lost to ensuing diagenetic and metamorphic changes, clays represent the first alteration products of basaltic glass under hydrous conditions and would be good candidates for aiding simple abiotic molecules, such as those found in meteorites, in undertaking evolutionary steps of prebiotic significance. As detailed in Deamer and Weber (2010), these minerals are known to adsorb organic molecules and actively participate as catalysts in their syntheses and reactions (Williams et al. 2005). In particular, the smectite group of expandable clays, such as montmorillonite and saponite, can undergo surface energy changes during diagenesis that will affect their surface H-bonding at key sites and form complex aromatic and polyaromatic hydrocarbons of up to C_{20} from methanol (e.g., Williams et al. 2005).

Also, as mentioned above, conditions of very low H_2O activities or elevated temperatures in aqueous solution can drive polymerization reactions that involve dehydration such as peptide formation. The latter possibility has inspired several experimental investigations of the

---

[4]Racemization, the reversal of configurations in water, involves the loss and reacquisition of hydrogen by the carbon adjacent to the carboxyl group, which is slightly acidic, and is not allowed when the H at C-2 is substituted with a methyl group.

potential for amino acid polymerization under hydrothermal conditions (Shock 1992b, 1993). Starting with amino acids, it has repeatedly been shown that dipeptides and cyclic dipeptides form rapidly in hydrothermal experiments (Imai et al. 1999b; Alargov et al. 2002; Li and Brill 2003; Lemke et al. 2009; Cleaves et al. 2009). Occasionally, these experimental studies also obtain small concentrations of tripeptides and longer oligomers (Imai et al. 1999a; Tsukahara et al. 2002), but the formation of cyclic dipeptides, which is thermodynamically favored (Shock 1992b), often dominates. It has also been shown that if experiments are started with somewhat larger oligomers, say three or four amino acids in length, then the peptides can be lengthened by hydrothermal reactions involving the monomers (Kawamura et al. 2005), and that polymers containing up to 20 amino acids can be generated hydrothermally from glutamic acid or aspartic acid, which do not form cyclic dipeptides (Kawamura and Shimahashi 2008). In addition, hydrothermal dehydration reactions involving alkanoic acids and glycerol produce lipid-like molecules capable of self-assembly (Simoneit et al. 2007). Taken together, these recent results show that condensation, polymerization, and peptide bond formation may commonly occur in hydrothermal conditions. If so, planetary processing of materials supplied from meteorites may have been integral to the emergence of living systems.

CHALLENGES AND FUTURE RESEARCH DIRECTIONS

Carbonaceous chondrites are natural samples of limited and unpredicted availability that, once reaching the Earth, are under the constant threat of terrestrial contamination. Their study has obviously met with challenges of material preservation, designing of analytical methodologies, identification of indigenous materials, and more; these will remain, *mutatis mutandis*, much the same in the future. Nonetheless, these meteorites have been analyzed successfully in great detail and their studies have been invaluable in determining the prebiotic possibilities of cosmochemical environments; however, they have not answered the basic exobiological question of whether extraterrestrial organic compounds contributed to molecular evolution on the early Earth and to the emergence of life.

That answer may never be possible, but, if we believe with Eschenmoser (2008) that life's origin " . . . *cannot be discovered, as other things in science, it can only be re-invented*", meteorite analyses will offer realistic molecular tools to attempt just that and much still can be done. After forty years of studying Murchison-type meteorites, a new group of Antarctic finds has shown that within the diverse cosmic environments may reside the capabilities of forming organic suites enriched in biomolecule precursors and of high prebiotic appeal. The CR2 organic compounds are still poorly characterized but new studies will define their extent and distribution. Many small molecules that could be useful for initiating molecular evolution could have escaped detection in earlier studies of these pristine meteorites and have not yet been targeted for analyses: glycolaldehyde, glyceraldehydes (detected but not quantified or unpublished), HCN, formamide, urea, and small peptides are all "stuff" required for modeling early evolutionary biology. Hopefully, we shall know soon their distribution in space also.

REFERENCES

Alargov DK, Deguchi S, Tsujii K, Horikoshi K. 2002. Reaction behaviors of glycine under super- and subcritical water conditions. *Origins Life Evol Biosph* **32:** 1–12.

Amend JP, Shock EL. 1998. Energetics of amino acid synthesis in hydrothermal ecosystems. *Science* **281:** 1659–1662.

Amend JP, Shock EL. 2001. Energetics of overall metabolic reactions in thermophilic and hyperthermophilic Archaea and Bacteria. *FEMS Microbiology Reviews* **25:** 175–243.

Arteaga O, Canillas A, Crusats J, El-Achemi Z, Jellison JE, Llorca J, Ribó JM. Chiral biases in solids by effect of sheer gradients: A speculation on the deterministic origin of biological homochirality. *Orig Life Evol Biosph* **40:** 27–40.

Ceccarelli C, Caselli P, Herbst E, Tielens AGGM, Caux E. 2007. Extreme deuteration and hot Corinos: The earliest chemical signatures of low-mass star formation. In *Protostars and planets* (ed. Reipurth VB , Jewitt D, Keil K), pp. 47–62. The University of Arizona Press, Tucson, AZ.

Chaban G, Pizzarello S. 2007. Meteoritic amino acids: The product of still unknown cosmic and solar environments. Abstracts of papers, 39th Lunar and Planetary Conference, Houston, TX. Abstract # 1389.

Charnley SD, Rodgers SB. 2002. The end of interstellar chemistry and the origin of nitrogen in comets and meteorites. *A J* **569**: L133–L137.

Chyba CF, Sagan C. 1992. Endogenous production, exogenous delivery, and impact-shock synthesis of organic molecules: an inventory for the origins of life. *Nature* **355**: 125–132.

Cleaves HJ, Aubrey AD, Bada JL. 2009. An evaluation of the critical parameters for abiotic peptide synthesis in submarine hydrothermal systems. *Origins of Life and the Evolution of the Biosphere* **39**: 109–126.

Cooper GW, Cronin JR. 1995. Linear and cyclic aliphatic carboxamides of the Murchison meteorite: Hydrolyzable derivatives of amino acids and other carboxylic acids. *Geochim Cosmochim Acta* **59**: 1003–1015.

Crick FHC, Orgel LE. 1973. Directed panspermia. *Icarus* **19**: 341–346.

Crisma M, Moretto A, Formaggio F, Kaptein B, Broxterman QB, Toniolo C. 2004. Meteoritic C^α-methylated α-amino acids an the homochirality of life: Searching for a link. *Angew Chem Int Ed* **43**: 6695–6699.

Cronin JR, Pizzarello S. 1997. Enantiomeric excesses in meteoritic amino acids. *Science* **275**: 951–955.

Deamer D. 1985. Boundary structures are formed by organic components of the Murchison carbonaceous chondrite. *Nature* **317**: 792–794.

Deamer D, Weber AL. 2010. Bioenergetics and life's origins. *Cold Spring Harb Perspect Biol* **2**: a004929.

Eschenmoser A. 2008. Fundamental questions: An interview with Albert Eschenmoser. In *The Scripps Research Institute News and Views*: 8. http://www.scripps.edu/newsandviews/e_20080421/.

Helgeson HC, Delany JM, Nesbitt WH, Bird DK. 1978. Summary and critique of the thermodynamic properties of rock forming minerals. *Am J Sci* **278A**: 1–229.

Imai E-I, Honda H, Hatori K, Brack A, Matsuno K. 1999a. Elongation of oligopeptides in a simulated submarine hydrothermal system. *Science* **283**: 831–833.

Imai E-I, Honda H, Hatori K, Matsuno K. 1999b. Autocatalytic synthesis of oligoglycine in a simulated submarine hydrothermal system. *Origins of Life and Evolution of the Biosphere* **29**: 249–259.

Johnson JW, Oelkers EH, Helgeson HC. 1992. SUPCRT92: A software package for calculating the standard molal thermodynamic properties of minerals, gases, aqueous species, and reactions from 1 to 5000 bar and 0 to 1000°C. *Computers & Geosciences* **18**: 932–934.

Kawamura K, Shimahashi M. 2008. One-step formation of oligopeptide-like molecules from Glu and Asp in hydrothermal environments. *Naturwissenschaften* **95**: 449–454.

Kawamura K, Nishi T, Sakiyama T. 2005. Consecutive elongation of alanine oligopeptides at the second time range under hydrothermal conditions using a microflow reactor system. *J Am Chem Soc* **127**: 522–523.

Kawasaki T, Hatase K, Fujii Y, Jo K, Soai K, Pizzarello S. 2006. The distribution of chiral asymmetry in meteorites: An investigation using asymmetric autocatalytic chiral sensors. *Geochim Cosmochim Acta* **70**: 5395–5402.

Lazcano A. 2010. Historical development of origins research. *Cold Spring Harb Perspect Biol* **2**: a002089.

Leman L, Orgel L, Ghadiri MR. 2004. Carbonyl sufide-mediated prebiotic formation of peptides. *Science* **306**: 283–286.

Lemke KH, Rosenbauer RJ, Bird DK. 2009. Peptide synthesis in early Earth hydrothermal systems. *Astrobiology* **9**: 141–146.

Levison HF, Bottke WF, Gounelle M, Morbidelli A, Nesvorn D, Tsiganis K. 2009. Contamination of the asteroid belt by primordial trans-Neptunian objects. *Nature* **460**: 364–366.

Li J, Brill TB. 2003. Spectroscopy of hydrothermal reactions. 27. Simultaneous determination of hydrolysis rate constants of glycylglycine to glycine and glycylglycine-diketopiperazine equilibrium constants at 310-330°C and 275 bar. *J Phys Chem A* **107**: 8575–8577.

Loinard S, Castets A, Ceccarelli C, Tielens AGGM, Faure A, Caux E, Duvert G. 2000. Double deuterated molecular species in protostellar environments. *Astron Astrophys* **359**: 1169–1174.

Luisi PL. 2007. Presented at: International School of Complexity-4th course: Basic questions on the Origin of Life; Ettore Majorana Foundation and Centre for Scientific Culture, Erice, Italy, 1–6 October 2006.

Mautner MN, Leonard R, Deamer DW. 1995. Meteorite organics in planetary environments: Hydrothermal release, surface activity and microbial utilization. *Planet Space Sci* **43**: 139–147.

Miller SL, Urey HC, Oró J. 1976. Origin of organic compounds on the primitive earth in meteorites. *J Mol Evol* **9**: 59–72.

Oro' J, Guidry CL. 1961. Direct synthesis of polypeptide I. Polycondensation of glycine in aqueous ammonia. *Arch Biochem Biophys* **93**: 166–171.

Pearson VK, Kearsley AT, Sephton MA, Gilmour I. 2007. The labelling of meteoritic organic material using osmium tetroxide vapour impregnation. *Planet Space Sci* **55**: 1310–1318.

Peltzer ET, Bada JL. 1978. α-Hydroxycarboxylic acids in the Murchison meteorite. *Nature* **272**: 443–444.

Pizzarello S. 2009. The chemistry that preceded life's origin: When is an evolutionary story an emergent story? *Orig Life Evol Biosph* (in press).

Pizzarello S, Cooper GW. 2001. Molecular and chiral analyses of some protein amino acid derivatives in the Murchison and Murray meteorites. *Meteorit Planet Sci* **36**: 897–909.

Pizzarello S, Holmes W. 2009. Nitrogen-containing compounds in two CR2 meteorites: ^{15}N composition, molecular distribution and precursor molecules. *Geochim Cosmochim Acta* **73**: 2150–2162.

Pizzarello S, Huang Y. 2005. The deuterium enrichment of individual amino acids in carbonaceous meteorites: A case for the presolar distribution of biomolecules precursors. *Geochim Cosmochim Acta* **69**: 599–605.

Pizzarello S, Weber AL. 2004. Prebiotic amino acids as asymmetric catalysts. *Science* **303**: 1151.

Pizzarello S, Weber AL. 2010. Stereoselective syntheses of pentose sugars under realistic prebiotic conditions. *Orig Life Evol Biosph* **40:** 3–10.

Pizzarello S, Zolensky M, Turk KA. 2003. Non racemic isovaline in the Murchison meteorite: chiral distribution and mineral association. *Geochim Cosmochim Acta* **67:** 1589–1595.

Pizzarello S, Huang Y, Alexandre MR. 2008. Molecular asymmetry in extraterrestrial chemistry: Insights from a pristine meteorite. *Proc Natl Acad Sci* **105:** 3700–3704.

Rodgers SB, Charnley SB. 2004. Interstellar diazenylium recombination and nitrogen isotopic fractionation. *Mont Not R Astron Soc* **352:** 600–604.

Rodgers SB, Charnley SB. 2008. Nitrogen superfractionation in dense cloud cores. *Mont Not R Astron Soc* **569:** L48–L52.

Roueff E, Gerin M. 2003. Deuterium in molecules of the interstellar medium. *Space Scie Rev* **106:** 61–72.

Shimoyama A, Ogasawara R. 2002. Dipeptides and diketopiperazines in the Yamato-791198 and Murchison carbonaceous chondrites. *Orig Life Evol Biosph* **32:** 165–179.

Shock EL. 1990. Geochemical constraints on the origin of organic compounds in hydrothermal systems. *Origins Life Evol Biosph* **20:** 331–367.

Shock EL. 1992a. Chemical environments in submarine hydrothermal systems. In: *Marine hydrothermal systems and the origin of life*, (ed. N Holm) a special issue of *Origins Life Evol Biosph* **22:** 67–107.

Shock EL. 1992b. Stability of peptides in high temperature aqueous solutions. *Geochim Cosmochim Acta* **56:** 3481–3491.

Shock EL. 1993. Hydrothermal dehydration of aqueous organic compounds. *Geochim Cosmochim Acta* **57:** 3341–3349.

Shock EL, Canovas P. 2010. The potential for abiotic organic synthesis and biosynthesis at seafloor hydrothermal systems. *Geofluids* (in press).

Shock EL, Helgeson HC. 1990. Calculation of the thermodynamic and transport properties of aqueous species at high pressures and temperatures: Standard partial molal properties of organic species. *Geochim Cosmochim Acta* **54:** 915–945.

Shock EL, Schulte MD. 1990. Amino acid synthesis in carbonaceous meteorites by aqueous alteration of polycyclic aromatic hydrocarbons. *Nature* **343:** 728–731.

Shock EL, Schulte MD. 1998. Organic synthesis during fluid mixing in hydrothermal systems. *J Geophys Res* **103:** 28513–28527.

Shock EL, Amend JP, Zolotov MYu. 2000. The early Earth vs. the origin of life. In *The origin of the Earth and Moon* (ed. R Canup, K Righter), pp. 527–543. University of Arizona Press.

Shock EL, Helgeson HC, Sverjensky DA. 1989 Calculation of the thermodynamic and transport properties of aqueous species at high pressures and temperatures: Standard partial molal properties of inorganic neutral species. *Geochim Cosmochim Acta* **53:** 2157–2183.

Shock EL, McCollom T, Schulte MD. 1998 The emergence of metabolism from within hydrothermal systems. In *Thermophiles: The keys to molecular evolution and the origin of life?* (ed. Adams Wiegel), pp. 59–76. Taylor & Francis, London, UK.

Shock EL, Oelkers EH, Johnson JW, Sverjensky DA, Helgeson HC. 1992. Calculation of the thermodynamic properties of aqueous species at high pressures and temperatures: Effective electrostatic radii, dissociation constants, and standard partial molal properties to $1000°C$ and 5 kb. *J Chem Soc, Faraday Trans* **88:** 803–826.

Shulte MD, Shock EL. 1993. Aldehydes in hydrothermal solutions: Standard partial molal thermodynamic properties and relative stabiirties at high-temperatures and pressures. *Geochim Cosmochim Acta* **57:** 3835–3846.

Simoneit BRT, Rushdi AI, Deamer DW. 2007. Abiotic formation of acylglycerols under simulated hydrothermal conditions and self-assembly properties of such lipid products. *Adv Space Res* **40:** 1649–1656.

Tsukahara K, Imai E-I, Honda H, Hatori K, Matsuno K. 2002. Prebiotic oligomerization on or inside lipid vesicles in hydrothermal environments. *Origins Life Evol Biosph* **32:** 13–21.

Wannier PG, Linke RA, Penzias AA. 1981. Observations of N-14N-15 in the galactic disk. *Ap J* **247:** 522–529.

Weber AL, Pizzarello S. 2006. The peptide catalyzed stereospecific synthesis of tetroses: A possible model for prebiotic molecular evolution. *Proc Natl Acad Sci* **103:** 12713–12717.

Yabuta H, Wiliams LB, Cody GD, Alexander CMO'D, Pizzarello S. 2007. The insoluble carbonaceous material of CM chondrites: A possible source of discrete organic compounds under hydrothermal conditions. *Meteor Planet Sci* **42:** 37–48.

Ribonucleotides

John D. Sutherland

School of Chemistry, The University of Manchester, Oxford Road, Manchester M13 9PL, United Kingdom
Corespondence: john.sutherland@manchester.ac.uk

It has normally been assumed that ribonucleotides arose on the early Earth through a process in which ribose, the nucleobases, and phosphate became conjoined. However, under plausible prebiotic conditions, condensation of nucleobases with ribose to give β-ribonucleosides is fraught with difficulties. The reaction with purine nucleobases is low-yielding and the reaction with the canonical pyrimidine nucleobases does not work at all. The reasons for these difficulties are considered and an alternative high-yielding synthesis of pyrimidine nucleotides is discussed. Fitting the new synthesis to a plausible geochemical scenario is a remaining challenge but the prospects appear good. Discovery of an improved method of purine synthesis, and an efficient means of stringing activated nucleotides together, will provide underpinning support to those theories that posit a central role for RNA in the origins of life.

Whether RNA first functioned in isolation, or in the presence of other macromolecules and small molecules is still an open question, and a question that is addressable through chemistry (Borsenberger et al. 2004). If synergies are found between RNA assembly chemistry and that associated with the assembly of lipids and/or peptides, the purist RNA world concept (Woese 1967; Crick 1968; Orgel 1968) might have to be loosened to allow other such molecules a role in the origin of life. Metabolism, or the roots of metabolism, could also potentially have coevolved with RNA if organic chemistry happened to work in a particular way on a set of plausible prebiotic feedstock molecules in a dynamic geochemical setting. Such considerations point to the need for an open mind when considering the chemical derivation of RNA. Notwithstanding these caveats, however, the self-assembly of polymeric RNA on the early Earth most likely involved activated monomers (Verlander et al. 1973; Ferris et al. 1996). These activated monomers could either have come together sequentially to make RNA one monomer at a time, or short oligoribonucleotides formed by such a process could have joined together by ligation in what would thus amount to a two-stage assembly of RNA polymers. Replication of RNA would then have involved template-directed versions of these or related chemistries (Orgel 2004).

The details of the polymerization processes that might plausibly have given rise to the first RNA molecules can only be investigated when there is some evidence as to the specific chemical nature of the activated monomers. Broadly speaking, however, it is possible to differentiate different polymerization chemistries on the basis of the bonds formed in the polymerization step. P–O bond forming polymerization chemistry is reasonable to consider first because of the simplicity of P–O retrosynthetic disconnections of RNA (Corey 1988). In the case of the simplest P–O bond forming polymerization chemistry, the monomeric products of preceding prebiotic chemistry would either be activated ribonucleoside-5′-phosphates **1**—activated through having a leaving group attached to the phosphate—or ribonucleoside-2′,3′-cyclic phosphates **2**, wherein the activation is intrinsic to the cyclic phosphate (Fig. 1). If all potential routes from prebiotic feedstock molecules to such monomers were to be investigated experimentally without success, then the potential for prebiotic self-assembly of monomers associated with more complicated polymerization chemistries, would additionally have to be investigated (this would include alternative P–O bond forming polymerization chemistry, as well as C–O and C–C bond forming chemistries). However, continuing with the simplest P–O bond forming polymerization chemistry, and the assumption that it seems reasonable to follow the simplest retrosynthetic disconnections first, **1** and **2** can then be conceptually reduced to ribose **3**, a nucleobase and phosphate (Joyce 2002; Joyce and Orgel 2006). Ribose **3** can then be disconnected to glycolaldehyde **4** and glyceraldehyde **5** through aldol chemistry, and the nucleobases disconnected to simpler carbon and nitrogen containing molecules—the pyrimidines to cyanamide **6** and cyanoacetylene **7** (conventionally through the hydration products urea **8** and cyanoacetaldehyde **9**), and the purines to hydrogen cyanide and a C(IV) oxidation level molecule such as **6** or **8**. This retrosynthetic analysis ultimately breaks ribonucleotides down into molecules that are sufficiently simple so that they can be deemed prebiotically plausible feedstock molecules (Fig. 2) (Sanchez et al. 1966; Pasek and Lauretta 2005; Bryant and Kee 2006; Thaddeus 2006).

It is not just because the simplest retrosynthetic disconnections of ribonucleotides proceed by way of ribose, nucleobases, and phosphate that people have tried to synthesize them via these three building blocks under prebiotically plausible conditions for the last 40–50 years—in terms of their appearance to the human eye, ribonucleotides undoubtedly *consist* of these three building blocks. Experimentally, there have been several notably successful reactions that ostensibly support this nucleobase ribosylation approach: Orgel's and Miller's syntheses of cytosine **10** (Ferris et al. 1968; Robertson and Miller 1995); Benner's and Darbre's syntheses of ribose **3** by aldolization of glycolaldehyde **4** and glyceraldehyde **5** (Ricardo et al. 2004; Kofoed et al. 2005); Pasek's and Kee's demonstration of phosphate synthesis by disproportionation of meteoritic metal phosphides (Pasek and Lauretta 2005; Bryant and Kee 2006); and Orgel's urea-catalyzed phosphorylation of nucleosides (eg., **11** (B=C) → **2** (B=C)) (Lohrmann and Orgel 1971). Indeed, for many years, a prebiotically plausible synthesis of ribonucleotides from ribose **3**, the nucleobases, and phosphate has been tantalizingly close but for one step of the assumed synthesis—the joining of ribose to the nucleobases.

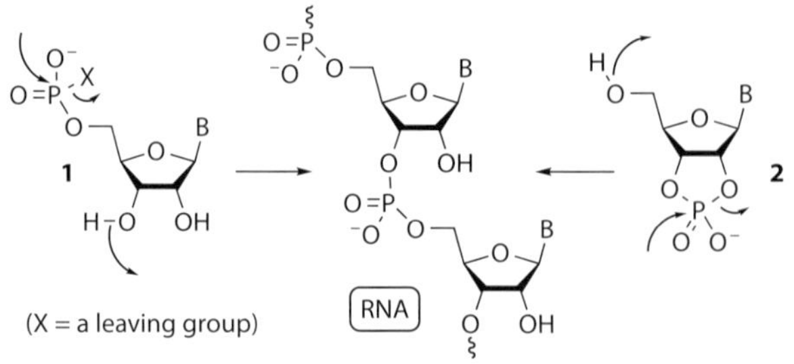

Figure 1. Activated ribonucleotides in the potentially prebiotic assembly of RNA. Potential P–O bond forming polymerization chemistry is indicated by the curved arrows.

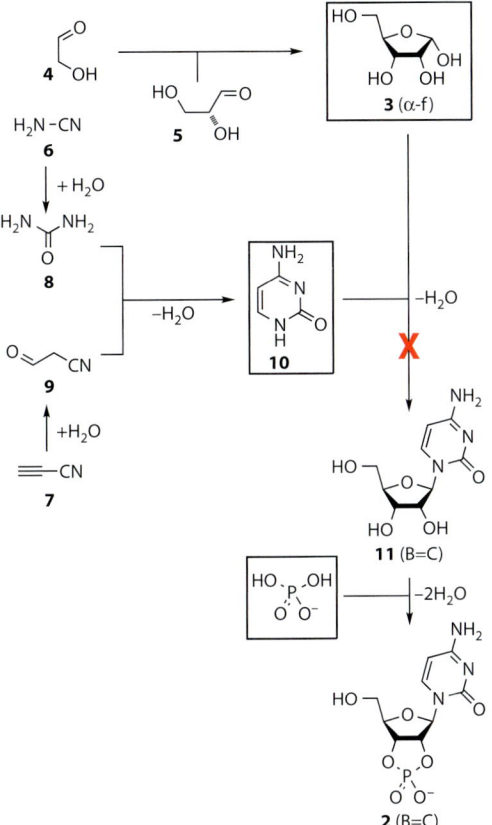

Figure 2. One of the synthetic routes to β-ribocytidine-2′,3′-cyclic phosphate 2 (B=C) implied by the assumption that nucleosides can self-assemble by nucleobase ribosylation. The general synthetic approach has been supported by the experimental demonstration of most of its steps. However, prebiotically plausible conditions under which the key nucleobase ribosylation step works have not been found despite numerous attempts over several decades.

This reaction works extremely poorly for the purines and not at all in the case of the pyrimidines (Fuller et al. 1972a, 1972b; Orgel 2004).

So, why does the ribosylation chemistry not work with free nucleobases and ribose 3 when, using the protecting and controlling groups of conventional synthetic chemistry, nucleobase ribosylation is possible? The reasons are predominantly kinetic and can be appreciated by consideration of the structure and reactivity of ribose 3 and representative nucleobases (Fig. 3). Ribose 3 exists as an equilibrating mixture of different forms in aqueous solution (Fig. 3A) (Drew et al. 1998). The mixture is dominated by β- and α-pyranose isomers (3 [β–p] and 3 [α–p]) with lesser amounts of β- and α-furanose isomers (3 [β–f] and 3 [α–f]). The various hemiacetal ring forms equilibrate via the open chain aldehyde form 3 (a), which is a very minor component along with an open chain hydrate. The purine nucleobase adenine 12 also exists in various equilibrating forms in aqueous solution (Fig. 3B) (Fonseca Guerra et al. 2006). In this case, the isomers differ in the position of protonation, the major tautomer, 12 has N9 protonated, but other tautomers such as 13—in which N1 is protonated—exist at extremely low concentration. To connect adenine 12 to ribose 3 to give a natural RNA ribonucleoside 11 (B=A) it is necessary for N9 of adenine to function as a nucleophile and C1 of 3 (α–f) to function as an electrophile. The latter is possible under acidic conditions when a small amount of 14—a selectively protonated form of 3 (α–f)—is present at equilibrium. The protonation converts the anomeric hydroxyl group into a better leaving group and enhances the electrophilicity of C1. The major tautomer of adenine, 12, is not nucleophilic at N9 because the lone pair on that atom is delocalized throughout the bicyclic ring structure. N9 of several minor tautomers such as 13 is nucleophilic because the nitrogen lone pair is localized, and so reaction with 14 is possible, though slow because of the low concentrations of the productively reactive species. To compound this sluggishness, the reaction is plagued with additional problems. First, the acid needed to activate 3 (α–f) also substantially protonates adenine, giving the cation 15, which is not nucleophilic on N9 (Christensen et al. 1970; Zimmer and Biltonen 1972; Major et al. 2002). Second, the most nucleophilic nitrogen—the 6-amino group—of the major tautomer of adenine 12 reacts with 3 (a)—the most reactive form of 3 despite its scarcity—resulting in N6-ribosyl adducts as by far the major products (Fuller et al. 1972a, 1972b). Third, the other isomeric forms of ribose, 3 (β–p), 3 (α–p), and 3 (β–f), can also react with 13 when they are protonated at

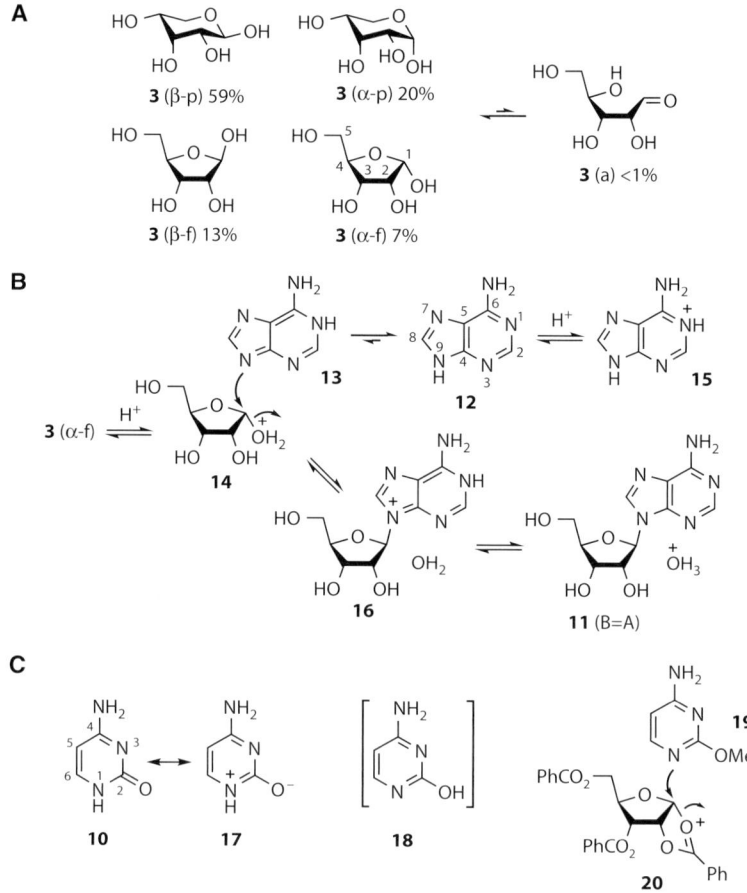

Figure 3. The difficulties of assembling β-ribonucleosides by nucleobase ribosylation. (*A*) The many different forms of ribose **3** adopted in aqueous solution. The pyranose (p) and furanose (f) forms interconvert via the open-chain aldehyde (a), which is also in equilibrium with an open-chain aldehyde hydrate (not shown). (*B*) Adenine tautomerism and the ribosylation step necessary to make the adenosine **11** (B=A) thought to be needed for RNA assembly. The low abundance of the reactive entities **13** and **14** is partly responsible for the low yield of **11** (B=A). (*C*) The reason for the lower nucleophilicity of N1 of the pyrimidines, and the conventional synthetic chemist's solution to the problems of ribosylation.

their anomeric hydroxyl groups. Fourth, N9 of the minor adenine tautomer **13** is not the only nucleophilic ring nitrogen of adenine; N1, N3, and N7 of the major tautomer **12** are also nucleophilic. These latter two points mean that the small amount of protonated adenosine **16**, that is formed when N9 of **13** reacts with **14**, is accompanied by a multitude of isomeric products. The final problem with the synthesis is reversibility. Any adenosine **11** (B=A) that is produced is formed in acid at equilibrium, and the equilibrium in aqueous solution lies in favor of **3** and **15**. The only way round this is to carry out the reaction in the dry-state in the presence of acidic catalysts. The best that has been achieved—a 4% yield of **11** (B=A)—involves such a dry-state reaction with an excess of ribose, followed by heating with concentrated ammonium hydroxide solution to hydrolyse N6-ribosyl adducts (Fuller et al. 1972a).

The situation with prebiotic pyrimidine ribosylation is even worse (Fig. 3c). Thus, for example, N1 of cytosine **10** is not nucleophilic because the lone pair is delocalized round the ring and into the carbonyl group (as indicated by the resonance canonical structure **17**). Experimentally, cytosine **10** cannot be ribosylated on N1 even using the conditions established for unselective ribosylation of adenine **12** (Fuller et al. 1972a). If there is any tautomeric form such as **18** with a localized N1 lone pair present at equilibrium, it must be in such a low concentration as to be effectively unreactive. The same considerations hold true for uracil.

In conventional synthetic chemistry, the aforementioned difficulties in nucleobase ribosylation can be overcome with directing, blocking, and activating groups on the nucleobase and ribose (Ueda and Nishino 1968). Thus, the cytosine derivative **19** is directed to function as a nucleophile at N1 by alkylation of O2. The ribose-derived intermediate **20** is constrained to the furanose form by benzoylation of the C5-hydroxyl group, and neighboring group participation from an O2-benzoyl group directs β-glycosylation. These molecular interventions are synthetically ingenious, but serve to emphasize the enormous difficulties that must be overcome if ribonucleosides are to be efficiently produced by nucleobase ribosylation under prebiotically plausible conditions. This impasse has led most people to abandon the idea that RNA might have assembled abiotically, and has prompted a search for potential pre-RNA informational molecules (Joyce et al. 1987; Eschenmoser 1999; Schöning et al. 2000; Zhang et al. 2005; Sutherland 2007). However, we realized that there were other possible synthetic approaches that although less obvious, still had the potential to make the ribonucleotides **1** and **2** (Anastasi et al. 2007). Furthermore, and as pointed out earlier, there are also alternative bond forming polymerization chemistries imaginable. Our plan was to work through these possibilities by systematic experimentation before deciding whether the abiogenesis of RNA is possible or not.

RECENT RESULTS

As building blocks for pyrimidine ribonucleotides, we have investigated the chemistry of glycolaldehyde **4**, glyceraldehyde **5**, cyanamide **6**, cyanoacetylene **7**, and phosphate—the same feedstock molecules ultimately invoked in the nucleobase ribosylation approach (Fig. 2) (in ongoing work on the prebiotic synthesis of purine nucleotides, the systems chemistry of mixtures containing hydrogen cyanide, is additionally being investigated). However, we have not presumed a particular order of assembly, and have systematically investigated many options over the last decade or so (Sutherland and Whitfield 1997; Ingar et al. 2003; Smith et al. 2004; Smith and Sutherland 2005; Anastasi et al. 2007, 2008). Furthermore, aware of the potential for certain molecules to act as catalysts of reactions in which other molecules are joined, an important aspect of our approach has been the study of the chemistry of multicomponent chemical systems, including mixed oxygenous and nitrogenous chemistry (Szostak 2009). It has long been held in prebiotic chemistry that the aldol chemistry of aldehydes potentially required to make sugars should not be mixed with, for example, the oligomerization of hydrogen cyanide and cyanamide potentially required to make purine nucleobases (Shapiro 1988). This is because the two chemistries have been assumed to interfere with each other, suggesting the potential for a "combinatorial explosion" of minor by-products. However, given the impasse of preformed nucleobase ribosylation, and the conceptual difficulties with which a transition from a pre-RNA world to the RNA world is fraught, taking a few synthetic risks seemed justified.

Cutting straight to the chase, we found that mixed oxygenous-nitrogenous chemistry of glycolaldehyde **4**-cyanamide **6** mixtures can be tamed by the inclusion of phosphate such that the heterocycle 2-aminooxazole **21** is obtained in >80% yield (Fig. 4) (Powner et al. 2009). The phosphate functions both as a pH buffer and as a general acid-base catalyst. The heterocycle **21** readily sublimes, potentially offering a prebiotically plausible purification through sublimation and precipitation or rain-in. In a smooth C–C bond forming reaction in water at neutral pH, **21** then adds to glyceraldehyde **5**, giving the four pentose aminooxazolines, but with high stereoselectivity for the *ribo*- and

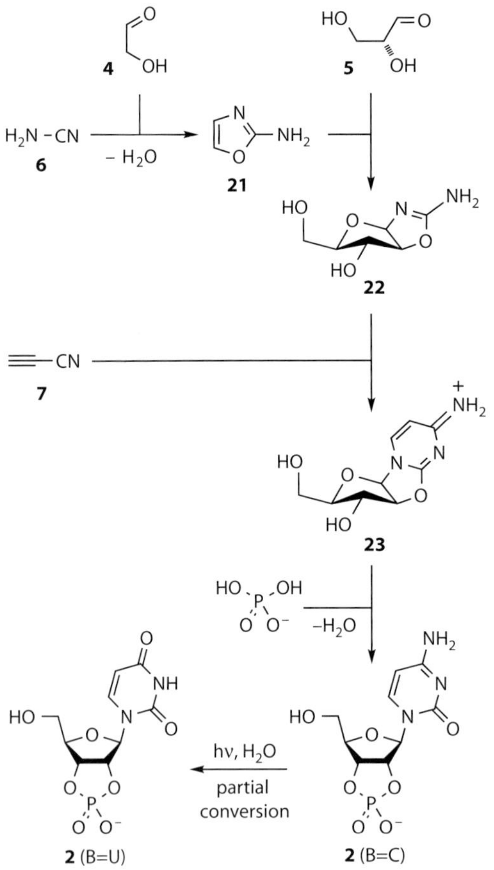

Figure 4. The recently uncovered route to activated pyrimidine nucleotides **2**. The nucleobase ribosylation problem is circumvented by the assembly proceeding through 2-aminooxazole **21**, which can be thought of as the chimera of half a pentose sugar and half a nucleobase. The second half of the pentose—glyceraldehyde **5**—and the second half of the nucleobase—cyanoacetylene **7**—are then added sequentially to give the anhydronucleoside **23**. Phosphorylation and rearrangement of **23** then furnishes **2** (B=C), and UV irradiation effects the partial conversion of **2** (B=C) to **2** (B=U).

urea **8** in the dry-state or in formamide solution then gives the activated ribonucleotide **2** (B=C). Finally, irradiation of **2** (B=C) with UV light in solution, followed by heating (Powner and Sutherland 2008), converts it to a mixture of **2** (B=C) and **2** (B=U), the two activated pyrimidine nucleotides needed for RNA synthesis.

The presumed mechanisms of the various reactions that make up the self-assembly sequence (Fig. 5) reveal some remarkable aspects of selectivity, and show the synthetic power of systems chemistry (Anastasi et al. 2006; Powner et al. 2009). In the first step, we found that the reaction of glycolaldehyde **4** and cyanamide **6** generated a series of dead end adducts and a small amount of 2-aminooxazole **21** in the absence of phosphate, but in its presence **21** was produced cleanly in extremely high yield. This improvement in the yield of **21** is attributed to general acid–base catalysis of two steps—that in which the nitrile carbon undergoes attack by a hydroxyl group nucleophile, and the C–H deprotonation step that generates the aromatic ring of **21** (Fig. 5A). Phosphate is an ideal general acid-base catalyst in this context as the reaction is ideally conducted at near neutral pH values close to the second pK_a of phosphoric acid. Phosphate is also known to catalyze both the dimerization and the hydrolysis of cyanamide **6** (Lohrmann and Orgel 1968), but these reactions are slower than the formation of **21**. However, if an excess of **6** is allowed to react with glycolaldehyde **4** in the presence of phosphate, the formation of **21** is followed by the formation of cyanoguanidine and urea **8**.

In the second step of the self-assembly process, 2-aminooxazole **21** adds to glyceraldehyde **5**, giving all four pentose aminooxazolines but, as mentioned previously, with high stereoselectivity for the *ribo*-isomer, and the key *arabino*-isomer **22** (Fig. 5B). The reaction initially follows the course of an electrophilic aromatic substitution reaction with the formation of a cationic intermediate. However, rather than rearomatizing by C–H deprotonation, this intermediate instead undergoes intramolecular addition of a hydroxyl group (Anastasi et al.

arabino-configured materials (Anastasi et al. 2006). Remarkably, the *ribo*-isomer then selectively crystallizes out of solution, leaving the *arabino*-isomer **22** the most abundant in solution. Reaction of **22** with cyanoacetylene **7** furnishes the anhydronucleoside **23** if the pH of the reaction is maintained at pH=6.5 through the use of phosphate as a buffer (Powner et al. 2009). Heating **23**, inorganic phosphate, and

2006). Although the C5′-hydroxyl group is intrinsically the most nucleophilic, it is the C4′-hydroxyl group that adds kinetically because its addition proceeds via a five-membered (cf., six-membered) transition state. The chemistry is partly reversible, however, especially in the presence of phosphate, and the product of attack by the C5′-hydroxyl group—a pyranose aminooxazoline—can be accessed, as can the electrophilic aromatic substitution product— an open sugar chain aminooxazole isomer— when the key C–H deprotonation step is general base catalysed by phosphate (Fig. 5B, red arrows). In the absence of phosphate, the ribo-arabino- and xylo-aminooxazolines exist in the furanose form, and it is only the lyxo-aminooxazoline that exists partly in the pyranose form, presumably because of steric encumbrance in the furanose form (Saewan et al. 2005). Under partial thermodynamic control in the presence of phosphate, the formation of aminooxazoline versus substituted aminooxazole is finely balanced and depends on the stability of the different aminooxazoline stereoisomers. Intriguingly, the ribo-, arabino-, and xylo-products preferentially exist as aminooxazolines but the lyxo-product preferentially exists as the substituted aminooxazole. Phosphate

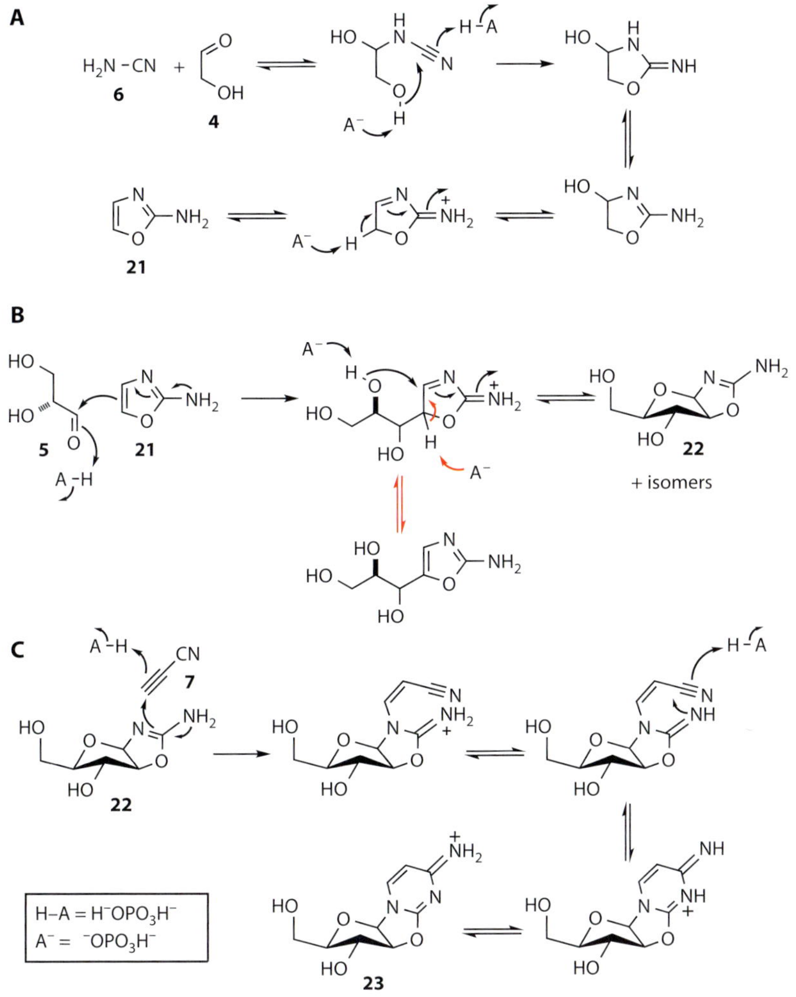

Figure 5. Mechanistic details of each of the five steps—(A) through (E)—of the recently uncovered route to activated pyrimidine nucleotides **2**. In E), resonance arrows are shown in red and reaction arrows in black.

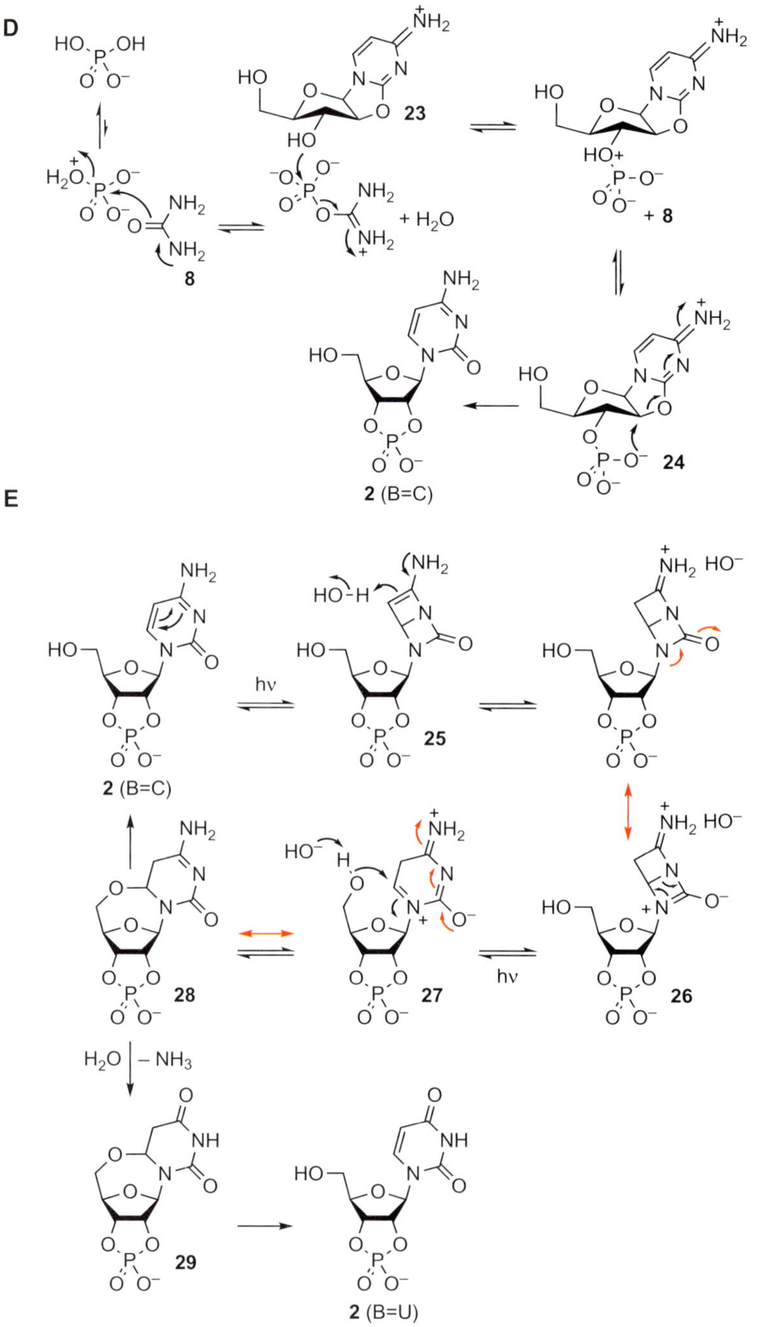

Figure 5. *Continued.*

thus makes the reaction of 2-aminooxazole **21** and glyceraldehyde **5** more selective for the formation of the desired *arabino*-aminooxazoline **22** (Powner et al. 2009).

In addition to its role as a pH buffer in the addition of **22** to cyanoacetylene **7** (at pH values higher than 6.5, or in the absence of phosphate, the anhydronucleoside **23** undergoes hydrolysis of the C2–O2′ bond giving β-arabinocytidine [Sanchez and Orgel 1970]), phosphate almost certainly functions as a general acid catalyst for the first addition step, thereby allowing the reaction to proceed by a path that does not require the intermediacy of a high-energy cyanovinyl anion (Fig. 5C). Furthermore, reaction of excess cyanoacetylene with the hydroxyl groups of the anhydronucleoside product **23** is prevented by phosphate since HPO_4^{2-} is more nucleophilic than these hydroxyl groups. In the course of acting as a chemical buffer in this way, the phosphate is partly transformed into cyanovinyl phosphate, and this latter compound, over time, reacts with more phosphate to give pyrophosphate (Ferris et al. 1970).

The urea **8** produced as a by-product in the reaction of glycolaldehyde **4** with cyanamide **6**, if the latter is in excess, then acts as a catalyst for the incorporation of phosphate in the fourth step of the self-assembly sequence (Fig. 5D). At high temperatures in the dry-state (Lohrmann and Orgel 1971), or in formamide solution (Schoffstall 1976), the weakly nucleophilic urea oxygen atom attacks a minor tautomer of $H_2PO_4^-$, displacing water via a dissociative, monomeric metaphosphate-like transition state to give an imidoyl phosphate. Although this reaction is reversible in principle, water is lost at high temperature, driving the formation of the high-energy intermediate, which can then phosphorylate the hydroxyl groups of nonvolatile compounds such as the anhydronucleoside **23**. Again, this reaction is reversible in principle, though it is unlikely to be fully so in urea melts because the viscosity of the reaction medium likely prevents free diffusion. It transpires that **23** adopts a conformation that potentially makes the C5′-hydroxyl group more hindered than the C3′-hydroxyl group; thus, phosphorylation of the latter is apparently preferred kinetically (Powner et al. 2009). The dianionic form of the phosphate group of the O3′-phosphorylated anhydronucleoside **24** can then attack C2′ in an intramolecular S_N2 reaction, thereby inverting the configuration of that carbon, and giving the ribocytidine-2′,3′-cyclic phosphate **2** (B=C) (Tapiero and Nagyvary 1971). Alternatively, if kinetic selectivity is not fully operative, the O5′- and O3′-phosphorylation products of **23** could both be sampled at equilibrium, and the equilibrium could then be displaced to the detriment of the former through irreversible conversion of the latter to **2** (B=C). Pyrophosphate—the product of reaction of the cyanovinyl phosphate by-product in the preceding step with additional phosphate (Ferris et al. 1970)—can also function as a phosphorylating agent in urea melts. In this case, the imidoyl phosphate is generated by nucleophilic displacement of phosphate (cf., water) via a dissociative, monomeric metaphosphate-like transition state. Pyrophosphate is not soluble in formamide, but there is another twist to phosphorylation of **23** with inorganic phosphate with urea in formamide solution, and that is that formamide is also weakly nucleophilic on its oxygen atom, and so can also react to give imidoyl phosphate intermediates—indeed, phosphorylation can be accomplished in formamide in the absence of urea (Schoffstall 1976). Formamide can be considered to be prebiotically available because it is the hydration product of hydrogen cyanide. The latter is inevitably also produced when cyanoacetylene **7** is made by high-energy atom recombination chemistry (Sanchez et al. 1970). Given that the purine nucleobases are constitutionally derivable from hydrogen cyanide, the case for the involvement of this compound and its hydration product in the prebiotic synthesis of pyrimidine and purine nucleotides seems extremely strong (Sanchez et al. 1967).

In the last step of the self-assembly sequence, **2** (B=C) is partly converted to **2** (B=U) by UV irradiation. Remarkably, the irradiation conditions destroy other cytidine nucleotides, and since some such compounds will likely arise from similar assembly chemistry proceeding from aminooxazolines other than **22**, the irradiation might also serve a sanitizing function such that **2** could be produced in the absence of stereoisomeric contaminants. The proposed mechanism of the photochemistry (Fig. 5E) is speculative, but supported by a significant body of fact. It has long been known that irradiation of other cytidine nucleosides and nucleotides leads to C5-photohydrates that eliminate water to regenerate the original cytosine derivative (Miller and Cerutti 1968). During prolonged irradiation, such cytosine derivatives go through many cycles of photoexcitation and thermal relaxation. Deleterious photochemistry—nucleobase loss, C1′- and C2′-epimerization, hydrolysis to uracil derivatives, etc.—increases with increased irradiation, presumably because high energy intermediates are thus accessed for longer (Powner et al. 2007; Powner and Sutherland 2008). The molecular details of these processes have been elusive, but recent work has suggested plausible mechanisms that can be further tested. It is thought that irradiation of the cytosine ring leads to the formation of a Dewar pyrimidine **25** through a disrotatory 4π electrocyclic process (Shaw and Shetlar 1990). Steric inhibition of resonance is then expected to make C5-protonation facile, giving an amidinium ion that can undergo an electrocyclic ring-opening—most easily visualized, as shown, for the resonance canonical form **26** (Fig. 5E). The consequence of this second pericyclic process is the formation of an N1,C6-iminium ion **27**. In most molecular contexts, an N1,C6-iminium ion would be expected to undergo attack by hydroxide anion, giving a cytidine photohydrate, but in the irradiation of **2** (B=C), the 2′,3′-cyclic phosphate allows ready access to otherwise inaccessible western sugar ring conformers (Saenger 1984) in which the C5′-hydroxyl group is proximal to C6 and can thus add to it instead. Thus, the specific nature of the cyclic phosphate of **2** (B=C) controls the cytosine photochemistry through enforced intramolecular nucleophilic addition. The resultant C5,O5′-anhydronucleotide **28** is thought to undergo thermal elimination back to **2** (B=C) more slowly than other cytidine photohydrates eliminate water because of conformational restriction about the C5–C6 bond. This has the effect that **2** (B=C) undergoes fewer photochemical excitation and thermal relaxation cycles than other cytidine nucleosides and nucleotides. The consequences of this are twofold. Firstly, deleterious chemistry from high-energy (charged) intermediates is reduced because these intermediates are accessed less often. Secondly, the persistence of the C5,O5′-anhydronucleotide allows for greater hydrolysis to the corresponding uracil derivative **29**. Relaxation of this uracil derivative is then responsible for the high yield of **2** (B=U) relative to the yield of the corresponding uridine nucleosides and nucleotides in the irradiation of other cytidine nucleosides and nucleotides. Indeed, most other pyrimidine nucleosides and nucleotides are destroyed by the prolonged irradiation that partly converts **2** (B=C) to **2** (B=U). This has the beneficial effect that **2** (B=C) and **2** (B=U) can be effectively sanitized by destruction of nucleoside and nucleotide by-products of the assembly process described herein.

TOWARD A GEOCHEMICALLY PLAUSIBLE SYNTHESIS

The contextually specific chemoselectivity, and the systems chemistry aspects of this newly uncovered self-assembly route to activated pyrimidine nucleotides suggest that the chemistry should be viewed as predisposed. However, the route as operated thus far in the laboratory is associated with several steps, and the conditions for these steps are different. Furthermore, purification in between certain steps was carried out to make analysis of the chemistry easier. Clearly, these issues need to be addressed before the synthesis can be seen as geochemically plausible.

First, the sequence of different conditions needs considering. Miller's iconic experiment (Miller 1953) has apparently conditioned many in the field to think that a prebiotic synthesis needs to occur under one set of conditions. But, think of chemistry occurring now on the Earth. There is no doubt that in many locations, conditions vary with time—there are light periods and dark periods, hot periods and cold periods, dry periods and wet periods. If the same were true on the primordial Earth, then why should we expect prebiotic synthesis just to occur in one particular period? Furthermore, the frequent impacts of asteroids, meteorites, and comets on the early Earth would surely have exacerbated the changeability of geochemical conditions. Given that there are so many geochemically plausible conditions, and sequences of conditions, it is difficult to try and predict solely from geochemistry the actual conditions that pertained for the prebiotic synthesis of any particular compound or class of compounds. Surely, it makes more sense to first find predisposed chemical routes to molecules of interest and then ask whether the sequence of conditions is geochemically plausible. If it then turns out that this sequence of conditions additionally supports the synthesis of other important molecules, then evidence might accumulate that the sequence of conditions actually did take place during the origin of life on Earth. Furthermore, the search for further relevant prebiotic synthesis would then be expedited. From what is now known about pyrimidine nucleotide self-assembly, one geochemically plausible sequence of conditions—coincidentally redolent of Darwin's suggestion (Darwin 1871)—involves a warm pond that evaporates and dries out, stays dry and is heated for a time, is then filled again by rain, and subsequently bathed in sunshine.

Secondly, one has to wonder if any purification steps could reasonably be invoked in such a scenario, and ask if the other chemistry would work in the absence of purification. 2-Aminooxazole **21** sublimes readily on being warmed at atmospheric pressure, and resolidifies on cooling surfaces. This property suggests that **21** could plausibly be purified prebiotically—if necessary—and be relocated by precipitation. If delivered to glyceraldehyde **5** and phosphate, then reaction of **21** and **5** giving the pentose aminooxazolines, including the key *arabino*-isomer **22**, certainly seems plausible. Subsequent delivery of cyanoacetylene **7**—produced by atmospheric chemistry—by rain-in also seems plausible, though there is no doubt that it would be accompanied by hydrogen cyanide. Will hydrogen cyanide interfere with the reaction of **22** and **7**, giving the anhydronucleoside **23**? The answer awaits experimental evaluation. If the hydrogen cyanide does not interfere, but slowly hydrolyses to formamide, then when the water in the pond evaporates, a formamide solution would remain. If this solution gets heated, the phosphorylation and rearrangement of **23** to **2** (B=C) will take place whether the pond additionally happened to contain urea or not. The issue seems not to be whether the chemistry could take place or not, but the yield and purity of products. Because of the other aminooxazolines, there would doubtless be other stereoisomeric products at this point, but this would not matter as subsequent dilution with water and irradiation would not just partly convert **2** (B=C) to **2** (B=U), but also destroy these other products. So, there is more work to be done, and although it is premature to conclude that the newly discovered self-assembly route to activated pyrimidine nucleotides (Powner et al. 2009) is geochemically plausible, the signs are looking good! What is now needed on top of further experimental work to assess the geochemical plausibility of the synthesis, is a thorough investigation of the synthesis of activated purine nucleotides.

FUTURE CHALLENGES

If further experimentation supports nucleoside-2′,3′-cyclic phosphates **2** as the activated monomers for prebiotic RNA synthesis, then an obvious next question is how such monomers oligomerise. Pioneering work from the 1970s partly addresses this and suggests that general acid-base catalysis is the key (Verlander et al. 1973). In the dry-state, mixtures of aliphatic amines and the corresponding ammonium salts are effective oligomerization catalysts. It is thought that the amines deprotonate the C5′-hydroxyl group of one nucleoside-2′,3′-cyclic phosphate **2**, enabling it to attack the phosphate group of another. The ammonium salts serve to protonate the oxygen leaving group generating a C2′- or C3′-hydroxyl group. Short oligomers are easily produced, but the chemistry generates both 3′,5′- and 2′,5′-internucleotide links, and further work is needed to address this selectivity issue.

The oligomerization of racemic mixtures of nucleoside-2′,3′-cyclic phosphates **2** clearly has the potential to generate extremely complex mixtures of diastereoisomers, so a major goal has to be the generation of these monomers in enantiopure form. This issue has not yet been solved, but it is apparent from the self-assembly chemistry that if glyceraldehyde **5** can be obtained in enantiopure form then the resultant pyrimidine nucleotides will be similarly enantiopure. How glyceraldehyde **5** might be obtained in enantiopure is a question that can only be answered when a prebiotically plausible synthesis of **5** is shown. The conditions potentially required to generate glyceraldehyde **5** by the aldolization of glycolaldehyde **4** and formaldehyde are the same conditions that favour the conversion of **5** to the more stable dihydroxyacetone. A synthesis of **5**—and indeed **4**—from formaldehyde involving umpolung would therefore be preferable, but prebiotically plausible carbonyl umpolung has not yet been shown.

Finally, the potential for simultaneous self-assembly of RNA and other molecules such as lipids and peptides must be investigated. Compartmentalization of RNA leading to protocells is seen as a crucial step towards systems that can undergo Darwinian evolution (Szostak et al. 2001), so can lipids be produced alongside nucleotides? Peptides, even dipeptides, offer additional scope for catalysis, and their prebiotic synthesis and use could help drive the evolution of the ribosome and the genetic code, so can peptides also be produced under the same conditions? Experimental investigation of such questions will rely heavily on analytical chemistry because of the complexity of the multicomponent systems that will have to be investigated. There is cause for optimism, however, as prebiotic systems chemistry, though still in its infancy, is already suggesting new solutions to old problems.

REFERENCES

Anastasi C, Buchet FF, Crowe MA, Helliwell M, Raftery J, Sutherland JD. 2008. The search for a potentially prebiotic synthesis of nucleotides via arabinose-3-phosphate and its cyanamide derivative. *Chem Eur J* **14**: 2375–2388.

Anastasi C, Buchet FF, Crowe MA, Parkes AL, Powner MW, Smith JM, Sutherland JD. 2007. RNA: Prebiotic product or biotic invention? *Chem Biodiversity* **4**: 721–739.

Anastasi C, Crowe MA, Powner MW, Sutherland JD. 2006. Direct assembly of nucleoside precursors from two- and three-carbon units. *Angew Chem Int Ed* **45**: 6176–6179.

Borsenberger V, Crowe MA, Lehbauer J, Raftery J, Helliwell M, Bhutia K, Cox T, Sutherland JD. 2004. Exploratory studies to investigate a linked prebiotic origin of RNA and coded peptides. *Chem Biodiversity* **1**: 203–246.

Bryant DE, Kee TP. 2006. Direct evidence for the availability of reactive, water soluble phosphorus on the early Earth. H-phosphinic acid from the Nantan meteorite. *Chem Commun* **14**: 2344–2346.

Christensen JJ, Rytting JH, Izatt RM. 1970. Thermodynamic pK, $\Delta H°$, $\Delta S°$, and $\Delta C_p°$ values for proton dissociation from several purines and their nucleosides in aqueous solution. *Biochemistry* **9**: 4907–4913.

Corey EJ. 1988. Retrosynthetic thinking - essentials and examples. *Chem Soc Rev* **17**: 111–133.

Crick FHC. 1968. The origin of the genetic code. *J Mol Biol* **38**: 367–379.

Darwin C. 1871. Letter to JD Hooker of 1st February. Cambridge University Library MS. DAR 94: 188–189 (*Calendar* 7471).

Drew KN, Zajicek J, Bondo G, Bose B, Serianni AS. 1998. ^{13}C-labeled aldopentoses: detection and quantitation of cyclic and acyclic forms by heteronuclear 1D and 2D NMR spectroscopy. *Carbohydrate Res* **307**: 199–209.

Eschenmoser A. 1999. Chemical etiology of nucleic acid structure. *Science* **284**: 2118–2124.

Ferris JP, Goldstein G, Beaulieu DJ. 1970. Chemical evolution, IV. An evaluation of cyanovinyl phosphate as a prebiotic phosphorylating agent. *J Am Chem Soc* **92**: 6598–6603.

Ferris JP, Sanchez RA, Orgel LE. 1968. Studies in prebiotic synthesis: III. Synthesis of pyrimidines from cyanoacetylene and cyanate. *J Mol Biol* **33**: 693–704.

Ferris JP, Hill AR Jr, Liu R, Orgel LE. 1996. Synthesis of long prebiotic oligomers on mineral surfaces. *Nature* **381**: 59–61.

Fonseca Guerra C, Bickelhaupt FM, Saha S, Wang F. 2006. Adenine tautomers: relative stabilities, ionization energies, and mismatch with cytosine. *J Phys Chem A* **110**: 4012–4020.

Fuller WD, Sanchez RA, Orgel LE. 1972a. Studies in prebiotic synthesis VI. Synthesis of purine nucleosides. *J Mol Biol* **67**: 25–33.

Fuller WD, Sanchez RA, Orgel LE. 1972b. Studies in prebiotic synthesis. VII Solid-state synthesis of purine nucleosides. *J Mol Evol* **1**: 249–257.

Ingar A-A, Luke RWA, Hayter BR, Sutherland JD. 2003. Synthesis of cytidine ribonucleotides by stepwise assembly of the heterocycle on a sugar phosphate. *ChemBioChem* **6**: 504–507.

Joyce GF. 2002. The antiquity of RNA-based evolution. *Nature* **418**: 214–221.

Joyce GF, Orgel LE. 2006. In *The RNA world*, 3rd ed. (ed. Gesteland R.F., Cech T.R., Atkins J.F.), pp. 23–56. Cold Spring Harbor Laboratory Press.

Joyce GF, Schwartz AW, Miller SL, Orgel LE. 1987. The case for an ancestral genetic system involving simple analogues of the nucleotides. *Proc Natl Acad Sci* **84:** 4398–4402.

Kofoed J, Reymond J-L, Darbre T. 2005. Prebiotic carbohydrate synthesis: Zinc-proline catalyses direct aqueous aldol reactions of α-hydroxy aldehydes and ketones. *Org Biomol Chem* **3:** 1850–1855.

Lohrmann R, Orgel LE. 1968. Prebiotic synthesis: Phosphorylation in aqueous solution. *Science* **161:** 64–66.

Lohrmann R, Orgel LE. 1971. Urea-inorganic phosphate mixtures as prebiotic phosphorylating agents. *Science* **171:** 490–494.

Major DT, Laxer A, Fischer B. 2002. Protonation studies of modified adenine and adenine nucleotides by theoretical calculations and ^{15}N NMR. *J Org Chem* **67:** 790–802.

Miller SL. 1953. A production of amino acids under possible primitive earth conditions. *Science* **117:** 528–529.

Miller N, Cerutti P. 1968. Structure of the photohydration products of cytidine and uridine. *Proc Natl Acad Sci* **59:** 34–38.

Orgel LE. 1968. Evolution of the genetic apparatus. *J Mol Biol* **38:** 381–393.

Orgel LE. 2004. Prebiotic chemistry and the origin of the RNA world. *Crit Rev Biochem Mol Biol* **39:** 99–123.

Pasek MA, Lauretta DS. 2005. Aqueous corrosion of phosphide minerals from iron meteorites: A highly reactive source of prebiotic phosphorus on the surface of the early Earth. *Astrobiology* **5:** 515–535.

Powner MW, Sutherland JD. 2008. Potentially prebiotic synthesis of pyrimidine β-D-ribonucleotides by photoanomerization/hydrolysis of β-D-cytidine-2'-phosphate. *ChemBioChem* **9:** 2386–2387.

Powner MW, Gerland B, Sutherland JD. 2009. Synthesis of activated pyrimidine nucleotides in prebiotically plausible conditions. *Nature* **459:** 239–242.

Powner MW, Anastasi C, Crowe MA, Parkes AL, Raftery J, Sutherland JD. 2007. On the prebiotic synthesis of ribonucleotides: Photoanomerisation of cytosine nucleosides and nucleotides revisited. *ChemBioChem* **8:** 1170–1179.

Ricardo A, Carrigan MA, Olcott AN, Benner SA. 2004. Borate minerals stabilize ribose. *Science* **303:** 196.

Robertson MP, Miller SL. 1995. An efficient prebiotic synthesis of cytosine and uracil. *Nature* **375:** 772–774.

Saenger W. 1984. *Principles of nucleic acid structure*. Springer-Verlag, New York.

Saewan N, Crowe MA, Helliwell M, Raftery J, Chantrapromma K, Sutherland JD. 2005. Exploratory studies to investigate a linked prebiotic origin of RNA and coded peptides. 4th Communication: Further observations concerning pyrimidine nucleoside synthesis by stepwise nucleobase assembly. *Chem Biodiversity* **2:** 66–83.

Sanchez RA, Orgel LE. 1970. Studies in prebiotic synthesis V. Synthesis and photoanomerization of pyrimidine nucleosides. *J Mol Biol* **47:** 531–543.

Sanchez RA, Ferris JP, Orgel LE. 1966. Cyanoacetylene in prebiotic synthesis. *Science* **154:** 784–785.

Sanchez RA, Ferris JP, Orgel LE. 1967. Studies in prebiotic synthesis II. Synthesis of purine precursors and amino acids from aqueous hydrogen cyanide. *J Mol Biol* **30:** 223–253.

Schoffstall AM. 1976. Prebiotic phosphorylation of nucleosides in formamide. *Origins Life* **7:** 399–412.

Schöning K-U, Scholz P, Guntha S, Wu X, Krishnamurthy R, Eschenmoser A. 2000. Chemical etiology of nucleic acid structure: the α-threofuranosyl-(3'→2') oligonucleotide system. *Science* **290:** 1347–1351.

Shapiro R. 1988. Prebiotic ribose synthesis: A critical analysis. *Origins Life Evol Biosphere* **18:** 71–85.

Shaw AA, Shetlar MD. 1990. 3-Ureidoacrylonitriles: Novel products from the photoisomerization of cytosine, 5-methylcytosine, and related compounds. *J Am Chem Soc* **112:** 7736–7742.

Smith JM, Sutherland JD. 2005. Aldolisation of bis(glycolaldehyde) phosphate and formaldehyde. *Chem Bio Chem* **6:** 1980–1982.

Smith JM, Borsenberger V, Raftery J, Sutherland JD. 2004. Exploratory studies to investigate a linked prebiotic origin of RNA and coded peptides. 2nd Communication: Derivation and reactivity of xylose phosphates. *Chem Biodiversity* **1:** 1418–1451.

Sutherland JD. 2007. Looking beyond the RNA structural neighborhood for potentially primordial genetic systems. *Angew Chem Int Ed* **46:** 2354–2356.

Sutherland JD, Whitfield JN. 1997. Studies on a potentially prebiotic synthesis of RNA. *Tetrahedron* **53:** 11595–11626.

Szostak JW. 2009. Systems chemistry on early Earth. *Nature* **459:** 171–172.

Szostak JW, Bartel DP, Luisi PL. 2001. Synthesizing life. *Nature* **409:** 387–390.

Tapiero CM, Nagyvary J. 1971. Prebiotic formation of cytidine nucleotides. *Nature* **231:** 42–43.

Thaddeus P. 2006. The prebiotic molecules observed in the interstellar gas. *Phil Trans R Soc B* **361:** 1681–1687.

Ueda T, Nishino H. 1968. On the Hilbert-Johnson procedure for pyrimidine nucleoside synthesis. *J Am Chem Soc* **90:** 1678–1679.

Verlander MS, Lohrmann R, Orgel LE. 1973. Catalysts for the self-polymerization of adenosine cyclic 2',3'-phosphate. *J Mol Evol* **2:** 303–316.

Woese C. 1967. *The genetic code*. Harper & Row, New York.

Zhang L, Peritz A, Meggers E. 2005. A simple glycol nucleic acid. *J Am Chem Soc* **127:** 4174–4175.

Zimmer S, Biltonen R. 1972. Thermodynamics of proton dissociation of adenine. *J Solution Chem* **1:** 291–298.

The Origin of Biological Homochirality

Donna G. Blackmond

Department of Chemistry, The Scripps Research Institute, La Jolla, California 92037

Correspondence: blackmond@scripps.edu

The single-handedness of biological molecules has fascinated scientists and laymen alike since Pasteur's first painstaking separation of the enantiomorphic crystals of a tartrate salt more than 150 yr ago. More recently, a number of theoretical and experimental investigations have helped to delineate models for how one enantiomer might have come to dominate over the other from what presumably was a racemic prebiotic world. This article highlights mechanisms for enantioenrichment that include either chemical or physical processes, or a combination of both. The scientific driving force for this work arises from an interest in understanding the origin of life, because the homochirality of biological molecules is a signature of life.

INTRODUCTION

Homochirality as a Signature of Life

For centuries, symmetry concepts have fascinated scientists as well as artists, mathematicians and writers, laymen and children. The property of chirality—nonsuperimposable forms that are mirror images of one another, as are left and right hands—is manifest in both molecular and macroscopic objects As early as 1874, and a quarter century after Pasteur showed that salts of tartaric acid exist as mirror image crystals, van't Hoff and Le Bel independently postulated the existence of chiral molecules (Heilbronner and Dunitz 1993) (Fig. 1). Chirality held Alice's attention as she pondered the macroscopic world she glimpsed through the looking glass, and her musings over whether looking-glass milk would be good to drink presaged our quest to understand the molecular importance of chirality.

The two forms of a chiral molecule, called enantiomers, have identical physical and chemical properties, but they manner in which each interacts with other chiral molecules may be different, just as a left hand interacts differently with left- and right-hand gloves. Chiral molecules in living organisms in Nature exist almost exclusively as single enantiomers, a property that is critical for molecular recognition and replication processes and would thus seem to be a prerequisite for the origin of life. Yet left and right-handed molecules of a compound will form in equal amounts (a racemic mixture) when we synthesize them in the laboratory in the absence of some type of directing template.

The fact of the single chirality of biological molecules—exclusively left-handed amino acids and right-handed sugars—presents us with two questions: First, what served as the original template for biasing production of one enantiomer over the other in the chemically

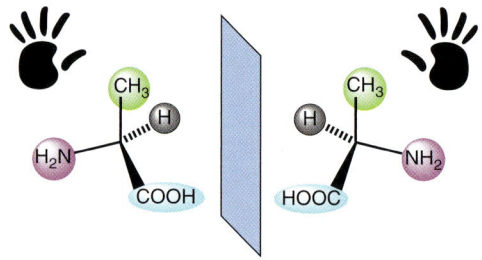

Figure 1. The two mirror-image enantiomers of the amino acid alanine.

austere, and presumably racemic, environment of the prebiotic world? And second, how was this bias sustained and propagated to give us the biological world of single chirality that surrounds us?

Short of constructing a Time Machine, we have no way of elucidating precisely the chain of events that led to life on earth today. What we may do instead is highlight the modern experimental and theoretical work that has attempted to probe these questions. After a brief discussion of the first question, this review focuses primarily on the second of the two questions raised earlier: the discussion of plausible mechanisms for the evolution of molecular homochirality as exemplified by the D-sugars and L-amino acids found in living organisms today.

BACKGROUND

How It All Got Started: Chance versus Determinism

"Symmetry breaking" is the term used to describe the occurrence of an imbalance between left and right enantiomeric molecules. This imbalance is traditionally measured in terms of the *enantiomeric excess*, or ee, where $ee = (R-S)/(R+S)$ and R and S are concentrations of the right and left hand molecules, respectively. Proposals for how an imbalance might have come about may be classified as either terrestrial or extraterrestrial, and then subdivided into either random or deterministic (sometimes called "de facto" and "de lege" respectively). Evidence of small enantiomeric excesses in amino acids found in chondritic meteor deposits (Pizzarello 2006) (see Zahnle et al. 2010) allows the hypothesis that the initial imbalance is not of our world (although it begs the question of where and how it did originate). Discussions of how an imbalance could have originated here on earth often debate the question of whether life was preordained to be based on D-sugars and L-amino acids or whether this happened by chance, implying that a life form based on the opposite chirality might have been just as likely at the outset. Here, physics enters the picture: the discovery of parity violation and the elaboration of one of its consequences, that the two enantiomers have a very small energy difference between them, led many to consider the implications for biological homochirality (Quack 2002). Quantitative estimates of this energy difference have been made and revised in the intervening years, but it is clear that it is very, very small; whereas experimental and theoretical work is ongoing, and the question is not yet settled, a relationship between biological homochirality and parity violation is not yet supported by experimental findings.

Proponents on the "chance" side of this question point out that absolute asymmetric synthesis—defined as the production of enantiomerically enriched products in the absence of a chemical or physical chiral directing force—could occur stochastically (Mislow 2003). A trivial example is that any collection of an odd number of enantiomeric molecules has, by definition, broken symmetry. Fluctuations in the physical and chemical environment could result in transient fluctuations in the relative numbers of left- and right-handed molecules. However, any small imbalance created in this way should average out as the racemic state unless some process intervenes to sustain and amplify it. Thus, whether or not the imbalance in enantiomers came about by chance, arising on earth or elsewhere, an amplification mechanism remains the key to increasing enantiomeric excess and ultimately to approaching the homochiral state. A description of mechanisms for how this imbalance might be amplified is the main subject of this review.

Amplifying the Imbalance

Theoretical models for how a small initial imbalance in enantiomer concentrations might ultimately be turned into the subsequent production of a single enantiomer have been discussed for more than half a century (Frank 1953; Calvin 1969), but only more recently have experimental studies begun to address this question directly. In the past two decades several distinct models with strikingly different features have emerged, leading to the comment that scientists are now "spoilt for choice" (Ball 2007) among possible explanations for how one enantiomer came to dominate over the other in biological molecules. These models draw on both the chemical and the physical behavior of chiral molecules, and they may be classified according to their relative emphasis on kinetics vs. thermodynamics of the processes involved. "Far-from-equilibrium" models involving autocatalytic chemical reactions or crystallization processes lie at one end of the spectrum. At the other end, a model based on equilibrium phase behavior proposes a physical explanation. And in between lies a model that invokes an interplay between thermodynamics and kinetics to explain how a combination of physical and chemical processes can drive a system of near-equal numbers of enantiomeric molecules to the left or to the right.

CHEMICAL MODELS

Homochirality via Autocatalysis

More than 60 yr ago, Frank developed a mathematical model for an autocatalytic reaction mechanism for the evolution of homochirality. The model is based on a simple idea: a substance that acts as a catalyst in its own self-production and at the same time acts to suppress synthesis of its enantiomer enables the evolution of enantiopure molecules from a near-racemic mixture. The experimental challenge to discover a reaction with these features was posed in the last sentence of this purely theoretical paper: "A laboratory demonstration may not be impossible" (Frank 1953; Wynberg 1989).

Mutual Antagonism

Frank's proposal serves to highlight the critical role played by an inhibition mechanism in autocatalytic models for the evolution of homochirality. Figure 2 illustrates this point using an example of a small group of L and D enantiomers that act as autocatalysts in an unlimited pool of substrate molecules. Each enantiomer is capable of reproducing itself in a reaction with a substrate molecule. In addition, there is "mutual antagonism" between L and D such that when they react together, both become deactivated and lose their capacity to self-replicate. The original pool shown in Figure 2 has an imbalance of one extra L molecule compared with the total number of D molecules, or 3:2 for this simple example. Let's say that one L and one D meet by chance and deactivate, whereas the remaining molecules feed on the pool of substrate and reproduce themselves. The active pool of enantiomers has now become 4:2. When these same processes repeat, the active pool becomes 6:2, then, then 10:2, and so on. One L and one D enantiomer are paired off in each mutual antagonism event, and hence the self-production of enantiomers will cause the ratio of L:D to grow as long as an initial imbalance was present at the beginning of the process. Together, autocatalysis and mutual antagonism propagate and amplify the imbalance in enantiomers. The only catch is that the smaller the initial imbalance, the greater the number of L and D molecules lost in the deactivation process before significant enantioenrichment can occur. If the substrate pool is large enough, however, productivity can remain high, and the selectivity of autocatalytic production of one enantiomer will eventually dominate.

Proof of Concept

This model might be considered trivial, and no experimental system is known to follow the process in Figure 2. But Frank's understated exhortation to experimental chemists to discover an autocatalytic reaction with these key features captivated several generations of chemists. More than forty years later, the first

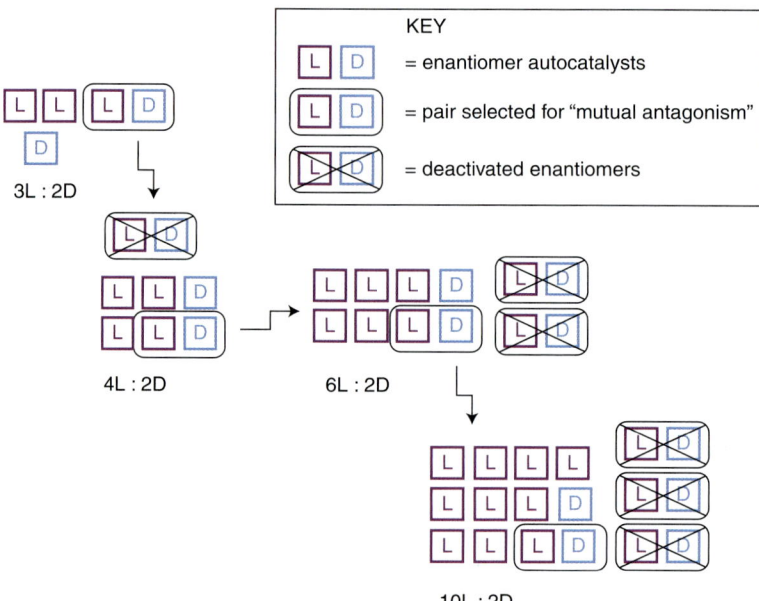

Figure 2. Schematic representation of the Frank model for the evolution of homochirality based on autocatalytic replication and mutual antagonism of enantiomers.

experimental proof of this concept was found when Soai and coworkers reported the autocatalytic alkylation of pyrimidyl aldehydes with dialkylzincs (Soai et al. 1995) (Scheme 1), in which the reaction rate is accelerated by addition of catalytic amounts of its alcohol product. In addition, and most strikingly, this reaction was shown to yield the autocatalytic product in very *high* enantiomeric excess starting from a very *low* enantiomeric excess in the original catalyst.

Since this initial discovery, Soai's group has gone on to present remarkable further observations of asymmetric amplification in the reaction that now bears his name. Enantiomeric excesses as high as 85% were reported for a reaction initiated with an initiator produced at 0.1% ee from exposure to circularly polarized light

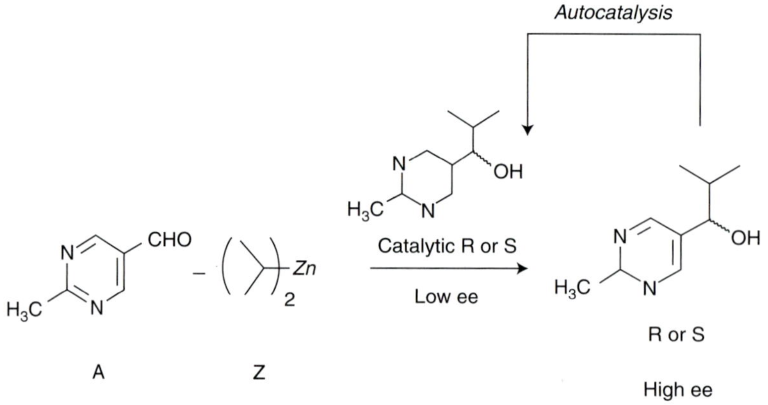

Scheme 1. The Soai autocatalytic reaction, in which the product catalyzes its own formation. The $-CH_3$ group in the pyrimidyl aldehyde may be replaced with other groups such as alkynyl groups.

(Shibata et al. 1998). Asymmetric amplification has also been observed for the reaction initiated by inorganic chiral materials such as quartz (Soai et al. 1999). Most recently Soai has shown that the reaction may be selectively triggered solely by the minute mirror-image difference provided by $^{12}C/^{13}C$ carbon isotope chirality of an initiator molecule (Kawasaki et al. 2009), demonstrating that the reaction needs only an extremely small nudge to direct it consistently to the left or to the right.

Mechanistic Corroboration of the Frank Model

Soai's observations continued to amaze and confound the community for several years before the first mechanistic rationalization of the reaction was reported by Blackmond and Brown in 2001 (Blackmond et al. 2001). A kinetic model was developed based on highly accurate in-situ measurements of the reaction's progress. What's more, the kinetic model independently predicted both the temporal degree of asymmetric amplification, confirmed by compositional analysis, as well as the relative concentrations of the catalyst species, confirmed by NMR spectroscopy. Figure 3 shows how the kinetic model compares with experimental data for asymmetric amplification in the Soai reaction performed with two different initial catalyst ee values. This work showed that the Soai reaction couples autocatalysis with a form of "mutual antagonism," thus evoking the main features of the Frank model, albeit in a more sophisticated chemical scenario.

The Blackmond/Brown model rationalizes asymmetric amplification in the autocatalytic Soai reaction based on an extension of Kagan's model for nonlinear effects in catalytic reactions (Girard and Kagan 1998), that is, cases in which the reaction product ee does not scale linearly with the catalyst ee. Such behavior may ensue when the catalyst molecules aggregate to form higher order species. The ML_2 model describes the formation of homochiral (RR and SS, Eqs. 1 and 2) and heterochiral (SR, Eq. 3) dimers from monomeric R and S molecules. The relative concentration of these dimers depends on an equilibrium constant, K_D (Eq. 4).

$$R + R \xrightarrow{K_{homo}} RR \quad (1)$$

$$S + S \xrightarrow{K_{homo}} SS \quad (2)$$

$$R + S \xrightarrow{K_{hetero}} SR \quad (3)$$

$$K_D = \frac{[SR]}{[RR] \cdot [SS]} = \left(\frac{K_{homo}}{K_{hetero}}\right)^2 \quad (4)$$

Kagan's model proposes that the two homochiral dimers act as enantiomeric catalysts, giving opposite product ee values with identical rate constants, whereas the heterochiral dimer catalyst produces racemic product and may show a rate constant different from that of the homochiral dimer catalysts. Some degree of asymmetric amplification will result for any system in which the heterochiral dimer catalyst is less active than its homochiral counterparts. A value of $K_D = 4$ indicates a stochastic or nonselective distribution of dimers, and larger values of K_D skew the distribution toward a preference for the heterochiral species.

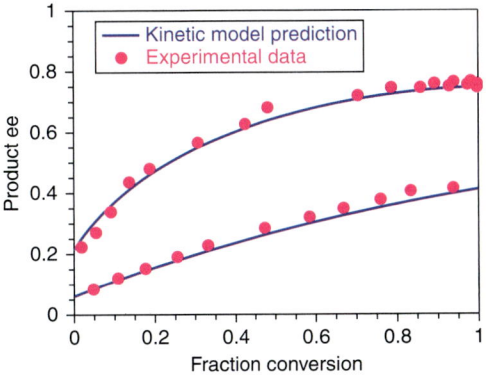

Figure 3. Product enantiomeric excess as a function of reaction progress in the Soai reaction of Scheme 1. Reactions catalyzed by 10 mol% of the reaction product with an initial ee of 6% and 22% (Buono and Blackond, 2003). Prediction of the Blackmond/Brown kinetic model (solid blue lines) and experimental values from HPLC analysis (filled magenta circles).

The upper limit attainable of ee amplification attainable in a catalytic reaction following this model occurs for when the heterochiral dimer is inactive and is dictated by the magnitude of K_D: thus for a 1% ee catalyst showing a stochastic distribution of dimers, the maximum amplification in the product ee is only twofold, to ca. 2% ee. Stronger amplification of ee may be realized only by creating a specific stereochemical bias toward a stable, inactive heterochiral dimer, that is, for systems showing K_D values that are significantly higher than that for a stochastic distribution. For example, a 1% ee catalyst would require a K_D value a millionfold higher than the stochastic distribution to approach a perfectly enantioselective reaction.

Selection Without Bias

Although this model provides a means for asymmetric amplification, it presents a quandary for the prebiotic evolution of homochirality. Today we can construct a strong bias in modern asymmetric catalysts to provide a high K_D value by drawing on the extensive natural chiral pool of molecules for complex building blocks, but this resource would not have been available to prebiotic molecules. The simple chiral molecules present in the prebiotic soup would not necessarily be expected to form dimers with highly unequal homo/heterochiral stabilities. How could asymmetric autocatalysts have achieved the strong stereochemical bias in dimer formation presumably needed to ensure strong asymmetric amplification?

The simple answer is that they didn't! The basic finding of Blackmond and Brown's studies is that the Soai reaction R and S products form a *stochastic* distribution of homochiral and heterochiral dimers, with essentially no stereochemical bias between the dimers ($K_D = 4$), and that the heterochiral dimer is inactive as a catalyst. The key to how this allows an approach to homochirality is provided by consideration of basic differences among catalytic reactions, which accelerate the formation of a species that is not the same as the catalyst, and autocatalytic reactions, in which the catalyst accelerates its own formation. Because the catalyst produces more of itself in autocatalysis, and because mutual antagonism allows the minor enantiomer to be siphoned off as an inactive heterochiral dimer (serving the role of "mutual antagonism" in the Frank model), the relative concentrations of the two enantiomers is not fixed at its initial value, as it is in a static catalytic system. Instead, in an autocatalytic system following this dimer model, catalyst concentration increases, and relative concentration of the two homochiral dimers changes, with reaction turnover. The ultimate product enantiomeric excess that may be achieved in such an autocatalytic reaction is limited only by the size of the substrate pool, not by the magnitude of K_D. Returning to our example of catalyst 1% ee with $K_D = 4$, comparison with an autocatalyst with identical initial ee and K_D shows that the autocatalytic system approaches homochirality after just 5000 cycles; strong robustness trumps mild intrinsic selectivity. Amplification of ee in autocatalysis requires *not* sophisticated stereoselection but *only* higher activity for the homochiral dimers, repeated over many autocatalytic cycles.

Homochirality: Chance Aided by Luck

Thus the dimer model provides an elegant and simple solution to one mystery of the evolution of homochirality (Blackmond 2006). If, as in the Soai reaction, the relative dimer reactivities happen to give the edge to the homochiral species, amplification, and ultimately homochirality, is ensured even for nonselective dimer formation. Statistics (stochastic dimer formation) and one stroke of luck (lower activity of the heterochiral dimer) are sufficient prerequisites to account for the evolution of our homochiral world today.

PHYSICAL MODELS

Kinetics versus Thermodynamics

The Soai autocatalytic reaction is essentially irreversible under the conditions used; the reaction proceeds faster as more product (catalyst!) is formed, as long as it continues to receive the nutrients it requires. In the time scale of the laboratory, the system never approaches the equilibrium state that would allow the reverse reactions to participate and ultimately erode selectivity. Analogous "far-from-equilibrium" processes showing amplified enantioselectivity as described earlier for the Soai reaction have been observed in physical processes such as the crystallization of molecules that form chiral solids. In these cases, kinetics must win out over thermodynamics in order for homochirality to evolve. The interconnection between kinetics and thermodynamics is addressed further in our consideration of two additional models for homochirality invoking the physical phase behavior of chiral solids in equilibrium with their solution phase molecules. An intrinsic feature of both of these models is the dominant role of the equilibrium condition, in contrast to the "far from equilibrium" cases invoking either chemical reactions and crystallization processes discussed earlier. These phase behavior models are underpinned by the fundamental concepts of the Gibbs Phase Rule as articulated by ternary phase diagrams of L + D + solvent.

Phase Behavior of Chiral Solids

Before discussing these two models in detail, it is instructive to review the types of solution-solid behavior commonly found for enantiomeric molecules such as amino acids (Jacques et al. 1994). Chiral compounds crystallize most commonly in one of two forms: (a) as a *racemic compound*, in which crystals contain a 1:1 ratio of D:L molecules; or (b) as a *conglomerate*, in which each crystal is comprised of molecules of a single enantiomer, and the crystals themselves are mirror images (as were those Pasteur separated with his tweezers), and there is no direct molecular interaction between D and L molecules. These types of compounds are illustrated schematically in Figure 4. The type of crystal a chiral molecule forms in the solid phase is a fundamental property of that molecule at a given temperature and pressure. Racemic compounds are more prevalent than conglomerates by ca. 10:1 on planet Earth, including all but two of the 19 proteinogenic amino acids that are chiral (the twentieth, glycine, is an achiral molecule).

Eutectic Composition

When unequal numbers of molecules that form either type of crystal solid are partially dissolved in water, the Gibbs phase rule teaches us that the solution composition at equilibrium is fixed at what is called the "eutectic composition." One interesting contrasting feature of the two types of compounds is that for conglomerates, this solution contains equal numbers of D and L molecules, giving a eutectic $ee^{eut} = 0$, whereas racemic compounds will show a nonzero ee^{eut}. This is easily rationalized by considering the solubility characteristics of each type of compound.

Conglomerates

Because the separate D and L crystals of a conglomerate are enantiomeric, they have identical properties, including solubility. A saturated solution of D crystals thus has the same solution

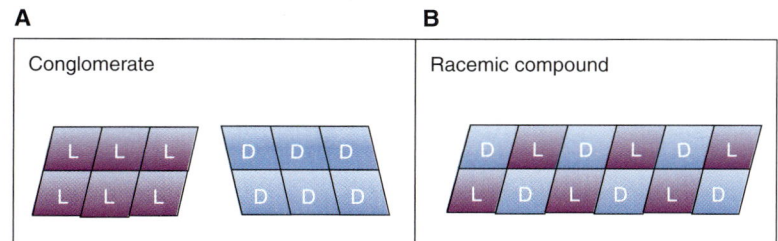

Figure 4. Two types of crystalline solids formed by chiral compounds. Rectangles represent solid phase enantiomeric molecules. (*A*) conglomerates form separate crystals of each enantiomer; (*B*) racemic compounds form mixed crystals in a 1:1 ratio of the two enantiomers.

concentration as does a saturated solution of L crystals. If a mixture of D and L crystals is allowed to equilibrate together, each enantiomorphic solid shows its own (identical) solubility independently, and the solution phase at equilibrium will contain equal numbers of of D and L molecules, and twice as many in total as for either system separately (Meyerhoffer 1904) This remains true even if the system contains an unequal number of D and L molecules in total, as shown in Figure 5A.

Racemic Compounds

The equilibrium phase behavior for a species forming a racemic compound is more complex. For the case of an unequal total number of D and L molecules, mixed 1:1 D:L crystals form preferentially, pairing up with all of the minor enantiomer in the solid phase, and any molecules of the excess enantiomer remaining in the solid phase form homochiral crystals, as shown in Figure 5B. Thus at equilibrium this system will contain two separate solid phases, one enantiopure, and one racemic, along with the solution phase. The D:L mixed solid contributes equal numbers of D and L molecules to the solution according to its characteristic solubility, and the enantiopure solid does the same with its one enantiomer, according to its solubility. The solution composition at equilibrium will depend on the relative solubilities of the pure and mixed crystals, with the result that the solution will show a eutectic ee value somewhere between 0 and 100% (Fig. 5B). This value is a characteristic of the particular chiral compound and is not known a priori.

Enantiomer Partitioning

In both cases, as Figure 5 shows, conglomerates and racemic compounds present in a nonracemic composition in equilibrium with a solvent show a partitioning of enantiomers between the solution and solid phases that is dictated by the characteristics of the type of solid formed. Solution ee does not equal solid phase ee, and neither value will be the same as the overall ee. The models discussed later make use of this concept of phase partitioning in different ways to provide enantioenrichment in either the solid phase or the solution phase. Figure 5 suggests that solution phase enantioenrichment might be the goal when racemic compounds are considered, whereas a mechanism for solid phase enantioenrichment might make more sense for conglomerates. This is exactly what has been found, as the sections later describe.

PHASE BEHAVIOR I

"Eve Crystal" Model for Conglomerates

It has been known for more than one hundred years that certain achiral molecules such as

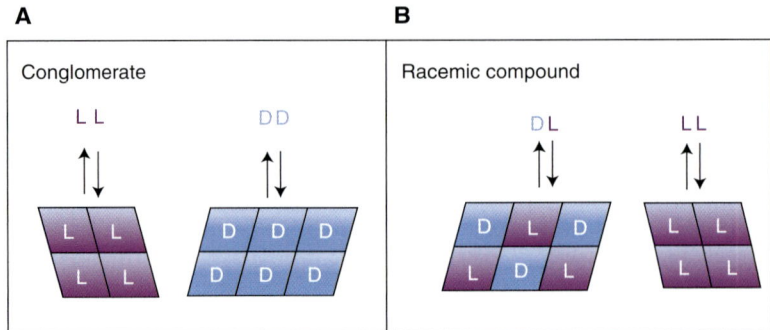

Figure 5. Depiction of equilibrium between chiral crystalline solids and their aqueous solution phases for nonracemic mixtures of enantiomers. Rectangles represent solid phase enantiomeric molecules; colored letters represent solution phase molecules in equilibrium with the solid phases. The solution phase composition is known as the eutectic. (A) conglomerates show $ee^{eut} = 0$; (B) racemic compounds show nonzero ee^{eut}.

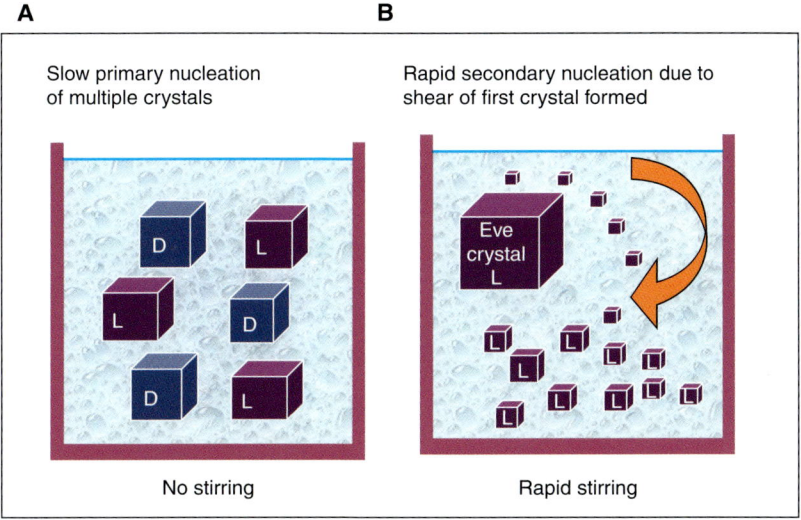

Figure 6. Crystallization from supersaturated solution of achiral molecules that form mirror enantiomorphic solid crystals. (A) Without rapid stirring, equal quantities of left- and right-handed crystals are formed; (B) under rapid stirring conditions, all crystals are grown from the fragments of a single primary crystal ("Eve crystal"), resulting in formation of only one enantiomorph.

$NaClO_3$ crystallize as chiral solids. Work from the late 19th century noted that these crystallizations often resulted in a formation of two separate chiral solid phases (Kipping and Pope 1898). Most often, equal amounts of left and right-handed crystals are formed, but under some conditions the system breaks symmetry during the crystallization. The most striking example of this phenomenon was reported nearly two decades ago by Kondepudi (Kondepudi et al. 1990) who showed that when the crystallization process was accompanied by rapid stirring, crystals of single chirality could be formed, randomly left-handed or right-handed in repeated experiments. This was rationalized by considering the dynamics of crystallization: formation of the first crystals (primary nucleation) from a homogeneous supersaturated solution is slower than the process of crystal growth by adding molecules to crystals already formed (secondary nucleation). Under rapid stirring, the first crystal formed, or "Eve" crystal, may be broken by shear into thousands of smaller crystals of the same chiral form. These "daughter" crystals then grow rapidly by drawing on solution molecules, and the solution rapidly becomes depleted. If this occurs more rapidly than formation of any new primary crystals, a single chiral solid state may result (McBride and Carter 1991) (Fig. 6). This phenomenon bears resemblance to "far-from-equilibrium" autocatalytic reaction processes such as the Soai reaction discussed earlier in this article.

PHASE BEHAVIOR II

Near Equilibrium Systems

Most recently, this $NaClO_3$ system that has fascinated scientists since van't Hoff's days was back in the news with a report by Viedma (2005) of a remarkable experimental finding. Viedma's experiment is not strictly a crystallization; it starts at what would be the end of a typical crystallization performed without the rapid stirring and shear described earlier, so that the system is equilibrated with an equal number of right- and left-hand crystals of $NaClO_3$ in saturated aqueous solution, as in Figure 6A. Under these conditions, no new crystals nucleate; the only processes occurring in the flask are the continual dissolution and reaccretion of $NaClO_3$ molecules to and from existing crystals, and these rates are balanced at equilibrium. No net change is expected, and that is exactly what has been found in such experiments for more than one hundred years. Viedma then added glass beads to the gently stirred vial, which enhanced the attrition of the crystals as they stirred. What he observed under these conditions is that the system evolved from equal quantities of left- and right-handed crystals to a single enantiomorphic solid. One chiral solid converted completely to the other over time, in a random fashion, sometimes to the left and sometimes to the right, with equal probability. The striking fact is that the system moved inexorably from an apparent equilibrium between two enantiomorphic solid states to a single chiral solid state, simply by virtue of the application of mechanical energy via the grinding of glass beads.

Viedma reasoned that the continual abrasion of the crystals by stirring with small glass beads enhanced both halves of the cycle of repetitive dissolution/crystallization that occurs at equilibrium. According to the Gibbs-Thomson rule, small crystals dissolve more readily than large crystals. Attrition by glass beads produces a greater number of smaller crystals, whose increased dissolution in turn causes a slight supersaturation of $NaClO_3$ in solution. Not sufficiently supersaturated to support primary nucleation, the system strives to redress the balance between solid and solution by increasing the rate of reaccretion of solution phase $NaClO_3$ onto existing crystals. A key point is that once a molecule of $NaClO_3$ dissolves from a crystal, it no longer possesses chirality and it retains no memory of the chiral form it previously showed as part of a crystal. Solution-phase $NaClO_3$ thus has no preference for re-accreting to a left- or right-handed crystal. What solution phase molecules do have, however, is a preference for adding to larger crystals over smaller ones, a phenomenon known as Ostwald ripening. If, by chance, the system contains a predominance of large crystals of one hand, solution phase $NaClO_3$ will preferentially add to these crystals, and the quantity of this enantiomorphic solid will increase relative to its mirror image form.

Chiral AMnesia

The mechanical energy imparted to the system by the enhanced attrition thus triggers an overall process that is cyclical: The action of the glass beads truncates the Ostwald ripening process, continually breaking up crystals as they attempt to grow in size. The growth of the crystals occurs in response to an increased concentration driving force created by slight supersaturation in solution, which in turn is the result of dissolution of small crystals created by the glass beads breaking up larger crystals. Evolution of solid phase homochirality in this case requires only an initial imbalance in the crystal size of left- versus right-hand crystals.

This proposed rationalization implies that the system never departs very far from equilibrium conditions and that system strives continually to restore the balance between the physical processes of dissolution and reaccretion. The key to the evolution of one chiral solid state is the ability of an achiral solution phase molecule to choose to add to either hand of its solid phase crystals. The solution phase serves as the conduit through which the molecules that form one hand of the crystal forget their solid-state chiral history and are free to choose a new solid-state chiral destiny, a process that has been given the name "chiral amnesia" (Blackmond 2007; Viedma 2007).

Extension to Chiral Molecules

These results led many to consider a possible extension of this process from the achiral $NaClO_3$ to intrinsically chiral molecules that form conglomerate solids, but this concept faces an important challenge. A chiral molecule in solution equilibrium with its two separate enantiomorphic solid phases would not have the ability to choose to add to either solid; a D-amino acid molecule may add only to a D-crystal, and an L-amino acid molecule to an L-crystal. Although this does not affect the Gibbs-Thomson and Ostwald ripening processes for dissolution and growth of crystals, conversion of crystals of one hand into the other would be frustrated without a process allowing

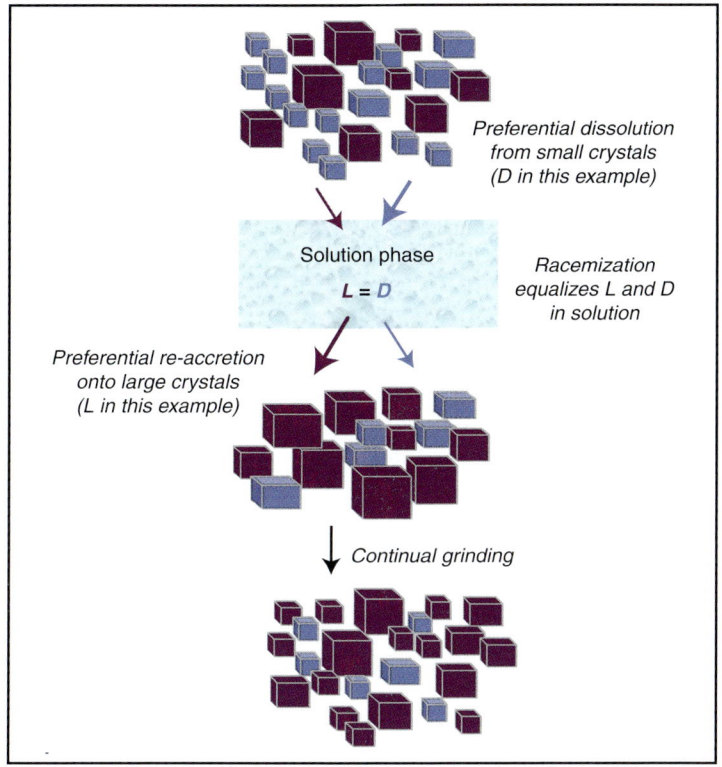

Figure 7. "Chiral amnesia" process for the evolution of solid-phase homochirality for chiral molecules that form conglomerate solids. In this example, an initial imbalance toward larger L crystals helps to drive the dissolution/re-accretion process from D crystals to L crystals. The process is aided by solution phase racemization, which converts the "excess" dissolved D molecules to L molecules, equalizing the solution composition and enabling molecules that were formerly part of a D crystal to add as L molecules to L crystals.

the molecules to forget their chiral signature in solution. However, organic chemists engaged in asymmetric synthesis know well—and are often themselves frustrated by—solution reactions in which chiral molecules can be made to convert between their left and right forms, known as racemization. Long considered a bane in organic synthesis, solution-phase racemization provides the key to extending the Viedma model for homochirality to intrinsically chiral molecules, as shown schematically in Figure 7. This was successfully shown recently for an amino acid derivative (Noorduin at al. 2008), for the proteinogenic amino acid aspartic acid (Viedma et al. 2008) (Scheme 2), and for a Mannich reaction product (Tsogoeva et al. 2009).

The studies of aspartic acid revealed that the inexorable move to one homochiral solid could be accomplished in the absence of glass beads,

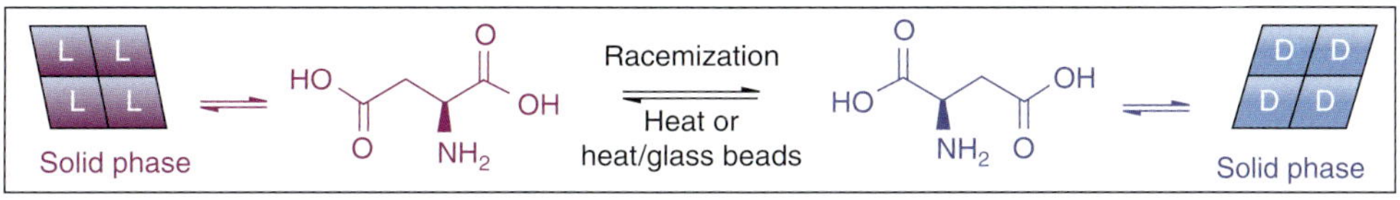

Scheme 2. Transformation of aspartic acid crystals from one enantiomorphic solid to the other via solution phase racemization. Mechanical or thermal energy input drives the dissolution/re-accretion process.

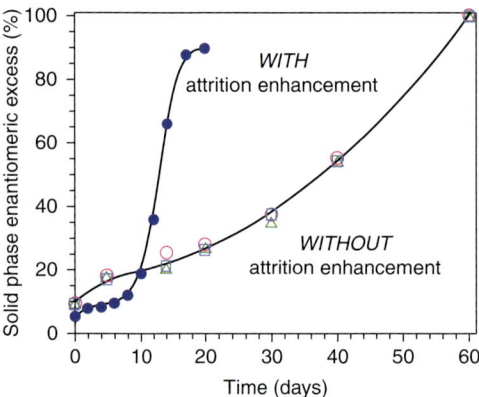

Figure 8. Evolution of solid-phase homochirality for aspartic acid via solution phase racemization. Energy input from grinding the crystals in the presence of enhanced attrition because of glass beads leads to a sigmoidal profile (filled blue circles), whereas thermal energy input drives the process in a linear fashion (open symbols).

simply by application of thermal energy rather than mechanical energy as the means to nudge the system away from equilibrium. The crystal ee profiles lose the exponential shape they show under the grinding conditions (Fig. 8), suggesting that Ostwald ripening may allow more extensive crystal growth when the energy input is thermal rather than mechanical.

PHASE BEHAVIOR III

Thermodynamic Model for Enantioenrichment

The phase properties of chiral compounds discussed earlier have been understood for more than one hundred years. More recently, a model for the origin of biological homochirality based on solution phase enantioenrichment of amino acids that form racemic compounds has been developed based on these concepts. It may be recognized from Figure 5B that in a heterogeneous mixture that contains unequal numbers of enantiopure D and L molecules, the minor enantiomer is present in the solution phase only to the extent that it can dissolve from the D:L crystal. The lower the solubility of the D:L crystals, the more strongly "trapped" in the solid phase will be the minor enantiomer, and the higher the resulting solution phase partitioning of the major enantiomer, manifested as a high ee^{eut}. Several of the proteinogenic amino acids form relatively insoluble D:L crystals and therefore show high eutectic ee values. For example, serine's eutectic occurs at >99% ee. This means that when a sample with nearly equal numbers of D and L serine molecules is partially dissolved, a virtually enantiopure solution results. Table 1 provides eutectic values for a number of amino acids (Klussmann et al. 2006).

Enantioenrichment in solution is thus dictated by thermodynamics for chiral compounds that happen to form relatively insoluble racemic compounds. This concept was first recognized by Morowitz (1969) 40 yr ago and was more recently elaborated by Blackmond's work probing eutectic composition for a variety of amino acids and other chiral compounds Klussmann et al. 2006), and by Breslow (Breslow and Levine 2006) who recently also reported high eutectic ee values for several nucleosides of prebiotic importance (Breslow and Cheng 2009). A recent theoretical treatment based on a two-dimensional lattice model successfully predicts the ternary phase behavior of amino acids based on the interactions that stabilize the racemic crystal, providing molecular level insight into the observed enantiomer partitioning (Lombardo et al. 2009). These studies suggest a general and facile route to homochirality that may have prebiotic relevance. Cycles of rain and evaporation establishing solid phase-solution phase equilibrium in pools containing a small initial imbalance of amino acid enantiomers could result in a solution of enantioenriched molecules that might then serve as efficient asymmetric catalysts or as building blocks themselves for construction of the complex molecules required for recognition, replication and ultimately for the chemical basis of life.

Sublime Partitioning

The phase partitioning of enantiomers that forms the basis of this model for solution phase enantioenrichment might be expected to

Table 1. Eutectic ee values for a number of proteinogenic amino acids, identified by their chemical structures and their three-letter names.

Ser	$ee^{eut} > 99\%$	His	$ee^{eut} = 93\%$
Leu	$ee^{eut} = 87\%$	Met	$ee^{eut} = 85\%$
Phe	$ee^{eut} = 88\%$	Ala	$ee^{eut} = 60\%$
Val	$ee^{eut} = 46\%$	Thr	$ee^{eut} = 0\%$

predict other phase behavior of enantiomers (Blackmond and Klussmann 2007a). Indeed, eutectic ee values correlate with fusion temperature for a number of amino acids. In addition, reports of enantioenrichment of amino acids via sublimation of enantioimpure D/L mixtures are also well correlated with the eutectic ee values for amino acids: for example, the sublimate from serine is nearly enantiopure, whereas that from threonine, which forms a conglomerate, gives 0% ee (Perry et al. 2007). Sublimate ee values of 72%–89% were observed for leucine, in accordance with it measured eutectic value of 88% ee (Fletcher et al. 2007). Enantiomer partitioning via sublimation raises tantalizing speculation about amino acids in space and an extraterrestrial origin of enantioenrichment.

Crystal Engineering for Tuning Eutectics

Because eutectic ee is a characteristic property of a compound, it would appear that enantioenrichment by this approach is limited to chiral compounds such as serine that happen to show high eutectic ee values. Thermodynamics suggests that other amino acids, such as valine with $ee^{eut} = 47\%$, may be stuck with the hand that Nature has dealt. However, further work revealed the exciting discovery of a route to solution phase enantioenrichment that may be achieved even for chiral compounds with low intrinsic ee^{eut} values. Eutectic ee composition may be "tuned" in many cases by through incorporation of a variety of small, achiral molecules into the solid phase structure of amino acid crystals via hydrogen bonding (Klussmann et al. 2007). The existence of cocrystals known as solvates or polymorphs is well known, but effects on solubility and the accompanying implications for solution phase enantioenrichment were not recognized, until Blackmond and coworkers (Klussmann et al. 2007a) showed that enhanced eutectic ee values may be obtained in a number of such cases. If the incorporated molecule reduces the solubility of the racemic crystal relative to that of the enantiopure crystal,

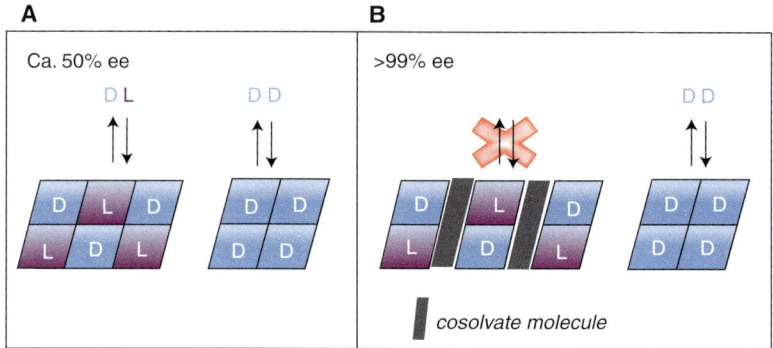

Figure 9. Manipulation of eutectic ee value by formation of a solvate that reduces the solubility of the racemic compound.

enhanced eutectic composition will result, as shown in Figure 9. For example, D:L proline incorporates CHCl$_3$ into its structure with a concomitant rise in ee^{eut} from 50% to >99% ee, and D:L valine and phenylalanine each form crystals incorporating fumaric acid; ee^{eut} rose from 47% and 88%, respectively, to >99% in both cases. The structure of the D:L compound of proline with chloroform is shown in Figure 10. Manipulation of the eutectic composition by additives may be thought of as an analogy to clathrate compounds, although here it is the amino acid enantiomers themselves that are trapped in the solvate-racemate structure, causing them to dissolve much less readily.

The finding that the enantiomeric excess of an amino acid in solution may be significantly enhanced via solvate formation enables an approach to enantioenrichment for a wide range of chiral compounds. A particularly appealing feature of this model is that it is based on an equilibrium mechanism, in contrast to the far-from-equilibrium environment invoked in kinetically induced amplification via autocatalytic reactions discussed earlier. Prebiotic pools containing nearly racemic amino acids could exist over long periods of time awaiting an influx of appropriate hydrogen-bonding partner molecules to form solvates that help provide enantiopure amino acids in solutions

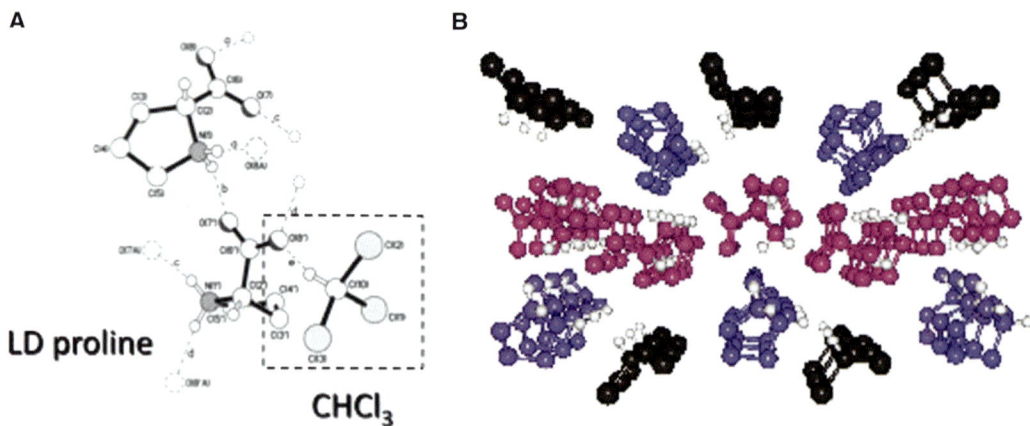

Figure 10. Crystal structure of LD proline incorporating one molecule of chloroform. (*A*) five independent hydrogen bonds are shown; (*B*) long range structure with proline enantiomers in blue and magenta, chloroform in black (Klussmann et al. 2006).

where the chemical reactions leading to life might begin to occur (Klussmann and Blackmond 2007b).

CHALLENGES AND FUTURE DIRECTIONS

Homochirality First?

Amino acids and sugar molecules are produced as single enantiomers in biological processes on earth today, and these molecules provide the building blocks for homochiral polymers such as peptide chains, RNA and DNA. However, it is far from certain that homochirality on the molecular level was required before nascent biopolymers began to play a role in the growing chemical complexity that led to life. This leaves open the possibility that the prebiotic molecular pool need not have been enantioenriched. It has been suggested that in competition for growth, heterochiral chains containing both enantiomers would readily give way to homochiral polymers (Wald 1957; Kuhn 1972, 2007), and therefore selection of one hand of an amino acid over the other could have come with oligopeptides rather than at the molecular level (Zepk et al. 2002). However, such a scenario still needs to invoke a mechanism—either chance or deterministic—for symmetry breaking of the mirror image homochiral polymer chains that would evolve. A plausible proposal is that partial enantioenrichment could have taken place at the molecular level, but that prebiotic amino acids and sugars need not to have evolved completely to single chirality before the formation of the first biopolymer chains. The greater the pre-enrichment on the molecular level, the more efficient would be the construction of homochiral chains, but the burden of chiral selectivity might have been shared as complexity increased.

The Future for Autocatalysis

The experimental conditions of the Soai reaction preclude it from being of direct prebiotic importance, because it is unlikely that the dialkylzinc chemistry involved would thrive an aqueous, aerobic prebiotic environment. However, the reaction has been particularly instructive in helping understand how autocatalysis coupled with inhibition could lead to a homochiral state. A number of experimental aspects of the Soai autocatalytic reaction are still under study, and a number of other theoretical kinetic models have been proposed, but the Blackmond/Brown model remains the only proposal that provides an adequate rationalization of the experimental data for the Soai reaction (Blackmond 2004, 2006a). Most recently, reports by Tsogoeva (Mauksch et al. 2007) of a purely organic reaction showing similar properties of autocatalysis and amplification of ee (Scheme 3) have excited the community, because the chemistry involved is much closer to what could be prebiotically plausible. Further work to understand the mechanism of this reaction is currently underway in a number of laboratories.

Phase Behavior Models

Comparison of the chiral amnesia and crystal engineering phase behavior models for the origin of homochirality reveal that they are complementary in many ways: the former produces solid-phase homochirality whereas the latter provides enantioenrichment of the solution phase; the chiral amnesia model converts

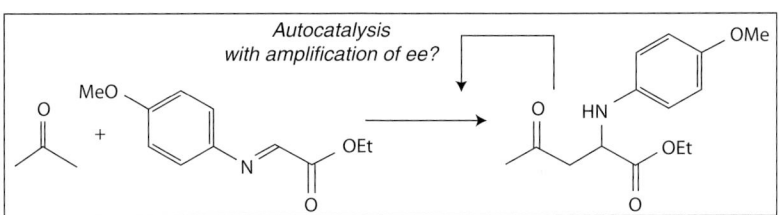

Scheme 3. Autocatalytic Mannich reaction reported by Tsogoeva.

one enantiomer to the other, whereas the crystal engineering model simply partitions the existing molecules between phase. Chiral amnesia may be applied only to molecules that form conglomerates, which means that only about 10% of known chiral compounds are candidates for enantioenrichment by this model. On the other hand, about 85% of chiral compounds might be amenable to the selective partitioning provided by the crystal engineering model. Perhaps some combination of the two led to the initial enantioenrichment of biologically relevant molecules. Both models provide reasonable prebiotic scenarios, and further work to understand the mechanism of enantioenrichment in each case is underway.

Where Do We Go from Here?

The pathway to life may be seen as a saga of increasing chemical and physical complexity (see, for example, Hazen 2010). The modern field of "systems chemistry" (von Kiedrowski 2005) seeks to understand the chemical roots of biological organization by studying the emergence of system properties that may be different from those showed individually by the components in isolation. The implications of the single chirality of biological molecules may be viewed in this context of complexity. Whether or not we will ever know how this property developed in the living systems represented on Earth today, studies of how single chirality might have emerged will aid us in understanding the much larger question of how life might have, and might again, emerge as a complex system.

REFERENCES

Ball P. 2007. Giving life a hand. *Chem World* **4**: 30–31.

Blackmond DG. 2004. Asymmetric autocatalysis and its implications for the origin of homochirality. *PNAS* **101**: 5732–5736.

Blackmond DG. 2006. Mechanistic study of the Soai autocatalytic reaction informed by kinetic analysis. *Tetrahedron: Asymmetry* **17**: 584–589.

Blackmond DG. 2009. The double solubility rule holds for racemizing enantiomers. *Chem Eur J* **15**: 3065–3068.

Blackmond DG, Klussmann M. 2007a. Spoilt for choice: Assessing phase behavior models for the evolution of homochirality. *Chem Commun* 3990–3996.

Blackmond DG, Klussmann M. 2007b. Investigating the evolution of biomolecutar homochirality, *AICHE J* **53**: 2–8.

Blackmond DG, McMillan CR, Ramdeehul S, Schorm A, Brown JM. 2001. Origins of asymmetric amplification in autocatalytic alkylzinc additions. *J Am Chem Soc* **123**: 10103–10104.

Breslow R, Cheng Z-L. 2009. On the origin of terrestrial homochirality for nucleosides and amino acids. *PNAS* **106**: 9144–9146.

Breslow R, Levine M. 2006. Amplification of enantiomeric concentrations under credible prebiotic conditions. *PNAS* **103**: 12979–12980.

Buono FG, Blackmond DG. 2003. Kinetic evidence for a tetrameric transition state in the asymmetric autocatalytic alkylation of pyrimidyl aldehydes. *J Am Chem Soc* **125**: 8978.

Calvin. 1969. *Molecular evolution*. Oxford University Press, Oxford, UK.

Fletcher SP, Jagt RBC, Feringa BL. 2007. An astrophysically relevant mechanism for amino acid enantiomer enrichment. *Chem Commun* 2578–2580.

Frank FC. 1953. On spontaneous asymmetric synthesis. *Biochim Biophys Acta* **11**: 459–463.

Girard C, Kagan HB. 1998. Nonlinear effects in asymmetric synthesis and stereoselective reactions: Ten years of investigation. *Angew Chemie Int Ed* **37**: 2923–2959.

Hazen RM. 2010. Mineral surfaces, geochemical complexities, and the origins of life. *Cold Spring Harb Perspect Biol* **2**: a002162.

Heilbronner E, Dunitz JD. 1993. *Reflections on symmetry in chemistry... and elsewhere.* Verlag Helvetica Chimica Acta, Basel.

Jacques J, Collet A, Wilen SH. 1994. *Enantiomers, racemates and resolution*, 2nd ed. Krieger Publishing Company, FL.

Kawasaki T, Matsumura Y, Tsutsumi T, Suzuki K, Ito M, Soai K. 2009. Asymmetric autocatalysis triggered by carbon isotope (13C/12C) chirality. *Science* **324**: 492–495.

Kipping WS, Pope WJ. 1898. Enantiomorphism. *J Chem Soc Trans* **73**: 606–617.

Klussmann M, Iwamura H, Mathew SP, Wells DH Jr, Pandya U, Armstrong A, Blackmond DG. 2006. Thermodynamic control of asymmetric amplification in amino acid catalysis. *Nature* **441**: 621–623.

Klussmann M, Izumi T, White AJP, Armstrong A, Blackmond DG. 2007. Emergence of solution-phase homochirality via crystal engineering of amino acids. *J Am Chem Soc* **123**: 7657–7660.

Klussmann K, White AJP, Armstrong A, Blackmond DG. 2006. Rationalization and prediction of solution enantiomeric excess in ternary phase systems. *Angew Chem Int Ed* **47**: 7985–7989.

Kondepudi DK, Kaufman RJ, Singh N. 1990. Chiral symmetry breaking in sodium chlorate crystallization. *Science* **250**: 975–976.

Kuhn H. 1972. Self-organization of molecular systems and evolution of the genetic apparatus. *Angew Chem Int Ed* **11**: 798–820.

Kuhn H. 2008. Origin of life—symmetry breaking in the universe: Emergence of homochirality. *Current Opinion Colloid Interface Sci* **13**: 3–11.

Lombardo TG, Stillinger FH, Debenedetti PG. 2009. Thermodynamic mechanism for solution phase chiral amplification. A lattice model. *PNAS* doi: 10.1073/pnas.0812867106.

Mauksch M, Tsogoeva SB, Martynova IM, Wei S. 2007. Evidence of asymmetric autocatalysis in organocatalytic reactions. *Angew Chem Int Ed* **46**: 393–396.

McBride JM, Carter RL. 1991. Spontaneous resolution by stirred crystallization. *Angew Chem Intl Ed* **30**: 293–295.

Morowitz M. 1969. A mechanism for the amplification of fluctuations in racemic mixtures. *J Theor Biol* **25**: 491–494.

Meyerhoffer W. 1904. Stereochemical notes, Pasteur's method of resolving by means of active compounds. Solubility of a tartrate compared with that of a racemate. *Ber Dtsch Chem Ges* **37**: 2604–2610.

Mislow K. 2003. Absolute asymmetric synthesis: A commentary. *Collect Czech Chem Commun* **68**: 849–864.

Noorduin WL, Izumi T, Millemaggi A, Leeman M, Meekes H, Van Enckevort WJP, Kellogg RM, Kaptein B, Vlieg E, Blackmond DG. 2008. Emergence of a single solid chiral state from a nearly racemic amino acid derivative. *J Am Chem Soc* **130**: 1158–1159.

Perry RH, Chunping W, Nefliu M, Cooks RG. 2007. Serine sublimes with spontaneous chiral amplification. *Chem Commun* 1071–73.

Pizzarello S. 2006. The chemistry of life's origin: A carbonaceous meteorite perspective. *Acc Chem Res* **39**: 231–237.

Quack M. 2002. How important is parity violation for molecular and biomolecular chirality? *Angew Chem Int Ed* **41**: 4618–4630.

Shibata J, Yamamoto T, Matsumoto N, Yonekubo S, Osanai S, Soai K. 1998. Amplification of a slight enantiomeric imbalance in molecules based on asymmetric autocatalysis. The first correlation between high enantiomeric enrichment ina chiral molecule and chircularly polarized light. *J Am Chem Soc* **120**: 12157–12158.

Soai K, Osanai S, Kadowaki K, Yonekubo S, Shibata T, Sato I. 1999. D- and L-Quartz-promoted Highly Enantioselective Synthesis of a Chiral Compound. *J Am Chem Soc* **121**: 11235–11236.

Soai K, Shibata T, Morioka H, Choji K. 1995. Asymmetric autocatalysis and amplification of enatiomeric excess of a chiral molecule. *Nature* **378**: 767–768.

Tsogoeva SB, Wei S, Freund M, Mauksch M. 2009. Generation of highly enantioenriched crystalline products in reversible asymmetric reactions with racemic or achiral catalysts. *Angew Chemie Int Ed* **48**: 598–602.

Viedma C. 2005. Chiral symmetry breaking during crystallization: Complete chiral purity induced by nonlinear autocatalysis and recycling. *Phys Rev Lett* **94**: 065504.

Viedma C. 2007. Chiral symmetry breaking and complete chiral purity by thermodynamic-kinetic feedback near equilibrium: Implications for the origin of biochirality. *Astrobiol* **7**: 312–319.

Viedma C, Ortiz JE, de Torres T, Izumi T, Blackmond DG. 2008. Evolution of solid phase homochirality for a proteinogenic amino acid. *J Am Chem Soc* **130**: 15274–15275.

von Kiedrowski G. Ruhr-Universität Bochum, Germany; editor of a new open-access journal "Systems Chemistry."

Wald G. 1957. The origin of optical activity. *Ann NY Acad Sci* **69**: 352–368.

Wynberg H. 1989. Asymmetric autocatalysis: Facts and fancy. *J Macromol Sci Chem* **A26**: 1033–1041.

Zahnle K, Schaefer L, Fegley B. 2010. Earth's early atmospheres. *Cold Spring Harb Perspect Biol* **2**: a004895.

Zepik H, Shavit E, Tang M, Jensen TR, Kjaer K, Bolbach G, Leiserowitz L, Weissbuch I, Lahav M. 2002. Chiral amplification of oligopeptides in two-dimensional crystalline self-assemblies on water. *Science* **295**: 1266.

Bioenergetics and Life's Origins

David Deamer[1] and Arthur L. Weber[2]

[1]Department of Biomolecular Engineering, Baskin School of Engineering, University of California, Santa Cruz, California 95064

[2]SETI Institute, NASA Ames Research Center, Mountain View, California 94043

Correspondence: deamer@soe.ucsc.edu

Bioenergetics is central to our understanding of living systems, yet has attracted relatively little attention in origins of life research. This article focuses on energy resources available to drive primitive metabolism and the synthesis of polymers that could be incorporated into molecular systems having properties associated with the living state. The compartmented systems are referred to as protocells, each different from all the rest and representing a kind of natural experiment. The origin of life was marked when a rare few protocells happened to have the ability to capture energy from the environment to initiate catalyzed heterotrophic growth directed by heritable genetic information in the polymers. This article examines potential sources of energy available to protocells, and mechanisms by which the energy could be used to drive polymer synthesis.

Previous research on life's origins has for the most part focused on the chemistry and energy sources required to produce the small molecules of life—amino acids, nucleobases, and amphiphiles—and to a lesser extent on condensation reactions by which the monomers can be linked into biologically relevant polymers. In modern living cells, polymers are synthesized from activated monomers such as the nucleoside triphosphates used by DNA and RNA polymerases, and the tRNA-amino acyl conjugates that supply ribosomes with activated amino acids. Activated monomers are essential because polymerization reactions occur in an aqueous medium and are therefore energetically uphill in the absence of activation.

A central problem therefore concerns mechanisms by which prebiotic monomers could have been activated to assemble into polymers. Most biopolymers of life are synthesized when the equivalent of a water molecule is removed to form the ester bonds of nucleic acids, glycoside bonds of polysaccharides, and peptide bonds in proteins. In life today, the removal of water is performed upstream of the actual bond formation. This process involves the energetically downhill transfer of electrons, which is coupled to either substrate-level oxidation or generation of a proton gradient that in turn is the energy source for the synthesis of anhydride pyrophosphate bonds in ATP. The energy stored in the pyrophosphate bond is then distributed throughout the cell to drive most other energy-dependent reactions. This is a complex and highly evolved process, so here we consider simpler ways in which energy could have been captured from the environment and made available for primitive versions of metabolism and polymer synthesis. Because the atmosphere of the primitive Earth did not contain appreciable oxygen, this review of primitive bioenergetics is limited to anaerobic sources of energy.

BACKGROUND

Fundamental Considerations from Thermodynamics and Kinetics

In general, life is characterized by the fact that the catalytic and genetic polymers exist in a steady state far from equilibrium. It follows that the origin of life can be understood in terms of a process in which the flow of energy through relatively simple systems of molecules produced a more complex set of polymeric molecules that had specific physical and chemical properties. The origin of life occurred when a subset of these molecules was captured in a compartment and could interact with one another to produce the properties we associate with the living state.

Energy flow, and the changes it produces, are described by the fundamental laws of thermodynamics and kinetics. These concepts are familiar to most readers, but it is less obvious how they can be applied to our understanding of the prebiotic environment and the increase in chemical complexity driven by energy flux. We will briefly recapitulate them here in relation to the origin of life.

1. The amount of energy released as a reaction proceeds toward equilibrium and is referred to as free energy, which has components of enthalpy and entropy. Both must be taken into account to understand how systems of molecules can become more complex. On the prebiotic Earth, immense numbers and varieties of chemical reactions were taking place because the Earth itself was in a state of thermodynamic disequilibrium. To understand the origin of life, it is essential to sort out which of the many energy sources were primary factors in driving the increasing complexity from which life emerged.

2. All reactions in principle are reversible and can approach equilibrium from either direction. This means that a potentially destructive reaction such as hydrolysis of polymers can proceed toward polymer synthesis if there is a way to remove water molecules so that covalent bonds can form. Their enzyme-catalyzed biosynthesis requires an input of metabolic energy, primarily delivered as pyrophosphate bonds of ATP, but the linking bonds are also thermodynamically unstable, which means that enzymes can also catalyze spontaneous hydrolysis of the bonds. The result is that life incorporates a continuous and controlled synthesis and breakdown of its polymeric components. Similar reactions were presumably occurring in the prebiotic environment, so it is essential to establish plausible mechanisms by which polymers could be synthesized and accumulate despite hydrolytic back reactions.

3. Because the concentration of reactants strongly affects reaction rates, it seems likely that for a prebiotic reaction to proceed at a useful rate, there must have been concentrating processes such as evaporation or adsorption. A dilute solution of activated monomers is unlikely to form polymers because at low concentrations the rate of monomer condensation is less than the rate of their hydrolytic deactivation. However, adsorption of activated monomers on mineral surfaces could enhance their polymerization by increasing their concentration and possibly orienting them in a way that favors condensation.

4. The reactants in a given reaction must overcome an energy barrier called activation energy that limits the rate at which the reaction can occur. Elevated temperatures provide activation energy to a potentially reactive system of molecules. The global temperature when life began is estimated to be in the range of 55 and 85°C (Kauth and Lowe 2003) but there would be considerable variability around this mean, ranging from cold polar regions to extensive high temperature vulcanism. It is reasonable to consider that thermal activation energy was likely to be abundant, so that the major hurdles to be overcome in promoting polymerization reactions would be to concentrate, organize, and chemically activate monomers.

5. Catalysts reduce the activation energy barrier so that a reaction can proceed more rapidly toward equilibrium. There were no protein enzymes on the prebiotic Earth, so simpler

catalysts like mineral surfaces, metal ions, and small polymers presumably acted as catalysts in primitive metabolic pathways.

6. Chemical kinetics defines the rates at which a given reaction occurs, and allows thermodynamically unstable molecular structures to exist far from equilibrium. A protein or nucleic acid in water, for instance, will ultimately hydrolyze to its component amino acids. However, in the absence of a catalyst, this is a slow reaction, so that faster catalyzed reactions of biosynthesis can keep up with the slower degradative rate of hydrolysis. The difference in reaction rates is referred to as a kinetic trap. On the early Earth, if there was a relatively fast process that could produce chemical bonds between monomers, kinetic traps would allow the resulting polymers to have a transient existence even if they were thermodynamically unstable.

Overview of Contemporary Bioenergetics

The bioenergetic pathways of contemporary anaerobic life are well understood. As depicted in Figure 1, the light energy captured by photolithotrophs is used to activate and release electrons from inorganic donors like H_2S, S^o, or H_2, thereby producing the electrochemical energy that reduces carbon dioxide and yields the organic molecules required by life. Similarly, chemolithotrophs transfer high energy electrons from H_2 and H_2S to lower energy states in electron acceptors such as CO_2 and NO_3^-, and use the derived free energy to reduce carbon dioxide to organic molecules. The organic products of such reactions are subsequently used by heterotrophic life as electron donors in their energy metabolism based on either anaerobic respiration or fermentation.

In anaerobic photolithotrophy, chemolithotrophy, and respiration, the acquired electrochemical energy is used to pump protons across membranes in such a way that an electrochemical proton gradient is produced, equivalent to approximately 0.2 volts of electrical potential. This energy of the proton potential is coupled to ATP synthesis catalyzed by an ATP synthase embedded in the membrane. The pyrophosphate bond energy in the ATP is then transferred by diffusion to the rest of the cell where it drives a variety of essential metabolic reactions, motor molecule functions, ion

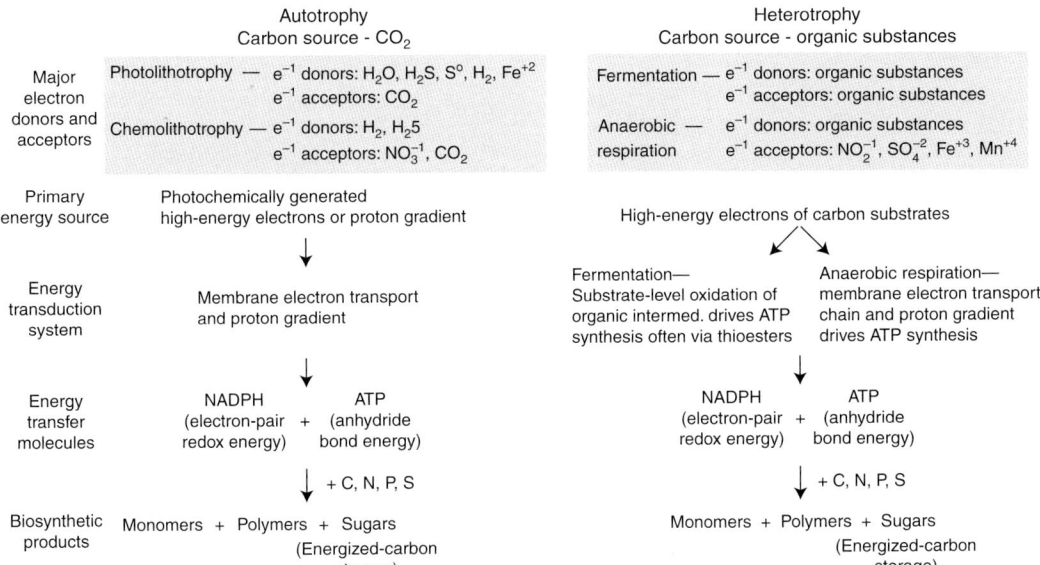

Figure 1. Energy sources and processing by anaerobic autotrophs and heterotrophs. See text for discussion.

transport processes, and polymerization reactions of biosynthesis.

In contrast, in fermentation, the electrochemical energy stored in organic substrates activates the transfer of high energy electrons from carbon groups (electron donors) to lower energy states provided by other carbon groups (electron acceptors). This substrate-level electron transfer oxidizes the electron donor to create an anhydride intermediate (usually a thioester) that in turn drives the synthesis of ATP (Gottschalk 1986)

The coupling of ATP synthesis to electron transfer and proton gradients is clearly a highly evolved system. The bioenergetic processes used by the first forms of cellular life must have been very different even though the same sources of energy were available, with the exception of electron transport to molecular oxygen. To understand the origin of life, we need to establish plausible sources of energy that would drive synthetic reactions.

Energy Sources on the Prebiotic Earth

Three kinds of energy are considered in this article. As shown in Figure 2, the first is the relatively high energy required to synthesize small molecules that have the potential to serve as monomers for polymerization reactions and feedstock for a primitive metabolism. These include photochemical energy available in ultraviolet light, atmospheric electric discharge, and geological electrochemical energy. The second is a series of relatively low energy reactions that incorporate condensation processes by which monomers are assembled into random polymers. These include anhydrous heat, mineral-catalyzed synthesis, and sugar-driven reactions. The third is the energy flow in metabolic networks in which an energy source in the local environment is captured and then transferred to standardized energy carriers (like ATP and NAD[P]H). These are catalytically coupled to a series of intracellular reactions that ultimately activate monomers required for catalyzed polymerization.

Table 1 shows the kinds and relative amounts of energy on today's Earth. It is a reasonable assumption that similar energy sources were available 4 billion years ago at the time of life's origin. Light energy is by far the most abundant, and in fact photochemistry drives virtually all life today. Could light have been a primary energy source for the first forms of life? This is an obvious possibility, yet there is a major conceptual problem: In modern life, capturing visible light requires a pigment system and a mechanism for transducing the energy content of photons into chemical energy to be used in metabolism, and there is as yet no plausible way to do this in a prebiotic scenario.

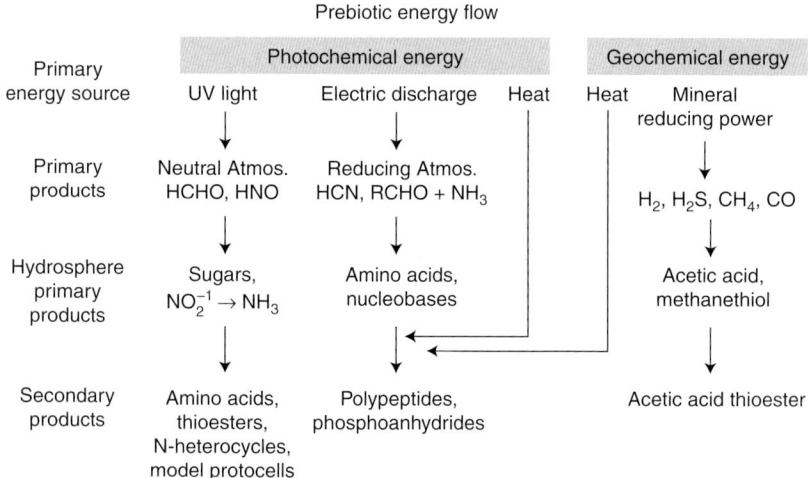

Figure 2. Possible pathways for synthetic prebiotic reactions. See text for discussion.

Table 1. Most of the energy flux on the early Earth was in the form of light energy from the sun, just as it is today.

Energy sources on the early Earth (kilojoules $m^{-2} yr^{-1}$)	
Solar radiation (UV<250 nM)	24,000
Shock waves from impacts	200
Radioactivity	117
Electrical discharges	2.9
Volcanoes	5.4
Chemical energy	Significant for the origin of life, not yet estimated.

The value shown in the table is the energy of ultraviolet radiation in a wavelength range that is absorbed by common organic compounds such as PAH, and therefore capable of activating photochemical reactions. Table adopted from estimates given by Chyba and Sagan (1992) and Miller and Urey (1959). There is considerable uncertainty in these estimates, which are presented only to illustrate order-of-magnitude approximations for energy flux in meter-scale localized environments.

However, ultraviolet light is clearly a potential source of high energy for small molecule synthesis, together with the energy released by shock waves generated by impacting comets and asteroid-sized objects (Chyba and Sagan 1992). The electrical discharge used in the original Miller-Urey experiments, and elevated temperature and pressures associated with geothermal environments (>200°C) can also do chemical work, but these sources represent a tiny fraction of the total energy flux. Electrical discharge is meant to simulate lightning in reducing gas mixtures, and Miller (1953) found that small amounts of highly reactive HCN and HCHO were produced in the discharge, which afterwards underwent Strecker reactions to produce several amino acids. This experiment revolutionized origins of life research, because for the first time, biologically relevant molecules were synthesized in a simulated prebiotic environment using a plausible source of energy.

Formaldehyde can also be synthesized by UV light interacting with water vapor and carbon dioxide in the upper atmosphere, and a plausible photochemical reaction was proposed by Pinto et al. (1980). From reasonable assumptions about UV flux and composition of the early atmosphere, these workers calculated that formaldehyde was added to the oceans at a rate of 10^{11} moles/year. It has long been known that formaldehyde (HCHO) in alkaline solutions readily reacts to form a variety of carbohydrates, as is discussed later. Cyanide (HCN) is another common product of UV and electrical energy impinging on mixtures containing nitrogen and gases such as CO and CO_2, and cyanide readily reacts with itself to produce other biologically relevant molecules. For instance, a few years after the Miller experiment was published, John Oro (1961) discovered that at high concentration, HCN could undergo pentamerization to form adenine.

Elevated pressures and temperatures associated with geothermal conditions can also promote significant chemical reactions. In particular, the Fischer-Tropsch (FT) reaction uses elevated temperatures to provide activation energy for a reaction in which carbon monoxide and hydrogen combine to produce long-chain hydrocarbons. Nooner et al. (1987) were among the first to test this reaction as a potential prebiotic source of hydrocarbon derivatives, using meteoritic iron as a catalyst, which lowers the activation energy of the reaction.

Although it is a reasonable assumption that the synthesis of organic compounds required for the origin of life was driven by energy available in the prebiotic environment, another source of potential chemical energy was available in organic compounds delivered by comets and meteoritic infall during late accretion. Chyba, Sagan, and co-workers (1990, 1992) estimated the total amount of organic carbon compounds that could be delivered in this way, and within an order of magnitude, it was in the range of the amounts estimated to be synthesized by Miller-Urey reactions under the most favorable conditions. The energy content would be present in the form of reduced carbon compounds that could undergo chemical modification if they were exposed to mineral surfaces in geological settings of appropriate fugacity (Shock 1990) (See also Pizzarello and Shock 2010).

This possibility is largely unexplored and is likely to be a fruitful direction for future research.

In summary, early investigations, beginning with Stanley Miller's experiments, showed that the energy available in electrical discharge, UV light, and elevated temperatures, when impinging on simple gas mixtures, can produce reactive compounds like HCN and HCHO, which in turn react to form amino acids, purine and pyrimidine bases, lipid-like amphiphiles, and carbohydrates. Assuming that inorganic phosphate was also available, all of the monomers required for the synthesis of major polymers of life would be available for the next stage of increasing complexity in the prebiotic envi-ronment.

Thermal Energy

Moderate heating at temperatures below 200°C is not a direct source of useful chemical energy. Heat speeds up the rate at which a reaction approaches chemical equilibrium, but does not change the position of equilibrium to favor one side of a reaction. However, heat can move the equilibrium of a reaction if it changes the concentration of the reactants by removing solvent, or by evaporating (volatilizing) a product of one side of a reaction, such as water from a condensation reaction. Under these conditions, net polymer synthesis becomes favorable because the monomer concentration is increased, and once the solvent water has evaporated, continued heating favors polymerization by removing the water produced by condensation reactions.

Early researchers investigating the origin of life attempted, with some success, to drive condensation reactions by drying potential reactants such as amino acids and nucleotides at elevated temperatures. The advantage of using anhydrous heating to drive condensation is that it is by far the simplest way to produce the polymers required for life to begin. Although the amino acid polymerization studied by Fox and colleagues is a prominent example (Fox and Harada 1958), there were also early attempts to drive phosphodiester bond formation under these conditions. For instance, Verlander and Orgel (1974) found that oligonucleotides up to six nucleotides long could be synthesized from cyclic 2′,3′-AMP when it was dried and heated. Usher (1977) suggested that the presence of a template should promote such reactions. Earlier McHale and Usher (1976) had shown that 6-mers of adenylic acid could form a phosphodiester bond to give a 12-mer, if a polyuridylic acid template was present.

Although anhydrous heat would appear to be a plausible source of energy to drive prebiotic condensation reactions, there are two problems with this approach. The first is obvious: In a dry state, potential reactants are trapped in a solid and diffusion is minimized, so that bond formation is limited to interactions between neighboring molecules. The second problem is that multiple reactions become activated as temperature increases, with the result that nonspecific chemical bonds begin to form, ultimately producing polymerized tars. But at temperatures between 60° and 80°C, it may be possible to organize reactants in such a way that bond formation is more specific. Cycling between hydrated and anhydrous conditions could then drive the synthesis of specific kinds of polymers. Furthermore, if the dry phase of a condensation reaction is cycled repeatedly through a hydrated phase, the reaction can lead to the accumulation of more complex products, as is discussed in the next section.

It seems reasonable to expect that small amounts of organic compounds would occur in the prebiotic environment, but they would be present as complex mixtures with inorganic anions and cations. The prebiotic environment was likely to have thousands of different organic compounds dissolved in a global ocean and lacustrine bodies of fresh water that accumulated on volcanic land masses rising above the ocean. The mixture would contain biologically relevant compounds such as amino acids and sugars, largely as racemic mixtures of their D and L enantiomers. Because the organic compounds would be present as very dilute solutions, it is essential to discover processes by which such compounds could be concentrated and organized to participate in intermolecular reactions such as polymer synthesis and protocell assembly needed for the origin of life.

There are two obvious ways in which dilute solutions can be concentrated. The simplest is the input of heat energy to evaporate the solution onto mineral surfaces. For instance, evaporating a volume of water containing one micromole of an organic solute such as an amino acid evenly spread onto a mineral surface area of 100 cm^2 will form a film approximately 10 molecules thick. During the last stages of evaporation, the concentration of solutes pass through molar concentrations, thereby promoting any chemical reactions that could not occur spontaneously in dilute solutions. Subsequent wetting cycles would release the products into the environment for further processing.

The second process, originally proposed by Bernal (1951), is that clay surfaces strongly adsorb organic compounds from dilute solutions and thereby concentrate them. Furthermore, the orderly arrangement of charged groups in the crystal structure of the clay can impose order on the adsorbed solutes and thereby promote polymerization reactions. A good example is the polymerization of activated nucleotides into short molecules of RNA, to be discussed later in this article (for review, see Ferris 1999, 2002).

It is interesting to note that simple freezing is also a process that concentrates potential reactants in an otherwise dilute solution. As a solution freezes, the microscopic crystals that initially form are nearly pure ice, so that solutes are concentrated into fluid eutectic phases between the crystals. Freezing has been used to promote nucleobase synthesis in frozen cyanide solutions (Miyakawa et al. 2002) and Kanavarioti et al. (2001) found that oligomers of RNA were synthesized from activated nucleotides frozen at $-18°C$. (See Orgel 2004 for review.) Although extensive ice on a hot early Earth seems unlikely, there is still uncertainty about actual temperatures available in specific sites, so ice eutectics cannot be dismissed as a means to concentrate dilute reactants in solution.

Thermal Energy Storage in Pyrophosphate Bonds

Most chemical energy in cells today circulates in the cytoplasm as ATP. Although cells use ATP to drive synthetic reactions, ATP is not a primary energy source, but rather is an energy transfer molecule that picks up energy from an energy source and then delivers it to energy-requiring reactions. This constant resynthesis (cycling) of ATP inside the cell is revealed by estimates showing that to synthesize 1 g of cell mass requires the energy of about 20–100 g of ATP (Stouthamer 1977). The chemical energy content of ATP is present in the pyrophosphate bonds that link the second and third phosphate groups of ATP. These are anhydride bonds, and their chemical energy is released by energetically downhill group transfer reactions of the phosphate group to other molecules, an activating process called phosphorylation. The second molecule gains chemical energy and can in turn undergo reactions that otherwise will not occur. Classic examples include the formation of aminoacyl-tRNA in protein synthesis, or acetyl-CoA in fatty-acid synthesis.

The question is whether pyrophosphate bond energy could have been a significant source of chemical energy in the reactions leading to the origin of life. In fact, phosphate is such an integral part of all contemporary life that phosphorylation reactions must have been incorporated in primitive metabolic pathways. Baltscheffsky (1996) has argued that this is plausible, in part because pyrophosphate and polyphosphates are readily produced simply by heating inorganic phosphate under anhydrous conditions, a process known to occur under natural conditions. Pyrophosphate-containing minerals, canaphite and wooldridgeite, have been discovered in quarries, albeit in minute quantities and in the form of microscopic crystals. Furthermore, Baltscheffsky found that the coupling membranes of a photosynthetic bacterial species—*Rhodospirillum rubrum*—synthesize pyrophosphate instead of ATP. The *R. rubrum* use the pyrophosphate as an energy source, much as other organisms use ATP.

Despite the ease of capturing thermal dehydration energy as pyrophosphate bonds, a plausible pathway for incorporating it as an energy source in early life has not yet been established. This is well worth further study, as is discussed in the last section of this article.

Sunlight and Photochemical Energy

Light only becomes a useful energy source if there is a pigment to absorb the photons. When photons are absorbed by a pigment molecule, they interact with the electronic structure of the molecule and add energy to the bonds to produce an excited state. This seemingly simple photochemical reaction is the energetic foundation of most life on the Earth, because this is what happens when a chlorophyll molecule absorbs light. Starting with chlorophyll in its ground state, a photon of red light is absorbed and increases the energy content of chlorophyll. The added energy causes it to go to an excited state that then donates an electron through multiple reactions to carbon dioxide, which ends up after further reactions as a carbohydrate. In addition, the transfer of the electron to lower energy states is coupled to the generation of a proton gradient across the membrane that is used to drive ATP synthesis. In this way, the original light energy is conserved in the form of chemical energy. After it loses the electron, chlorophyll is positively charged and the electron is replaced from a water molecule in the "water splitting reaction," which releases oxygen. This is the source of virtually all of the oxygen in the Earth's atmosphere.

If photochemistry is essential for life today, could it also have provided energy for the origin of life? Perhaps, but there is a problem: What pigment molecules were available? Certainly not chlorophyll, which is a very complex molecule requiring multiple enzymatic steps for its synthesis. Furthermore, even if primitive pigments were present, the capture of light energy would not be possible unless there were membranes available that could contain the pigment molecules. To process the absorbed light energy, a primitive cell would also require a system of electron transport molecules to transduce the light energy into chemical energy. Future research might someday discover a mechanism by which this seemingly complex series of reactions could arise, but until then it seems plausible that an energy source other than light was used by the first living cells. In other words, the first life was likely to be heterotrophic.

Geological Electrochemical Energy

The electrochemical reactions used by aerobic life today require oxygen produced by plants. In mitochondria and aerobic bacteria, an electron transfer from reduced substrates to molecular oxygen produces 0.2 volts and enough protonic current to synthesize ATP. It is improbable that a complete chemiosmotic system was available to the first forms of life, but simpler electrochemical reactions are still conceivable mechanisms in which donors and acceptors of electrons provide an energy source. Several potential donors were likely to be available in the prebiotic environment. Perhaps the most plausible is hydrogen gas itself, as well as hydrogen sulfide (H_2S) and methane. A variety of microorganisms today use these gases as a source of electrons, a good example being the abundant bacteria present in hydrothermal vents. There is a consensus that little or no oxygen was present in the early atmosphere, so what electron acceptors might have served instead of oxygen? A useful list of potential anaerobic donors and acceptors was presented by Gaidos et al. (1999), who considered the interesting question of energy sources that might be available in a putative ocean beneath Europa's icy surface. Terrestrial anaerobes use sulfate, nitrate, iron (III), and manganese (IV) as electron acceptors, with CO, nitrite, and hydrogen sulfide as electron donors. Sulfate would be abundant on the early Earth, as it is today. Nitrate and nitrite could have been made on the prebiotic Earth from nitrogen gas by photochemical synthesis and aqueous processes (Summers and Khare 2007) and in high pressure geochemical processes (Brandes et al. 1998). In addition, Cody et al. (2000, 2004) have shown how sulfide minerals could have been involved in primitive metabolic pathways.

Energy and Sulfur Chemistry

Sulfur chemistry is of particular relevance to the origin of life, because the minimal amount of molecular oxygen in the prebiotic environment would allow sulfur to be abundant as an element, as a reduced gas (hydrogen sulfide), and as iron sulfide minerals. Taking into consideration this prebiotic context, Wächtershäuser (1988) proposed that life did not arise by assembly of pre-existing organic compounds, but instead as two-dimensional synthetic chemistry on an iron sulfide mineral surface called pyrite. According to this idea, when hydrogen sulfide reacts with iron in solution to produce insoluble iron sulfide mineral, the reaction generates electrons that can be donated to the bound compounds and thereby drive a series of energetically uphill chemical reactions that otherwise cannot take place in solution. Wächtershäuser sees these reactions as the beginning metabolism, which occurred on a mineral surface rather than in the volume of a cell. To test this idea, Huber and Wächtershäuser (1997) heated a dispersion of iron and nickel sulfides together with a source of carbon and analyzed the products. There was no evidence of a long chain of integrated reactions, but using CO and H_2S, acetic acid was produced, and from CO and CH_3SH the thioester CH_3-CO-SCH_3 was synthesized. In later work in which amino acids were added to the simulation (Huber and Wächtershäuser 1998; Huber et al. 2003), the amino acids were observed to be chemically activated and formed peptide bonds.

DeDuve (1991) has proposed another version of sulfur chemistry as a source of chemical energy related to the origin of life, which involves thioesters that could have served to activate energy-dependent steps in primitive metabolic pathways before the advent of ATP. Thioesters are synthesized when a water molecule is lost during the reaction of an organic acid and a thiol: R-COOH+R'-SH –> R-CO-SR'. Unlike the relatively low-energy content of an ester bond, the thioester bond has an energy content equivalent to the pyrophosphate bond of ATP. Earlier Weber (1984, 1998) had shown that α-hydroxy acid and α-amino acid thioesters could be synthesized by reacting sugars (or sugar precursors) with ammonia and a thiol under plausible prebiotic conditions.

Several investigators have incorporated sulfur chemistry in their research. Wieland (1953) originally showed that amino acid thioesters spontaneously form peptide bonds in aqueous solution, and Maurel and Orgel (2000) showed the elongation of a decapeptide of oligoglutamic acid up to 15 mers in the presence of thioglutamic acid. Zepik et al. (2006) extended this reaction to protocells by encapsulating decaglutamic acid in liposomes composed of dimyristoylphosphatidylcholine, and found that externally added thioglutamic acid was able to permeate the vesicle membrane at a rate sufficient to support elongation of the decapeptide within the vesicles. It is clear that sulfur chemistry has considerable potential for expanding our understanding of early bioenergetic processes.

Chemiosmotic Energy Conversion to Anhydride Energy

Life today uses either chemiosmosis or substrate-level phosphorylation to convert electrochemical energy to anhydride chemical energy of ATP (Mitchell 1961, 1966). Was chemiosmosis an ancient process used by the first cells, or a later discovery? Arthur Koch (1985) first addressed this question and speculated that chemiosmosis may have arisen in the first forms of primitive cellular life, and this conjecture is worth considering here. Three relevant postulates of chemiosmosis are listed below:

- Coupling membranes maintain a proton gradient, and the gradient must be of sufficient magnitude to drive ATP synthesis.

- Coupling membranes pump protons by electron transport using either light or electron-transfer as an energy source.

- Coupling membranes contain an ATPase that is also a proton pump.

The lipid composition of biological membranes today contains specific lipids that are products of highly evolved metabolic pathways. A stable hydrophobic lipid bilayer barrier composed of hydrocarbon chains 16–18 carbons in length is required to maintain ionic concentration gradients, particularly protons in the case of coupling membranes. Lipid bilayers are notoriously leaky to protons (Nichols and Deamer 1980; Paula et al. 1996), so the first self-assembled membranes composed of relatively short-chain amphiphiles would be unable to maintain significant proton gradients, as recently shown by Chen and Szostak (2004). Mansy (2010) discusses the role of membrane permeability in primitive cellular systems. Even if a plausible primitive barrier membrane can be discovered, an electron transport system and ATP synthase would need to be incorporated in the bilayer for chemiosmosis to be a source of chemical energy. In our judgment, the complex requirements for a functioning chemiosmotic system weighs against the proposition that primitive life used chemiosmosis for converting electrochemical energy to the anhydride energy of ATP.

Chemical Energy of Organic Substrates: Carbohydrates

The environment of the prebiotic Earth was far from equilibrium, so that a variety of chemical reactions were occurring simultaneously. The problem is to gain some understanding of which of these was relevant to the origin of life, and how they were incorporated. Living systems today use chemical reactions to release energy in small steps called metabolism, which can be defined as a series of chemical reactions linked in a molecular system that provides energy and small molecules required for growth. Each step is catalyzed by a specific enzyme, and the reaction rates are controlled by feedback loops in which a product is an allosteric inhibitor of the enzyme to be regulated. If the first life was heterotrophic, what nutrients might have been available as a source of chemical energy?

Of all the organic substrates, sugars are by far the most attractive organic energy substrate of primitive anaerobic life, because they are able to provide all the energy and carbon needed for the growth and maintenance of a fermentative metabolism. In fact, the sugars that are the first substrates of the glycolytic pathway can be considered to be optimal biosynthetic substrates because they contain mainly alcohol groups that have maximum self-transformation energy, and a single carbonyl group (aldehyde or ketone) that makes them reactive and able to form covalent adducts to enzyme active sites (Weber 2004). Moreover, in fermentation, the energy content of sugars is converted to the anhydride energy of ATP by substrate-level oxidation phosphorylation, a process that does not require the organized membrane structures of phosphorylation coupled to electron transfer. As discussed later, the energy content and reactivity of sugars also allows them to act as substrates for chemically spontaneous synthetic processes that yield many of the molecular products required for the origin of life. Such sugar-driven syntheses require no external source of chemical energy (Weber 2000).

RECENT EXPERIMENTAL STUDIES

Sugar-driven Prebiotic Synthesis

In addition to being the sole energy and carbon source of fermentative organisms today, sugars have chemical properties that make them very attractive substrates for synthetic processes needed for the origin of life. First, sugars can be synthesized under plausible prebiotic conditions from formaldehyde and glycolaldehyde by the formose reaction (Schwartz and de Graaf 1993; See also Benner et al. 2010). Second, sugars are reactive and contain considerable self-transformation energy, properties that allow them to react with ammonia, yielding many types of molecules needed for the origin of life. These sugar-driven syntheses require no additional source of chemical energy (Weber 2000).

The synthetic versatility of sugars is shown by their spontaneous reactions in the presence of ammonia that yield catalytic amines, biomonomers (amino acids), metabolites (pyruvate, glycolate), energy molecules (hydroxy and amino acid thioesters), alternative nucleobases (2-pyrazinones that resemble uracil), heterocyclic molecules (furans, pyrroles, imidazoles, pyridines, and pyrazines), polymers (polypyrroles and polyfurans), and cell-like organic microspherules (Weber 2001-2008, refs. therein). Sugars have also been shown to drive the prebiotic synthesis of ammonia from nitrite. Remarkably, these prebiotic synthetic processes based on sugar chemistry can evolve directly into modern sugar-driven biosynthesis without violating the principle of evolutionary continuity.

Finally, sugar synthesis from formaldehyde and glycolaldehyde, and the subsequent conversion of sugar products to carbonyl-containing products can be catalyzed by small molecules (ammonia and amines including amino acids and peptides). In fact, small L-dipeptides (the isomer found in proteins) stereoselectively catalyzed the formation of D-ribose (Pizzarello and Weber 2010). These ammonia and amine-catalyzed reactions yielded aldotriose (glyceraldehyde), ketotriose (dihydroxyacetone), aldotetroses (erythrose and threose), ketotetrose (erythrulose), pyruvaldehyde, acetaldehyde, glyoxal, pyruvate, glyoxylate, and several unidentified carbonyl products. The uncatalyzed control reaction yielded no pyruvate or glyoxylate, and only trace amounts of pyruvaldehyde, acetaldehyde, and glyoxal. With L-alanine, the rates of triose and pyruvaldehyde synthesis were about 15-times and 1200-times faster, respectively, than the uncatalyzed reaction (Weber 2001). Because amines are also products of sugar–ammonia reactions, these studies suggested that the sugar–ammonia reaction could be autocatalytic. This possibility was tested in a later study, which showed that reaction of the triose sugar (glyceraldehyde) with ammonia yielded a crude product mixture capable of catalyzing a 10-fold acceleration of the same sugar–ammonia reaction that produced the catalytic products (Weber 2007).

Amphiphile Synthesis Using Geothermal Energy

Franz Fischer and Hans Tropsch discovered in the 1920s that a gaseous mixture of carbon monoxide and hydrogen, when passed over a hot iron catalyst, produced excellent yields of hydrocarbons. Oro and coworkers (Nooner et al. 1987) first showed that the Fischer-Tropsch type synthesis (FTT) also worked with meteoritic iron-nickel as a catalyst, and proposed that long-chain fatty acids may have been produced this way on the early Earth. McCollom et al. (1999) and Rushdi et al. (2001) found that the FTT reaction also worked simply by treating oxalic acid to elevated temperatures $150°–250°C$ and corresponding pressures in a stainless steel chamber, producing a mixture of fatty acids and alcohols as products. At elevated temperatures, oxalic breaks down into a mixture of CO, CO_2, and H_2, the reactive gases required for FTT synthesis. In a later paper, Simoneit et al. (2006) found that if stoichiometric glycerol was present in the mixture along with a variety of fatty acids, the same conditions produced good yields of monoglycerides.

These results are significant because the majority of membrane-forming lipids today are glycerol esters. Furthermore, monoglycerides by themselves can assemble into lipid bilayers, and when mixed with fatty acids are able to form stable membranes (Monnard et al. 2002; Mansy and Szostak 2008). Future research in this area should be directed toward characterizing such membranes in terms of long-term stability and capacity for maintaining ionic concentration gradients of protons and other cations.

Phosphodiester Bond Synthesis Driven by Anhydrous Cycles

Clay mineral is a common example of an organizing surface. Clays have a surprisingly large crystalline surface area, over 100 m^2/g, and have a multilamellar structure that seems likely to adsorb, concentrate, and organize potential monomers on and between the lamellae. Ferris and coworkers (1996) have made extensive studies of montmorillonite clay as an organizing agent, and established that activated monomers of mononucleotides can in fact polymerize on clay surfaces into RNA-like polymers up to 50 nucleotides in length containing both $3'–5'$ and $2'–5'$ phosphodiester bonds. However, the resulting polymers are tightly bound to the clay surfaces, and if they are to participate in the origin of cellular life, the RNA products must somehow be associated with membranous vesicles to form a protocellular system. Significantly, Hanczyc et al. (2003) found that clay particles with bound RNA were readily encapsulated in fatty-acid vesicles.

It is not generally realized that lipids also form multilamellar structures that have the potential to organize and concentrate monomers. The most common image of lipids is in the form of microscopic vesicles bounded by a lipid bilayer membrane, usually referred to as liposomes. However, in the dry state, lipids are present as multilamellar matrices consisting of stacked bilayers, or less often as hexagonal phases in which the lipids form indefinitely long tubes that tend to have a hexagonal packing, rather than lamellar (Reiss-Husson and Luzzati 1967; Deamer et al. 1970). Furthermore, if lipid vesicles undergo drying, the bilayers fuse into the multilamellar phase, and any solutes present are trapped and concentrated between the lipid head groups of bilayers. Unlike the solid matrix of clay surfaces, the bilayers are liquid crystals, which means that trapped solutes have diffusional mobility.

Rajamani et al. (2008) investigated the possibility that the order imposed on mononucleotides by multilamellar lipid matrices could promote polymerization. The nucleotides used were ordinary 5′-ribonucleotides that had not been chemically activated. Instead, the chemical potential for ester bond formation was provided by cycling through anhydrous conditions at moderately elevated temperatures in the range of 60°–90°C. Thermal cycling of the mononucleotides in mass ratios from 2:1 to 1:2 with the phospholipids was shown to yield relatively long chains of linear RNA-like polymers, ranging from 20 to 100 nucleotides in length. The linearity was determined by nanopore analysis of the products, and the RNA-like character was established by ^{32}P-end labeling with T7 RNA kinase. A variety of phospholipids could promote the polymerization, but in the absence of lipids, only short nucleotide oligomers were detected.

These results show that there is sufficient chemical potential available in anhydrous conditions to drive phosphate ester synthesis as long as the nucleotide monomers are concentrated and organized within a fluid liquid crystalline matrix that permits diffusional mobility. Furthermore, at the end of the reaction when the lipid matrix is rehydrated, the products of the reaction are encapsulated in lipid vesicles. This appears to be a plausible pathway to protocells, which could be generated at the edges of volcanic geothermal pools where wet–dry cycles would be common.

Carbonyl Sulfide as a Plausible Prebiotic Condensing Agent

At some point, polymerization reactions must have evolved from simple processes driven by an input of physical energy to a more complex mechanism involving primitive versions of activation occurring in an aqueous environment. For many years, research has focused on discovering a plausible condensing agent that can perform this feat, and recent discoveries suggest that carbonyl sulfide is a likely candidate. Carbonyl sulfide (COS) is a reactive compound that has been detected in volcanic gas and mineral ash (Rasmussen et al. 1982), along with its chemical relatives, carbon disulfide and carbon dioxide. Leman et al. (2004) found that if COS is present in an aqueous solution of amino acids, di- and tripeptides are synthesized with yields up to 80%. In a second paper, Leman et al. (2006) reported that amino-acyl phosphate anhydrides up to 30% yields were synthesized in mixtures of amino acids, phosphate, and COS.

These results offer a strong clue to the manner in which phosphate was initially incorporated into primitive metabolic pathways, particularly those leading to peptide bond formation and the synthesis of small oligopeptides that may have served as catalysts and structural components of early life.

Novel Prebiotic Synthesis of Nucleotides

The nucleotide monomers of nucleic acids consist of a purine or pyrimidine linked through a nitrogen–carbon bond to a pentose sugar, which in turn is phosphorylated through an ester bond on the 5′ hydroxyl group. Each of the component molecules of nucleotides was presumably present in organic mixtures on the prebiotic Earth, but must be linked into the more complex molecular structure of nucleotides before they can be incorporated into nucleic acid polymers. One might imagine that phosphorylation of a ribose sugar could occur, because ester bonds are not difficult to synthesize by simple condensation reactions, but synthesis of the C-N bond between a sugar and a nucleobase has been much more challenging.

A recent paper by Powner et al. (2009) (see also Sutherland 2010) reported a remarkably efficient series of reactions that leads to mononucleotide synthesis. Instead of attempting to produce C-N bonds between an existing pyrimidine such as cytosine and ribose, the reaction uses sequential additions of cyanamide, cyanoacetylene, glycolaldehyde, and glyceraldehyde, all in the presence of phosphate. Under these conditions, spontaneous reactions first form arabinose aminooxazoline and anhydronucleoside intermediates, which then add phosphate and condense into cytosine monophosphate. Although it is unlikely that such a complex mixture of reactants and phosphate might occur in the prebiotic environment, the fact that the reaction can occur at all in aqueous solution, using only the chemical energy of the reactants, opens a new direction for future investigations that may reveal a simpler process.

CHALLENGES AND FUTURE RESEARCH DIRECTIONS

Identifying a Source of Condensation Energy

In this article, we briefly touched on processes by which certain kinds of energy could have driven chemical reactions related to the origin of life. There is a consensus that electrical discharge and ultraviolet light could drive the synthesis of reactive molecules like cyanide and formaldehyde, which in turn would react to produce more complex molecules that could serve as potential monomers for primitive forms of life. However, a plausible energy source for polymerization remains an open question. Condensation reactions driven by cycles of anhydrous conditions and hydration would seem to be one obvious possibility, but seem limited by the lack of specificity of the chemical bonds that are formed. On the other hand, there may be conditions yet to be discovered that organize monomers in such a way that the formation of specific chemical bonds is promoted. The organizing effect of clay mineral surfaces is one well-known example, and incorporation of monomers into orderly lipid matrices also promotes specific polymerization reactions. Future studies of conditions that can add order to reactive monomers are likely to reveal new clues to plausible polymerization processes that could occur in the prebiotic environment.

An alternative to anhydrous heat is a condensing agent such as carbonyl sulfide, which can activate peptide bond and phosphanhydride bond synthesis in aqueous solutions. This discovery certainly deserves further attention. So far, only short oligomers have been produced using COS and dilute monomers, but conditions that can concentrate and organize the reactants are likely to produce much longer polymers. Related to carbonyl sulfide as a condensation agent is the energy available in thioester bonds (DeDuve 2005). Although molecular oxygen was virtually absent from the prebiotic atmosphere, its neighbor in the periodic table—sulfur—was abundant, both as an element and in common gases such as H_2S and COS. Therefore, it seems probable that sulfur was incorporated into a variety of organic molecules, and further studies of the thioester bond as a plausible intermediate in primitive metabolic pathways should be fruitful.

A third source of energy for polymerization processes is the chemical potential of simple carbohydrates. Sugars, when heated near $70°C$ in the presence of ammonia or amines, are known to produce polymers containing furan and pyrrole residues, and cell-like microspherules (Weber 2005, refs. therein). Sugars also drive the synthesis of α-hydroxyacid thioesters and α-amino acid thioesters that are known to oligomerize forming polyesters and peptides, respectively (Weber 1998, refs. therein). In addition, in his description of the sugar-driven synthesis of uracil-like 2-pyrazinone, Weber (2008) proposed that sugars also have the potential to form pyrazinone monomers with α-hydroxycarbonyl side-chain groups (CH_2OH-CO-CH_2-pyrazinone) that could spontaneously polymerize to give oligomers joined by enol ether linkages or dehydrated aldol linkages. In contrast to other prebiotic syntheses, the previously cited sugar-driven processes are "one-pot" reactions that yield reactive monomers capable of spontaneous oligomerization without being coupled to an additional source of chemical energy, like ATP.

Phosphate Reactions

There are good reasons why phosphate is central to energy metabolism today. These were described by Westheimer (1987) in an excellent review that should be required reading for students interested in the origins of life. An important, still unanswered question concerns how phosphate might have first become involved in life processes. Phosphate today is mostly present in the form of a mineral called apatite, the same combination of calcium and phosphate that composes tooth enamel and bones. Apatite has a very low solubility, so what was the original source of a soluble form of phosphate? There are no convincing explanations yet, but it is possible that life began in a low pH environment similar to that of volcanic hydrothermal springs. Calcium phosphate readily dissolves at acidic pH ranges to release phosphate anions. The presence of free phosphate in solution may have permitted incorporation into organic compounds as phosphate esters, followed by a second set of chemical reactions that initiated primitive metabolic pathways involving phosphate. For instance, Prabahar and Ferris (1997) reported that adenine itself is able to activate phosphate under certain conditions. Identifying a plausible mechanism for prebiotic phosphorylation represents an important problem for future research.

Pigments and Photosynthesis

Sunlight drives virtually the entire biosphere today, but when and how did life first begin to capture light as an energy source? Although today's photosynthetic process seems much too complex to have been a source of energy for early life, there must have been some sort of pigment available that could begin capturing light energy in a useful way and initiate the evolutionary path toward modern photosynthesis. Pteridines are examples of relatively simple pigment molecules that can be generated from amino acids exposed to anhydrous heat. This reaction and potential roles of pteridines in the origin of life have been extensively explored by Kritsky and coworkers and deserve further study (for review, see Kritsky and Telegina 2004).

An alternative scenario is that pigments existed in the environment that partitioned into lipid bilayers of membranes in such a way that light energy could be transduced into chemical energy. Were organic molecules present in the prebiotic environment that might serve as membrane-bound pigments? Polycyclic aromatic hydrocarbons (PAH) are abundant forms of organic carbon, and most PAH species absorb light in the near UV and blue region of the spectrum. After accepting photons, the excited states can act as reducing agents and release protons, thereby generating chemiosmotic pH gradients (for review, see Deamer et al. 1994). Another interesting photochemical reaction of PAH derivatives is photocarboxylation. For instance, when exposed to near UV light, phenanthrene reacts directly with carbon dioxide to produce phenanthrene carboxylic acid, probably the simplest possible example of carbon dioxide fixation (Tazuke et al. 1986; See Deamer 1997 for review). These properties of PAH compounds have yet to be explored thoroughly and surely represent a fruitful area for future research.

REFERENCES

Baltscheffsky H. 1996. *Origin and evolution of biological energy conversion*. New York: Wiley VCH.

Benner SA, Kim H-J, Kim M-J, Ricardo A. 2010. Planetary organic chemistry and the origins of Biomolecules. *Cold Spring Harb Perspect Biol* **2:** a003467.

Bernal JD. 1951. *The physical basis of life*. London: Routledge and Kegan Paul.

Brandes JA, Boctor NZ, Cody GD, Cooper BA, Hazen RM, Yoder HS. 1998. Abiotic nitrogen reduction on the early Earth. *Nature* **395:** 365–336.

Chen IA, Szostak JW. 2004. A kinetic study of the growth of fatty acid vesicles. *Biophys J* **87:** 988–998.

Chyba CF, Sagan C. 1992. Endogenous production, exogenous delivery and impact-shock synthesis of organic molecules: An inventory for the origin of life. *Nature* **355:** 125–113.

Chyba CF, Thomas PJ, Brookshaw, Sagan C. 1990. Cometary delivery of organic molecules to the early Earth. *Science* **249:** 366–373.

Cody GD. 2004. Transition metal sulfides and the origin of metabolism. *Ann Rev Earth Planetary Sci* **32:** 569–599.

Cody GD, Boctor NZ, Filley TR, Hazen RM, Scott JH, Sharma A, Yoder HS. 2000. Primordial carbonylated iron-sulfur compounds and synthesis of pyruvate. *Science* **289:** 1337–1340.

Deamer DW. 1991. Polycyclic aromatic hydrocarbons: Primitive pigment systems in the prebiotic environment. *Adv Space Res* **12:** 183–189.

Deamer DW. 1997. The first living systems: A bioenergetic perspective. *Microbiol Mol Biol Rev.* **61:** 239–262.

Deamer DW, Harang-Mahon E, Bosco G. 1994. Self-assembly and function of primitive membrane structures. In: *Early life on Earth.* Nobel Symposium No. 84. Bengston S. ed.

Deamer DW, Leonard R, Tardieu A, Branton D. 1970. Lamellar and hexagonal lipid phases visualized by freeze-etching. *Biochim Biophys Acta* **219:** 47–60.

DeDuve C. 1991. *Blueprint for a cell: The nature and origin of life.* New York: Neil Patterson Publishers.

DeDuve C. 2005. *Singularities: Landmarks on the pathway of life.* Cambridge University Press.

Ferris JP. 1999. Prebiotic synthesis on minerals: Bridging the prebiotic and RNA worlds. *Biol Bull* **196:** 311–314.

Ferris J. 2002. Montmorillonite catalysis of 30–50 mer oligonucleotides: Laboratory demonstration of potential steps in the origin of the RNA world. *Orig Life Evol Biospheres* **32:** 311–332.

Ferris JP, Hill AR, Liu R, Orgel LE. 1996. Synthesis of long prebiotic oligomers on mineral surfaces. *Nature* **381:** 59.

Fox SW, Harada K. 1958. Thermal copolymerization of amino acids to a product resembling protein. *Science* **128:** 1214.

Gottschalk G. 1986. *Bacterial metabolism.* New York: Springer-Verlag, New York, pp. 1–11, 210–317.

Gaidos EJ, Nealson KH, Kirschvink JL. 1999. Life in ice-covered oceans. *Science* **284:** 1631–1633.

Hanczyc MM, Fujikawa SM, Szostak JW. 2003. Experimental models of primitive cellular compartments: Encapsulation, growth, and division. *Science* **302:** 618–622.

Hargreaves WR, Deamer DW. 1978. Liposomes from ionic, single-chain amphiphiles. *Biochemistry* **17:** 3759–3768.

Huber C, Wächtershäuser G. 1997. Activated acetic acid by carbon fixation on (Fe,Ni)S under primordial conditions. *Science* **276:** 245–247.

Huber C, Wächtershäuser G. 1998. Peptides by activation of amino acids with CO on (Ni,Fe)S surfaces: Implications for the origin of life. *Science* **281:** 670–672.

Huber C, Eisenreich W, Hecht S, Wächtershäuser G. 2003. A possible primordial peptide cycle. *Science* **301:** 938–940.

Kanavarioti A, Monnard P-A, Deamer DW. 2001. Eutectic phases in ice facilitate nonenzymatic nucleic acid synthesis. *Astrobiology* **1:** 271–281.

Knauth LP, Lowe DR. 2003. High Archean climatic temperature inferred from oxygen isotope geochemistry of cherts in the 3.5 Ga Swaziland Supergroup, South Africa. *GSA Bulletin* **115:** 566–580.

Koch A. 1985. Primeval cells: Possible energy-generating and cell-division mechanisms. *J Mol Evol* **21:** 270–77.

Koch AL, Schmidt TM. 1991. The first cellular bioenergetic process: Primitive generation of a proton motive force. *J Mol Evo.* **33:** 297–304.

Kritsky MS, Telegina TA. 2004. Role of nucleotide-like coenzymes in primitive evolution. In *Cellular origin, life in extreme environments and astrobiology.* J. Seckbach, Ed. Springer Netherlands.

Leman L, Orgel L, Ghadiri MR. 2004. Carbonyl sulfide-mediated prebiotic formation of peptides. *Science* **306:** 283–286.

Leman LJ, Orgel LE, Ghadiri MR. 2006. Amino acid dependent formation of phoshate anhydrides in water mediated by carbonyl sulfide. *J Am Chem Soc* **128:** 20–21.

Lohrmann R, Bridson PK, Orgel LE. 1980. Efficient metal-ion catalyzed template-directed oligonucleotide synthesis. *Science* **208:** 1464–1465.

Mansy SS. 2010. Membrane transport in primitive cells. *Cold Spring Harb Perspect Biol* **2:** a002188.

Mansy SS, Szostak JW. 2008. Thermostability of model protocell membranes. *Proc Natl Acad Sci* **105:** 13351–13355.

Maurel M-C, Orgel LE. 2000. Oligomerization of α-thioglutamic acid. *Orig Life Evol Biospheres* **30:** 423–430.

McCollom TM, Ritter G, Simoneit BRT. 1999. Lipid synthesis under hydrothermal conditions by Fischer-Tropsch-type reactions. *Orig Life Evol Biospheres* **29:** 153–166.

Miller SL. 1953. Production of amino acids under possible primitive Earth conditions. *Science* **117:** 528–529.

Miller AL, Urey HC. 1959. Organic compound synthesis on the primitive Earth. *Science* **130:** 245–51.

Mitchell P. 1961. Coupling of phosphorylation to electron and hydrogen transfer by a chemi-osmotic type of mechanism. *Nature* **191:** 144–148.

Mitchell P. 1966. Chemiosmotic coupling in oxidative and photosynthetic phosphorylation. *Biol Rev* **41:** 445–501.

Miyakawa S, Cleaves HJ, Miller SL. 2002. The cold origin of life: Implications based on pyrimidines and purines pruduced from frozen ammonium cyanide solutions. *Orig Life Evol Biosph* **32:** 209–218.

Monnard PA, Apel CL, Kanavarioti A, Deamer DW. 2002. Influence of ionic inorganic solutes on self-assembly and polymerization processes related to early forms of life: Implications for a prebiotic aqueous medium. *Astrobiology* **2:** 139–152.

Nichols JW, Deamer DW. 1980. Net proton-hydroxide permeability of large unilamellar liposomes measured by an acid-base titration technique. *Proc Natl Acad Sci* **77:** 2038–2042.

Nooner DW, Oro J. 1987. Synthesis of fatty acids by a closed system Fischer-Tropsch process. *Adv Chem* **178:** 159–171.

Orgel L. 2004. Prebiotic adenine revisited: Eutectics and photochemistry. *Orig Life Evol Biospheres* **34:** 361–369.

Oro J. 1961. Mechanism of synthesis of adenine from hydrogen cyanide under possible primitive Earth conditions. *Nature* **191:** 1193–1194.

Paula S, Volkov AG, Van Hoek AN, Haines TH, Deamer DW. 1996. Permeation of protons, potassium ions, and small polar molecules through phospholipid bilayers as a function of membrane thickness. *Biophys J* **70:** 339–348.

Pinto JP, Gladstone GR, Yung YL. 1980. Photochemical production of formaldehyde in Earth's primitive atmosphere. *Science* **210:** 183–185.

Pizzarello S, Shock E. 2010. The organic composition of carbonaceous meteorites: The evolutionary story ahead of biochemistry. *Cold Spring Harb Perspect Biol* **2**: a002105.

Pizzarello S, Weber AL. 2010. Stereoselective syntheses of pentose sugars under realistic prebiotic conditions. *Orig Life Evol Biosph* (in press).

Powner MW, Gerland B, Sutherland JD. 2009. Synthesis of activated pyrimidine ribonucleotides in prebiotically plausible conditions. *Nature* **459**: 239–242.

Prabahar KJ, Ferris JP. 1997. Adenine derivatives as phosphate-activating groups for the regioselective formation of 3′,5′-linked oligoadenylates on montmorillonite: possible phosphate-activating groups for the prebiotic synthesis of RNA. *J Am Chem Soc* **119**: 4330–4337.

Rajamani S, Vlassov A, Benner S, Coombs A, Olasagasti F, Deamer D. 2008. Lipid-assisted synthesis of RNA-like polymers from mononucleotides. *Orig Life Evol Biosphere* **38**: 57–74.

Rasmussen RA, Khalil MAK, Dalluge RW, Penkett SA, Jones B. 1982. Carbonyl sulfide and carbon disulfide from the eruptions of Mount St. Helens. *Science* **215**: 665–667.

Reiss-Husson F, Luzzati V. 1967. Phase transitions in lipids in relation to the structure of membranes. *Adv Biol Med Phys* **11**: 87–107.

Rushdi AI, Simoneit BRT. 2001. Lipid formation by aqueous Fischer-Tropsch-type synthesis over a temperature range of 100 to 400°C. *Orig Life Evol Biosphere* **31**: 103–118.

Schwartz AW, de Graaf RM. 1993. The prebiotic synthesis of carbohydrates: A reassessment. *J Mol Evolution* **35**: 101–106.

Shock E. 1990. Geochemical constraints on the origin of organic compounds in hydrothermal systems. *Orig Life Evol Biospheres* **20**: 331–367.

Simoneit BRT, Rushdi AI, Deamer DW. 2006. Abiotic formation of acylglycerols under simulated hydrothermal conditions and self-assembly properties of such lipid products. *Adv Space Res* **11**: 1649–1656.

Sleep NH, Zahnle K, Kasting JF, Morowitz HJ. 1989. Annihilation of ecosystems by large asteroid impacts on the early Earth. *Nature* **342**: 139–142.

Stouthamer AH. 1977. Energetic aspects of the growth of microorganisms. In Microbial Energetics. BA Haddock and WA Hamilton, Eds. Cambridge University Press London. pp. 285–315.

Stribling R, Miller SL. 1987. Energy yields for hydrogen cyanide and formaldehyde synthesis: The HCN and amino acid concentration in the primitive ocean. *Orig Life Evol Biospheres* **17**: 261–273.

Summers DR, Khare B. 2007. Nitrogen fixation on early Mars and other terrestrial planets: Experimental demonstration of abiotic fixation reactions to nitrite and nitrate. *Astrobiology* **7**: 333–341.

Sutherland JD. 2010. Ribonucleotides. *Cold Spring Harb Perspect Biol* **2**: a005439.

Szostak JW, Bartel DP, Luisi PL. 2001. Synthesizing life. *Nature* **409**: 387–390.

Tazuke S, Kazama S, Kitamura N. 1986. Reductive photocarboxylation of aromatic hydrocarbons. *J Org Chem* **51**: 4548–4553.

Usher DA. 1977. Early chemical evolution of nucleic acids: a theoretical model. *Science* **196**: 311–313.

Usher DA, McHale AH. 1976. Nonenzymic joining of oligoadenylates on a polyuridylic acid template. *Science* **192**: 53–54.

Verlander MS, Orgel LE. 1974. Analysis of high molecular weight material from the polymerization of adenosine cyclic 2′, 3′-phosphate. *J Mol Evol* **3**: 115–120.

Wachtershauser G. 1988. Before enzyme and template: Theory of surface metabolism. *Microbiol Rev* **52**: 452–484.

Walde P, Wick R, Fresta M, Mangone A, Luisi PL. 1994. Autopoietic self-reproduction of fatty acid vesicles. *J Am Chem Soc* **116**: 11649–11654.

Weber AL. 1998. Prebiotic amino acid thioester synthesis: Thiol-dependent amino acid synthesis from formose substrates (formaldehyde and glycolaldehyde) and ammonia. *Orig Life Evol Biospheres* **28**: 259–270.

Weber AL. 1984. Nonenzymatic formation of "energy-rich" lactoyl and glyceroyl thioesters from glyceraldehyde and a thiol. *J Mol Evol* **20**: 157–166.

Weber AL. 2000. Sugars as the optimal biosynthetic carbon substrate of aqueous life throughout the universe. *Orig Life Evol Biospheres* **30**: 33–43.

Weber AL. 2001. The Sugar Model: Catalysis by amines and amino acid products. *Orig Life Evol Biospheres* **31**: 71–86.

Weber AL. 2004. Kinetics of organic transformations under mild aqueous conditions: Implications for the origin of life and its metabolism. *Orig Life Evol Biosphere* **34**: 473–495.

Weber AL. 2005. Growth of organic microspherules in sugar-ammonia reactions. *Orig Life Evol Biospheres* **35**: 523–536.

Weber AL. 2007. The Sugar Model: Autocatalytic activity of the triose-ammonia. *Orig Life Evol Biospheres* **37**: 105–111.

Weber AL. 2008. Sugar-driven prebiotic synthesis of 3,5(6)-dimethylpyrazin-2-one: A possible nucleobase of a primitive replication process. *Orig Life Evol Biospheres* **38**: 279–292.

Westheimer FH. 1987. Why nature chose phosphates. *Science* **235**: 1173–1178.

Wieland T, Bokelmann E, Bauer L, Lang HU. 1953. Polypeptide syntheses. 8. Formation of sulfur containing peptides by the intramolecular *Liebigs. Ann Chem* **582**: 129–149.

Zepik HH, Rajamani S, Maurel MC, Deamer D. 2007. Oligomerization of thioglutamic acid: encapsulated reactions and lipid catalysis. *Orig Life Evol Biospheres* **37**: 495–505.

Mineral Surfaces, Geochemical Complexities, and the Origins of Life

Robert M. Hazen[1] and Dimitri A. Sverjensky[2]

[1]Geophysical Laboratory, Carnegie Institution of Washington, Washington, DC 20015
[2]Department of Earth and Planetary Sciences, Johns Hopkins University, Baltimore, Maryland 21218

Correspondence: rhazen@ciw.edu

Crystalline surfaces of common rock-forming minerals are likely to have played several important roles in life's geochemical origins. Transition metal sulfides and oxides promote a variety of organic reactions, including nitrogen reduction, hydroformylation, amination, and Fischer-Tropsch-type synthesis. Fine-grained clay minerals and hydroxides facilitate lipid self-organization and condensation polymerization reactions, notably of RNA monomers. Surfaces of common rock-forming oxides, silicates, and carbonates select and concentrate specific amino acids, sugars, and other molecular species, while potentially enhancing their thermal stabilities. Chiral surfaces of these minerals also have been shown to separate left- and right-handed molecules. Thus, mineral surfaces may have contributed centrally to the linked prebiotic problems of containment and organization by promoting the transition from a dilute prebiotic "soup" to highly ordered local domains of key biomolecules.

The question of life's origin is in essence a problem of information transfer from a geochemical environment to a highly localized volume. Earth's prebiotic environment possessed a varied inventory of raw materials—an atmosphere, oceans, rocks and minerals, and a diverse suite of small organic molecules. The processes by which the Hadean Earth was transformed to a living world required the selection, concentration, and organization of specific organic molecules into successively more information-rich localized assemblages. In this view, life's origins can be modeled as a problem in emergent chemical complexification (Morowitz 1992; de Duve 1995; Lahav 1999; Hazen 2005; Zaikowski and Friedrich 2007).

At least five aspects of Hadean geochemical environments contributed to Earth's prebiotic complexity and thus may have played significant roles in the emergence of life.

1. *Chemical Complexity*: The simplest chemical models for life's origins use only four essential elements—C, H, O, and N (e.g., Oparin 1938; Urey 1951; Morowitz 1992; Morowitz et al. 2000)—with the possible addition of S (Wächtershäuser 1990; de Duve 1995) and/or P (Westheimer 1987; Kornberg et al. 1999). Experiments to probe origins-of-life chemistry have often used correspondingly simple chemical systems (Miller 1953; Fox and Harada 1958; Oró 1961; Sanchez et al.

1967; Shapiro 1988; Hennet et al. 1992; Marshall 1994; Eschenmoser 1999; Bernstein et al. 2002). Geochemical environments, in contrast, typically incorporate a dozen or more major and minor elements, with dozens more trace elements (Turekian 1968; Albarède and Hoffmann 2003; McSween et al. 2003; Steele et al. 2009). Therefore, it is important to consider the roles of a wider chemical spectrum in essential prebiotic reaction pathways.

2. *Interfaces*: Even given the most optimistic assessments of exogenous and endogenous sources of prebiotic organic molecules (Chyba and Sagan 1992), the Hadean oceans or large terrestrial bodies of water would have been extremely dilute (Pinto et al. 1980; Stribling and Miller 1987; Cohn et al. 2001). Mechanisms for the selection and concentration of essential biomolecules are thus required. In this regard, numerous authors have focused on the effectiveness of interfaces between minerals and aqueous solutions (Goldschmidt 1952; Cairns-Smith and Hartman 1986; Lahav 1994; Ertem and Ferris 1996; Ferris 1999, 2005; Ertem 2002; Schoonen et al. 2004; Hazen 2006), or among immiscible fluids (Lasaga et al. 1971; Deamer and Pashley 1989; Morowitz 1992; Dobson et al. 2000; Tuck 2002; Monnard et al. 2002). Such surfaces provide loci where organic molecules can be selected and concentrated from more dilute solutions.

3. *Gradients*: Gradients were important disequilibrium features of the Hadean Earth. Thermal gradients were sustained both by solar radiation and by geothermal heat, whereas chemical gradients were produced by mineral dissolution and the mixing of different fluid reservoirs. Thermal and chemical gradients are striking characteristics of hydrothermal systems, both on the deep ocean floor and in near-surface continental environments. At modern-day deep sea hydrothermal vents, for example, thermal gradients may reach 300°C at scales of a few centimeters, whereas significant gradients in pH, oxidation state, and in dissolved cationic and anionic species occur at scales significantly less than a meter (Shock 1992; Carl 1995; Van Dover 2000; Kelley et al. 2005). Less severe gradients also occur at off-axis hydrothermal systems, and during mixing of fluids of differing salinities, for example where river waters enter the oceans.

4. *Fluxes*: The dynamic circulation and mixing of fluids through such varied processes as hydrothermal venting, ocean currents, stream and groundwater flow, winds, and tides is another ubiquitous disequilibrium feature of Earth's near-surface environment.

5. *Cycles*: The prebiotic geochemical characteristics perhaps most critical for life's origins were pervasive cycling of environmental conditions at or near Earth's Hadean surface. Periodic cycles such as day-night, high tide-low tide, hot-cold, and wet-dry subjected prebiotic chemicals to repeated selective pressures and thus winnowed the pool of available chemical species. Episodic events, including asteroid impacts and volcanic eruptions, added additional selective pressures to the near-surface organic inventory.

These five geochemical characteristics must be considered individually and collectively in origins-of-life models. This article focuses primarily on just one of these aspects, the possible roles of crystalline surfaces of common rock-forming minerals, which provided ubiquitous crystal-fluid interfaces for a variety of molecular processes. The mineral-water interface is a dynamic, energetic environment that can selectively adsorb biomolecules, increase the thermal stability of organic species, promote chemical reactions, and facilitate the type of molecular concentration and organization that must have preceded life's origins.

BACKGROUND

Chemical interactions at crystal-water interfaces are crucial to a wide range of scientific and technological topics, including corrosion, heterogeneous catalysts, chemical sensors, teeth and bones, titanium implants and other prosthetic medical devices, and myriad commercial products including paints, glues, dyes, lubricants, solvents, and cleaners. Geochemists pay special attention to reactions between mineral surfaces and aqueous species—interactions central to weathering, soil formation, hydrothermal ore-forming fluids, biomineralization, biofilm formation, uptake and release of chemicals that affect water quality, and many other natural processes (Davis and Kent 1990; Stumm 1992; Vaughan 1995; Hochella 1995; Drever 1997; Langmuir 1997; Brown et al. 1998; Brown and Parks 2001; Davis et al. 2004; De Yoreo and Dove 2004; Lee et al. 2006, 2007; Glamoclija et al. 2009). Studies of mineral-molecule interactions related to origins of life build on this vast geochemical literature.

On the Nature of Mineral Surfaces

The idealized crystalline surface terminates in an arrangement of atoms that approximates the planar truncation of a periodic three-dimensional crystal structure (Fig. 1A). In real crystals this ideal situation is altered in several ways (e.g., Hochella and White 1990; Somorjai 1994; Hochella 1995; Vaughan 1995; Brown et al. 1998). First, surface atoms reside in an environment quite different from those below the surface, and thus undergo relaxation owing to boundary effects—typically slight deviations from their formal crystallographic positions (Hochella 1990; Stipp and Hochella 1991; Wright et al. 2001). Second, mineral surfaces in air or an aqueous medium are commonly subject to chemical alteration through oxidation, hydration, or hydroxylation (Guevremont et al. 1998; Biino et al. 1999; Stipp 2002). Third, crystals invariably have defects and impurities that alter local surface physical properties and chemical reactivity (Hochella 1990; Cygan et al. 2002).

The topology of real crystal surfaces also represents an important deviation from ideality because crystal surfaces are seldom flat. In the intensively studied case of cubic close-packed metal surfaces, including Pt, Ag, Au, and Cu, ideally flat terraces can only exist for (100), (110), and (111). All other surfaces of these metals must incorporate steps and/or kinks (Fig. 1B and C). Kink sites are intrinsic to all high-index surfaces, whether on metals or minerals, and they dictate molecular adsorption behavior on these faces (McFadden et al. 1996; Sholl 1998; Power and Sholl 1999; Ahmadi et al. 1999; Gellman et al. 2001).

Mineral surfaces, with their lower symmetries and multiple crystallographically distinct atomic sites, present additional complexities compared with metals (Lasaga 1990; Hazen 2004). Although some common surfaces of rock-forming minerals can be ideally planar at the atomic scale (e.g., the [100] plane of quartz [SiO_2], the [001] planes of graphite [C] and molybdenite [MoS_2], and the [001] planes of varied layer silicates such as micas and

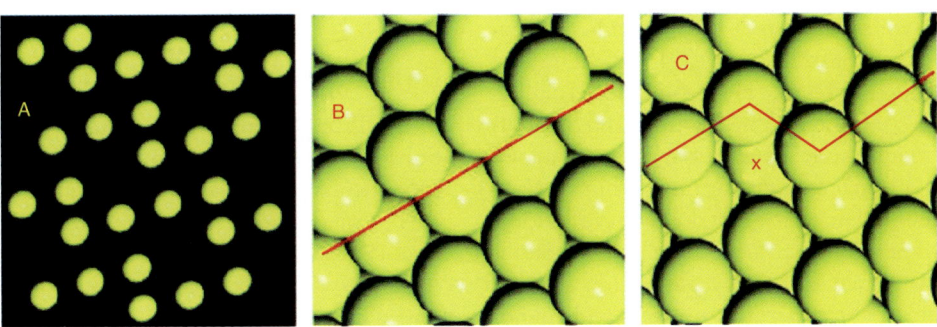

Figure 1. Crystal surfaces display a variety of atomic surface features. (*A*) The surface of an ideal crystal may be represented as a periodic two-dimensional arrangement of atoms; these atoms may be coplanar or they may occur at slightly different heights. Real crystals feature surfaces that are typically stepped (*B*) or kinked (*C*). Kink sites provide chiral (left- or right-handed) centers (X). Experimental and theoretical studies reveal that molecular adsorption is enhanced at such surface irregularities.

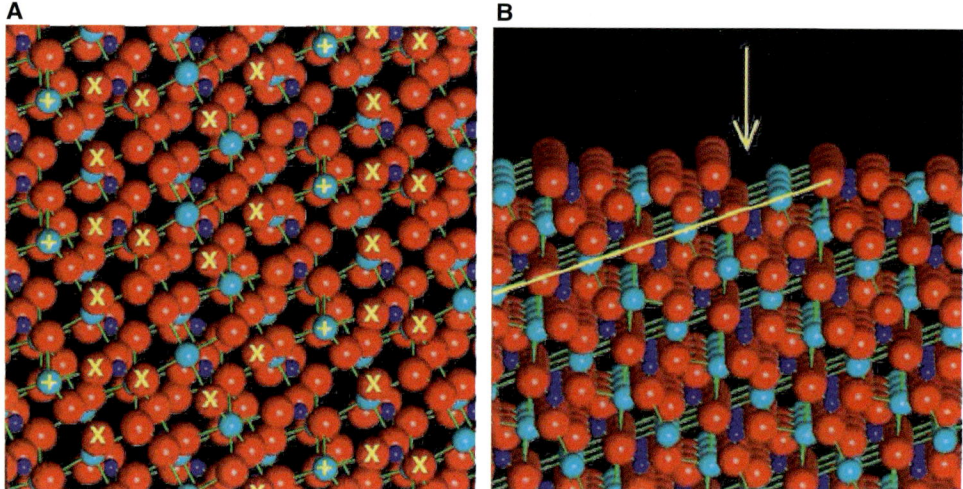

Figure 2. The common calcite form of $CaCO_3$ often displays chiral surfaces. (*A*) The structure of the (21–34) face of calcite features a chiral arrangement of positive (+) and negative (X) charge centers near the crystal termination. Ca, C, and O atoms are turquoise, blue, and red, respectively. In this 20×20 Å view the (01–8) axis is vertical—an orientation that provides a useful image of the surface structure. (*B*) A view of this surface tilted 3° from horizontal (projected almost down the [01–8] axis) reveals the irregular surface topology, including 2-Å-deep steps (yellow arrow) that result from the oblique intersection of layers of Ca and rigid CO_3 groups with the surface (yellow line).

chlorites), most surfaces are intrinsically irregular, as shown by the ~2 Å relief on the common (21–34) surfaces of calcite ($CaCO_3$) (Fig. 2) (Hazen 2004). Mineral surfaces also commonly incorporate growth defects, including step edges and kink sites, which provide promising docking loci for organic molecules (Lasaga 1990; Teng and Dove 1997; Teng et al. 1998; Orme et al. 2001; Hazen and Sholl 2003; De Yoreo and Dove 2004). For example, Teng et al. (2006) showed the step-dependent adsorption of succinic acid (1,4-dicaboxlyic acid) on irregular (10–14) growth surfaces of calcite. The presence of succinic acid in solution blocks certain growth directions and thus dramatically modifies calcite surface growth morphology.

Additional complexities arise from a variety of geological materials that do not have periodic two-dimensional surfaces, notably amorphous materials such as basaltic glass from seafloor volcanoes. Mesoporous zeolites and nanoparticulate clays (including layer phyllosilicates and hydroxide minerals) interact with organic molecules in complex three-dimensional environments (Smith 1998; Greenwell and Coveney 2006; Benetoli et al. 2007). For example, Pitsch et al. (1995) showed that double-layer hydroxide minerals such as hydrotalcite efficiently adsorb glycoaldehyde phosphate and formaldehyde, presumably into their relatively spacious inter-layer regions, and promote condensation reactions to tetrose and hexose sugar phosphates. Given these complexities, any realistic modeling of interactions between biomolecules and mineral surfaces must take into account the geometries of both molecules and surfaces.

On the Nature of the Mineral–Water Interface

Mineral surfaces undergo significant modifications in an aqueous environment, especially in an electrolyte solution such as sea water (Parks 1990; Davis and Kent 1990; Van Cappellen et al. 1993; de Leeuw and Parker 1997; de Leeuw et al. 1999; Wright et al. 2001; Stipp 2002). In the classic electrical double layer (EDL) model, a crystal surface in pure water directly contacts a compact quasi-periodic layer of H^+ or OH^- ions, whereas a second diffuse layer of mobile ions extends from the compact layer a few Ångstroms into the fluid. The nature of the

EDL is strongly dependent on pH: At lower pH, the surface is typically protonated and thus positively charged, whereas at higher pH, the surface is negatively charged with hydroxyls. The crossover pH at which the surface is electrically neutral is called the point of zero charge (pH_{PZC}).

It is important to recognize that crystallographically distinct faces on the same crystal will have different surface structures, and thus may have significantly different pH_{PZC}s, surface reactivities, and other properties (Guevremont et al. 1998; Hung et al. 2003; Churchill et al. 2004). For this reason, powdering a crystal sample to increase the reactive surface area in an experiment may destroy significant information regarding specific reaction mechanisms associated with specific crystallographic faces (Hazen 2006).

On the Nature of Mineral–Molecule Interactions

Adsorbed ionic and molecular species modify the electrical double layer of mineral surfaces by displacing OH^-, H^+, and H_2O at the solid–fluid interface. Recent developments in surface complexation modeling enable accurate modeling of surface speciation. Sverjensky and Fukushi (2006) developed an extended triple-layer (ETL) model to account for the process of inner-sphere surface complexation by ligand exchange. The ETL model treats a previously neglected phenomenon integral to ligand exchange reactions: the electrostatic work during desorption of water dipoles from a charged surface. The magnitude of this work is substantial and depends only on the stoichiometry of the surface reaction. When structures of adsorbed anions established in spectroscopic studies are used to calibrate an ETL model of bulk adsorption data, the model can independently predict proportions of inner- to outer-sphere surface complexes as functions of pH, ionic strength, and surface coverage—proportions that have been confirmed in experiments, for example in studies of aspartate and glutamate adsorption on titanium dioxide (e.g., Jonsson et al. 2009).

More than a century of experimental and theoretical research has explored the interaction of dissolved aqueous chemical species with mineral surfaces (for reviews see, e.g., Parsons 1990; Hochella and White 1990; Brown et al. 1998). Most of this extensive literature focuses on dissolved ions and inorganic complexes; however, many of the principles developed for mineral–ion interactions also apply to biomolecules.

All surface-promoted reactions require at least one molecular species to interact with the surface. These interactions can be mediated by water molecules, protons, or hydroxyl groups through relatively weak interactions (outer-sphere adsorption, or "physisorption"). Alternatively, one or more chemical bonds can form (inner-sphere adsorption, or "chemisorption"). Chemisorbed ions typically bond to one or two surface atoms, whereas larger molecules can adopt a variety of surface topologies with multiple attachments (Fig. 3), as discussed later in this article (Davis and Kent 1990; Zhang et al. 2004; Sverjensky et al. 2008; Jonsson et al. 2009).

Details of molecular adsorption are dependent on several variables, most notably pH, the nature and concentrations of molecular solutes, and the identities and concentrations of electrolytes (Schindler 1990; Sverjensky 2005; Sverjensky and Fukushi 2006; Jonsson et al. 2009). Additional complexities arise when organic molecules interact with crystal surface irregularities (Teng and Dove 1997; Teng et al. 1998, 2000; Orme et al. 2001; De Yoreo and Dove 2004; Elhadj et al. 2006). Such interactions can be strikingly revealed during crystal growth or dissolution in the presence of organic molecules, which can preferentially dock along crystallographically distinct edges and kinks. Such binding may inhibit crystal growth in certain directions and thus result in unusual crystal morphologies (e.g., Teng et al. 2006). For example, Cody and Cody (1991) showed strikingly varied skewed growth of macroscopic gypsum ($CaSO^4H_2O$) crystals in the presence of 28 different chiral organic solutes, including amino acids and carboxylic acids. Such exquisite molecular control of crystal growth and orientation points to strategies for nano-engineering and underscores the intricacies of modeling real-world prebiotic organic selection and organization (De Yoreo and Dove 2004).

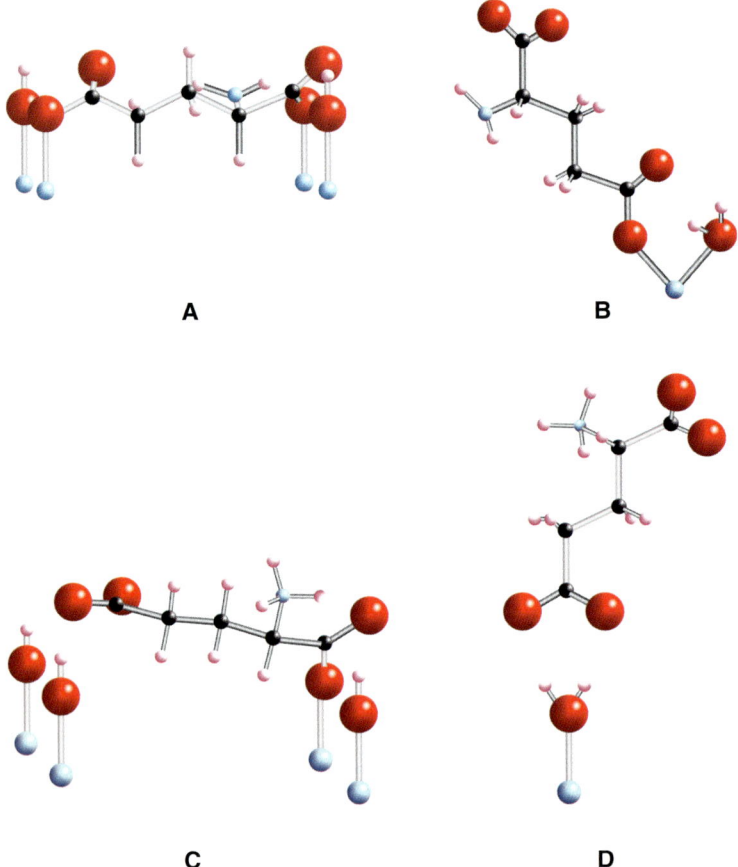

Figure 3. Amino acids bind in different ways to mineral surfaces. Numerous possible modes of attachment exist for glutamate adsorbed to rutile (TiO_2) surface sites, consistent with surface complexation calculations (Sverjensky et al. 2008; Jonsson et al. 2009). Large red spheres indicate oxygen atoms, small black spheres carbon, small pink or blue spheres hydrogen or nitrogen, respectively, and the lowermost blue spheres titanium at the rutile surface. (*A*) Bridging-bidentate species with four points of attachment involving one inner-sphere Ti-O-C bond and one Ti-OH...O=C hydrogen bond for each carboxylate. (*B*) Chelating species with two points of attachment involving one inner-sphere Ti-O-C bond and one Ti-OH$_2^+$...O=C to a single titanium. (*C*) Alternative to the bridging-bidentate species in (*A*). This bridging-bidentate species has four points of attachment involving one inner-sphere Ti-O-C bond and one Ti-OH...O=C hydrogen bond of the a-carboxylate, and one Ti-OH...⁻O-C hydrogen bond and one Ti-OH...O=C hydrogen bond of the g-carboxylate (stabilized through resonance). (*D*) Alternative to the chelating species in (*B*), outer-sphere or hydrogen bonded to the surface. After Jonsson et al. 2009.

Mineral Evolution

Any consideration of minerals and the origins of life must take into account which mineral species were available on the Hadean Earth. Recent studies have shown that the diversity and distribution of minerals at or near Earth's surface has changed dramatically over geological time (Hazen et al. 2008). Perhaps two-thirds of the approximately 4400 known mineral species represent weathering products owing to a biologically oxidized atmosphere, and thus are the indirect consequence of life. The nature and abundances of hydroxides, zeolites, and clay minerals, which are often invoked in origins-of-life models, were undoubtedly very different from today (Schoonen et al. 2004; Hazen et al. 2008). Many other minerals that concentrate rare pegmatophile elements such as Li, Be, Cs, Ta, and U may have taken a billion years or more to form, and so their first appearances would have postdated the origins of life (Grew and Hazen 2009; Hazen et al. 2009). Thus, origin-of-life scenarios that invoke rare or exotic minerals such as uraninite (UO_2) (Adam 2007) or colemanite [$Ca_2B_6O_{11}5(H_2O)$] (Ricardo et al. 2004) may be untenable.

RECENT RESULTS

The first emergent step in life's origin, the prebiotic synthesis of biomolecules, is also the best understood. Since the pioneering synthesis studies of Stanley Miller and Harold Urey (Miller 1953; Miller and Urey 1959; see also Wills and Bada 2000), facile production of organic molecules has been shown for several plausible prebiotic environments (Chyba and Sagan 1992; Lahav 1999; Fry 2000; Hazen 2005). More problematic is the second emergent step by which life's idiosyncratic subset of biomolecules was selected and then concentrated from the dilute, diverse suite of prebiotic molecular species that must have accumulated in the primitive oceans.

Consider, for example, the problem of amino acids. More than 70 different amino acids have been extracted from the Murchison meteorite and other carbonaceous chondrites (Cronin and Chang 1993; Sephton 2002; Pizzarello and Weber 2004; Pizzarello 2006; Martins 2007). This inventory, moreover, contains a mixture of both left- and right-handed amino acids, although by some accounts concentration of L-amino acids exceeds D-amino acids in some meteorites (Pizzarello and Cronin 2000; Pizzarello 2006). Life, in contrast, uses only 20 principal amino acids and these molecules occur almost exclusively in their so-called "left-handed" forms. Life's sugars, lipids, and polycyclic molecules show similar molecular selectivity (Nelson and Cox 2004; Hazen 2005). By what process did this molecular selection and concentration occur?

Mineral surfaces have long been recognized as having the potential to select and organize organic molecules in the path from geochemistry to biochemistry (Bernal 1951; Goldschmidt 1952; Lahav 1994). The following sections thus consider recent results on aspects of mineral-mediated molecular synthesis, selection, and organization.

Mineral-"Catalyzed" Organic Synthesis

The synthesis of small organic molecules from inorganic precursors, including mineral-mediated synthesis, is perhaps the best understood aspect of life's origins. For example, recent experiments by several groups have shown production of ammonia from reactions of N_2 or nitrate with a wide range of oxides or sulfides (Brandes et al. 1998, 2008; Schoonen and Xu 2001; Dorr et al. 2003); thiols from CO_2 and sulfides (Heinen and Lauwers 1996); and amino acids and other compounds from CO or CO_2, NH_3, and H_2O and sulfides (Nakajima et al. 1975; Huber and Wächtershäuser 1997, 2006; Schoonen et al. 1999; Cody et al. 2000, 2001, 2004). Transition metal oxide or sulfide minerals are not true catalysts in these reactions because the surface is altered; rather, the minerals participate as reactants through a redox couple, for example:

$$CO_2 + FeS + H_2S \rightarrow HCOOH + FeS_2 + H_2O \, (e.g., Cody\,2004), \text{ or}$$

$$4Cu_2O + 2H_2O + NO_3^- \rightarrow NH_3 + 8CuO + OH^-$$
(Brandes et al. 2008).

Alternatively, many common minerals have been shown to catalyze carbon addition reactions, including Fischer-Tropsch-type synthesis and hydroformylation (Heinen and Lauwers 1996; McCollum et al. 1999; Cody et al. 2004). Varied roles of minerals in the synthesis of small organic molecules have been reviewed recently by Cody (2004) and Schoonen et al. (2004) and are not considered further here.

Molecular Adsorption and Stabilization

Experimental and theoretical investigations have documented the nature of molecular adsorption for numerous mineral-molecule combinations (e.g., Lambert 2008; Lambert et al. 2009). Dozens of recent studies reveal complexity in structure and speciation of adsorbed organics, particularly through in situ FTIR and other spectroscopic studies (Somasundaran and Krishnakumar 1994; Rodriguez et al. 1996; Holmen et al. 1997; Roddick-Lanzilotta et al. 1998; Fitts et al. 1999; Klug and Forsling 1999; Kubicki et al. 1999; Roddick-Lanzilotta and McQuillan 1999, 2000; Duckworth and Martin 2001; Sheals et al. 2002; Lackovic et al. 2003; Rosenqvist et al. 2003; Yoon et al. 2004, 2005; Johnson et al. 2004a, 2004b, 2004c, 2005a, 2005b; Trout and Kubicki 2004; Persson and Axe 2005; Perezgasga et al. 2005; Benetoli et al. 2007; Arora and Kamaluddin 2009; Kitadai et al. 2009; Pászti and Guczi 2009). Important conclusions of these studies include the recognition of multiple adsorption configurations, including both single and multiple inner- and outer-sphere binding, for a given mineral-molecule pair.

Integration of experiments and theoretical molecular methods has been performed for amino acids on metal surfaces (Chen et al. 2002; Toomes et al. 2003; Efstathiou and Woodruff 2003; Barlow and Raval 2003; Jones and Baddeley 2004), for anions adsorbed on oxide surfaces (Collins et al. 1999; Kubicki et al. 1999; Kwon and Kubicki 2004; Peacock and Sherman 2004; Yoon et al. 2004; Bargar et al. 2005; Persson and Axe 2005), and for varied biomolecules on calcite (Thomas et al. 1993; Orme et al. 2001; Asthagiri and Hazen 2007). In addition, for the calcite-water interface, many theoretical simulations and observations exist to guide surface chemical models (Stipp and Hochella 1991; de Leeuw and Parker 1997, 1999; Teng et al. 1998, 2000; Fenter et al. 2000; Wright et al. 2001; Stipp 2002; de Leeuw and Cooper 2004; Geissbuhler et al. 2004; Kristensen et al. 2004).

Recent studies of organic anion adsorption on oxide surfaces show complex surface speciation: both inner- and outer-sphere species vary in relative importance over a range of pH, ionic strength, and surface coverage. Many investigations document significant variations in proportions of inner- to outer-sphere species as a function of pH or surface coverage (e.g., Hug and Sulzberger 1994; Fitts et al. 1999; Nowack and Stone 1999; Roddick-Lanzilotta and McQuillan 2000; Lackovic et al. 2003; Sheals et al. 2003; Persson and Axe 2005). Proton titrations of oxide surfaces in electrolyte solutions, both with and without an organic adsorbate, provide powerful constraints on the possible reactions responsible for adsorption, particularly when used in combination with in situ attenuated total reflection Fourier transform infrared (ATR-FTIR) studies (Holman and Casey 1996; Nordin et al. 1997; Boily et al. 2000a, 2000b, 2000c, 2005; Sheals et al. 2002, 2003; Lackovic et al. 2003; Lindegren et al. 2005). However, with few exceptions (Gisler 1981; Whitehead 2003; Vlasova and Golovkova 2004; Vlasova 2005; Jonsson et al. 2009). Most adsorption studies of amino acids on oxide surfaces have been limited to systems without control of pH and ionic strength (e.g., Holm et al. 1983; Matrajt and Blanot 2004).

The work of Jonsson et al. (2009, 2010), who studied the adsorption of L-glutamate and L-aspartate on the surface of rutile (α-TiO$_2$, pH$_{PZC}$ = 5.4) in NaCl solutions using potentiometric titrations and batch adsorption experiments over a wide range of pH, ligand-to-solid ratio, and ionic strength, illustrates the need for such integrated studies. Not only did they find that adsorption depends strongly on ionic strength and glutamate concentration, but the extended triple-layer surface complexation model of all the experimental results also indicated the existence of at least two surface glutamate complexes. For example, one possible mode of glutamate attachment involves a bridging-bidentate species binding through both carboxyl groups, which can be thought of as "lying down" on the surface (as found previously for amorphous titanium dioxide and hydrous ferric oxide) (Fig. 3A). Another adsorption mode involves a chelating species, which binds only through the γ-carboxyl group, i.e., "standing up" at the surface (Fig. 3B). The calculated proportions of these two surface glutamate species vary strongly, particularly with pH and glutamate concentration (Fig. 4). Any model of prebiotic interactions between mineral surfaces

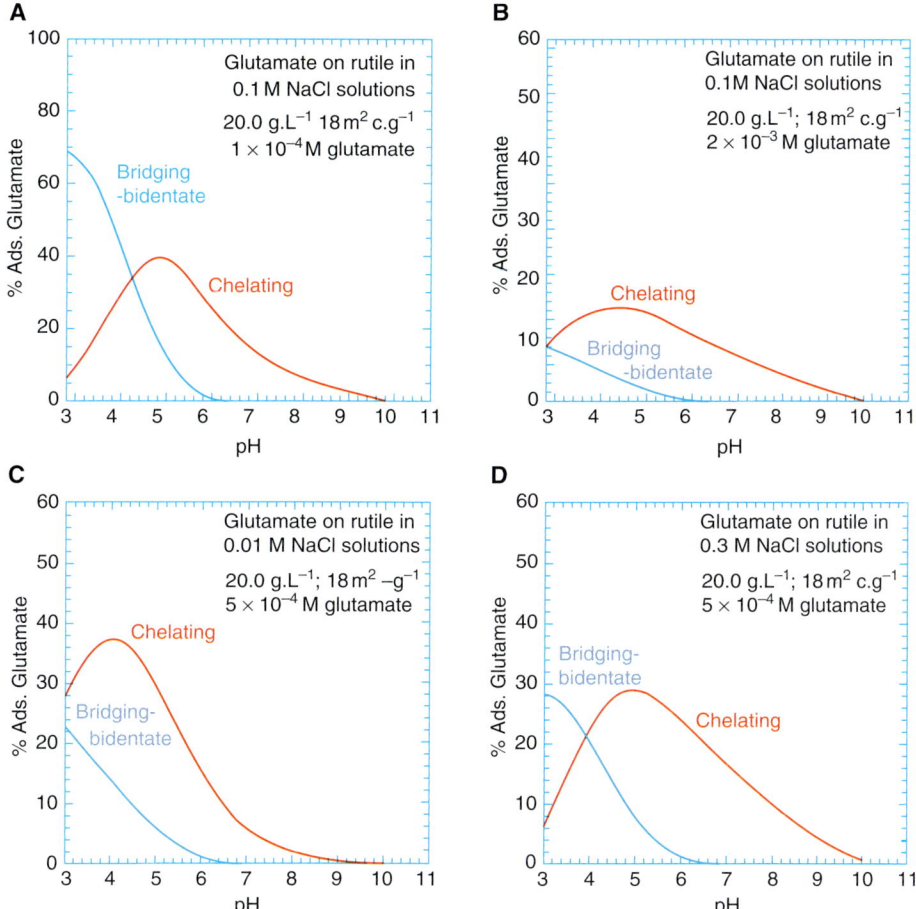

Figure 4. Predicted surface speciation of glutamate on rutile as a function of environmental conditions. The species names refer to the pictures in Figure 3. After Jonsson et al. 2009.

and biomolecules must take these added complexities into account.

Chiral Molecular Adsorption

A key attribute of life, and an important consideration in origins-of-life models, is life's molecular handedness, or chirality. Chiral crystalline surfaces provide effective environments for discrimination of left- and right-handed molecules in both natural and industrial contexts (Hazen and Sholl 2003). A chiral crystal surface is defined as any terminal arrangement of atoms that cannot be superimposed on its reflection in a mirror perpendicular to the surface. Such surfaces have long been cited in reference to their possible role in the origins of biochemical homochirality (Tsuchida et al. 1935; Karagounis and Coumoulos 1938; Amariglio et al. 1968; Bonner et al. 1975; Lahav 1999; Hazen et al. 2001). More recently, these crystal surfaces have received attention for their potential applications in the selection and purification of chiral pharmaceuticals and other molecular products (Soai et al. 1999; Kahr and Gurney 2001; Jacoby 2002; Rouhi 2004).

Most studies have focused on the behavior of chiral surfaces of cubic close-packed (CCP) metals, including copper, silver, gold, and platinum. Theoretical studies of these metal surfaces have shown the potential for significant differences in adsorption energies of right- versus left-handed molecules (McFadden et al. 1996; Sholl 1998; Šljivančanin et al. 2002), whereas experiments provide indirect evidence for chiral selectivity (Ahmadi et al. 1999; Attard 2001; Sholl et al. 2001; Horvath and Gellman 2001, 2002; Kühnle et al. 2002). Less attention has been devoted to the wide variety of chiral oxide and silicate mineral surfaces, which are ubiquitous in Earth's crust. Such surfaces provide the most abundant and accessible local chiral geochemical environments, and thus represent logical sites for the prebiotic chiral selection and organization of essential biomolecules (Hazen and Sholl 2003; Hazen 2004, 2005, 2006; Castro-Puyana et al. 2008). Hazen et al. (2001) first showed chiral-selective mineral adsorption, specifically of aspartate on the common (21–34) surfaces of calcite ($CaCO_3$), whereas alanine displayed no such selection. Subsequent calculations (Asthagiri et al. 2004; Hazen 2006; Asthagiri and Hazen 2007) rationalize these results by demonstrating that aspartate, but not alanine, binds to the calcite surface with three noncolinear attachments (Fig. 5)—a prerequisite for chiral selection (Davankov 1997).

Numerous other common rock-forming minerals, including quartz (SiO_2), alkali feldspar [$(Na,K)AlSi_3O_8$], and clinopyroxene [$(Ca,Mg,Fe)SiO_3$], possess chiral crystal surfaces (Hazen 2004). One or more of these minerals is present in most rocks in Earth's crust, as well as on the Moon, Mars, and other terrestrial bodies, so chiral crystal environments are correspondingly ubiquitous. Furthermore, any irregular mineral fracture surface will provide an additional variety of local chiral environments. These natural chiral surface environments occur in both left- and right-handed variants in approximately equal proportions (Frondel 1978; Evgenii and Wolfram 1978). Nevertheless, the widespread occurrence of local chiral environments provided the prebiotic Earth with innumerable sites for experiments in chiral selection and organization—experiments that may have led, through a process of chiral amplification, to a fortuitous, self-replicating homochiral entity (Bonner 1991, 1995; Lippmann and Dix 1999; Zepik et al. 2002; Klussman et al. 2006, 2007; Noorduin et al. 2008).

Mineral-induced Polymerization

Many of life's essential macromolecules, including proteins, carbohydrates, and DNA, form from water-soluble monomeric units—amino acids, sugars, and nucleic acids, respectively. Under some (but not all) conditions, these polymers tend to break down rather than form in an aqueous medium (Shock 1993). Mineral surfaces provide a means to concentrate and assemble these bio-monomers. Lahav et al. (1978) showed that amino acids concentrate and polymerize on clay minerals to form small, protein-like molecules. Such reactions occur when a solution containing amino acids evaporates in

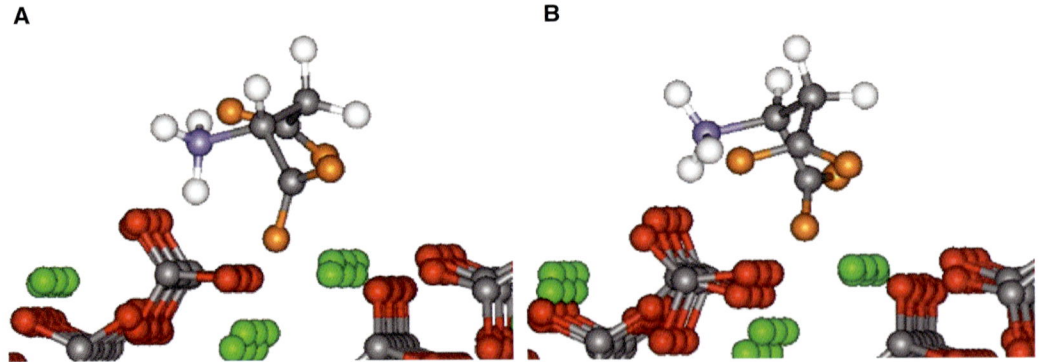

Figure 5. The most stable configurations for L- and D-aspartate on the calcite (21–34) surface (A and B, respectively). The D enantiomer, which requires significantly less calcite surface relaxation and aspartate distortion, is favored by 8 Kcal/mol—the largest known enantiospecific effect.

the presence of clays—a situation not unlike the evaporation that might have repeatedly dried up shallow prebiotic ponds or tidal pools. Numerous subsequent studies have elucidated the adsorption and polymerization of amino acids on varied crystalline surfaces (Zamaraev et al. 1997; Hill et al. 1998; Liu and Orgel 1998; Lambert 2008; Lambert et al. 2009; Rimola et al. 2009).

Ferris and colleagues (Holm et al. 1993; Ferris et al. 1996; Ertem and Ferris 1996, 1997; Ertem 2002; Ferris 1999, 2005; Ertem et al. 2007, 2008) induced clays to act as scaffolds in the formation of RNA oligomers up to 50-mers. Despite these advances, Orgel (1998) concluded that as more molecules are added to a lengthening polymer the strand becomes more tightly bound to the mineral surface. Such strong adhesion might prove problematic in the production of useful biologically active macromolecules. One possible solution was described by Hanczyc et al. (2003), who mixed clays, RNA nucleotides, and lipids in a single experiment. The clays adsorb RNA while hastening the formation of lipid vesicles. In the process, RNA-decorated clay particles are incorporated into the vesicles. This spontaneous self-assembly of RNA-containing vesicles represents a plausible pathway to the emergence of a self-replicating cell-like entity.

Smith and coworkers (Smith 1998; Parsons et al. 1998; Smith et al. 1999) have explored an alternative to adsorption on clays by invoking reactions within the channels of mesoporous zeolite minerals. Prebiotic molecules concentrated and aligned inside zeolite channels might undergo polymerization reactions.

The most elaborate mineral-based origins scenario posits that self-replicating clay minerals were, themselves, the first living entities (Cairns-Smith 1968, 1977, 1982, 1985a, 1985b, 1988; Cairns-Smith and Hartman 1986). According to this model, crystal growth defects, aperiodic cation distributions, or random layer stacking sequences constitute a kind of genetic information, analogous to the sequence of nucleotides (A, T, G, and C) in DNA. Cairns-Smith speculated that clay minerals could "replicate" by cleaving, whereas more favorable (i.e., stable) sequences evolve at the expense of less favorable sequences through the selective processes of growth and dissolution of individual particles. Ultimately, according to this hypothesis, organic molecules used clay-life as scaffolding for the evolution of modern biochemistry. Greenwell and Coveney (2006) have proposed a similar model using layered double hydroxides as "information storage and transfer compounds." Recent tests of the crystal gene model by Bullard and coworkers (2007) underscore the experimental difficulties inherent in testing such a model.

CHALLENGES AND FUTURE RESEARCH DIRECTIONS

A central objective of this collection is to move away from classical chemical scenarios that emphasize synthesis of simple organic molecules, and instead treat the origin of life as a pathway extending from the smallest organic molecules to the emergence of molecular systems that can be contained in some form of semi-permeable compartment. An important added emphasis is the consideration of realistic physical and chemical environments of early Earth. Life's origins occurred in a geochemical milieu in which chemically complex constituents at solid-fluid and fluid-fluid interfaces were subjected to chemical and thermal gradients, fluid fluxes, and a variety of cyclic processes.

Recent experiments underscore the potential for these geochemical complexities to enhance significantly organic reaction rates and pathways, as well as the rates and mechanisms of molecular selection, concentration, and self-organization. Thus, these nonequilibrium aspects of dynamic geochemical environments may have been critical to promoting key steps in prebiotic chemical evolution. This article has emphasized possible roles of mineral surfaces in such information transfer.

Opportunities in Mineral Surfaces Research

Studies of mineral-molecule interactions are still in their infancy, but several intriguing observations should inform future work.

1. *Differential adsorption:* It is well known that electrolytes, metals, and organic molecules compete for surface binding sites (e.g., Schindler 1990), but few studies have

addressed the question of competitive or cooperative biomolecular adsorption. For example, Pontes-Buarques et al. (2001) discovered that while adenosine monophosphate (AMP) alone does not easily adsorb onto pyrite (FeS_2) in the absence of divalent cations, the addition of acetate strongly enhances AMP binding. Such interactive molecular adsorption phenomena could have important implications for life's chemical origins.

Competitive adsorption may also play an important role. Churchill et al. (2004) observed that amino acid adsorption may be highly dependent on mineral surface charge. Quartz ($pH_{PZC} \sim 2.8$) tends to adsorb amino acids most strongly when the mineral pH_{PZC} and the isoelectric point (pI) of the amino acid differ significantly. Thus, quartz adsorbs lysine (pI = 9.74) more strongly than amino acids with lower pI (glycine, alanine, aspartate, glutamate, tyrosine, and leucine). In contrast, although calcite ($pH_{PZC} = 9.5$) interacts most strongly with aspartate (pI = 2.98) and glutamate (pI = 3.08), it also adsorbs a variety of other amino acids with $6 < pI < 10$. Calcite may thus represent a more plausible template than quartz for prebiotic selection and organization of homochiral polypeptides. Similarly, ribose is selectively concentrated on rutile from equimolar solutions of the isomeric pentose sugars: arabinose, lyxose, ribose, and xylose (Hazen 2006; Bielski and Tencer 2006; Cleaves et al. 2009). Additional competitive adsorption experiments, including studies that incorporate realistic sea water salinity, are needed, even though divalent cations in solution may inhibit molecular self-organization in some systems (Monnard et al. 2002).

2. *Molecular Organization*: In the special case of highly planar crystal surfaces, it is sometimes possible to image individual adsorbed molecules or clusters of molecules. In some cases, adsorbed molecules have been found to organize into periodic two-dimensional surface structure—structures that might have played a role in prebiotic chemical evolution. Sowerby et al. (1996, 1998a, 1998b, 2002) used scanning tunneling microscopy to document structures of self-assembled monolayers of adenine and guanine on graphite (C) and molybdenite (MoS_2), Kühnle et al. (2002) imaged individual cysteine dimers adsorbed on gold, and Uchihashi et al. (1999) observed adenine adsorption on graphite with noncontact AFM. AFM studies and molecular calculations also suggest that amino acids can adsorb selectively along linear surface steps of calcite (Orme et al. 2001; de Yoreo and Dove 2004), perhaps providing an alignment conducive to homochiral polymerization (Hazen et al. 2001).

These results point to the need for in situ studies not just of adsorption, but also of molecular organization and polymerization on mineral surfaces. The potential roles of irregular surface topologies on polymerization (e.g., Elhadj et al. 2006) should also be explored.

3. *Molecular Stabilization*: A significant recurrent objection to any role of hydrothermal systems in life's origins is the rapid decomposition of many important biomolecules at elevated temperatures (Bada et al. 1995; Wills and Bada 2000; Bada and Lazcano 2002; see, however, Shock 1993; Seewald et al. 2006). However, inner-sphere bonding of organic molecules to mineral surfaces may play a significant role in enhancing the thermal stability of these molecules (Hazen 2006; Lambert 2008). For example, it is well known that the inner-sphere (calcium-oxygen) bonding of proteins such as osteocalcin and collagen to hydroxylapatite in bones (e.g., Hoang et al. 2004) can lead to the preservation of these proteins for periods in excess of a million years (Collins et al. 2000; Nielsen-Marsh et al. 2005; Schweitzer et al. 2007). Consequently, much more research needs to be performed on mineral-induced thermal stabilization of biomolecules.

Experimental Design and the Origin of Life

Perhaps the greatest challenge facing origins-of-life researchers is conducting experiments that effectively mimic the complex prebiotic geochemical environment. Origins investigators have long recognized the importance of non-equilibrium characteristics of the prebiotic world and, accordingly, have incorporated some aspects of these geochemical realities into their experiments. The transformational experiments of Miller and Urey (Miller 1953; Miller and Urey 1959) used thermal gradients and fluid fluxes in association with spark discharges to achieve organic synthesis. Subsequent theoretical models (Wächtershäuser 1988, 1990, 1992; de Duve 1995; Russell and Hall 1997, 2006) and experiments (Lahav et al. 1978; Ferris et al. 1996; Huber and Wächtershäuser 1997; McCollum and Simoneit 1999; Whitfield 2009) have incorporated aspects of molecular complexity that may emerge from multi-component geochemical systems with gradients, fluxes, cycles, and interfaces. For example, Budin et al. (2009) have found that lateral thermal gradients in a narrow capillary environment enhance localized concentrations of lipids by orders of magnitude—conditions that foster vesicle formation in the low-temperature zones of the experiment.

Of special note are experiments in molecular evolution, for example of selectively binding RNA aptamers (Ellington and Szostak 1990; Wilson and Szostak 1999) and peptides (Seelig and Szostak 2007). These experiments incorporate chemical complexity, fluxes, interfaces, and cycles of molecular selection, each of which adds information to the system and quickly leads to highly functional molecules (Szostak 2003; Carothers et al. 2004; Hazen et al. 2007). Such selective, cycling environments likely represent an essential aspect of life's origins and evolution.

In the case of experiments on mineral-molecule interactions in an aqueous environment, most experiments to date have focused on a single well-characterized mineral with one solute in water with at most a single electrolyte at room conditions. Such experiments are essential to obtain baseline information on the magnitude and geometry of adsorption for various mineral-molecule pairs. Nevertheless, these studies do not replicate prebiotic complexities, including the multiple electrolytes of seawater, numerous competing organic species in the prebiotic soup, and numerous competing mineral phases and surfaces, all present over a range of temperature, pressure, pH, and solute concentrations.

Added to these challenges are the daunting efforts required to reproduce nonequilibrium geochemical complexities in the laboratory. For example, any attempt to impose a thermal gradient on a chemical experiment adds at least three experimental variables (T_{max}, T_{min}, and distance) that must be specified and controlled throughout an experiment. Imposition of cycles is even more challenging, for it requires control of experimental conditions for the two end-member states of the system, as well as the temporal variables related to cycle lengths and rates of change between these two states. Experiments to simulate geochemical environments in a closed laboratory system may encounter unanticipated problems. For example, initial attempts to mimic the interaction of mineral-rich hydrothermal vents with colder ocean water at the Geophysical Laboratory and elsewhere have been thwarted, because metal sulfides dissolve in regions of hotter fluids and then precipitate and clog the system in cooler portions.

Nevertheless, despite the added experimental complexity of incorporating gradients, fluxes and cycles, these aspects of natural geochemical environments were probably essential to the emergence of biochemical complexity and thus must inform the design of future origins-of-life experiments.

ACKNOWLEDGMENTS

We thank R.J. Hemley for thoughtful comments and suggestions, David Deamer and Jack Szostak for spearheading this collection, and NASA's Astrobiology Institute, the National Science Foundation, the Alfred P. Sloan Foundation, and the Carnegie Institution of Washington for support of research on life's origins.

REFERENCES

Adam Z. 2007. Actinides and life's origins. *Astrobiology* **7:** 852–872.

Ahmadi A, Attard G, Feliu J, Rodes A. 1999. Surface reactivity at "chiral" platinum surfaces. *Langmuir* **15:** 2420–2424.

Albarède F, Hoffmann AW. 2003. *Geochemistry: An introduction*. Cambridge University Press, Cambridge, UK.

Amariglio A, Amariglio H, Duval X. 1968. Asymmetric reactions on optically active quartz. *Helv Chim Acta* **51:** 2110.

Arora AK, Kamaluddin. 2009. Role of metal oxides in chemical evolution: Interaction of ribose nucleotide with alumina. *Astrobiology* **9:** 165–171.

Asthagiri A, Hazen RM. 2007. An *ab initio* study of adsorption of alanine on the chiral calcite (21-31) surface. *Molec Sim* **33:** 343–351.

Asthagiri A, Downs RT, Hazen RM. 2004. Density functional theory modeling of interactions between amino acids and chiral mineral surfaces. *Geol Soc Am Abstr with Prog* (Denver CO):

Attard GA. 2001. Electrochemical studies of enantioselectivity at chiral metal surfaces. *J Phys Chem B* **105:** 3158–3167.

Bada JL, Lazcano A. 2002. Some like it hot, but not the first biomolecules. *Science* **296:** 1982–1983.

Bada JL, Miller SL, Zhao M. 1995. The stability of amino acids at submarine hydrothermal vent temperatures. *Orig Life Evol Biosph* **25:** 111–118.

Bargar JR, Kubicki JD, Reitmeyer R, Davis JA. 2005. ATR-FTIR spectroscopic characterization of coexisting carbonate surface complexes on hematite. *Geochim Cosmochim Acta* **69:** 1527–1542.

Barlow SM, Raval R. 2003. Complex organic molecules at metal surfaces: Bonding, organisation and chirality. *Surf Sci Repts* **50:** 201–341.

Benetoli LOB, de Souza CMD, da Silva KL, de Souza IG Jr, de Santana H, Paesano A Jr, da Costa ACS, Zaia CTBV, Zaia DAM. 2007. Amino acid interaction with and adsorption on clays: FT-IR and Mossbauer spectroscopy and X-ray diffractometry investigations. *Orig Life Evol Biosph* **37:** 479–493.

Bernal JD. 1951. *The physical basis of life*. Routledge and Kegan Paul, London, UK.

Bernstein MP, Dworkin JP, Sandford SA, Cooper GW, Allamandola LJ. 2002. Racemic amino acids from the ultraviolet photolysis of interstellar ice analogues. *Nature* **416:** 401–403.

Bielski R, Tencer M. 2006. A possible path to the RNA World: Enantioselective and diastereoselective purification of ribose. *Orig Life Evol Biosph* **37:** 167–175.

Biino GG, Mannella N, Kay A, Mun B, Fadley CS. 1999. Surface chemical characterization and surface diffraction effects of real margarite (001): An angle-resolved XPS investigation. *Am Mineral* **84:** 629–638.

Boily J-F, Persson P, Sjöberg S. 2000a. Benzenecarboxylate complexation at the goethite-water interface: I. A mechanistic description of pyromellitate surface complexes from the combined evidence of infrared spectroscopy, potentiometry, adsorption data and surface complexation modeling. *Langmuir* **16:** 5719–5729.

Boily J-F, Persson P, Sjöberg S. 2000b. Benzenecarboxylate complexation at the goethite-water interface: II. Linking IR spectroscopic observations to mechanistic-like surface complexation models. *Geochim Cosmochim Acta* **64:** 3453–3470.

Boily J-F, Persson P, Sjöberg S. 2000c. Benzenecarboxylate complexation at the goethite-water interface: III. The significance of modelling parameters and the influence of particle surface area. *J Colloid Interface Sci* **227:** 132–140.

Boily J-F, Sjöberg S, Persson P. 2005. Structures and stabilities of Cd and Cd-phthalate complexes at the goethite-water interface. *Geochim Cosmochim Acta* **69:** 3219–3236.

Bonner WA. 1991. The origin and amplification of biomolecular chirality. *Orig Life Evol Biosph* **21:** 59–111.

Bonner WA. 1995. Chirality and life. *Orig Life Evol Biosph* **25:** 175–190.

Bonner WA, Kavasmaneck PR, Martin FS, Flores JJ. 1974. Asymmetric adsorption of alanine by quartz. *Science* **186:** 143–144.

Bonner WA, Kavasmaneck PR, Martin FS, Flores JJ. 1975. Asymmetric adsorption by quartz: A model for the prebiotic origin of optical activity. *Orig Life* **6:** 367–376.

Brandes JA, Hazen RM, Yoder HS Jr. 2008. Inorganic nitrogen reduction and stability under hydrothermal conditions. *Astrobiology* **8:** 1113–1126.

Brandes JA, Boctor NZ, Cody GD, Cooper BA, Hazen RM, Yoder HS Jr. 1998. Abiotic nitrogen reduction on the early Earth. *Nature* **395:** 365–367.

Brown GE Jr, Parks GA. 2001. Sorption of trace elements on mineral surfaces: Modern perspectives from spectroscopic studies, and comments on sorption in the marine environment. *Int Geol Rev* **43:** 963–1073.

Brown GE Jr, Heinrich VE, Casey WH, Clark DL, Eggleston C, Felmy A, Goodman DW, Grätzel M, Maciel G, McCarthy MI, et al. 1998. Metal oxide surfaces and their interactions with aqueous solutions and microbial organisms. *Chem Rev* **99:** 77–174.

Budin I, Bruckner R, Szostak J. 2009. Formation of protocell-like vesicles in a thermal diffusion column. *J Am Chem Soc* **131:** 9628–9629.

Bullard T, Freudenthal J, Avagyan S, Kahr B. 2007. Tests of Cairn-Smith's 'crystals-as-genes' hypothesis. *Faraday Disc* **136:** 231–245.

Cairns-Smith AG. 1968. The origin of life and the nature of the primitive gene. *J Theor Biol* **10:** 53–88.

Cairns-Smith AG. 1977. Takeover mechanisms and early biochemical evolution. *Biosystems* **9:** 105–109.

Cairns-Smith AG. 1982. *Genetic takeover and the mineral origins of life*. Cambridge University Press, Cambridge, UK.

Cairns-Smith AG. 1985a. *Seven clues to the origin of life*. Cambridge University Press, Cambridge, UK.

Cairns-Smith AG. 1985b. The first organisms. *Sci Amer* **252:** 90–100.

Cairns-Smith AG. 1988. The chemistry of materials for artificial Darwinian systems. *Int Rev Phys Chem* **7:** 209–250.

Cairns-Smith AG, Hartman H. 1986. *Clay minerals and the origin of life*. Cambridge University Press, Cambridge, UK.

Carl DM. Editor. 1995. *Deep-sea hydrothermal vents*. CRC Press, Boca Raton, FL.

Carothers JM, Oestreich SO, Davis JH, Szostak JW. 2004. *J Am Chem Soc* **126:** 5130–5137.

Carter PW. 1978. Adsorption of amino acids-containing organic matter by calcite and quartz. *Geochim Cosmochim Acta* **42:** 1239–1242.

Castro-Puyana M, Salgado A, Hazen RM, Crego AL, Marina ML. 2008. Investigation of the enantioselective adsorption of 3-carboxy adipic acid on minerals by capillary electrophoresis. *Electrophoresis* **29:** 1548–1555.

Chen Q, Frankel DJ, Richardson NV. 2002. Chemisorption induced chirality: Glycine on Cu{110}. *Surf Sci* **497:** 37–46.

Churchill H, Teng H, Hazen RM. 2004. Measurements of pH-dependent surface charge with atomic force microscopy: Implications for amino acid adsorption and the origin of life. *Am Mineral* **89:** 1048–1055.

Chyba CF, Sagan C. 1992. Endogenous production, exogenous delivery, and impact-shock synthesis of organic molecules: An inventory for the origins of life. *Nature* **355:** 125–132.

Cleaves HJ, Jonsson CM, Jonsson CL, Sverjensky DA, Hazen RM. 2010. Adsorption of nucleic acid components on rutile (TiO_2) surfaces. *Astrobiology* (in press).

Cody GD. 2004. Transition metal sulfides and the origins of metabolism. *Annu Rev Earth Planet Sci* **32:** 569–599.

Cody AM, Cody RD. 1991. Chiral habit modifications of gypsum from epitaxial like adsorption of stereo-specific growth inhibitors. *J Cryst Growth* **113:** 508–529.

Cody GD, Boctor NZ, Brandes JA, Filley TR, Hazen RM, Yoder HSJr. 2004. Assaying the catalytic potential of transition metal sulfides for prebiotic carbon fixation. *Geochim Cosmochim Acta* **68:** 2185–2196.

Cody GD, Boctor NZ, Filley TR, Hazen RM, Scott JH, Yoder HSJr. 2000. The primordial synthesis of carbonylated iron-sulfur clusters and the synthesis of pyruvate. *Science* **289:** 1339–1342.

Cody GD, Boctor NZ, Hazen RM, Brandes JA, Morowitz HJ, Yoder HS Jr. 2001. Geochemical roots of autotrophic carbon fixation: Hydrothermal experiments in the system citric acid-H_2O-($\pm$FeS)-($\pm$NiS). *Geochim Cosmochim Acta* **65:** 3557–3576.

Cohn CA, Hansson TK, Larsson HS, Sowerby SJ, Holm NG. 2001. Fate of prebiotic adenine. *Astrobiology* **1:** 477–480.

Collins CR, Ragnarsdottir KV, Sherman DM. 1999. Effect of inorganic and organic ligands on the mechanism of cadmium sorption to goethite. *Geochim Cosmochim Acta* **63:** 2989–3002.

Collins MJ, Gernaey AM, Nielsen-Marsh CM, Vermeer C, Westbroeck P. 2000. Slow rates of degradation of osteocalcin: green light for fossil bone protein? *Geology* **26:** 1139–1142.

Cronin JR, Chang S. 1993. Organic matter in meteorites: Molecular and isotopic analyses of the Murchison meteorite. In *The chemistry of life's origins* (ed. JM Greenberg et al.), pp. 209–258. Kluwer, The Netherlands.

Cygan RT, Wright K, Fisler DK, Gale JD, Slater B. 2002. Atomistic models of carbonate minerals: Bulk and surface structures, defects, and diffusion. *Molec Sim* **28:** 475–495.

Davankov VA. 1997. The nature of chiral recognition: Is it a three-point interaction? *Chirality* **9:** 99–102.

Davis JA, Kent DB. 1990. Surface complexation modeling in aqueous geochemistry. In *Mineral-water interface geochemistry* (ed. MF Hochella Jr, AF White), *Rev Mineral* **23:** 177–260.

Davis JA, Meece DE, Kohler M, Curtis GP. 2004. Approaches to surface complexation modeling of uranium (VI) adsorption on aquifer sediments. *Geochim Cosmochim Acta* **68:** 3621–3642.

Deamer DW, Pashley RM. 1989. Amphiphilic components of the Murchison carbonaceous chondrite: Surface properties and membrane formation. *Orig Life Evol Biosph* **19:** 21–38.

de Duve C. 1995. *Vital dust: Life as a cosmic imperative.* Basic Books, New York.

de Leeuw NH, Cooper TG. 2004. A computer modeling study of the inhibiting effect of organic adsorbates on calcite crystal growth. *Crystal Growth Design* **4:** 123–133.

de Leeuw NH, Parker SC. 1997. Atomistic simulation of the effect of molecular adsorption of water on the surface structure and energies of calcite surfaces. *J Chem Soc, Faraday Trans* **93:** 467–475.

de Leeuw NH, Parker SC, Harding JH. 1999. Molecular dynamics simulation of crystal dissolution from calcite steps. *Phys Rev B* **60:** 13792–13799.

De Yoreo JJ, Dove PM. 2004. Shaping crystals with biomolecules. *Science* **306:** 1301–1302.

Dobson CM, Ellison GB, Tuck AF, Vaida V. 2000. Atmospheric aerosols as prebiotic chemical reactors. *Proc Natl Acad Sci* **97:** 11864–11868.

Dorr M, Kasbohrer J, Grunert R, Kreisel G, Brand WA, Werner RA, Geilmann H, Apfel C, Robl C, Weigand W. 2003. A possible prebiotic formation of ammonia from dinitrogen on iron oxide surfaces. *Angew Chem Int Ed Engl* **42:** 1540–1543.

Downs RT, Hazen RM. 2004. Chiral indices of crystalline surfaces as a measure of enantioselective potential. *J Mol Catal* **216:** 273–285.

Drever JI. 1997. *The geochemstry of natural waters.* Prentice Hall, New York.

Duckworth OW, Martin ST. 2001. Surface complexation and dissolution of hematite by C1-C6 dicarboxylic acids at pH=5.0. *Geochim Cosmochim Acta* **65:** 4289–4301.

Dzombak DA, Morel FMM. 1990. *Surface complexation modeling.* Wiley, New York.

Efstathiou V, Woodruff DP. 2003. Characterization of the interaction of glycine with Cu (100) and Cu (111). *Surf Sci* **531:** 304–318.

Ellington AE, Szostak JW. 1990. Selection *in vitro* of single-stranded DNA molecules that fold into specific ligand binding sites. *Nature* **346:** 818–822.

Elhadj S, Salter EA, Wierzbicki A, De Yoreo JJ, Han N, Dove PM. 2006. Peptide controls growth on calcite mineralization: Polyaspartate chain length affects growth kinetics and acts as a stereochemical switch on morphology. *Crystal Growth Design* **6:** 197–201.

Ertem G. 2002. Montmorillonite, oligonucleotides, RNA and the origin of life. *Orig Life Evol Biosph* **34:** 549–570.

Ertem G, Ferris JP. 1996. Synthesis of RNA oligomers on heterogeneous templates. *Nature* **379:** 238–240.

Ertem G, Ferris JP. 1997. Template-directed synthesis using the heterogeneous templates produced by montmorillonite catalysis: A possible bridge between the prebiotic and RNA worlds. *J Am Chem Soc* **119**: 7197–7201.

Ertem G, Hazen RM, Dworkin JP. 2007. Sequence analysis of trimer isomers formed by montmorillonite catalysis in the reaction of binary monomer mixtures. *Astrobiology* **7**: 715–724.

Ertem G, Hazen RM, Snellinger AM, Dworkin JP, Johnston MV. 2008. Sequence- and region-selective formation of RNA-like oligomers by montmorillonite catalysis. *Int J Astrobiology* **7**: 1–7.

Eschenmoser A. 1999. Chemical etiology of nucleic acid structure. *Science* **284**: 2118–2124.

Efstathiou V, Woodruff DP. 2003. Characterization of the interaction of glycine with Cu(100) and Cu(111). *Surf Sci* **531**: 304–318.

Evgenii K, Wolfram T. 1978. The role of quartz in the origin of optical activity on Earth. *Orig Life Evol Biosph* **30**: 431–434.

Fenter P, Geissbühler P, DiMasi E, Srajer G, Sorenson LB, Sturchio NC. 2000. Surface speciation of calcite observed in situ by high-resolution X-ray reflectivity. *Geochim Cosmochim Acta* **64**: 1221–1228.

Ferris JP. 1999. Prebiotic synthesis on minerals: bridging the prebiotic and RNA worlds. *Biol Bull* **196**: 311–314.

Ferris JP. 2005. Mineral catalysis and prebiotic synthesis: montmorillonite-catalyzed formation of RNA. *Elements* **1**: 145–149.

Ferris JP, Hill AR Jr, Liu R, Orgel LE. 1996. Synthesis of long prebiotic oligomers on mineral surfaces. *Nature* **381**: 59–61.

Fitts JP, Persson P, Brown GE Jr, Parks GA. 1999. Structure and bonding of Cu(II)-glutamate complexes at the γ-Al_2O_3-water interface. *J Colloid Surface Sci* **220**: 133–147.

Fox SW, Harada K. 1958. Thermal copolymerization of amino acids to a product resembling protein. *Science* **128**: 1214.

Frondel C. 1978. Characters of quartz fibers. *Am Mineral* **63**: 17–27.

Fry I. 2000. *The emergence of life on Earth: A historical and scientific overview*. Rutgers University Press, New Brunswick, NJ.

Geissbuhler P, Fenter P, DiMasi E, Srajer G, Sorensen LB, Sturchio NC. 2004. Three dimensional structure of the calcite-water interface by surface X-ray scattering. *Surf Sci* **573**: 191–203.

Gellman AJ, Horvath JD, Buelow MT. 2001. Chiral single crystal surface chemistry. *J Mol Catal A* **167**: 3–11.

Gisler A. 1981. Die adsorption von aminosauren an grenzflachen oxid-wasser. PhD Thesis, 124 pp., Universitat Bern.

Glamoclija M, Steele A, Fries M, Schieber J, Voytek MA, Cockell CS. 2009. Association of anatase (TiO_2) amd microbes: unusual fossilization effect or a potential biosignature? In *The ICDP-USGS Deep Drilling Project in the Chesapeake Bay Impact Structure: Results from the Eyreville Core Holes* (ed. Gohn GS, et al.) *Geol Soc Am Spec Pap* **458**: 965–975.

Goldschmidt VM. 1952. Geochemical aspects of the origin of complex organic molecules on the earth, as precursors to organic life. *New Biol* **12**: 97–105.

Greenwell HC, Coveney PV. 2006. Layered double hydroxide minerals as possible prebiotic information storage and transfer compounds. *Orig Life Evol Biosph* **36**: 13–37.

Grew ES, Hazen RM. 2009. Evolution of the minerals of beryllium: a quintessential crustal element [abstract]. *Geol Soc Am Abstr with Prog* (Portland OR).

Guevremont JM, Strongin DR, Schoonin MAA. 1998. Thermal chemistry of H_2S and H_2O on the (100) plane of pyrite: unique reactivity of defect sites. *Am Mineral* **83**: 1246–1255.

Guevremont JM, Elsetinow AR, Strongin DR, Bebie J, Schoonen MAA. 1998. Structure sensitivity of pyrite oxidation: comparison of the (100) and (111) planes. *Am Mineral* **83**: 1353–1356.

Hanczyc MM, Fujikawa SM, Szostak JW. 2003. Experimental models of primitive cellular compartments: encapsulation, growth, and division. *Science* **302**: 618–622.

Hazen RM. 2004. Chiral crystal faces of common rock-forming minerals. In *Progress in Biological Chirality* (Eds. G Palyi et al.), pp. 137–151. Elsevier, New York.

Hazen RM. 2005. *Genesis: The scientific quest for life's origins*. Joseph Henry Press, National Academy of Sciences, Washington, DC.

Hazen RM. 2006. Mineral surfaces and the prebiotic selection and organization of biomolecules. *Am Mineral* **91**: 1715–1729.

Hazen RM, Sholl DS. 2003. Chiral selection on inorganic crystalline surfaces. *Nature Mater* **2**: 367–374.

Hazen RM, Ewing RJ, Sverjensky DA. 2009. The evolution of uranium and thorium minerals. *Am Mineral* **94**: 1293–1311.

Hazen RM, Filley TR, Goodfriend GA. 2001. Selective adsorption of L- and D-amino acids on calcite: Implications for biochemical homochirality. *Proc Natl Acad Sci* **98**: 5487–5490.

Hazen RM, Griffin P, Carothers J, Szostak J. 2007. Functional information and the emergence of biocomplexity. *Proc Natl Acad Sci* **104**: 8574–8581.

Hazen RM, Papineau D, Bleeker W, Downs RT, Ferry JM, McCoy TJ, Sverjensky DA, Yang H. 2008. Mineral evolution. *Am Mineral* **93**: 1693–1720.

Heinen W, Lauwers A-M. 1996. Organic sulfur compounds resulting from the interaction of iron sulfide, hydrogen sulfide and carbon dioxide in an anaerobic environment. *Orig Life Evol Biosph* **26**: 131–150.

Hennet RJC, Holm NG, Engel MH. 1992. Abiotic synthesis of amino acids under hydrothermal conditions and the origin of life: a perpetual phenomenon? *Naturwissenschaften* **79**: 361–365.

Hill ARJr, Böhler C, Orgel LE. 1998. Polymerization on the rocks: Negatively-charged α-amino acids. *Orig Life Evol Biosph* **28**: 235–242.

Hoang QQ, Sicheri F, Howard AJ, Yang DSC. 2003. Bone recognition mechanism of porcine osteocalcin from crystal structure. *Nature* **425**: 977–980.

Hochella MF Jr. 1990. Atomic structure, microtopography, composition, and reactivity of mineral surfaces. In *Mineral-water interface geochemistry*, (ed. MF Hochella Jr, AF White), *Rev Mineral* **23**: 87–132.

Hochella MF Jr. 1995. Mineral surfaces: their characterization and their physical and reactive nature. In *Mineral surfaces* (ed. DJ Vaughan, RAD Pattrick), pp. 17–60. Chapman and Hall, New York.

Hochella MF Jr, White AF. Editors. 1990. *Mineral-Water Interface Geochemistry.* Rev Mineral 23, Mineralogical Society of America, Chantilly, VA.

Hochella MFJr, Eggleston CM, Eilings VB, Thompson MS. 1990. Atomic structure and morphology of the albite (010) surface: an atomic-force microscopy and electron diffraction study. *Am Mineral* **75:** 723–730.

Holm NG, Ertem G, Ferris JP. 1993. The binding and reactions of nucleotides and polynucleotides on iron oxide hydroxide polymorphs. *Orig Life Evol Biosph* **23:** 195–215.

Holm NG, Dowler MJ, Wadsten T, Arrhenius G. 1983. β-FeOOH.Cl$_n$ (akaganite) and Fe$_{1-x}$O (wustite) in hot brine from the Atlantis II Deep (Red Sea) and the uptake of amino acids by synthetic β-FeOOH.Cl$_n$. *Geochim Cosmochim Acta* **47:** 1465–1470.

Holmen BA, Casey WH. 1996. Hydroxamate ligands, surface chemistry, and the mechanism of ligand-promoted dissolution of goethite [FeOOH(s)]. *Geochim Cosmochim Acta* **60:** 4403–4416.

Horvath JD, Gellman AJ. 2001. Enantiospecific desorption of R- and ZS-propylene oxide from a chiral Cu (643) surface. *J Am Chem Soc* **123:** 7953–7954.

Horvath JD, Gellman AJ. 2002. Enantiospecific desorption of chiral compounds from chiral Cu (643) and achiral Cu (111) surfaces. *J Am Chem Soc* **124:** 2384–2392.

Huber C, Wächtershäuser G. 1997. Activated acetic acid by carbon fixation on (Fe,Ni)S under primordial conditions. *Science* **276:** 245–247.

Huber C, Wächtershäuser G. 2006. α-hydroxy and α-amino acids under possible Hadean, volcanic origin-of-life conditions. *Science* **314:** 630–632.

Hug SJ, Sulzberger B. 1994. In situ Fourier transform infrared spectroscopic evidence for the formation of several different surface complexes of oxalate on TiO$_2$ in the aqueous phase. *Langmuir* **10:** 3587–3597.

Hung A, Yarovsky I, Russo SP. 2003. Density-functional theory studies of xanthate adsorption on the pyrite FeS$_2$(110) and (111) surfaces. *J Chem Phys* **118:** 6022–6029.

Jacoby M. 2002. 2-D stereoselectivity. *Chem Eng News* **80** (March 25, 2002): 43–46.

Johnson BB, Sjoberg S, Persson P. 2004a. Surface complexation of mellitic acid to goethite: An attenuated total reflection Fourier transform infrared study. *Langmuir* **20:** 823–828.

Johnson SB, Brown GE Jr, Healy TW, Scales PJ. 2005b. Adsorption of organic matter at mineral/water interfaces: 6. Effect of inner-sphere versus outer-sphere adsorption on colloidal stability. *Langmuir* **21:** 6356–6365.

Johnson SB, Yoon TH, Brown GE Jr. 2005a. Adsorption of organic matter at mineral/water interfaces: 5. Effects of adsorbed natural organic matter analogues on mineral dissolution. *Langmuir* **21:** 2811–2821.

Johnson SB, Yoon TH, Kokar BD, Brown GE Jr. 2004b. Adsorption of organic matter at mineral/water interfaces: 2. Outer-sphere adsorption of maleate and implications for dissolution processes. *Langmuir* **20:** 4996–5006.

Johnson SB, Yoon TH, Kokar BD, Brown GE Jr. 2004c. Adsorption of organic matter at mineral/water interfaces: 3. Implications of surface dissolution for adsorption of oxalate. *Langmuir* **20:** 11480–11492.

Jones TE, Baddeley CJ. 2004. An investigation of the adsorption of (R,R)-tartaric acid on oxidised Ni{111} surfaces. *J Molec Catal A* **216:** 223–231.

Jonsson CM, Jonsson CL, Sverjensky DA, Cleaves HJ, Hazen RM. 2009. Attachment of L-glutamate to rutile (α-TiO$_2$): a potentiometric, adsorption and surface complexation study. *Langmuir* **25:** 12127–12135.

Jonsson CM, Jonsson CL, Estrada C, Sverjensky DA, Cleaves HJ, Hazen RM. 2010. Adsorption of L-aspartate to rutile (α-TiO$_2$): experimental and theoretical surface complexation studies. *Geochem Cosmochim Acta* **74:** 2356–2367.

Kahr B, Gurney RW. 2001. Dyeing crystals. *Chem Rev* **101:** 893–951.

Karagounis G, Coumoulos G. 1938. A new method for resolving a racemic compound. *Nature* **142:** 162–163.

Kelley DS, Karson JA, Früh-Green GL, Yoerger DR, Shank TM, Butterfield DA, Hayes JM, Schrenk MO, Olson EJ, Proskurowski G, et al. 2005. A serpentinite-hosted ecosystem: The Lost City hydrothermal field. *Science* **307:** 1428–1434.

Kitadai N, Yokoyama T, Nakashima S. 2009. ATR-IR spectroscopic study of L-lysine dsorption on amorphous silica. *J Colloid Interface Sci* **329:** 31–37.

Klug O, Forsling W. 1999. A spectroscopic study of phthalate adsorption on γ-aluminum oxide. *Langmuir* **15:** 6961–6868.

Klussmann M, Iwamura H, Mathew SP, Wells DHJr, Pandya U, Armstrong A, Blackmond DG. 2006. Thermodynamic control of asymmetric amplification in amino acid catalysis. *Nature* **441:** 621–623.

Klussmann M, Izumi T, White AJ, Armstrong A, Blackmond DG. 2007. Emergence of solution-phase homochirality via engineering of amino acids. *J Am Chem Soc* **129:** 7657–7660.

Koretsky CM, Sverjensky DA, Sahai N. 1998. A model of surface site types on oxide and silicate minerals based on crystal chemistry: Implications for site types and densities, multi-site adsorption, surface infrared spectroscopy, and dissolution kinetics. *Am J Sci* **298:** 349–438.

Kornberg A, Rao NN, Ault-Riché D. 1999. Inorganic polyphosphate: A molecule of many functions. *Annu Rev Biochem* **68:** 89–125.

Kristensen K, Stipp SLS, Refson K. 2004. Modeling steps and kinks on the surface of calcite. *J Chem Phys* **121:** 8511–8523.

Kubicki JD, Schroeter LM, Itoh MJ. 1999. Attenuated total reflectance Fouriertransform infrared spectroscopy of carboxylic acids adsorbed onto mineral surfaces. *Geochim Cosmochim Acta* **63:** 2709–2725.

Kühnle A, Linderoth TR, Hammer B, Besenbacher F. 2002. Chiral recognition in dimerization of adsorbed cysteine observed by scanning tunneling microscopy. *Nature* **415:** 891–893.

Kwon K, Kubicki JD. 2004. Molecular orbital study on surface complex structures of phosphates to iron hydroxides: Calculation of vibrational frequencies and adsorption energies. *Langmuir* **20:** 9249–9254.

Lackovic K, Angove MJ, Johnson BB, Wells JD. 2004. Modeling the adsorption of Cd(II) onto goethite in the presence of citric acid. *J Colloid Interface Sci* **267:** 49–59.

Lackovic K, Johnson BB, Angove MJ, Wells JD. 2003. Modeling the adsorption of citric acid onto Muloorina illite and related clay minerals. *J Colloid Interface Sci* **267:** 49–59.

Lahav N. 1994. Minerals and the origin of life: Hypotheses and experiments in heterogeneous chemistry. *Hetreo Chem Rev* **1:** 159–179.

Lahav N. 1999. *Biogenesis: Theories of life's origins.* Oxford University Press, New York.

Lahav N, White D, Chang S. 1978. Peptide formation in the prebiotic era: Thermal condensation of glycine in fluctuating clay environments. *Science* **201:** 67–69.

Lambert J-F. 2008. Adsorption and polymerization of amino acids on mineral surfaces: a review. *Orig Life Evol Biosph* **38:** 211–242.

Lambert J-F, Stievano L, Lopes I, Gharsallah, Piao L. 2009. The fate of amino acids adsorbed on mineral matter. *Planet Space Sci* **57:** 460–467.

Langel W, Menken L. 2003. Simulation of the interface between titanium oxide and amino acids in solution by first principles MD. *Surf Sci* **538:** 1–9.

Langmuir D. 1997. *Aqueous environmental geochemistry.* Prentice-Hall, New York.

Lasaga AC. 1990. Atomic treatment of mineral-water surface reactions. In *Mineral-water interface geochemistry* (ed. MF Hochella Jr, AF White), *Rev Mineral* **23:** 17–85.

Lasaga AC, Holland HD, Dwyer MJ. 1971. Primordial oil slick. *Science* **174:** 53–55.

Lee H, Lee BP, Messersmith PB. 2007. A reversible wet/dry adhesive inspired by mussels and geckos. *Nature* **448:** 338–342.

Lee H, Scherer NF, Messersmith PB. 2006. Single-molecule mechanics of mussel adhesion. *Proc Natl Acad Sci* **103:** 12999–13003.

Lindegren M, Loring JS, Redden G, Persson P. 2005. Citrate adsorption at the water-goethite interface: a spectroscopic evaluation of surface complexes. Goldschmidt Conf Abstr (Idaho), A369.

Lippmann DZ, Dix J. 1999. Possible mechanisms for spontaneous production of enantiomeric excess. In *Advances in biochirality* (ed. G Pályi, et al.), 85–98. Elsevier, New York.

Liu R, Orgel LE. 1998. Polymerization on the rocks: β-amino acids and arginine. *Orig Life Evol Biosph* **28:** 245–257.

Lorenzo MO, Baddeley CJ, Muryn C, Raval R. 2000. Extended surface chirality from supramolecular assemblies of adsorbed chiral molecules. *Nature* **404:** 376–379.

Lowenstam HA, Weiner S. 1989. *On biomineralization.* Oxford University Press, New York.

Marshall WL. 1994. Hydrothermal synthesis of amino acids. *Geochim Cosmochim Acta* **58:** 2099–2106.

Martin ZCTP. 2007. Chemical analysis of organic molecules in carbonaceous meteorites. PhD Thesis, Lisbon University, Portugal.

Matrajt G, Blanot D. 2004. Properties of synthetic ferrihydrite as an amino acid adsorbent and a promoter of peptide bond formation. *Amino Acids* **26:** 153–158.

McCollum TM, Ritter G, Simoneit BR. 1999. Lipid synthesis under hydrothermal conditions by Fischer-Tropsch-type reactions. *Orig Life Evol Biosph* **29:** 153–166.

McCollom TM, Simoneit BRT. 1999. Abiotic formation of hydrocarbons and oxygenated compounds during thermal decomposition of iron oxalate. *Orig Life Evol Biosph* **29:** 167–186.

McFadden CF, Cremer PS, Gellman AJ. 1996. Adsorption of chiral alcohols on "chiral" metal surfaces. *Langmuir* **12:** 2483–2487.

McSween HY, Richardson SM, Uhle M. 2003. *Geochemistry: Pathways and processes*, 2nd ed. Columbia University Press, New York.

Miller SL. 1953. A Production of amino acids under possible primitive Earth conditions. *Science* **117:** 528–529.

Miller SL, Urey HC. 1959. Organic compound synthesis on the primitive Earth. *Science* **130:** 245–251.

Monnard P-A, Apel CL, Kanavarioti A, Deamer DW. 2002. Influence of ionic inorganic solutes on self-assembly and polymerization processes related to early forms of life: Implications for a prebiotic aqueous medium. *Astrobiology* **2:** 139–152.

Morowitz HJ. 1992. *The beginnings of cellular life: metabolism recapitulates biogenesis.* Yale University Press, New Haven, CT.

Morowitz HJ, Kostelnik JD, Yang J, Cody GD. 2000. The origins of intermediary metabolism. *Proc Natl Acad Sci* **97:** 7704–7708.

Nakajima T, Yabushita Y, Tabushi I. 1975. Amino acid synthesis through biogenetic-type CO_2 fixation. *Nature* **256:** 60–61.

Nelson DL, Cox MM. 2004. *Lehninger's principles of biochemistry*, 4th ed. Worth Publishers, New York.

Nielsen-Marsh CM, Richards MP, Hauschka PV, Thomas-Oates JE, Trinkaus E, Pettitt PB, Karavanic I, Poinar H, Collins MJ. 2005. Osteocalcin protein sequences of Neanderthals and modern primates. *Proc Natl Acad Sci* **102:** 4409–4413.

Noorduin WL, Izumi T, Millemaggi A, Leeman M, Meekes H, Van Enckevort WJP, Kellogg RM, Kaptein B, Vlieg E, Blackmond DG. 2008. Emergence of a single solid chiral state from a nearly racemic amino acid derivative. *J Am Chem Soc* **130:** 1158–1159.

Nordin J, Persson P, Laiti E, Sjoberg S. 1997. Adsorption of o-phthalate at the water-boehmite (γ-AlOOH) interface: evidence for two coordination modes. *Langmuir* **13:** 4085–4093.

Nowack B, Stone AT. 1999. Adsorption of phosphonates onto the goethite-water interface. *J Colloid Interface Sci* **214:** 20–30.

Oparin AI. 1938. *The origin of life* (S Morgulis, translator). Macmillan, New York.

Orgel LE. 1998. Polymerization on the rocks: Theoretical introduction. *Orig Life Evol Biosph* **28:** 227–234.

Orme CA, Noy A, Wierzbicki A, McBride MT, Grantham M, Teng HH, Dove PM, DeYoreo JJ. 2001. Formation of chiral morphologies through selective binding of amino acids to calcite surface steps. *Nature* **411:** 775–778.

Oró J. 1961. Mechanism of synthesis of adenine from hydrogen cyanide under possible primitive Earth conditions. *Nature* **191**: 1193–1194.

Parks GA. 1990. Surface energy and adsorption at mineral-water interfaces: An introduction. In *Mineral-water interface geochemistry* (ed. MF Hochella Jr, AF White), *Rev Mineral* **23**: 133–175.

Parsons R. 1990. Electrical double layer: Recent experimental and theoretical developments. *Chem Rev* **90**: 813–826.

Parsons I, Lee MR, Smith JV. 1998. Biochemical evolution II: Origin of life in tubular microstructures in weathered feldspar surfaces. *Proc Natl Acad Sci* **95**: 15173–15176.

Pászti Z, Guczi L. 2009. Amino acid adsorption on hydrophilic TiO_2: a sum frequency generation vibrational spectroscopy study. *Vib Spec* **50**: 48–56.

Peacock CL, Sherman DM. 2004. Vanadium adsorption onto goethite (a-FeOOH) at pH 1.5 to 12: a surface complexation model based on ab initio molecular geometries and EAFS spectroscopy. *Geochim Cosmochim Acta* **68**: 1723–1733.

Perezgasga L, Serrato-Diaz A, Negron-Mendoza A, de Pablo Galan L, Mosqueira FG. 2005. Sites of adsorption of adenine, uracil, and their corresponding derivatives on sodium montmorillonite. *Orig Life Evol Biosph* **35**: 91–110.

Persson P, Axe K. 2005. Adsorption of oxalate and malonate at the water-goethite interface: molecular surface speciation from IR spectroscopy. *Geochim Cosmochim Acta* **69**: 541–552.

Pinto JP, Gladstone GR, Yung YL. 1980. Photochemical production of formaldehyde in Earth's primitive atmosphere. *Science* **210**: 183–185.

Pitsch S, Eschenmoser A, Gedulin B, Hui S, Arrhenius G. 1995. Mineral induced formation of sugar phosphates. *Orig Life Evol Biosph* **25**: 297–334.

Pizzarello S. 2006. The chemistry of life's origin: A carbonaceous meteorite perspective. *Accounts of Chemical Research* **39**: 231–237.

Pizzarello S, Cronin JR. 2000. Non-racemic amino acids in the Murray and Murchison meteorites. *Geochem Cosmochim Acta* **64**: 329–338.

Pizzarello S, Weber A. 2004. Prebiotic amino acids as asymmetric catalysts. *Science* **303**: 1151.

Pontes-Buarques M, Tessis AC, Bonapace JA, Monte MBM, Cortés-Lopez G, de Souza-Barros F, Vieyra A. 2001. Modulation of adenosine 5′-monophosphate adsorption onto aqueous resident pyrite: potential mechanisms for prebiotic reactions. *Orig Life Evol Biosph* **31**: 343–362.

Power TD, Sholl DS. 1999. Enantiospecific adsorption of chiral hydrocarbons on naturally chiral Pt and Cu surfaces. *J Vac Sci Technol A* **17**: 1700–1704.

Ricardo A, Carrigan MA, Olcott AN, Benner SA. 2004. Borate minerals stabilize ribose. *Science* **303**: 196–197.

Rimola A, Ugliengo P, Sodupe M. 2009. Formation versus hydrolysis of the peptide bond from a quantum-mechanical viewpoint: the role of mineral surfaces and implications for the origin of life. *Int J Mol Sci* **10**: 746–760.

Roddick-Lanzilotta AD, McQuillan AJ. 1999. An in situ infrared spectroscopic investigation of lysine peptide and polylysine adsorption to TiO_2 from aqueous solutions. *J Colloid Interface Sci* **217**: 194–202.

Roddick-Lanzilotta AD, McQuillan AJ. 2000. An in situ infrared spectroscopic study of glutamic acid and of aspartic acid adsorbed on TiO_2: implications for the biocompatibility of titanium. *J Colloid Interface Sci* **227**: 48–54.

Roddick-Lanzilotta AD, Connor PA, McQuillan AJ. 1998. An in situ infrared spectroscopic study of the adsorption of lysine to TiO_2 from an aqueous solution. *Langmuir* **14**: 6479–6484.

Rodriguez R, Blesa MA, Regazzoni AE. 1996. Surface complexation at the TiO_2 (anatase)/aqueous solution interface: chemisorptions of catechol. *J Colloid Interface Sci* **177**: 122–131.

Rouhi AM. 2004. Chiral chemistry. *Chem Eng News* **82** (June 14, 2004): 47–62.

Rosenqvist J, Axe K, Sjöberg S, Persson P. 2003. Adsorption of dicarboxylates on nano-sized gibbsite particles: effects of ligand structure on bonding mechanism. *Colloids Surfaces* **220**: 91–104.

Russell MJ, Hall AJ. 1997. The emergence of life from iron monosulphide bubbles at a submarine hydrothermal redox and pH front. *J Geol Soc London* **154**: 377–402.

Sanchez R, Ferris J, Orgel LE. 1967. Studies in prebiotic synthesis. II. Synthesis of purine precursors and amino acids from aqueous hydrogen cyanide. *J Molec Biol* **30**: 223–253.

Schindler PW. 1990. Co-adsorption of metal ions and organic ligands: formation of ternary surface complexes. In *Mineral-Water Interface Geochemistry* (ed. MF Hochella Jr, AF White), *Rev Mineral* **23**: 281–307.

Schoonen MAA, Xu Y. 2001. Nitrogen reduction under hydrothermal vent conditions: Implications for the synthesis of C-H-O-N compounds. *Astrobiology* **1**: 133–140.

Schoonen MAA, Xu Y, Bebie J. 1999. Energetics and kinetics of the prebiotic synthesis of simple organic acids and amino acids with the FeS/FeS_2 redox couple as reductant. *Orig Life Evol Biosph* **29**: 5–32.

Schoonen MAA, Smirnov A, Cohn C. 2004. A perspective on the role of minerals in prebiotic synthesis. *Ambio* **33**: 539–551.

Schweitzer MH, Suo Z, Avci R, Asara JM, Allen MA, Arce FT, Horner JR. 2007. Analyses of soft tissue from *Tyrannosaurus rex* suggest the presence of protein. *Science* **316**: 277–280.

Seelig B, Szostak JW. 2007. Selection and evolution of enzymes from a partially randomized non-catalytic scaffold. *Nature* **448**: 828–833.

Seewald JS, Zolotov MY, McCollom T. 2006. Experimental investigation of single carbon compounds under hydrothermal conditions. *Geochim Cosmochim Acta* **70**: 446–460.

Sephton MA. 2002. Organic compounds in carbonaceous meteorites. *Natural Products Report* **19**: 292–311.

Shapiro R. 1988. Prebiotic ribose synthesis: a critical analysis. *Orig Life Evol Biosph* **18**: 71–85.

Sheals J, Sjoberg S, Persson P. 2002. Adsorption of glyphosate on goethite: molecular characterization of surface complexes. *Environ Sci Tech* **36**: 3090–3095.

Sheals J, Granstrom M, Sjoberg S, Persson P. 2003. Coadsorption of Cu(II) and glyphosate at the water-goethite (α-FeOOH) interface: molecular structures from FTIR and EXAFS measurements. *J Colloid Interface Sci* **262:** 38–47.

Shock EL. 1992. Chemical environments of submarine hydrothermal systems. *Orig Life Evol Biosph* **22:** 67–107.

Shock EL. 1993. Hydrothermal dehydration of aqueous organic compounds. *Geochim Cosmochim Acta* **57:** 3341–3349.

Sholl DS. 1998. Adsorption of chiral hydrocarbons on chiral platinum surfaces. *Langmuir* **14:** 862–867.

Sholl DS, Asthagiri A, Power TD. 2001. Naturally chiral metal surfaces as enantiospecific adsorbents. *J Phys Chem B* **105:** 4771–4782

Sljivancanin Z, Gothelf KV, Hammer B. 2002. Density functional theory study of enantiospecific adsorption at chiral surfaces. *J Am Chem Soc* **124:** 14789–14798.

Smith JV. 1998. Biochemical evolution. I. Polymerization on internal, organophilic silica surfaces of dealuminated zeolites and feldspars. *Proc Natl Acad Sci* **95:** 3370–3375.

Smith JV, Arnold FP Jr, Parsons I, Lee MP. 1999. Biochemical evolution III: Polymerization on organophilic silica-rich surfaces, crystal-chemical modeling, formation of first cells, and geological clues. *Proc Natl Acad Sci* **96:** 3479–3485.

Soai K, Osanai S, Kadowaki K, Yonekubo S, Shibata T, Sato I. 1999. d- and l-quartz-promoted highly enantioselective synthesis of a chiral organic compound. *J Am Chem Soc* **121:** 11235–11236.

Somasundaran P, Krishnakumar S. 1994. In situ spectroscopic investigations of adsorbed surfactant and polymer layers in aqueous and non-aqueous systems. *Colloids Surfaces* **93:** 79–95.

Somorjai GA. 1994. *Introduction to surface chemistry*. Wiley, New York.

Sowerby SJ, Heckl WM, Petersen GB. 1996. Chiral symmetry breaking during the self assembly of monolayers from achiral purine molecules. *J Molec Evol* **43:** 419–424.

Sowerby SJ, Edelwirth M, Heckl WM. 1998a. Self-assembly at the prebiotic solid-liquid interface: Structures of self-assembled monolayers of adenine and guanine bases formed on inorganic surfaces. *J Phys Chem* **102:** 5914–5922.

Sowerby SJ, Edelwirth M, Reiter M, Heckl WM. 1998b. Scanning tunneling microscopy image contrast as a function of scan angle in hydrogen-bonded self-assembled monolayers. *Langmuir* **14:** 5195–5202.

Sowerby SJ, Petersen NG, Holm NG. 2002. Primordial coding on amino acids by adsorbed purine bases. *Orig Life Evol Biosph* **32:** 35–46.

Steele JH, Thorpe SA, Turekian KK. Editors. 2009. *Encyclopedia of ocean sciences*, 2nd ed. Elsevier, Burlington, MA.

Stipp SL. 2002. Where the bulk terminates: Experimental evidence for restructuring, chemibonded OH- and H+, adsorbed water and hydrocarbons on calcite surfaces. *Molec Sim* **28:** 497–516.

Stipp SL, Hochella MF Jr. 1991. Structure and bonding environments at the calcite surface as observed with X-ray photoelectron spectroscopy (XPS) and low energy electron diffraction (LEED). *Geochim Cosmochim Acta* **55:** 1723–1736.

Stribling R, Miller SL. 1987. Energy yields for hydrogen cyanide and formaldehyde synthesis: the HCN and amino acid concentrations in the primitive ocean. *Orig Life* **17:** 261–273.

Stumm W. 1992. *Chemistry of the solid-water interface*. Wiley, New York.

Sverjensky DA. 2005. Prediction of surface charge on oxides in salt solutions: revisions for 1:1 (M^+L^-) electrolytes. *Geochim Cosmochim Acta* **69:** 225–257.

Sverjensky DA, Fukushi K. 2006. Anion adsorption on oxide surfaces: Inclusion of the water dipole in modeling the electrostatics of ligand exchange. *Environ Sci Tech* **40:** 263–271.

Sverjensky DM, Jonsson CM, Jonsson CL, Cleaves HJ, Hazen RM. 2008. Glutamate surface speciation on amorphous titanium dioxide and hydrous ferric oxide. *Environ Sci Tech* **42:** 6034–6039.

Szostak J. 2003. Functional information. *Nature* **423:** 689.

Teng HH, Dove PM. 1997. Surface site-specific interactions of aspartate with calcite during dissolution: Implications for biomineralization. *Am Mineral* **82:** 878–887.

Teng HH, Chen Y, Pauli E. 2006. Direction specific interactions of 1,4-dicarboxylic acid with calcite surfaces. *J Am Chem Soc* **128:** 14482–14484.

Teng HH, Dove PM, DeYoreo JJ. 2000. Kinetics of calcite growth: analysis of surface processes and relationships to macroscopic rate laws. *Geochim Cosmochim Acta* **64:** 2255–2266.

Teng HH, Dove PM, Orme C, DeYoreo JJ. 1998. The thermodynamics of calcite growth: a baseline for understanding biomineral formation. *Science* **282:** 724–727.

Thomas MM, Clouse JA, Longo JM. 1993. Adsorption of organic compounds on carbonate minerals. I. model compounds and their influence on mineral wettability. *Chem Geol* **109:** 201–213.

Toomes RL, Kang J-H, Woodruff DP, Polcik M, Kittel M, Hoeft J-T. 2003. Can glycine form homochiral structural domains on low-index copper surfaces? *Surface Science* **552:** L9–L14.

Trout CC, Kubicki JD. 2004. UV resonance Raman spectra and molecular orbital calculations of salicylic and phthalic acids complexed to Al(3+) in solution and on mineral surfaces. *J Phys Chem A* **108:** 11580–11590.

Tuck A. 2002. The role of atmospheric aerosols in the origin of life. *Surveys Geophys* **23:** 379–409.

Tsuchida R, Kobayashi M, Nakamura A. 1935. Asymmetric adsorption of complex salts on quartz. *J Chem Soc Japan* **56:** 1339.

Turekian KK. 1968. *Oceans*. Prentice-Hall, New York.

Uchihashi T, Okada T, Sugawara Y, Yokoyama K, Morita S. 1999. Self-assembled monolayer of adenine base on graphite studied by noncontact atomic force microscopy. *Phys Rev B* **60:** 8309–8313.

Van Cappellen P, Charlet L, Stumm W, Wersin P. 1993. A surface complexation model of the carbonate mineral-aqueous solution interface. *Geochim Cosmochim Acta* **57:** 3505–3518.

Van Dover CL. 2000. *The ecology of deep-sea hydrothermal vents*. Princeton University Press, Princeton, NJ.

Vaughan DJ. 1995. Mineral surfaces: an overview. In *Mineral Surfaces*, (ed. DJ Vaughan, RAD Pattrick), pp. 1–16. Chapman and Hall, New York.

Vlasova NN. 2005. Adsorption of copper-amino acid complexes on the surface of highly dispersed silica. *Colloid J* **67:** 537–541.

Vlasova NN, Golovkova LP. 2004. The adsorption of amino acids on the surface of highly dispersed silica. *Colloid J* **66:** 733–738.

Wächtershäuser G. 1988. Before enzymes and templates: theory of surface metabolism. *Microbiol Rev* **52:** 452–484.

Wächtershäuser G. 1990. Evolution of the first metabolic cycles. *Proc Natl Acad Sci* **87:** 200–204.

Wächtershäuser G. 1992. Groundworks for an evolutionary biochemistry: the iron-sulfur world. *Prog Biophys Molec Biol* **58:** 85–201.

Weiner S, Addadi L. 1997. Design strategies in mineralized biological materials. *J Mater Chem* **7:** 689–702.

Weissbuch I, Addadi L, Lahav M, Leiserowitz L. 1991. Molecular recognition at crystal interfaces. *Science* **253:** 637–644.

Westheimer FH. 1987. Why nature chose phosphates. *Science* **235:** 1173–1178.

Whitehead CF. 2003. (Amino)carboxylate coordination reactions with ferric (hydr)oxides: adsorption and ligand-assisted dissolution. PhD Thesis, 303 pp., Johns Hopkins University.

Whitfield J. 2009. Nascence man. *Nature* **459:** 316–319.

Wills C, Bada JL. 2000. *The spark of life: Darwin and the primeval soup*. Perseus, Cambridge, MA.

Wilson DS, Szostak JW. 1999. In vitro selection of functional nucleic acids. *Annu Rev Biochem* **68:** 611–647.

Wright K, Cygan RT, Slater B. 2001. Structure of the (10-14) surfaces of calcite, dolomite, and magnesite under wet and dry conditions. *Phys Chem Chem Phys* **3:** 839–844.

Yoon TH, Johnson SB, Brown GE Jr. 2005. Adsorption of organic matter at mineral/water interfaces: IV. Adsorption of humic substances at boehmite/water interfaces and impact on boehmite dissolution. *Langmuir* **21:** 5002–5012.

Yoon TH, Johnson SB, Musgrave CB, Brown GE Jr. 2004. Adsorption of organic matter at mineral/water interfaces: I. ATR-FTIR spectroscopic and quantum chemical study of oxalate adsorbed at boehmite/water and corundum/water interfaces. *Geochim Cosmochim Acta* **68:** 4505–4518.

Zaikowski L, Friedrich JM. Editors. 2007. *Chemical Evolution across Space and Time*. American Chemical Society Symposium Series, **981:** 1–430.

Zamaraev KL, Romannikov VN, Salganik RI, Wlassof WA, Khramtsov VV. 1997. Modeling of the prebiotic synthesis of oligopeptides: silicate catalysts help to overcome the critical stage. *Orig Life Evol Biosph* **23:** 325–337.

Zepik H, Shavit E, Tang M, Jensen TR, Kjaer K, Bolbach G, Leiserowitz L, Weissbuch I, Lahav M. 2002. Chiral amplification of oligopeptides in two-dimensional crystalline self-assemblies on water. *Science* **295:** 1266–1269.

Zhang Z, Fenter P, Cheng L, Sturchio NC, Bedzyk MJ, Machesky ML, Wesolowski DJ. 2004. Model-independent X-ray imaging of adsorbed cations at the crystal-water interface. *Surf Sci* **554:** L95–L100.

Zhao XY, Zhao RG, Yang WS. 2000. Scanning tunneling microscopy investigation of L-lysine adsorbed on Cu (001). *Langmuir* **16:** 9812–9818.

From Self-Assembled Vesicles to Protocells

Irene A. Chen[1] and Peter Walde[2]

[1]FAS Center for Systems Biology, Harvard University, Cambridge, Massachusetts 02138
[2]Department of Materials, ETH Zurich, CH-8093, Zurich, Switzerland

Correspondence: ichen@lsdiv.harvard.edu

Self-assembled vesicles are essential components of primitive cells. We review the importance of vesicles during the origins of life, fundamental thermodynamics and kinetics of self-assembly, and experimental models of simple vesicles, focusing on prebiotically plausible fatty acids and their derivatives. We review recent work on interactions of simple vesicles with RNA and other studies of the transition from vesicles to protocells. Finally we discuss current challenges in understanding the biophysics of protocells, as well as conceptual questions in information transmission and self-replication.

For synthetic biologists, a useful operational definition of life is "a self-sustaining chemical system capable of Darwinian evolution," which was adopted by the Exobiology program of NASA (Joyce 1994). In the quest to build a simple living system, much recent interest has focused on encapsulating a genetic or metabolic system inside membrane vesicles (Deamer and Dworkin 2005; Luisi et al. 1999; Morowitz et al. 1988; Ourisson and Nakatani 1994; Szostak et al. 2001). Vesicles are supramolecular aggregates containing an aqueous interior that is separated from the bulk solution by one or more bilayers of amphiphiles.

Why use vesicles? Although they are not strictly required in this definition of life, there are two major reasons why vesicle membranes are thought to be important. The first is that the membrane forms a semipermeable barrier that permits small molecules to pass into the cellular space and traps modified (e.g., phosphorylated or polymerized) products. The second reason is evolutionary: The membrane separates different genomes from one another and reduces the problem of inactive parasites (Szathmary and Demeter 1987; Szostak et al. 2001). During the origin of life, physical grouping is a plausible way for replicator enzymes (replicases) to interact nonrandomly. For example, replicases encapsulated in growing and dividing membrane vesicles would tend to be trapped with sequences related to their own sequence, and thus would preferentially copy those sequences. Because the vesicles separate different genomes from each other, poor replicases would not have access to active replicases, whereas mutants with improved replicase activity would benefit directly themselves, as their descendants remain in the same vesicle and copy each other. An occasional parasitic sequence would be separated from most of the active polymerases during vesicle division and could not poison the entire system (the "stochastic corrector" model) (Smith and Szathmary 1995; Szathmary and

Demeter 1987). Thus, a higher level of population organization, the cell, greatly facilitates the evolution of more efficient replicases (Cavalier-Smith 2001; Koch 1984; Matsuura et al. 2002; Szathmary and Demeter 1987; Szostak et al. 2001).

Membrane vesicles are not the only way to segregate different genomes. The attachment of molecules onto surfaces also creates a heterogeneous distribution of interactions based on spatial proximity, a scenario that has been investigated theoretically using cellular automata models (Szabo et al. 2002). Although they may not have been the initial means of achieving genomic segregation during the origin of life, membranes are the dominant means of separating cells today. Membranes presumably assumed this function very long ago, at least three to four billion years ago, at some time before the diversification from the last common ancestor.

Vesicle morphologies and topologies can cover a rich and diverse landscape (Fig. 1, top), although experimentalists tend to prefer unilamellar vesicles because data can be more easily interpreted in this context, and because these vesicles resemble contemporary cells, which use the plasma membrane to separate the cell's

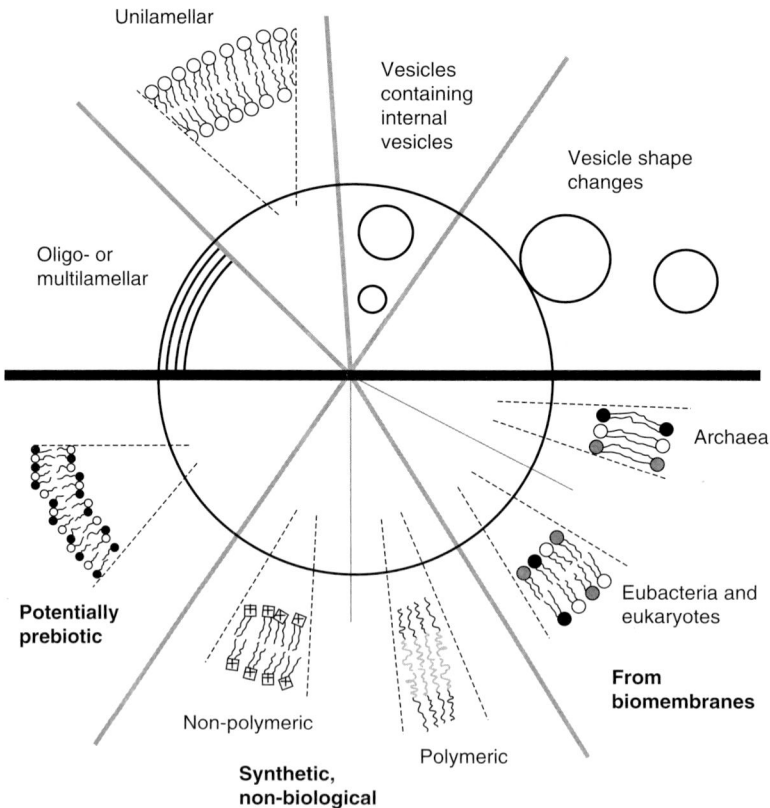

Figure 1. Diversity of morphology and composition of self-assembled vesicles. *Top*: Schematic representation of possible morphologies and shape changes. Vesicles may be multi-, oligo-, or unilamellar. They may also be multivesicular (containing smaller vesicles inside a large vesicle). Under certain conditions and for certain amphiphiles, vesicle shape changes can be induced (e.g., leading to vesicle budding and fission). Vesicles may also be nonspherical (e.g., tubular). The diameter of vesicles may vary between about 30 nm and more than 100 μm. *Bottom*: Vesicle formation occurs for a large number of chemically diverse amphiphiles, including those naturally occurring in biomembranes as well as completely synthetic amphiphiles. Of particular interest are single-chain amphiphiles (or mixtures of amphiphiles) that are potentially prebiotic.

interior from the outside environment. The plasma membrane is composed of roughly equal parts protein and lipid amphiphiles, so one might assume that a protocell membrane was also composed of amphiphilic lipids and/or peptides. However, for simplicity, most experimental work thus far has focused on vesicles made of only one or two prebiotically plausible components (e.g., fatty acid). Vesicles can be made using many different types of amphiphiles, either naturally occurring or synthetic (Fig. 1, bottom). Because of the general robustness of the formation of vesicles, this process has been called an "archetype of self-assembly" (Antonietti and Foerster 2003).

In this article, we first review some fundamentals of self-assembly and focus on important features of vesicles made from single chain amphiphiles. For further discussion of vesicles as model prebiotic experimental systems, we refer the reader to reviews already in the literature (Mansy 2009; Monnard and Deamer 2003; Walde and Ichikawa 2001; Walde et al. 2006) and references contained within these reviews. We then turn to recent results and current challenges in research related to vesicles in the origins of life.

BACKGROUND

Thermodynamics of Self-Assembly of Amphiphiles

Self-assembled structures of amphiphiles are the result of a balance of attractive and repulsive forces. Amphiphiles tend to aggregate because of the hydrophobic effect, which stems primarily from the strong attraction of water for itself (Tanford 1973). Nonpolar solutes disrupt the isotropic hydrogen bonding of water, causing an entropic loss at the solute-water interface. Therefore nonpolar molecules tend to aggregate to minimize the interface (direct attractive interactions, such as van der Waals forces, play a relatively small role). In the absence of a repulsive force, the hydrophobic effect would lead to bulk-phase separation. However, repulsion among amphiphiles resulting from sterics (e.g., among hydrated head groups) or electrostatics favors the formation of structured assemblages.

A minimum number of molecules is required for effective reduction of the water–hydrophobic surface interface in an aggregate relative to free solution. This leads to a highly cooperative transition in which the concentration of aggregates increases dramatically near a critical aggregation concentration (cac) and for micelles, the critical micellization concentration (cmc). This process is often modeled as a phase transition; although such models are not strictly correct, they often capture the essential behavior of the system within experimental error. This approximation is quite good when aggregates are large (i.e., containing >50 molecules), as is the case for most micelles and vesicles.

Micelles are self-assembled structures with a liquidlike hydrocarbon interior and polar headgroups on the exterior (Fig. 2A). Several geometries are possible, including spherical, cylindrical, and ellipsoidal. The free energy of transferring an amphiphile from water into a micelle depends linearly on chain length, so the cmc depends exponentially on chain length (Tanford 1973). Vesicles are closed bilayer structures with an aqueous compartment (Fig. 2B). Whether micelles or vesicles are formed can be explained largely by packing considerations (Israelachvili et al. 1976; Israelachvili et al. 1977). The balance of attractive interactions and repulsive interactions gives an optimal interfacial area per molecule. Amphiphiles self-assemble into structures having area exposed to solvent (a) close to this optimum. Spherical micelles have the greatest interfacial area, in which $v/a = l/3$ (where v = volume and l = hydrophobic chain length), cylindrical micelles have $v/a = l/2$, and vesicles have $v/a = l$. To a first approximation, the presence of two chains on an amphiphile doubles v whereas the optimal a and l are constant, so a consequence of these considerations is that single-chain amphiphiles tend to form micelles whereas double-chain amphiphiles tend to form cylindrical micelles or vesicles. Vesicle formation is also favored by factors that reduce the optimal interfacial area. For example, fatty acids (prebiotically plausible single-chain

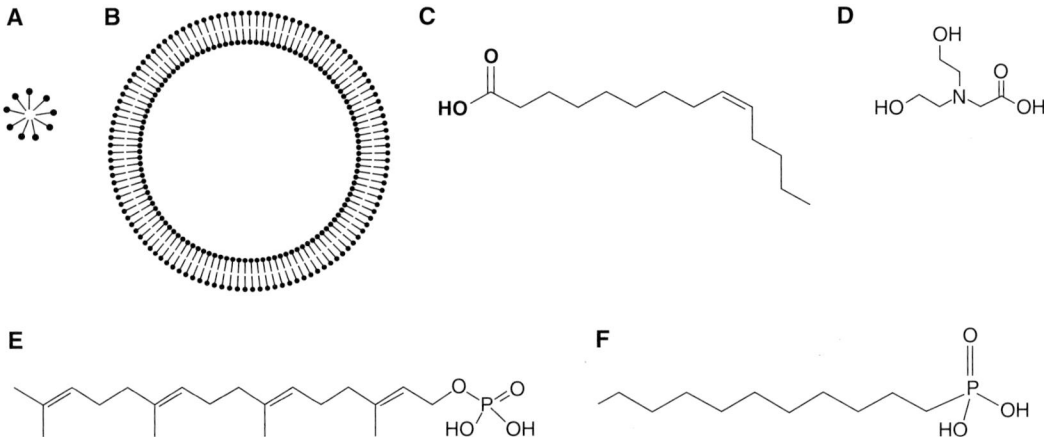

Figure 2. Structures of prebiotically plausible single chain amphiphiles and a commonly used buffer. (*A*) micelle; (*B*) vesicle; (*C*) myristoleic acid; (*D*) bicine; (*E*) geranylgeranyl phosphoric acid; (*F*) *n*-decylphosphonic acid.

amphiphiles) form micelles at high pH when their head groups are negatively charged, but they form vesicles at lower pH when their head groups interact more favorably (Gebicki and Hicks 1973).

Packing constraints also imply a minimum chain length required to form vesicles; for fatty acids (a single chain amphiphile), this length is around eight carbons (Monnard and Deamer 2003). Organic extracts of the Murchison meteorite (Lawless and Yuen 1979) show that abundances of fatty acids decreased as chain length increases, with a three-carbon increase corresponding to a ~10-fold drop in abundance. Roughly speaking, this drop in abundance is somewhat compensated by the decrease of cac as chain length increases (a three-carbon increase corresponds to a ~10-fold decrease in cac) (Chen et al. 2006). Although the details of these relationships depend on the synthetic pathway and buffer conditions, this general pattern suggests that, during prebiotic synthesis, each fatty acid would reach its cac at around the same time, leading to a relatively sharp transition to a vesicle world.

Kinetics of Self-Assembly of Amphiphiles

The kinetics of nucleation and growth of micelles are generally quite fast. Nucleation occurs on the order of microseconds to milliseconds, whereas exchange of monomers through the aqueous phase occurs on the order of nanoseconds to microseconds. These timescales depend on the association and dissociation rates of amphiphiles. As chain length increases, dissociation rates decrease (for alkyl sulphates, roughly 10-fold per three-carbon increase), whereas association rates decrease only slightly because of slower diffusion (within one order of magnitude from C6 to C14) (Aniansson et al. 1976; Hunter 2001).

Vesicle formation from a solution of micelles appears to have two relaxation times, a rapid mixing or aggregation of micelles, followed by slower growth and closure to form vesicles. For a mixture of anionic and zwitterionic surfactant micelles, the micelles mix within a few milliseconds and then coalesce into disclike micelles that grow and close into vesicles within a second (Weiss et al. 2005). For vesicles formed by increasing the pH of a solution of fatty-acid micelles, aggregation of micelles occurs very quickly (<12 ms) whereas relaxation into vesicles takes seconds to minutes (Bloechliger et al. 1998; Chen and Szostak 2004a). The kinetics of fatty-acid vesicle formation are particularly interesting with regard to the origins of life for two reasons: (1) the reaction can be autocatalytic, i.e., the presence of vesicles accelerates the formation of more vesicles (Walde et al. 1994b), and (2) vesicle formation is catalyzed by many types of surfaces (Hanczyc et al. 2003; Hanczyc et al. 2007a), including montmorillonite, which also catalyzes RNA polymerization from activated monomers (Ferris and Ertem 1992). Once formed, closed vesicles can be fairly stable as supramolecular structures, surviving for several days or longer, even if their components exchange relatively rapidly.

Amphiphiles for Protocells

The primary components of modern cells are double-chain amphiphiles, particularly phospholipids, which consist of a headgroup and two acyl chains. Because of their geometry, double-chain amphiphiles form stable vesicles very readily. However, they pose several fundamental problems for a protocell, which result from slow dynamics. Phospholipids are relatively impermeable to charged compounds (e.g., K^+ permeability coefficient: 10^{-10} to 10^{-12} cm/s (Paula et al. 1996)), possibly because the low mobility of amphiphiles between the leaflets prevents facilitation of transport. Slow flip-flop would also limit vesicle growth by incorporation of new lipid (which requires redistribution of lipid from the outer to inner leaflet), inhibiting a key feature of a self-reproducing system. Modern cells treat their membranes as relatively static barriers, using flipases and permeases to mediate flip-flop and transport, respectively.

On the other hand, single-chain amphiphiles appear to be excellent candidates for prebiotic protocells for several reasons. They have relatively high permeability to ions and small molecules (e.g., K^+ permeability coefficient: 10^{-6} cm/s (Chen and Szostak 2004b)), presumably due at least in part to increased mobility of amphiphiles flipping between the leaflets, facilitating ion passage and circumventing the need for specific transporters. Fast flip-flop also permits vesicle growth through addition of an amphiphile feedstock to the vesicle exterior (Berclaz et al. 2001; Chen and Szostak 2004a). Single-chain amphiphiles also have high dissociation rates and relatively high water solubility, so the membrane components can be exchanged among vesicles rapidly, allowing redistribution to "growing" protocells (Chen et al. 2004). The cac is higher for single-chain lipids compared with double-chain lipids of the same chain length and similar head group structure, yielding a greater reservoir of monomers in solution. A limitation of single-chain amphiphiles is that they often form micelles, depending on the molecular geometries and buffer conditions, so the amphiphiles considered for prebiotic vesicles are constrained to have relatively small headgroups compared with double-chain amphiphiles.

Although we do not know which type of amphiphiles constituted the membranes of protocells, much attention has focused on fatty acids (Fig. 2c) and mixtures of fatty acids with their derivatives (e.g., fatty alcohols, amines, and monoacylglycerols). Fatty acids can be synthesized abiotically in several ways. For example, Miller-Urey-type electrical discharge reactions in a solution of ammonia under a nitrogen and methane atmosphere yield fatty acids with a chain length up to C12 (Allen and Ponnamperuma 1967; Yuen et al. 1981). Abiotic syntheses generally yield decreasing amounts of fatty acids of longer chain lengths. A synthesis simulating hydrothermal vents yielded fatty acids up to C33, from an aqueous Fischer-Tropsch-type synthesis using a heated solution of oxalic acid (which disproportionates into H_2, CO_2, and CO) (McCollom et al. 1999; Rushdi and Simoneit 2001).

Direct evidence for the abiotic presence of fatty acids comes from the detection of fatty acids in the interior of the Murray and Murchison carbonaceous chondrite meteorites from Australia (up to C8), as well as an Asuka carbonaceous chondrite meteorite (A-881458) from Antarctica (up to C12) (Lawless and Yuen 1979; Naraoka et al. 1999; Yuen et al. 1984; Yuen and Kvenvolden 1973). Fatty acids are relatively abundant in these meteorites, being 20 times more abundant than amino acids in the organic extract of A-881458. Indeed, organic extracts from the Murchison meteorite form boundary membranes when rehydrated (Deamer 1985; Deamer and Pashley 1989). The presence of fatty acids is particularly suggestive because the chemical composition of these meteorites is believed to resemble that of the early solar system.

Depending on the solution pH, fatty acids self-assemble into different structures (Cistola et al. 1988; Small 1986). At low pH, the molecules are protonated and uncharged, resulting in an oil phase. At high pH, the molecules are deprotonated and negatively charged, resulting in the formation of micelles with a hydrophobic core and surface-exposed carboxylates that repel one another. Although the pK_a of a carboxylic acid is typically 4–5, the self-assembly of fatty acids leads to a cooperative effect that increases the pK_a. For example, oleic acid (C18:1) monomers have a pK_a of 4.5, but oleic acid assembled into a bilayer membrane in bicine buffer (Fig. 2D) has a pK_a of 8.5. Medium- and long-chain fatty acids incorporated into membranes have pK_as in the general range of 7–9 (Cistola et al. 1988). When the solution pH is near the pK_a, fatty acids assemble into bilayer membrane vesicles that are capable of entrapping solutes (Apel et al. 2001; Walde et al. 1994a). These vesicles have a net negative charge, with a formal surface charge density close to half of the molecular density of the membrane. As a result, cations are also associated with the surface, forming an electrical double layer (Grahame 1947; Hunter 2001). Several articles describe in detail the characteristics and limitations of fatty-acid systems (Deamer and Dworkin 2005; Mansy 2009; Monnard et al. 2002; Morigaki and Walde 2007; Walde et al. 2006).

Another category of single-chain amphiphiles that may be promising as a model for prebiotic vesicles uses a phosphate headgroup, such as the polyprenyl phosphates (e.g., geranylgeranyl phosphate; Fig. 2E). These amphiphiles can self-assemble into vesicles (Ourisson and Nakatani 1994; Pozzi et al. 1996). In an interesting parallel to fatty-acid membranes, which can be stabilized to pH changes by acyl alcohols (Monnard et al. 2002), polyprenyl phosphate membranes are stabilized by addition of polyprenyl alcohols (Streiff et al. 2007). Other phosphate-based systems under investigation include mono-n-alkyl phosphates and phosphonates (Walde et al. 1997); alkyl phosphonates in particular have been observed in organic extracts from the Murchison meteorite (e.g., decylphosphonate; Fig. 2F) (Cooper et al. 1992).

RECENT RESULTS

Membranes and Nucleic Acids

Vesicles could play a simple role as compartment boundaries for ribozymes (Chen et al. 2005), but they may contribute to an early cell in other ways as well. Because of the importance of the RNA world during origins of life, several groups are studying direct or indirect interactions between vesicle membranes and RNA. One mode of direct interaction between RNA and cationic vesicles is electrostatic attraction to form nucleic acid-lipid complexes (Thomas and Luisi 2005). Such complexes may also be formed with neutral or zwitterionic vesicles under the right buffer conditions (e.g., high divalent cation concentration). For anionic amphiphiles, the electrostatic repulsion between RNA (poly-U) and the vesicles can be overcome by covalently decorating the phopholipid headgroup with adenosine (Milani et al. 2007), analogous to the annealing of two complementary, negatively charged nucleic acid strands. Conversely, the nucleic acid can be conjugated to a hydrophobic moiety, such as cholesterol, which inserts into the membrane and brings the nucleic acid into the proximity of the vesicle (Banchelli et al. 2008). RNA aptamers for membranes have also been found by in vitro selection for RNAs that bound to phospholipid vesicles. These sequences appear to bind the membrane as aggregates (Janas and Yarus 2003). One such sequence was modified to include an RNA motif for tryptophan binding, and together these membrane-binding RNAs had a small but measurable effect on the rate of tryptophan permeation through the membrane (Janas et al. 2004).

Membrane vesicles can also have surprising consequences for the evolution of the RNA world. Membranes can accelerate the polymerization of RNA mononucleotides or amino acids during drying and wetting cycles (Deamer et al. 2006; Rajamani et al. 2008; Zepik et al. 2007). This effect may be driven by an increase in local concentration of monomers, through trapping of monomers between lamellae during drying and mass action from dehydration. Although the mechanism is unknown, the liquid crystalline structure of the dried lamellae is believed to be important for this effect.

On the theoretical side, fresh efforts have been made in the analysis of polymerization dynamics relevant to nucleic acids ("prelife"). This mathematical analysis describes the distribution of sequences with and without replication and the occurrence of a phase transition among these regimes (Manapat et al. 2009; Nowak and Ohtsuki 2008). One interesting direction for this research would be consideration in the context of a protocell, to determine how small numbers and competition among protocells might affect the prelife dynamics.

Systems-Level Properties of Protocells

How did interactions arise between the membrane and its encapsulated contents? Two interactions have been described based solely on the physicochemical properties of the system. The growth of vesicles consists of the incorporation of fatty-acid molecules into the outer leaflet of the membrane, followed by the flip-flop of about half of these molecules into the inner leaflet. Fatty acid flip-flop into the inner leaflet will also transport protons (Hamilton 1998; Kamp and Hamilton 1992), so growth transduces the energy from the micelle-vesicle transition into a pH gradient across the membrane (Chen and Szostak 2004b). However, proton gradients across fatty-acid membranes only last as long as cations are not transported in the opposite direction, so a cell would need to use the energy stored in the proton gradient quickly before it decayed (Kamp and Hamilton 1992; Zeng et al. 1998). Regardless, the capture of energy is one way the growth of one component, the membrane, could confer an advantage to another component (encapsulated contents).

The emergence of the cell can also be approached from the perspective of the encapsulated genome. A transmembrane gradient of any impermeable solute (Δc) exerts osmotic pressure on a semipermeable membrane ($\Delta \Pi = n\Delta c RT$ for small solutes in dilute solution, where $n = $ number of particles into which the solute dissociates). A charged solute generates more osmotic pressure, and the osmotic pressure of a charged polymer, such as RNA, will be due primarily to the counterions associated with it, in accordance with the Gibbs-Donnan equilibrium (Bartlett and Kromhout 1952; Tinoco et al. 2002). The replication of a genomic polymer would increase the osmotic pressure in a vesicle (assuming that the monomers equilibrate across the membrane). The resulting areal strain favors membrane growth, which relieves the strain by increasing the total area of the membrane. For amphiphiles with high rates of monomer exchange and flip-flop, like fatty acids, this leads to a competition among vesicles for amphiphiles, translating the fitness of the encapsulated replicase into the fitness of the whole cell (Chen et al. 2004). Darwinian competition between "cells" could have emerged simply on encapsulation of replicases or metabolic cycles inside vesicles; empty vesicles and vesicles encapsulating less efficient systems would shrink by losing membrane.

This competition has interesting implications for the natural selection of chemical traits in the genome and membrane. Because counterions cause most of the change in osmotic pressure after polymerization, protocells containing charged polymers as the genetic material, such as RNA or DNA, would outcompete those with neutral polymers (e.g., peptide nucleic acid). With respect to the membrane, the protocellular competition might favor the evolution of an enzyme that stabilized the membrane against loss of amphiphiles, such as by condensing two fatty-acid molecules together. In turn, the reduced permeability and growth dynamics of stabilized membranes might drive the coevolution of transporters and more enzymes. Thus the competition arising from the simple physical properties of an encapsulated replicase could result in a "snowball" effect on the evolution of cellular complexity. Such predictions may suggest further experimental exploration.

Computer Simulation of Protocell-Like Systems

In an era of increasingly powerful and accessible computation, molecular dynamics (MD) simulation is a promising technique for exploring our understanding of the behavior of relatively large chemical systems like vesicles. In general these simulations use a combination of empirical and *ab initio* information to describe energy potentials for molecular conformations and intermolecular forces (dispersion and electrostatic forces) and solve the Newtonian equations of motion numerically. The size of a vesicle system is roughly hundreds of thousands to a few millions of atoms (e.g., a single fatty acid is roughly 50 atoms, and thousands of fatty-acid molecules would compose a small vesicle, and water would also compose a substantial mole fraction of the system). Currently an MD simulation of this size could simulate about 100 nanoseconds, which would be enough to study the fast kinetics of self-assembly. An atomistic simulation of dipalmitoyl phosphatidylcholine (DPPC) in water recapitulated vesicle self-assembly (de Vries et al. 2004). In addition to testing our understanding of the system, ideally these simulations could be useful to experimentalists for building biophysical intuition and emphasizing features to be further explored. For example, the simulation of a DPPC vesicle indicated a large asymmetry in mass and ordering between the inner and outer bilayers, suggesting a potential mechanism for asymmetric insertion of molecules into the membrane.

Slower dynamics or larger systems could be modeled using coarse-grained rather than atomistic simulations (Klein and Shinoda 2008). One interesting but poorly understood phenomenon that might benefit from MD simulation is the recent observation that dried lipid membranes accelerate the polymerization of amino acids and nucleotide monophosphates, perhaps by preorganizing the reactants into the reactive geometry (Rajamani et al. 2008).

Another style of simulation of protocells focuses on the reaction (or reaction-diffusion) kinetics of the system without modeling the molecular details. Although much work, such as Manfred Eigen's classic model of hypercycles (Eigen 1971), dealt with metabolic cycles without explicit reference to a membrane, recently the potential for interactions between the membrane and encapsulated metabolic reactions has come under closer scrutiny (Mavelli and Ruiz-Mirazo 2007). Stochastic simulations are of particular interest because the volume of a small vesicle could be less than one attoliter, such that micromolar concentrations would give only a handful of molecules per vesicle. Although care is always required when dealing with assumptions in a model, simulations can provide a testing ground for our understanding, and may become a useful tool for understanding experimental results and generating interesting, experimentally testable predictions in protocell-sized systems.

CHALLENGES AND FUTURE DIRECTIONS

Physical Forces and Protocells

How would protocells interact with their environment? Although some interest has focused on interactions at a molecular level, such as the permeability and thermostability of vesicles (Mansy et al. 2008; Mansy and Szostak 2008), less work has been performed to relate colloidal processes to protocells. For example, transport phenomena might give us clues about protocell movement through a heterogeneous environment or even initially homogeneous environment. Hanczyc et al. showed how oleic anhydride oil droplets added to an alkaline solution spontaneously move, as exothermic hydrolysis of the anhydride sets up a convective current within the droplet; the direction of motion appeared to be the outcome of symmetry-breaking fluctuation (Hanczyc et al. 2007b). The theory of osmophoresis suggests that vesicles in a solute gradient will move in the direction of lower solute concentration (Anderson 1983; Nardi et al. 1999). Such effects, if experimentally relevant, could inform our understanding of how a protocell would move within its environment as a consequence of colloidal forces, without motor machinery.

Entropy might play a very interesting role in protocells. For example, theory predicts that two chromosomes in a cell tend to segregate from each other to maximize conformational entropy (Jun and Mulder 2006). This finding suggests that protocell genomes might not require extra machinery to segregate its genetic material. Along similar lines, cell division machinery (in particular, a contractile ring) was recently discovered to be unnecessary for reproduction and division of *Bacillus subtilis* (Leaver et al. 2009). The observed mode of division is reminiscent of a physical phenomenon, the Rayleigh instability (also known as "pearling" in tubular liposomes), lending credence to the idea that protocells would not require division machinery (Chen 2009). One might also expect protocells to interact with one another through entropic effects, such as the depletion effect, in which the entropy of the aqueous solution is maximized when large particles are close together, resulting in an effective attraction between the particles. Such effects could be interesting as leading to assembly of protocells into larger structures. Investigation of the role of physical forces in protocells may present many opportunities for collaboration between traditional soft condensed matter physicists and origins of life researchers.

Vesicles from Mixtures

Studies on vesicles formed from potentially prebiotic amphiphiles have so far been limited to systems containing one or two types of amphiphiles. This is in contrast to the output of simulated prebiotic chemical reactions, which typically produce very heterogeneous mixtures of compounds. In addition, little experimental work has been performed to integrate peptides and lipids. (A notable exception to this bias toward working with one or two component systems was a study using the heterogeneous organic extract from a carbonaceous chondrite meteorite to form vesicles [Deamer 1985].) The reason for this bias is mainly the avoidance of analytical difficulties, but clearly it is not very realistic.

What properties might vary with membrane composition? Several features characterize a particular vesicle system and differentiate it from others, including (1) lamellarity, (2) average membrane thickness and its fluctuations, (3) amphiphile dynamics in the membrane, (4) membrane permeability, i.e., the ability to keep solutes captured inside the vesicles and to stabilize concentration gradients across the membrane, (5) chemical stability of the amphiphiles, (6) membrane surface charge, (7) membrane fluidity, (8) possibility of membrane budding and fission (leading to vesicle divisions), (9) vesicle aggregation or vesicle adsorption to surfaces, (10) vesicle fusion, (11) colloidal stability at different temperatures and bulk solution compositions. All these properties ultimately depend on the chemical structure of the amphiphiles, the environmental conditions, and the process of vesicle formation. The process of vesicle formation is relevant because most vesicles are kinetically trapped systems (i.e., not at true thermodynamic equilibrium). Lamellarity and size depend strongly on how the vesicles are prepared, but the other features depend primarily on the vesicle composition and buffer conditions.

There are challenges to using vesicles based on fatty acids as protocell models (Monnard and Deamer 2002). One limitation for practical studies is the failure in preparing vesicles from long-chain fatty acids at room temperature, for example from hexadecanoic acid, because of the high melting temperature of saturated linear hydrocarbon chains. According to the phase diagram, a lamellar phase exists at $70°C$ (Skurtveit et al. 1989). However, recently it was shown that vesicles form from long-chain fatty acids at room temperature if guanidine hydrochloride is added (Douliez et al. submitted for publication). This illustrates how vesicles of mixed composition can actually have unanticipated properties beneficial to a protocell, such as robustness to ionic strength, lowered cac, or even propagation of information (as discussed later) (Monnard and Deamer 2003). One of the challenges for the future will be to further study these mixtures to understand the chemical basis for their effects on vesicle characteristics. Other mixtures to explore include mixtures of fatty acids with small molecule components, mixtures of different types of fatty acids, and mixtures of fatty acids with peptides.

Self-Assembled Structures and Information

The storage and propagation of information is a key feature of biological systems. Nucleic acid sequences fulfill this requirement in a straightforward fashion, particularly in the case of ribozymes, where the informational material itself has chemical activity. However, in a very broad sense, molecules other than nucleic acids carry information about their reactivity and physicochemical properties (Graham et al. 2004). Self-assembling molecules contain the information for building supramolecular structures, such as vesicles and micelles. For example, potassium decanoate, $CH_3(CH_2)_6COO^-K^+$, carries information on the thermodynamics of its self-assembly behavior. When dissolved in water at a concentration of 29.1 g/L (150 mM), micellar aggregates form (Namani and Walde 2005). Each micelle contains a hydrophobic core that allows some solubilization of molecules that are not soluble in water, so the self-assembly of the amphiphilic decanoate molecules confers new physicochemical properties to the aqueous solution (Luisi 2002). Could self-assembled, noncovalent systems also propagate their "information"?

An interesting model addressing this question was proposed recently (Segre et al. 2000). In a solution containing a mixture of amphiphiles, if amphiphiles of a certain type favored the incorporation of other amphiphiles in a selective fashion, an assemblage would reach a particular steady state whose composition contains some information. In simulations they observed that abrupt transitions could occur among different steady states, suggesting that different compositions could coexist and thereby provide variation on which natural selection might act. An experimental system corresponding to this model would require determining the on- and off- rates of different amphiphiles in vesicles spanning a range of compositions. A particularly interesting technical challenge associated with this theoretical problem is the identification of the composition of an individual vesicle.

Self-assembly as a Quantitative Trait

Although one working definition of life (a self-sustaining chemical reaction capable of Darwinian evolution) was given in the introduction to this article, alternative definitions also exist, often in the form of several criteria. The difficulties encountered in defining life suggest that life is a quantitative, not qualitative, trait, which could be measured by an appropriate metric. Chirikjian and coworkers developed metrics for the degree of self-replication and task complexity in a framework of self-replicating robots (Lee et al. 2008). For example, some robots copy themselves by joining two already complex building blocks (low self-replication) in a highly structured environment with many landmarks (low task complexity), whereas other robots copy themselves by joining several simple building blocks (high self-replication) in an unstructured environment (high task complexity). These calculations require knowledge of the components, their interconnections, and the environment. Application of this quantitative approach to chemical systems and to other properties (i.e., evolvability and life) poses an interesting intellectual challenge that might also help integrate our understanding of different model systems of protocells.

REFERENCES

Allen WV, Ponnamperuma C. 1967. A possible prebiotic synthesis of monocarboxylic acids. *Currents Mod Biol* **1:** 24–28.

Anderson JL. 1983. Movement of a semipermeable vesicle through an osmotic gradient. *Phys Fluids* **26:** 2871–2879.

Aniansson EAG, Wall SN, Almgren M, Hoffmann H, Kielmann I, Ulbricht W, Zana R, et al. 1976. Theory of kinetics of micellar equilibria and quantitative interpretation of chemical relaxation studies of micellar solutions of ionic surfactants. *J Phys Chem* **80:** 905–922.

Antonietti M, Foerster S. 2003. Vesicles and liposomes: A self-assembly principle beyond lipids. *Adv Materials* **15:** 1323–1333.

Apel CL, Deamer DW, Mautner MN. 2001. Self-assembled vesicles of monocarboxylic acids and alcohols: Conditions for stability and for the encapsulation of biopolymers. *Biochim Biophys Acta* **1559:** 1–9.

Banchelli M, Betti F, Berti D, Caminati G, Bombelli FB, Brown T, Wilhelmsson LM, et al. 2008. Phospholipid membranes decorated by cholesterol-based oligonucleotides as soft hybrid nanostructures. *J Phys Chem B* **112:** 10942–10952.

Bartlett JH, Kromhout RA. 1952. The Donnan equilibrium. *Bull Math Biophys* **14:** 385–391.

Berclaz N, Muller M, Walde P, Luisi PL. 2001. Growth and transformation of vesicles studied by ferritin labeling and cryotransmission electron microscopy. *J Phys Chem B* **105:** 1056–1064.

Bloechliger E, Blocher M, Walde P, Luisi PL. 1998. Matrix effect in the size distribution of fatty acid vesicles. *J Phys Chem B* **102:** 10383–10390.

Cavalier-Smith T. 2001. Obcells as proto-organisms: Membrane heredity, lithophosphorylation, and the origins of the genetic code, the first cells, and photosynthesis. *J Mol Evol* **53:** 555–595.

Chen IA. 2009. Cell division: Breaking up is easy to do. *Curr Biol* **19:** R327–R328.

Chen IA, Szostak JW. 2004a. A kinetic study of the growth of fatty acid vesicles. *Biophys. J.* **87:** 988–998.

Chen IA, Szostak JW. 2004b. Membrane growth can generate a transmembrane pH gradient in fatty acid vesicles. *Proc. Natl. Acad. Sci* **101:** 7965–7970.

Chen IA, Roberts RW, Szostak JW. 2004. The emergence of competition between model protocells. *Science* **305:** 1474–1476.

Chen IA, Salehi-Ashtiani K, Szostak JW. 2005. RNA catalysis in model protocell vesicles. *J. Am. Chem. Soc.* **127:** 13213–13219.

Chen IA, Hanczyc MM, Sazani PL, Szostak JW. 2006. Protocells: Genetic polymers inside membrane vesicles. *In*: Atkins J.F. ed. RNA World, 3rd ed., Cold Spring Harbor Laboratory Press.

Cistola DP, Hamilton JA, Jackson D, Small DM. 1988. Ionization and phase behavior of fatty acids in water: Application of the Gibbs phase rule. *Biochemistry* **27:** 1881–1888.

Cooper GW, Onwo WM, Cronin JR. 1992. Alkyl phosphonic-acids and sulfonic-acids in the Murchison meteorite. *Geochimica Et Cosmochimica Acta* **56:** 4109–4115.

de Vries AH, Mark AE, Marrink SJ. 2004. Molecular dynamics simulation of the spontaneous formation of a small DPPC vesicle in water in atomistic detail. *J Am Chem Soc* **126:** 4488–4489.

Deamer DW. 1985. Boundary structures are formed by organic components of the Murchison carbonaceous chondrite. *Nature* **317:** 792–794.

Deamer DW, Pashley RM. 1989. Amphiphilic components of the Murchison carbonaceous chondrite: Surface properties and membrane formation. *Orig Life Evol Biosph* **19:** 21–38.

Deamer DW, Dworkin JP. 2005. Chemistry and physics of primitive membranes, Pages 1–27. *In*: Walde P. ed. *Prebiotic chemistry: From simple amphiphiles to protocell models*. Topics in Current Chemistry.

Deamer D, Singaram S, Rajamani S, Kompanichenko V, Guggenheim S. 2006. Self-assembly processes in the prebiotic environment. *Phil Trans Roy Soc B-Biol Sci* **361:** 1809–1818.

Eigen M. 1971. Selforganization of matter and the evolution of biological macromolecules. *Naturwissenschaften* **58:** 465–523.

Ferris JP, Ertem G. 1992. Oligomerization of ribonucleotides on montmorillonite: reaction of the $5'$-phosphorimidazolide of adenosine. *Science* **257:** 1387–1389.

Gebicki JM, Hicks M. 1973. Ufasomes are stable particles surrounded by unsaturated fatty acid membranes. *Nature* **243:** 232–234.

Graham DJ, Malarkey C, Schulmerich MV. 2004. Information content in organic molecules: Quantification and statistical structure via Brownian processing. *J Chem Inf Computer Sci* **44:** 1601–1611.

Grahame DC. 1947. The electrical double layer and the theory of electrocapillarity. *Chem Rev* **41:** 441–501.

Hamilton JA. 1998. Fatty acid transport: Difficult or easy? *J Lipid Res* **39:** 467–481.

Hanczyc MM, Fujikawa SM, Szostak JW. 2003. Experimental models of primitive cellular compartments: encapsulation, growth, and division. *Science* **302:** 618–622.

Hanczyc MM, Mansy SS, Szostak JW. 2007a. Mineral surface directed membrane assembly. *Origins of Life and Evolution of the Biosphere* **37:** 67–82.

Hanczyc MM, Toyota T, Ikegami T, Packard N, Sugawara T. 2007b. Fatty acid chemistry at the oil-water interface: Self-propelled oil droplets. *J Am Chem Soc* **129:** 9386–9391.

Hunter RJ. 2001. Foundations of colloid science. New York, Oxford University Press.

Israelachvili JN, Mitchell DJ, Ninham BW. 1976. Theory of self-assembly of hydrocarbon amphiphiles into micelles and bilayers. *J Chem Soc Faraday Trans II* **72:** 1525–1568.

Israelachvili JN, Mitchell DJ, Ninham BW. 1977. Theory of self-assembly of lipid bilayers and vesicles. *Biochim Biophys Acta* **470:** 185–201.

Janas T, Yarus M. 2003. Visualization of membrane RNAs. *Rna-a Publication of the Rna Society* **9:** 1353–1361.

Janas T, Janas T, Yarus M. 2004. A membrane transporter for tryptophan composed of RNA. *Rna-a Publication of the Rna Society* **10:** 1541–1549.

Joyce G. 1994. Foreword. *In*: Deamer D.W., Fleischaker G.R. eds. *Origins of life: The central concepts*. Boston, Jones and Bartlett Publishers.

Jun S, Mulder B. 2006. Entropy-driven spatial organization of highly confined polymers: Lessons for the bacterial chromosome. *Proc Natl Acad Sci* **103:** 12388–12393.

Kamp F, Hamilton JA. 1992. pH gradients across phospholipid membranes caused by fast flip-flop of un-ionized fatty acids. *Proc Natl Acad Sci* **89:** 11367–11370.

Klein ML, Shinoda W. 2008. Large-scale molecular dynamics simulations of self-assembling systems. *Science* **321:** 798–800.

Koch AL. 1984. Evolution vs the number of gene copies per primitive cell. *J Mol Evol* **20:** 71–76.

Lawless JG, Yuen GU. 1979. Quantification of monocarboxylic acids in the Murchison carbonaceous meteorite. *Nature* **282:** 396–398.

Leaver M, Dominguez-Cuevas P, Coxhead JM, Daniel RA, Errington J. 2009. Life without a wall or division machine in *Bacillus subtilis*. *Nature* **457:** 849–853.

Lee K, Moses M, Chirikjian GS. 2008. Robotic Self-replication in structured environments: Physical demonstrations and complexity measures. *Int J Robotics Res* **27:** 387.

Luisi PL. 2002. Emergence in chemistry: Chemistry as the embodiment of emergence. *Foundations Chem* **4**: 183–200.

Luisi PL, Walde P, Oberholzer T. 1999. Lipid vesicles as possible intermediates in the origin of life. *Current Opinion Colloid Interface Sci* **4**: 33–39.

Manapat M, Ohtsuki H, Burger R, Nowak MA. 2009. Originator dynamics. *J Theoret Biol* **256**: 586–595.

Mansy SS. 2009. Model protocells from single-chain lipids. *Int J Mol Sci* **10**: 835–843.

Mansy SS, Szostak JW. 2008. Thermostability of model protocell membranes. *Proc Natl Acad Sci* **105**: 13351–13355.

Mansy SS, Schrum JP, Krishnamurthy M, Tobe S, Treco DA, Szostak JW. 2008. Template-directed synthesis of a genetic polymer in a model protocell. *Nature* **454**: p122–U110.

Matsuura T, Yamaguchi M, Ko-Mitamura EP, Shima Y, Urabe I, Yomo T. 2002. Importance of compartment formation for a self-encoding system. *Proc Natl Acad Sci* **99**: 7514–7517.

Mavelli F, Ruiz-Mirazo K. 2007. Stochastic simulations of minimal self-reproducing cellular systems. *Philosophical Trans Roy Soc B-Biological Sciences* **362**: 1789–1802.

McCollom TM, Ritter G, Simoneit BR. 1999. Lipid synthesis under hydrothermal conditions by Fischer-Tropsch-type reactions. *Orig. Life Evol. Biosph.* **29**: 153–166.

Milani S, Bombelli FB, Berti D, Baglioni P. 2007. Nucleolipoplexes: A new paradigm for phospholipid bilayer-nucleic acid interactions. *J Am Chem Soc* **129**: 11664–+.

Monnard PA, Deamer DW. 2002. Membrane self-assembly processes: Steps toward the first cellular life. *Anat Rec* **268**: 196–207.

Monnard PA, Deamer DW. 2003. Preparation of vesicles from nonphospholipid amphiphiles. *Meth Enzymol* **372**: 133–151.

Monnard PA, Apel CL, Kanavarioti A, Deamer DW. 2002. Influence of ionic inorganic solutes on self-assembly and polymerization processes related to early forms of life: Implications for a prebiotic aqueous medium. *Astrobiology* **2**: 139–152.

Morigaki K, Walde P. 2007. Fatty acid vesicles. *Current Opinion in Colloid Interface Sci* **12**: 75–80.

Morowitz HJ, Heinz B, Deamer DW. 1988. The chemical logic of a minimum protocell. *Orig Life Evol Biosph* **18**: 281–287.

Namani T, Walde P. 2005. From decanoate micelles to decanoic acid/dodecylbenzenesulfonate vesicles. *Langmuir* **21**: 6210–6219.

Naraoka H, Shimoyama A, Harada K. 1999. Molecular distribution of monocarboxylic acids in Asuka carbonaceous chondrites from Antarctica. *Orig Life Evol Biosph* **29**: 187–201.

Nardi J, Bruinsma R, Sackmann E. 1999. Vesicles as Osmotic Motors. *Phys Rev Lett* **82**: 5168–5171.

Nowak MA, Ohtsuki H. 2008. Prevolutionary dynamics and the origin of evolution. *Proc Natl Acad Sci* **105**: 14924–14927.

Ourisson G, Nakatani Y. 1994. The terpenoid theory of the origin of cellular life: the evolution of terpenoids to cholesterol. *Chem Biol* **1**: 11–23.

Paula S, Volkov AG, Van Hoek AN, Haines TH, Deamer DW. 1996. Permeation of protons, potassium ions, and small polar molecules through phospholipid bilayers as a function of membrane thickness. *Biophys J* **70**: 339–348.

Pozzi G, Birault V, Werner B, Dannenmuller O, Nakatani Y, Ourisson G, Terakawa S. 1996. Single-chain polyprenyl phosphates form "primitive" membranes. *Angewandte Chemie-International Edition* **35**: 177–180.

Rajamani S, Vlassov A, Benner S, Coombs A, Olasagasti F, Deamer D. 2008. Lipid-assisted synthesis of RNA-like polymers from mononucleotides. *Orig Life Evol Biosph* **38**: 57–74.

Rushdi AI, Simoneit BR. 2001. Lipid formation by aqueous Fischer-Tropsch-type synthesis over a temperature range of 100 to 400 degrees C. *Orig. Life Evol. Biosph.* **31**: 103–118.

Segre D, Ben-Eli D, Lancet D. 2000. Compositional genomes: Prebiotic information transfer in mutually catalytic noncovalent assemblies. *Proc Natl Acad Sci* **97**: 4112–4117.

Skurtveit R, Sjoblom J, Hoiland H. 1989. Emulsions under Elevated-Temperature and Pressure Conditions. 1. The Model System Water Hexadecanoic Acid Sodium Hexadecanoate Decane at 70-Degrees-C. *J Colloid Interface Sci* **133**: 395–403.

Small DM. 1986. Chapter 9, Pages 285–343. *In:* Small D.M. ed. The physical chemistry of lipids: from alkanes to phospholipids. *Handbook of lipid research*. New York, Plenum Press.

Smith JM, Szathmary E. 1995. *The major transitions in evolution*. Oxford, W.H. Freeman.

Streiff S, Ribeiro N, Wu ZY, Gumienna-Kontecka E, Elhabiri M, Albrecht-Gary AM, Ourisson G, et al. 2007. "Primitive" membrane from polyprenyl phosphates and polyprenyl alcohols. *Chem Biol* **14**: 313–319.

Szabo P, Scheuring I, Czaran T, Szathmary E. 2002. In silico simulations reveal that replicators with limited dispersal evolve towards higher efficiency and fidelity. *Nature* **420**: 340–343.

Szathmary E, Demeter L. 1987. Group selection of early replicators and the origin of life. *J Theor Biol* **128**: 463–486.

Szostak JW, Bartel DP, Luisi PL. 2001. Synthesizing life. *Nature* **409**: 387–390.

Tanford C. 1973. *The hydrophobic effect: formation of micelles and biological membranes*. New York, John Wiley and Sons, Inc.

Thomas CF, Luisi PL. 2005. RNA selectively interacts with vesicles depending on their size. *J Phys Chem B* **109**: 14544–14550.

Tinoco I, Sauer K, Wang JC, Puglisi JD. 2002. Physical Chemistry: Principles and applications in biological sciences. Upper Saddle River, New Jersey, Prentice Hall.

Walde P, Ichikawa S. 2001. Enzymes inside lipid vesicles: Preparation, reactivity and applications. *Biomol Eng* **18**: 143–177.

Walde P, Goto A, Monnard P-A, Wessicken M, Luisi PL. 1994a. Oparin's reactions revisited: enzymatic synthesis of poly(adenylic acid) in micelles and self-reproducing vesicles. *J Am Chem Soc* **116**: 7541–7547.

Walde P, Namani T, Morigaki K, Hauser H. 2006. Formation and Properties of Fatty Acid Vesicles (Liposomes), Pages

1–19. *In*: Gregoriadis G. ed. Liposome Technology. New York, Informa Healthcare.

Walde P, Wessicken M, Radler U, Berclaz N, CondeFrieboes K, Luisi PL. 1997. Preparation and characterization of vesicles from mono-n-alkyl phosphates and phosphonates. *J Phys Chem B* **101:** 7390–7397.

Walde P, Wick R, Fresta M, Mangone A, Luisi PL. 1994b. Autopoietic self-reproduction of fatty acid vesicles. *J Am Chem Soc* **116:** 11649–11654.

Weiss TM, Narayanan T, Wolf C, Gradzielski M, Panine P, Finet S, Helsby WI. 2005. Dynamics of the self-assembly of unilamellar vesicles. *Phys Rev Lett* **94:** 038303.

Yuen GU, Kvenvolden KA. 1973. Monocarboxylic acids in Murray and Murchison carbonaceous meteorites. *Nature* **246:** 301–303.

Yuen GU, Lawless JG, Edelson EH. 1981. Quantification of monocarboxylic acids from a spark discharge synthesis. *J Mol Evol* **17:** 43–47.

Yuen G, Blair N, Des Marais DJ, Chang S. 1984. Carbon isotope composition of low molecular weight hydrocarbons and monocarboxylic acids from Murchison meteorite. *Nature* **307:** 252–254.

Zeng Y, Han X, Schlesinger P, Gross RW. 1998. Nonesterified fatty acids induce transmembrane monovalent cation flux: Host-guest interactions as determinants of fatty acid-induced ion transport. *Biochemistry* **37:** 9497–9508.

Zepik HH, Rajamani S, Maurel MC, Deamer D. 2007. Oligomerization of thioglutamic acid: Encapsulated reactions and lipid catalysis. *Orig Life Evol Biosph* **37:** 495–505.

Membrane Transport in Primitive Cells

Sheref S. Mansy

Armenise-Harvard Laboratory of Synthetic and Reconstructive Biology, Centre for Integrative Biology (CIBIO), University of Trento, Italy

Correspondence: mansy@science.unitn.it

Although model protocellular membranes consisting of monoacyl lipids are similar to membranes composed of contemporary diacyl lipids, they differ in at least one important aspect. Model protocellular membranes allow for the passage of polar solutes and thus can potentially support cell-to functions without the aid of transport machinery. The ability to transport polar molecules likely stems from increased lipid dynamics. Selectively permeable vesicle membranes composed of monoacyl lipids allow for many lifelike processes to emerge from a remarkably small set of molecules.

Lipid bilayer membranes are an integral component of living cells, providing a permeability barrier that is essential for nutrient transport and energy production. It is reasonable to assume that a similar boundary structure would be required for the origin of cellular life (Szostak et al. 2001). Even though bilayer membranes are a cellular necessity, they also pose a significant obstacle to early cellular functions, the most obvious being that the permeability barrier would inhibit chemical exchange with the environment. Such an exchange is important not only for acquiring nutrient substrates for primitive metabolic processes, but also for the release of inhibitory side-products.

Contemporary cells circumvent the permeability problem by incorporating complex transmembrane protein machinery that provides specific transport capabilities. It is unlikely that Earth's first cells assembled bilayer membranes together with specific membrane protein transporters. Rather, intermediate evolutionary steps must have existed in which simple lipid molecules provided many of the characteristics of contemporary membranes without relying on advanced protein machinery. What seems to have been necessary was the appearance of a simple membrane system capable of retaining and releasing specific molecules. In short, a protocell needed to be selectively permeable.

BACKGROUND

Membrane Structure

The structure of biological membranes has been extensively characterized, but extending this knowledge to protocellular membranes is problematic. Insight can be gained by comparing what is known of the structure and permeability of biological membranes with those derived from laboratory models of protocells. The lipid bilayers of present-day biological membranes consist primarily of diacyl lipids, such as glycerophospholipids and sphingolipids, in addition to other potential components including glycolipids (derivatives of either sphingo- or glycerophospholipids), sterols, and proteins (Fig. 1). A single bilayer membrane is approximately 5 nm thick (Wiener and White 1992), which corresponds to seven to nine amino acids of a β-strand or 20–25 amino acids of an α-helix (Nelson and Cox 2008). Within the membrane, lipids are arranged so that their hydrophilic head groups extend outward toward the bulk aqueous phase and their hydrophobic tails are buried in the nonpolar interior of the bilayer. The spatial organization of carbonyl ester dipole moments within the membrane produces a significant membrane dipole potential estimated to be +240 mV (Gawrisch et al. 1992; Peterson et al. 2002; Wang et al. 2006). The lipids that constitute biological membranes usually have hydrocarbon chains between 16 and 24 carbons long and have either zwitterionic or anionic head groups (Tanford 1980; Evans and Wennerstrom 1999). Unsaturated lipids have a *cis* double bond at carbon 9, which is the position within the hydrocarbon chain that optimally disrupts inter-acyl packing (Menger et al. 1988). The distribution of membrane lipids is asymmetric, with certain

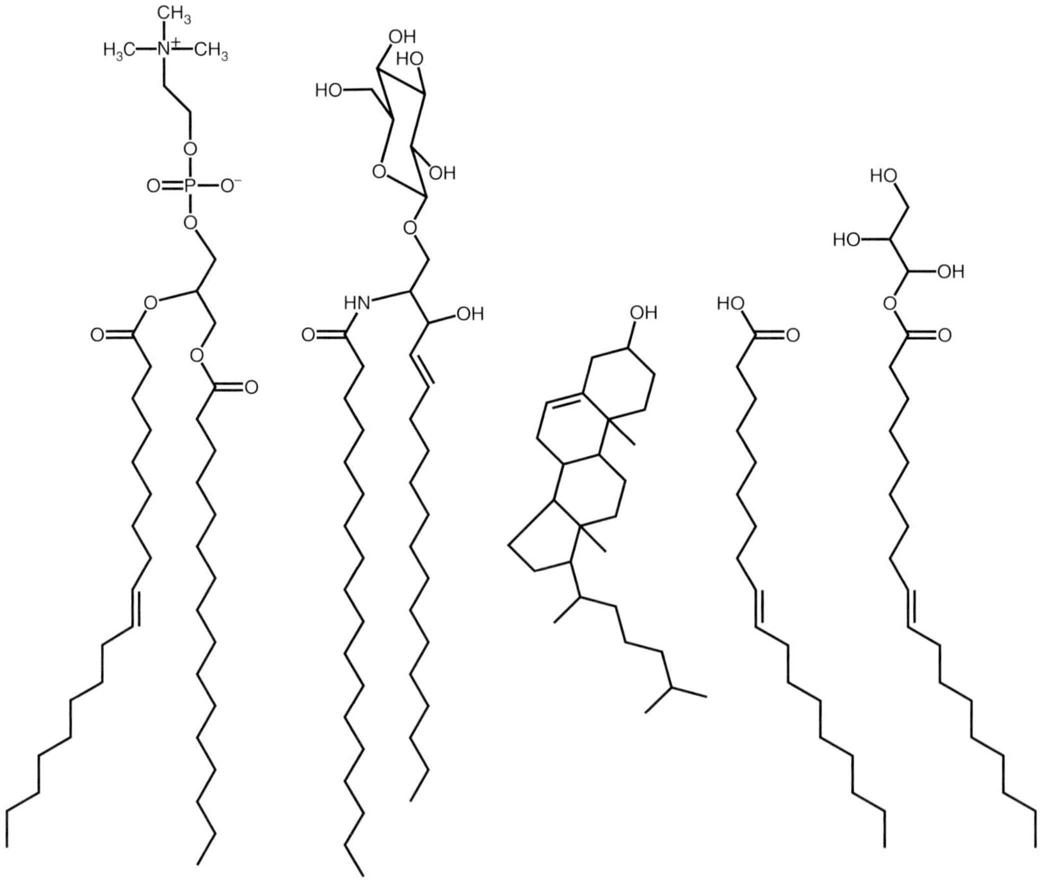

Figure 1. Lipid chemical structures. From *left* to *right*, phospholipid (1-palmitoyl-2-oleoyl-*sn*-glycero-3-phosphocholine), sphingolipid (glucosyl ceramide), sterol (cholesterol), fatty acid (oleate), and glycerol monoester of fatty acid (monoolein). Phospholipids, spingolipids, and sterols are common components of biological membranes, whereas fatty acids and glycerol monoesters of fatty acids are constituents of model protocellular membranes.

classes of lipids, e.g., glycolipids, predominating in one leaflet. Water molecules extensively interact with the hydrophilic exterior of the membrane (Gawrisch et al. 2007). The depth of water penetration extends past the head group and glycerol to the carbonyl groups, yielding a hydrophobic membrane interior of approximately 3 nm (Wiener and White 1992; Gawrisch et al. 2007). It should be emphasized that biological membranes also contain nonlipid components, with the mass of protein components often exceeding that of lipids (Singer 2004). In summary, for a solute to cross a lipid bilayer membrane of 5 nm, the solute must penetrate a 3-nm hydrophobic barrier.

The prevalence of phosphate in lipid head groups is striking. With few exceptions (van Mooy et al. 2009), diacyl membrane lipids contain phosphate head groups, even though amphiphilic molecules that form bilayer membranes do not require the presence of phosphate (Israelachvili 1991; Evans and Wennerstrom 1999). Indeed, the importance of phosphate is a recurring theme in biology, with examples in lipids, nucleic acids (Benner and Hutter 2002), metabolism (Westheimer 1987; Deamer 1997), and cellular signaling (Tarrant and Cole 2009).

Prebiotic membrane composition and structure

The lipid composition of protocellular membranes is not known. However, inferences can be drawn from simulated prebiotic synthesis, geological and astronomical analyses, as well as data from constructed laboratory model systems. Thus far, interest has centered on fatty acid and fatty acid derivatives as plausible prebiotic membrane constituents. First, they can be synthesized under simulated prebiotic conditions via Fischer-Tropsch-type reactions (McCollom et al. 1999). Second, analysis of carbonaceous meteorites, such as the Murchison meteorite, has identified fatty acids as one of many complex organic molecules (Lawless and Yuen 1979). The latter should not necessarily be interpreted as evidence for seeding life on Earth through the extraterrestrial delivery of organic material. Rather, it shows that abiotic synthesis using molecules that were available on the prebiotic Earth can generate appreciable quantities of amphiphilic compounds such as fatty acids (Pizzarello 2006; Pizzarello 2007). Interestingly, organic extracts of the Murchison meteorite that contained a mixture of abiotically synthesized amphiphilc molecules self-assembled into vesicles in aqueous solution (Deamer 1985; Deamer and Pasley 1989).

Laboratory vesicle systems have shown that fatty acids alone can form bilayer membrane structures that are similar to the membranes composed of diacyl phospholipids (Gebicki and Hicks 1973; Gebicki and Hicks 1976; Hargreaves and Deamer 1978). From a structural perspective, the only difference is the presence of a single acyl chain in the interior, hydrophobic core of the membrane per hydrophilic head group. Other properties of membranes composed of fatty acids and phospholipids are similar. For example, both types of lipids generate similarly sized vesicles (Hargreaves and Deamer 1978) and require hydrocarbon chains of at least 10 carbons to form stable bilayers (Deamer 1997). Fatty acid and phospholipid membranes also are similar in thickness (Cistola et al. 1986), thermal stability (Mansy and Szostak 2008), and tensile strength (Chen et al. 2004). Finally, both kinds of membranes are capable of entrapping polymers (Walde et al. 1994) and retaining the activity of encapsulated enzymes (Walde et al. 1994; Chen et al. 2005). It should be noted that whereas meteorite samples show the presence of short, saturated fatty acids, laboratory model systems are often based on longer, unsaturated fatty acids. The data thus far show that many of the structural properties of fatty acid and phospholipid membranes are similar.

Lipid Dynamics

Membranes are highly fluid structures showing a wide range of lipid motions (Singer and Nicolson 1972; Gawrisch 2004). Within membranes, lipids can undergo *trans*-gauche isomerizations, rotate, wobble, laterally diffuse, and flip from one leaflet to another (Table 1). The dynamics of the carbons of a single lipid vary with respect to their position within the

Table 1. Lipid dynamics within liquid-crystalline bilayer membranes.

Motion type	Diacyl phospholipid	Monoacyl fatty acid[a]	Reference
Trans-gauche isomerization	ps–ns		(Venable et al. 1993; Gawrisch 2004)
Wobble	ns		(Klauda et al. 2008)
Axial rotation	ns		(Klauda et al. 2008)
Lateral diffusion	ns–ms		(Rubenstein et al. 1979; Gawrisch 2004; Kahya and Schwille 2006)
Flip-flop	days	ms–s	(Kamp and Hamilton 1992; Chen and Szostak 2004b)
Escape	days	s	(Roseman and Thompson 1980; Fujikawa et al. 2005)

[a]*Trans*-gauche isomerization, wobble, axial rotation, and lateral diffusion correlation times are presumed to be similar for fatty acid and phospholipid membranes.

hydrocarbon chain, with greater dynamic motion observed toward the center of the membrane, i.e., near the terminal methyl groups, and less motility measured for the carbons near the surface of the membrane (McFarland and McConnell 1971). Lipid dynamics are influenced by acyl chain packing, head group interactions, and sterol content. For example, the presence of cholesterol can decrease lateral mobility by an order of magnitude (Rubenstein et al. 1979), decrease *trans*-gauche isomerization of the carbons near the head group, and increase the dynamics of terminal methyls for long acyl chain lipids (McIntosh 1978; Gawrisch 2004). Many of these smaller scale motions, such as *trans*-gauche isomerizations, are presumed to be similar between diacyl and monoacyl lipids, although differences likely exist.

Larger scale lipid motions, such as flip-flop, are influenced by acyl chain and head group compositions. Decreases in inter-acyl packing through unsaturation or branching increase lipid flip-flop rates (Armstrong et al. 2003; Chen and Szostak 2004a). Similarly, membranes in the more fluid, liquid-crystalline phase contain lipids that flip-flop faster than membranes in the gel phase. However, diacyl phospholipid flip-flop is greatest at the phase transition temperature in which packing defects presumably more strongly diminish acyl chain interactions (John et al. 2002). The lipid head group also influences flip-flop rates with highly polar head group containing lipids being incapable of flipping. Because fatty acids possess fewer acyl chains and a less polar head group than phospholipids, fatty acids flip-flop more quickly than phospholipids (Hamilton 2003; Chen and Szostak 2004b).

Lipid escape (exchange) dynamics represent an additional type of large-scale motion that is strongly influenced by hydrophobicity. Diacyl phospholipids have a large hydrophobic surface area that results in a substantial energetic barrier for the acyl chains of the lipid to move from the hydrophobic environment of the membrane's interior to aqueous solution. Therefore, diacyl phospholipids do not readily exchange between vesicles or micelles and instead form kinetically trapped structures. The decreased hydrophobic surface area of monoacyl lipids poses a much lower energetic barrier for lipid escape, and so fatty acids rapidly equilibrate between vesicles and micelles. Therefore, fatty acids form highly dynamic aggregate structures that behave more like a system under thermodynamic rather than kinetic control (Luisi 2001). Aggregate systems containing mixtures of mono- and diacyl-lipids confirm this difference in dynamics (Fujikawa et al. 2005), which can be exploited for liposome generation by the detergent depletion method (Evans and Wennerstrom 1999). In short, monoacyl lipids, such as fatty acids, have higher escape and flip-flop dynamics than diacyl phospholipids, even though both monoacyl and diacyl lipids form structurally similar membranes.

RECENT RESULTS

Permeability Mechanisms

Although contemporary cells facilitate transport by using protein channels or carriers that provide less energetically costly paths for the solute to pass through the hydrophobic interior of the membrane, many small, neutral molecules such as water and carbon dioxide are able to cross the membrane without the aid of proteins (Inoue 1996). Permeability decreases with increasing molecular size, for instance, for solutes ranging in size from urea and glycerol up to glucose. The precise mechanism by which this occurs is not yet fully understood, as the route taken by the solute likely depends on a complex set of factors deriving from both the nature of the solute and the membrane. However, insight into the process has progressed considerably so that experimentally measured solute permeation rates can often be corroborated by mechanistically based computational modeling. Current permeability models take into account solute hydrophobicity, size, and shape along with membrane fluidity and packing density.

Solubility-Diffusion Model

Our current understanding of the barrier posed by lipid membranes is largely due to the work of Ernest Overton. His interpretation of permeability data is known as Overton's rule and is simply a correlation between solute hydrophobicity, that is the solute's ability to partition into the oily interior of the membrane, and permeation rates (Kleinzeller 1999). This remarkably simple correlation accurately predicts permeability trends for a large number of solutes and is incorporated into most currently used permeability models.

Overton's rule forms the foundation of the solubility-diffusion permeability model (Fig. 2A). In essence, it is a combination of Overton's hydrophobicity correlation with Fick's first law of diffusion, which describes how molecules move down concentration gradients at a rate directly proportional to the magnitude of the gradient (Eisenberg and Crothers 1979). The mathematical expression of Fick's first law is $J = -D(dc/dx)$, in which J is the solute flux, D is the diffusion coefficient, c is the solute concentration, and x is the distance that the solute must traverse when it crosses the membrane. If the rate of diffusion through the interior of the membrane is taken to be the rate limiting step for solute flux through a membrane, then D represents the diffusion coefficient within the hydrophobic interior of the membrane and x becomes the hydrophobic thickness of the bilayer membrane (Paula and Deamer 1999). Also, because the total flux of a solute through a vesicle membrane is directly proportional to

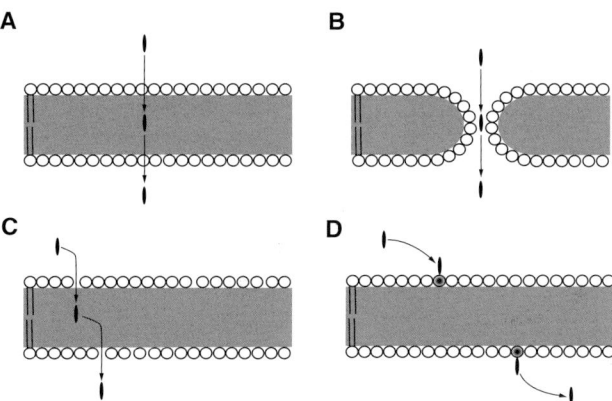

Figure 2. Membrane transport mechanisms. (*A*) Solubility-diffusion model. (*B*) Transient pore model. (*C*) Head-group gated model. (*D*) Lipid flipping-carrier model. Circles represent lipid head groups and the gray area indicates the hydrophobic region occupied by lipid acyl chains. The black oval is the solute molecule.

the surface area of the membrane, a term for the surface area (SA) of the membrane is added. Permeability is usually expressed as a permeability coefficient (P), rather than as solute flux, which more clearly shows the relationship between permeability and the properties of the solute and the membrane. P and J are related by $P = -J/(c_2-c_1)$, in which c_2-c_1 represents the solute concentration gradient. Finally, to account for the ability of a molecule to dissolve in the hydrophobic region of the membrane, a partition coefficient (K) is included, which gives the more familiar form of the solubility-diffusion equation $P = (K \cdot D \cdot SA)/x$.

This equation predicts a dependence on solute hydrophobicity, as more hydrophobic solutes can partition more effectively into the membrane, and a dependence on membrane thickness with thicker membranes having lower permeability coefficients. The model is consistent with a number of observations, working particularly well for neutral solutes like glycerol, urea, and ethanol up to hexoses. Membrane fluidity is taken into account by the diffusion term (D), because diffusion is dependent on solvent viscosity. Although the solubility-diffusion model appears well suited to describe a variety of solute permeability data, particularly for nonpolar molecules, the model does not take into consideration dynamic motion of the bilayer membrane structure (Paula et al. 1998).

Transient Pore Model

Although the elegantly simple solubility-diffusion model accurately predicts permeability trends for a large number of solutes, some inconsistencies between experimental data and theoretical modeling exist. For instance, ion permeability data are not consistent with the solubility-diffusion model. The major factor influencing ion permeation rates is their inability to partition into the nonpolar membrane phase, which is inhibited by the ion's "self energy," i.e., Born energy (Parsegian 1969). It is interesting that the encountered barrier is not the same for cations and anions. Because of the membrane's intrinsic dipole potential, anions more readily cross the initial barrier of head groups than cations, resulting in greater anion permeability coefficients (Fig. 3).

To account for this discrepancy, alternate models have been developed for ion permeation. One of these is known as the transient pore model (Fig. 2B), which invokes the opening of pores because of thermal fluctuations in membrane structure (Paula and Deamer 1999). If the transient pore is lined by lipid head groups or water molecules, then the Born energy barrier for ion permeation is removed, resulting in increased permeability coefficients (Paula et al.

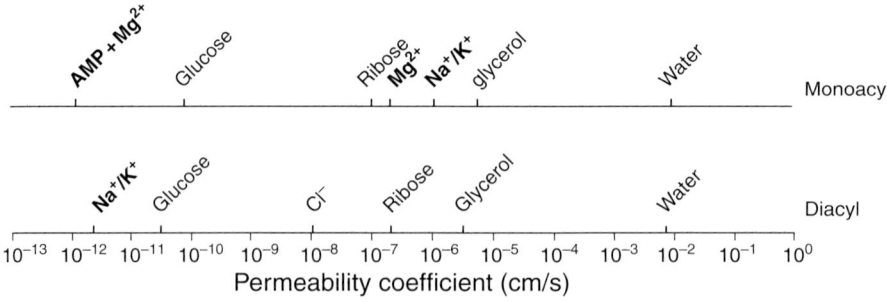

Figure 3. Comparison between solute permeability coefficients of liquid-crystalline diacyl phospholipid and monoacyl fatty acid membranes. Solutes in bold type permeate differently through both membranes. AMP permeation depends on the presence of Mg^{++}. AMP and Mg^{++} permeability through diacyl phospholipid membranes does not occur within the time scale represented by the x-axis. Differences in permeability between monoacyl and diacyl lipid membranes are small for nonpolar solutes and large for polar solutes. (Data were compiled from Paula et al. 1996; Paula et al. 1998; Paula and Deamer 1999; Chen and Szostak 2004b; Chen et al. 2005; Sacerdote and Szostak 2005; Mansy et al. 2008.)

1996). This alternate theory makes a variety of testable predictions that can be used in comparison with the solubility diffusion model. For instance, the creation of pores more strongly depends on membrane thickness because the required membrane thinning defects are more readily populated in membranes composed of lipids with short acyl chains (Paula et al. 1998). Permeability coefficients versus membrane thickness data for Na^+ and K^+ were consistent with a transient pore model rather than a solubility-diffusion model for phospholipids with short acyl chains, whereas acyl chains ≥ 18 carbons, showed permeation consistent with the solubility-diffusion model, particularly for anions (Paula et al. 1996).

Packing Defect Model

Packing defects have also been invoked as a mechanism by which solutes can pass through membranes. The differences, if any, are not clear as to what distinguishes transient pores from packing defects. Typically, packing defect mechanisms are used to describe the permeability of membranes composed of saturated lipids, whereas transient pores are typically used to describe processes occurring in membranes composed of lipids with at least one unsaturated acyl chain. Inconsistencies in permeability between these two types of membranes often have been observed (Xiang and Anderson 1998; Paula and Deamer 1999; Nagle et al. 2008). Nevertheless, packing defects are believed to exist for saturated lipid membranes at their phase transition temperature and result in dramatic increases in permeability (Nagle and Scott 1978; Jansen and Blume 1995). For example, even large ionic molecules, such as nucleotides, can traverse diacyl phospholipid membranes at the phase transition temperature of the membrane (Monnard and Deamer 2001; Cisse et al. 2007; Monnard et al. 2007).

Lipid Head Group-Gated Model

Despite the power of the earlier approaches in explaining solute permeability, inconsistencies continue to persist. For example, none of the models adequately take into account solute size and acyl chain ordering effects. This may in part be because of an overly simplified view of diffusion processes, particularly for solutes that diffuse through multiple phases. An insightful alternative perspective builds on concepts from the free volume model for diffusion in which solvent translational motions create cavities of free volume of sufficient size that can be occupied by an adjacent solute (Xiang and Anderson 1998). In other words, diffusion can be thought of as a statistical process of solute molecules finding nearby cavities to fill. This free volume approach can be adapted to better explain the processes of a solute passing through a barrier gated by lipid head groups. More specifically, solute passage through a membrane may depend on the probability of the solute finding free surface area to exploit (Xiang and Anderson 1998). The free surface area can be thought of as representing gaps within the hydrophilic head group region of the membrane (Fig. 2C). Two conclusions are obvious from such a perspective. First, the size and shape of the solute (described by its cross sectional area) influences its ability to find gaps, or free surface area, of sufficient size to allow for solute entry. Second, lipid packing density within the membrane similarly affects the size of free surface area and thus rates of permeation.

The head group gated model described earlier was combined with the solubility-diffusion model to give a model that takes into consideration both free surface area and solute size along with partitioning and diffusion through the membrane's hydrophobic interior (Mathai et al. 2008; Nagle et al. 2008). This three-slab model describes a process in which solutes must find free surface area for entry before partitioning into and diffusing through an oily interior, and then finally emerging through another head group-gated passage into aqueous solution. The model is particularly satisfying as it can readily explain differences in diffusion arising from solute size and shape, whereas also maintaining a strong solubility-diffusion component consistent with Overton's rule (Mathai et al. 2008; Nagle et al. 2008).

All of the models discussed thus far are based on data from diacyl phospholipid membranes.

For nonpolar solutes, the model appears to fit data from both monoacyl and diacyl lipid membranes. Even subtle changes in apolar solute–water interactions (Wei and Pohorille 2009) can result in dramatic changes in permeability coefficients for both mono- and diacyl lipid membranes (Sacerdote and Szostak 2005). Conversely, ionic solute permeabilities differ greatly between these two types of membranes (Fig. 3).

Lipid Carrier Model

Ionic solutes appear to operate via a different permeability mechanism when encountering a membrane that contains monoacyl lipids. Initial insight into this process was obtained from diacyl phospholipid membranes containing small amounts of monoacyl fatty acids, which revealed dramatic increases in proton and monovalent cation permeabilities (Kamp and Hamilton 1992). The proposed mechanism differed from previous models by invoking lipid flipping (Fig. 2D). More specifically, neutralized fatty acids, either through protonation or association of monovalent cations with the fatty acid's carboxylic acid head group, carry solute while flipping from one leaflet to the other before solute release (Kamp and Hamilton 1992). Similar studies with membranes composed solely of monoacyl fatty acids were consistent with this proposed carrier mechanism (Chen and Szostak 2004b).

The ability of monoacyl lipids to enhance the permeation rates of ionic solutes is not confined to alkali (Chen and Szostak 2004b) and alkaline Earth metals (Chen et al. 2005). Fatty acid membranes can similarly accommodate the passage of nucleotide mono- and diphosphates, if divalent cations are present to reduce the overall charge of the solute (Chen et al. 2005; Mansy et al. 2008). Although a mechanism for large, ionic, organic solute permeation has not been established, the data are consistent with a carrier-type process (Mansy et al. 2008).

Lipid Dynamics and Permeability

All of the described permeability mechanisms are dependent on lipid dynamics, such as membrane fluidity dynamics that facilitate diffusion through the hydrophobic region of the membrane or lipid dynamics at the surface of the membrane to create an entry space for solutes. The lipid dynamics that are exploited for solute passage depend on membrane composition and solute polarity. A well-documented example is the disparity between permeability data from saturated and unsaturated diacyl phospholipid membranes. It seems likely that such discrepancies reflect real differences in routes of solute passage and thus membrane dynamics. For example, the increased lipid flipping dynamics observed at the phase transition temperature of saturated diacyl phospholipid membranes (John et al. 2002) may facilitate the creation of hydrophilic pores (or packing defects lined with hydrophilic residues) that can be exploited for polar solute passage.

A revealing method to gain insight into the relationship between lipid dynamics and solute permeability is to compare solute permeabilities between monounsaturated diacyl phospholipids and fatty acid membranes. Many of the structural properties of the two membrane types are similar, but their associated lipid dynamics are significantly different, particularly those of lipid flip-flop and lipid escape (Table 1). Less polar molecules that are able to partition more effectively into the membrane permeate similarly through diacyl and monoacyl membranes. This suggests a similar dependence on the types of lipid dynamics, e.g., *trans*-gauche isomerizations, exploited for solute passage and in this case is consistent with a solubility-diffusion based mechanism. Conversely, highly polar molecules show much different behavior depending on the type of membrane encountered. Because polar molecules are less able to partition into the hydrophobic interior of the membrane, they must exploit alternate paths, and thus other available lipid dynamics, for solute passage. Unlike diacyl phospholipids, monoacyl fatty acids can rapidly flip-flop. It has been proposed previously that the flip-flop dynamics of fatty acid membranes may provide an alternate route for large, polar solute passage similar to the carrier mechanism proposed for monovalent cations (Mansy et al. 2008).

Computational studies are consistent with monoacyl lipid dynamics providing lower Born energy paths for ion permeation (Wilson and Pohorille 1996). Although definitive proof of this mechanism is lacking, results so far are consistent with a link between the increased lipid dynamics of fatty acid membranes and their increased permeability characteristics.

Although the carrier mechanism appears different from typical biological mechanisms of solute transport, similarities between the fatty acid carrier mechanism and the mechanism used by membrane transport proteins exist. Membrane transport proteins function similarly to enzymes in lowering the activation energy barrier by providing polar side-chain interactions that substitute for the solute's interactions with water. In effect, they are providing an alternate path for passage through the membrane that compensates for lost aqueous solvent interactions. This may not be much different than what occurs during the permeation of ionic solutes through fatty acid membranes. Solute interactions with carboxylate head groups may substitute for water interactions in the same way that protein-side chains substitute for hydration within a protein pore. Perhaps Earth's first cells exploited membrane lipid compositions that functioned both as a solute barrier and simultaneously as a solute transporter.

CHALLENGES AND FUTURE RESEARCH DIRECTIONS

Implications of Selective Permeability for the Origin of Life

Despite the presumed difficulty in nutrient exchange across membranes, permeability data from laboratory models of protocells reveal that selective permeability is an inherent property of fatty acid membranes. Vesicles composed of fatty acids are able to retain large polymers and highly charged molecules, while simultaneously allowing for the diffusion of smaller, less polar solutes (Chen et al. 2006). Even subtle differences in partition coefficients can lead to large permeability differences that could potentially be exploited for early metabolic processes.

For example, within a series of aldopentoses, ribose crosses monoacyl and diacyl lipid membranes approximately fivefold faster than diastereomers of equal molecular weight (Sacerdote and Szostak 2005). This remarkable kinetic selectivity apparently arises from less extensive solvent interactions for ribose than for other five-carbon sugars (Wei and Pohorille 2009) and is consistent with permeability models that incorporate Overton's rule. Because the preferential acceleration of a few reaction rates can result in directional flux through a complex set of potential chemical reactions (Copley et al. 2007), the kinetic availability of ribose in protocellular structures could have significantly influenced the resulting encapsulated protometabolic network.

The ability of model protocells with membranes composed of monoacyl lipids to uptake nucleotides provides for a route to replicate nucleic acid genomes in the absence of proteins. Building on work from the Orgel laboratory (Lohrmann and Orgel 1976; Tohidi et al 1987), a sequence-general, template-directed, nonenzymatic nucleic acid replication method was developed by the Szostak laboratory (Schrum et al. 2009). Briefly, the method exploits nucleotides activated with an imidazole leaving group in place of the pyrophosphate and a $2'$-amino nucleophile in place of the $3'$-hydroxyl. Although nucleotides likely were not prebiotically activated in precisely this way, it seems probable that nucleotides were activated differently than they are now in contemporary cells (Lohrmann and Orgel 1976). Because fatty acid membranes allow for the passage of nucleotides (Chen et al. 2005; Mansy et al. 2008), the encapsulation of a nucleic acid template inside of monoacyl, fatty acid vesicles resulted in efficient template copying when $2'$-amino, $5'$-phosphorimidazolide nucleotides were added externally to the vesicles. The same conditions did not result in template copying inside of diacyl phospholipid vesicles because of the impermeability of the membrane (Mansy et al. 2008). These studies highlight the possibility of simple, cell-like structures composed of leaky (Deamer 2008), monoacyl lipid membranes to replicate its own genome by feeding off of the environment.

The permeability and selectivity of fatty acid vesicles suggest that the simplest prebiotic cell-like structures may have been heterotrophic, in the sense that growth was achieved by taking up nutrient compounds from the external solution, rather than by developing encapsulated metabolic networks to synthesize nutrients. Conversely, an autotrophic protocell would likely have required diacyl lipids to retain synthesized small molecules and thus would have additionally faced a series of transport problems that required greater complexity in the form of membrane transport proteins. Nevertheless, protocell research is still in its infancy and much more work is needed to define possible routes to the emergence of protocells.

Early Transporters

The arguments presented here show that nutrient acquisition can occur across lipid membranes without the aid of specialized proteins or RNA molecules. However, cell-like structures that depend on passive diffusion mechanisms are limited in the complexity that they can achieve. To fuel active metabolic processes, membrane transport must be rapid enough to replenish consumed nutrients. Although the total amount of solute flux across a membrane can be increased by increasing vesicle size, i.e., membrane surface area (r^2 dependence), the accompanying changes in volume (r^3 dependence) result in a net decrease in nutrient availability. This is a significant limitation because compartment size is directly related to the potential complexity of the system. In other words, small compartments can only hold a small number of molecules (Morowitz 1992), thus placing an upper limit on the number of possible encoded ribozymes, for example. Similarly, smaller compartments are more susceptible to fluctuations in molecular composition; even the addition of a single proton within a small vesicle results in significant pH changes (Morowitz 1992). Although these compartment size considerations assume active metabolic processes that may not be relevant to the early steps in the emergence of protocells, progression toward cell-like structures with increased complexity and metabolic activity likely would have selected for larger compartments and in turn active transport mechanisms.

If protocellular evolution resulted in progressively less permeable membrane compositions (Szathmary 2007), then the simultaneous development of transmembrane solute transporters would have been necessary. The simplest biological examples of transporters are short, antibiotic peptides that can mediate the passage of cations across otherwise impermeable phospholipid membranes. Such peptides often contain D- and L-amino acids, are nonribosomally synthesized, and often display remarkable solute specificity (Eisenberg 1998). Even simple hydrophobic polypeptides, such as poly-alanine, can increase proton permeation rates (Oliver and Deamer 1994). It is easy to envisage how short peptide sequences could have been exploited for solute passage by a protocell. However, a mechanism for the propagation of the resulting phenotype from one generation to another would be problematic in the absence of genetic material.

As evidence continues to grow suggesting that life originated in or passed through a RNA world, an important question is whether RNA is able to provide the membrane transport functionality provided by proteins and short peptides. Although RNA mediated transport across membranes appears unlikely because of the high polarity of the phosphodiester backbone of RNA molecules, RNA could modulate the properties of a membrane through interactions with the hydrophilic, solvent exposed membrane surfaces. Lipid head-group organization is an important determinant of membrane stability and fluidity, and for the ability of solutes to gain access to the membrane's hydrophobic interior. Through in vitro selection experiments, RNA molecules have been identified that tightly associate with membrane surfaces and that may also partially penetrate the membrane (Khvorova et al. 1999; Janas and Yarus 2006). As some permeability mechanisms invoke transient pores and packing defects as routes for solute passage, it is plausible that RNA molecules that simply interact with membrane surfaces could influence the structure and dynamics of the membrane enough to effectively enhance solute permeation. In fact, a RNA chimera consisting of selected membrane and tryptophan binding motifs is able to facilitate tryptophan transport across diacyl phospholipid membranes (Janas and Yarus 2004). Although few examples exist of such RNA-membrane interactions, the ability of RNA to serve a wide range of functions in vitro and in contemporary cells suggests that facilitated solute transport may not have been impossibility during the origin of life (Janas et al. 2006).

Membrane Permeability Influences

Selective permeability influences processes beyond those of nutrient acquisition, including growth, shape, and division processes. Concentration gradients across vesicle membranes because of solute impermeability produce osmotic pressures that can drive chemical reactions, such as vesicle growth (Chen et al. 2004). A mixture of vesicles experiencing differing osmotic pressures arising from differences in RNA content, for example, can lead to the preferential growth of RNA containing vesicles at the expense of empty vesicles. In other words, osmotic gradients across monoacyl, fatty acid membranes create a competition for resources between model protocellular structures (Chen et al. 2004). Similarly, osmotic gradients across multilamellar membranes result in vesicle shape changes because of surface area–volume imbalances that predispose the vesicle to division during growth processes (Zhu and Szostak 2009). The requisite shape changes only occur in the presence of slowly permeable solutes, whereas multilamellar vesicle growth in the presence of highly permeable solutes does not result in vesicle shape changes (Zhu and Szostak 2009). Selectively permeable membranes endow vesicle systems with the lifelike properties of growth, competition, and division.

CONCLUSIONS

It appears that many of the desirable permeability properties of fatty acid membranes arise from the increased lipid dynamics of the vesicle system, whereas many of the early assumptions on the influences of membranes on protocellular evolution were based on less dynamic, diacyl phospholipid membranes. In addition to facilitating the selective absorption and release of molecular building blocks, the dynamic nature of monoacyl lipid membranes allow for growth (Hanczyc et al. 2003), competition (Chen et al. 2004), and division (Zhu and Szostak 2009). Importantly, many of these processes are coupled and so a complete cycle of growth, replication, and division appears achievable (Mansy and Szostak 2009). For example, the uptake of activated nucleotides can result in an increase in genomic content and thus an increase in osmotic pressures on the vesicle system that can be partially relieved by growth. Protocells incapable of genomic replication would not grow and would be selected against. Growing multilamellar vesicles with semipermeable membranes would then proceed through a series of shape changes resulting from surface area-volume imbalances and ultimately divide under mild agitation. Surprisingly, data thus far suggest that a complete replication cycle may only require four components, including two types of monoacyl lipids, activated nucleotides, and a nucleic acid template (Mansy and Szostak 2009).

Thus far a robust and complete cycle including both nucleic acid and compartment replication has not been shown, even though many of the necessary individual steps have been realized. Further efforts in developing nucleic acid replication mechanisms that properly coordinate strand separation, annealing, and copying with vesicle shape changes and division is needed (Mansy and Szostak 2009). Nearly every step in the growth and replication of protocellular structures is influenced by membrane permeability. Protocells with a fatty acid membrane can acquire polar nutrients without first evolving additional transport machinery. The demonstration of a complete cycle that can undergo Darwinian evolution will greatly deepen our understanding of the chemical and physical basis of life.

ACKNOWLEDGMENTS

I thank Cristina Del Bianco, members of CIBIO at the University of Trento, and the Szostak

Laboratory for insightful discussions, the Editors for helpful comments, and the Armenise-Harvard foundation for support.

REFERENCES

Armstrong VT, Brzustowicz MR, Wassall SR, Jenski LJ, Stillwell W. 2003. Rapid flip-flop in polyunsaturated (docosahexaenoate) phospholipid membranes. *Arch Biochem Biophys* **414:** 74–82.

Benner SA, Hutter D. 2002. Phosphates, DNA, and the search for nonterrean life: A second generation model for genetic molecules. *Bioorg Chem* **30:** 62–80.

Chen IA, Szostak JW. 2004a. A kinetic study of the growth of fatty acid vesicles. *Biophys J* **87:** 988–998.

Chen IA, Szostak JW. 2004b. Membrane growth can generate a transmembrane pH gradient in fatty acid vesicles. *Proc Natl Acad Sci* **101:** 7965–7970.

Chen IA, Roberts RW, Szostak JW. 2004. The emergence of competition between model protocells. *Science* **305:** 1474–1476.

Chen IA, Salehi-Ashtiani K, Szostak JW. 2005. RNA catalysis in model protocell vesicles. *J Am Chem Soc* **127:** 13213–13219.

Chen IA, Hanczyc MM, Sazani PL, Szostak JW. 2006. Protocells: Genetic polymers inside membrane vesicles. in *The RNA world* (ed. Gesteland R.F., Cech T.R., Atkins J.F.), pp. 57–88. Cold Spring Harbor Laboratory Press, Cold Spring Harbor.

Cisse I, Okumus B, Joo C, Ha T. 2007. Fueling protein-DNA interactions inside porous nanocontainers. *Proc Natl Acad Sci* **104:** 12646–12650.

Cistola DP, Atkinson D, Hamilton JA, Small DM. 1986. Phase behavior and bilayer properties of fatty acids: Hydrated 1:1 acid-soaps. *Biochemistry* **25:** 2804–2812.

Copley SD, Smith E, Morowitz HJ. 2007. The origin of the RNA world: Co-evolution of genes and metabolism. *Bioorg Chem* **35:** 430–443.

Deamer DW. 1985. Boundary structures are formed by organic components of the Murchison carbonaceous chondrite. *Nature* **317:** 792–794.

Deamer DW. 1997. The first living systems: A bioenergetic perspective. *Microbiol Mol Biol Rev* **61:** 239–261.

Deamer DW. 2008. Origins of life: How leaky were primitive cells? *Nature* **454:** 37–38.

Deamer DW, Pasley RM. 1989. Amphiphilic components of the Murchison carbonaceous chondrite: Surface properties and membrane formation. *Orig Life Evol Biosphere* **19:** 21–38.

Eisenberg D, Crothers D. 1979. *Physical chemistry with applications to the life sciences.* The Benjamin/Cummings publishing company, Inc., Menlo Park.

Eisenberg E. 1998. Ionic channels in biological membranes: Natural nanotubes. *Acc Chem Res* **31:** 117–123.

Evans DF, Wennerstrom H. 1999. *The colloidal domain: Where physics, chemistry, biology, and technology meet.* Wiley-VCH, New York.

Fujikawa SM, Chen IA, Szostak JW. 2005. Shrink-wrap vesicles. *Langmuir* **21:** 12124–12129.

Gawrisch K. 2004. The dynamics of membrane lipids. In *The structure of biologial membranes* (ed. Yeagle P.L.), pp. 147–172. CRC Press, Boca Raton.

Gawrisch K, Gaede HC, Mihailescu M, White SH. 2007. Hydration of POPC bilayers studies by 1H-PFG-MAS-NOESY and neutron diffraction. *Eur Biophys J* **36:** 281–291.

Gawrisch K, Ruston D, Zimmerberg J, Parsegian VA, Rand RP, Fuller N. 1992. Membrane dipole potentials, hydration forces, and the ordering of water at membrane surfaces. *Biophys J* **61:** 1213–1223.

Gebicki JM, Hicks M. 1973. Ufasomes are stable particles surrounded by unsaturated fatty acid membranes. *Nature* **243:** 232–234.

Gebicki JM, Hicks M. 1976. Preparation and properties of vesicles enclosed by fatty acid membranes. *Chem Phys Lipids* **16:** 142–146.

Hamilton JA. 2003. Fast flip-flop of cholesterol and fatty acids in membranes: Implications for membrane transport proteins. *Curr Opin Lipidol* **14:** 263–271.

Hanczyc MM, Fujikawa SM, Szostsak JW. 2003. Experimental models of primitive cellular compartments: Encapsulation, growth, and division. *Science* **302:** 618–622.

Hargreaves WR, Deamer DW. 1978. Liposomes from ionic, single-chain amphiphiles. *Biochemistry* **17:** 3759–3768.

Inoue T. 1996. Interaction of surfactants with phospholipid vesicles. In *Vesicles (Surfactant Science)*. Marcel Dekker, Inc, New York.

Israelachvili JN. 1991. *Intermolecular & surface forces.* Academic Press, London.

Janas T, Yarus M. 2004. A membrane transporter for tryptophan composed of RNA. *RNA* **10:** 1541–1549.

Janas T, Yarus M. 2006. Specific RNA binding to ordered phospholipid bilayers. *Nucleic Acids Res* **34:** 2128–2136.

Janas T, Janas T, Yarus M. 2006. RNA, lipids, and membranes. In *The RNA world* (ed. R.F. Gesteland, T.R. Cech, J.F. Atkins), pp. 207–225. Cold Spring Harbor Laboratory Press, Cold Spring Harbor.

Jansen M, Blume A. 1995. A comparative study of diffusive and osmotic water permeation across bilayers composed of phospholipids with different head groups and fatty acyl chains. *Biophys J* **68:** 997–1008.

John K, Schreiber S, Kubelt J, Herrmann A, Muller P. 2002. Transbilayer movement of phospholipids at the main phase transition of lipid membranes: Implications for rapid flip-flop in biological membranes. *Biophys J* **83:** 3315–3323.

Kahya N, Schwille P. 2006. How phospholipid-cholesterol interactions modulate lipid lateral diffusion, as revealed by fluorescence correlation spectroscopy. *J Fluoresc* **16:** 671–678.

Kamp F, Hamilton JA. 1992. pH gradients across phospholipid membranes caused by fast flip-flop of un-ionized fatty acids. *Proc Natl Acad Sci* **89:** 11367–11370.

Khvorova A, Kwak YG, Tamkun M, Majerfeld I, Yarus M. 1999. RNAs that bind and change the permeability of phospholipid membranes. *Proc Natl Acad Sci* **96:** 10649–10654.

Klauda JB, Roberts MF, Redfield AG, Brooks BR, Pastor RW. 2008. Rotation of lipids in membranes: molecular dynamics simulation, 31P spin-lattice relaxation, and rigid-body dynamics. *Biophys J* **94**: 3074–3083.

Kleinzeller A. 1999. Charles Ernest Overton's concept of a cell membrane. In *Membrane permeability: 100 years since Ernest Overton* (ed. D.W. Deamer, A. Kleinzeller, D.M. Fambrough), pp. 1–18. Academic Press, San Diego.

Lawless JG, Yuen GU. 1979. Quantification of monocarboxylic acids in the Murchison carbonaceous meteorite. *Nature* **282**: 396–398.

Lohrmann R, Orgel LE. 1976. Template-directed synthesis of high molecular weight polynucleotide analogues. *Nature* **261**: 342–344.

Luisi PL. 2001. Are micelles and vesicles chemical equilibrium systems? *J Chem Fd* **78**: 380–384.

Mansy SS, Szostak JW. 2008. Thermostability of model protocell membranes. *Proc Natl Acad Sci* **105**: 13351–13355.

Mansy SS, Szostak JW. 2009. Reconstructing the emergence of cellular life through the synthesis of model protocells. *Cold Spring Harb Symp Quant Biol* (in press).

Mansy SS, Schrum JP, Krishnamurthy M, Tobe S, Treco DA, Szostak JW. 2008. Template-directed synthesis of a genetic polymer in a model protocell. *Nature* **454**: 122–125.

Mathai JC, Tristram-Nagle S, Nagle JF, Zeidel ML. 2008. Structural determinants of water permeability through the lipid membrane. *J Gen Physiol* **131**: 69–76.

McCollom TM, Ritter G, Simoneit BR. 1999. Lipid synthesis under hydrothermal conditions by Fischer-Tropsch-type reactions. *Orig Life Evol Biosph* **29**: 153–166.

McFarland BG, McConnell HM. 1971. Bent fatty acid chains in lecithin bilayers. *Proc Natl Acad Sci* **68**: 1274–1278.

McIntosh TJ. 1978. The effect of cholesterol on the structure of phosphatidylcholine bilayers. *Biochim Biophys Acta* **513**: 43–58.

Menger FM, Wood MG, Zhou QZ, Hopkins HP, Fumero J. 1988. Thermotropic properties of synthetic chain-substituted phosphatidylcholines: Effect of substituent size, polarity, number, and location on molecular packing. *J Am Chem Soc* **110**: 6804–6810.

Monnard PA, Deamer DW. 2001. Nutrient uptake by protocells: A liposome model system. *Orig Life Evol Biosph* **31**: 147–155.

Monnard PA, Luptak A, Deamer DW. 2007. Models of primitive cellular life: Polymerases and templates in liposomes. *Philos Trans R Soc Lond B Biol Sci* **362**: 1741–1750.

Morowitz HJ. 1992. *Beginnings of cellular life. Metabolism recapitulates biogenesis.* Yale University Press, New Haven.

Nagle JF, Scott HL Jr. 1978. Lateral compressibility of lipid mono- and bilayers. Theory of membrane permeability. *Biochim Biophys Acta* **513**: 236–243.

Nagle JF, Mathai JC, Zeidel ML, Tristram-Nagle S. 2008. Theory of passive permeability through lipid bilayers. *J Gen Physiol* **131**: 77–85.

Nelson DL, Cox MM. 2008. *Lehninger: principles of biochemistry.* W. H. Freeman and Company, New York.

Oliver AE, Deamer DW. 1994. α-helical hydrophobic polypeptides form proton-selective channels in lipid bilayers. *Biophys J* **66**: 1364–1379.

Parsegian A. 1969. Energy of an ion crossing a low dielectric membrane: Solutions to four relevant electrostatic problems. *Nature* **221**: 844–846.

Paula S, Deamer DW. 1999. Membrane permeability barriers to ionic and polar solutes. In *Membrane permeability: 100 years since Ernest Overton* (ed. D.W. Deamer, A. Kleinzeller, D.M. Fambrough), pp. 77–95. Academic Press, San Diego.

Paula S, Volkov AG, Deamer DW. 1998. Permeation of halide anions through phospholipid bilayers occurs by the solubility-diffusion mechanism. *Biophys J* **74**: 319–327.

Paula S, Volkov AG, Van Hoek AN, Haines TH, Deamer DW. 1996. Permeation of protons, potassium ions, and small polar molecules through phospholipid bilayers as a function of membrane thickness. *Biophys J* **70**: 339–348.

Peterson U, Mannock DA, Lewis RN, Pohl P, McElhaney RN, Pohl EE. 2002. Origin of membrane dipole potential: Contribution of the phospholipid fatty acid chains. *Chem Phys Lipids* **117**: 19–27.

Pizzarello S. 2006. The chemistry of life's origin: A carbonaceous meteorite perspective. *Acc Chem Res* **39**: 231–237.

Pizzarello S. 2007. The chemistry that preceded life's orgin: A study guide from meteorites. *Chem Biodivers* **4**: 680–693.

Roseman MA, Thompson TE. 1980. Mechanism of the spontaneous transfer of phospholipids between bilayers. *Biochemistry* **19**: 439–444.

Rubenstein JL, Smith BA, McConnell HM. 1979. Lateral diffusion in binary mixtures of cholesterol and phosphatidylcholines. *Proc Natl Acad Sci* **76**: 15–18.

Sacerdote MG, Szostak JW. 2005. Semipermeable lipid bilayers exhibit diastereoselectivity favoring ribose. *Proc Natl Acad Sci* **102**: 6004–6008.

Schrum JP, Ricardo A, Krishnamurthy M, Blain JC, Szostak JW. 2009. Efficient and rapid template-directed nucleic acid copying using 2'-amino-2',3'-dideoxyribonucleoside-5'-phosphorimidazolide monomers. *J Am Chem Soc* **131**: 14560–14570.

Singer SJ. 2004. Some early history of membrane molecular biology. *Annu Rev Physiol* **66**: 1–27.

Singer SJ, Nicolson GL. 1972. The fluid mosaic model of the structure of cell membranes. *Science* **175**: 720–731.

Szathmary E. 2007. Coevolution of metabolic networks and membranes: the scenario of progressive sequestration. *Philos Trans R Soc Lond B Biol Sci* **362**: 1781–1787.

Szostak JW, Bartel DP, Luisi PL. 2001. Synthesizing life. *Nature* **409**: 387–390.

Tanford C. 1980. *The hydrophobic effect: Formation of micelles and biological membranes.* John Wiley & Sons, New York.

Tarrant MK, Cole PA. 2009. The chemical biology of protein phosphorylation. *Ann Rev Biochem* **78**: 797–825.

Tohidi M, Zielinski WS, Chen CH, Orgel LE. 1987. Oligomerization of 3'amino-3' deoxyguanosine-5' phosphorimidazolidate on a d(CpCpCpCpC) template. *J Mol Evol* **25**: 97–99.

van Mooy BA, Fredricks HF, Pedler BE, Dyhrman ST, Karl ST, Noblizek M, Lomas MW, Mincer TJ, Moor LR, Moutin T, et al. 2009. Phytoplankton in the ocean use non-phosphorus lipids in response to phosphorus scarcity. *Nature* **458**: 69–71.

Venable RM, Zhang Y, Hardy BJ, Pastor RW. 1993. Molecular dynamics simulations of a lipid bilayer and of hexadecane: An investigation of membrane fluidity. *Science* **262**: 223–226.

Walde P, Goto A, Monnard PA, Wessicken M, Luisi PL. 1994. Oparin's reactions revisited: Enzymatic synthesis of poly(adenylic acid) in micelles and self-reproducing vesicles. *J Am Chem Soc* **116**: 7541–7547.

Wang L, Bose PS, Sigworth FJ. 2006. Using cryo-EM to measure the dipole potential of a lipid membrane. *Proc Natl Acad Sci* **103**: 18528–18533.

Wei C, Pohorille A. 2009. Permeation of membranes by ribose and its diastereomers. *J Am Chem Soc* **131**: 10237–10245.

Westheimer FH. 1987. Why nature chose phosphates. *Science* **235**: 1173–1178.

Wiener MC, White SH. 1992. Structure of a fluid dioleoylphosphatidycholine bilayer determined by joint refinement of x-ray and neutron diffraction data. III. Complete structure. *Biophys J* **61**: 434–447.

Wilson MA, Pohorille A. 1996. Mechanism of unassisted ion transport across membrane bilayers. *J Am Chem Soc* **118**: 6580–6587.

Xiang TX, Anderson BD. 1998. Influence of chain ordering on the selectivity of dipalmitoylphosphatidylcholine bilayer membranes for permeant size and shape. *Biophys J* **75**: 2658–2671.

Zhu TF, Szostak JW. 2009. Coupled growth and division of model protocell membranes. *J Am Chem Soc* **131**: 5705–5713.

Primitive Genetic Polymers

Aaron E. Engelhart and Nicholas V. Hud

School of Chemistry and Biochemistry, Georgia Institute of Technology, Atlanta, Georgia 30332

Correspondence: hud@chemistry.gatech.edu

Since the structure of DNA was elucidated more than 50 years ago, Watson-Crick base pairing has been widely speculated to be the likely mode of both information storage and transfer in the earliest genetic polymers. The discovery of catalytic RNA molecules subsequently provided support for the hypothesis that RNA was perhaps even the first polymer of life. However, the de novo synthesis of RNA using only plausible prebiotic chemistry has proven difficult, to say the least. Experimental investigations, made possible by the application of synthetic and physical organic chemistry, have now provided evidence that the nucleobases (A, G, C, and T/U), the trifunctional moiety ([deoxy]ribose), and the linkage chemistry (phosphate esters) of contemporary nucleic acids may be optimally suited for their present roles—a situation that suggests refinement by evolution. Here, we consider studies of variations in these three distinct components of nucleic acids with regard to the question: Is RNA, as is generally acknowledged of DNA, the product of evolution? If so, what chemical and structural features might have been more likely and advantageous for a proto-RNA?

In contemporary life, nucleic acids provide the amino acid sequence information required for protein synthesis, while protein enzymes carry out the catalysis required for nucleic acid synthesis. This mutual dependence has been described as a "chicken-or-the-egg" dilemma concerning which came first. However, requiring that these biopolymers appeared strictly sequentially may be an overly restrictive preconception—nucleic acids and *noncoded* peptides may have arisen independently and only later become dependent on each other. Nevertheless, the requirements for the chemical emergence of life would appear simplified if one polymer was initially able to store and transfer information as well as perform selective chemical catalysis—two essential features of life.

The discovery of catalytic RNA molecules in the early 1980s (Kruger et al. 1982; Guerrier-Takada et al. 1983) created widespread interest in an earlier proposal (Woese 1967; Crick 1968; Orgel 1968) that nucleic acids were the first biopolymers of life, as nucleic acids transmit genetic information and could have once been responsible for catalyzing a wide range of reactions. The ever-increasing list of processes that involve RNA in contemporary life continues to strengthen this view (Mandal and Breaker 2004; Gesteland and Atkins 2006). Furthermore, the rule-based one-to-one pairing of complementary bases in a Watson-Crick duplex (Fig. 1) provides a robust mechanism for information transfer during replication that could have been operative from the advent of oligonucleotides. In contrast, there is no obvious and general mechanism by which the amino acid sequence of a polypeptide can be transferred to a new polypeptide as part of a replication process.

If we accept that nucleic acids must have appeared without the aid of coded proteins, we are still faced with the question of how the first nucleic acid molecules came to be. Broadly defined, there are two schools of thought regarding the origin of the earliest nucleic acids. In one school, it is proposed that abiotic chemical processes initially gave rise to nucleotides (i.e., phosphorylated nucleosides), which were then coupled together to yield polymers identical in chemical structure to contemporary RNA. In support of this model, Sutherland presents in his article current progress toward discovering possible chemical pathways for the prebiotic synthesis of RNA mononucleotides, as well as methods for their protein-free polymerization (Sutherland 2010).

A second school of thought, discussed in this article, considers RNA to be a product of evolution, and that a different RNA-like polymer (or proto-RNA) was used by the earliest forms of life. Just as the deoxyribose sugar of DNA was likely the product of Darwinian evolution (selected for the hydrolytic stability it provides this long-lived biopolymer), so, too, might the sugar, phosphate, and bases of RNA have been refined by evolution. In this scenario, a proto-RNA is more likely to have spontaneously formed than RNA, because a proto-RNA could have had more favorable *chemical* characteristics (e.g., greater availability of precursors and ease of assembly), but such a polymer was eventually replaced, through evolution, by RNA (potentially after several incremental changes), based on *functional* characteristics (e.g. nucleoside stability, versatility in forming catalytic structures). Thus, contemporary RNA may possess chemical traits that, although optimally suited for contemporary life, may have been ill-suited for the earliest biopolymers, with the converse being true for proto-RNA.

BACKGROUND

Challenges to "Reinventing" Proto-RNA

If proto-RNA came before RNA, possibly comprised of different bases and a different backbone, then the chemical space of proto-RNA candidates seems almost limitless. Leslie Orgel once called this scenario a "gloomy prospect" with regard to solving the origin of life (Orgel 1998). However, we do not see the prospect of

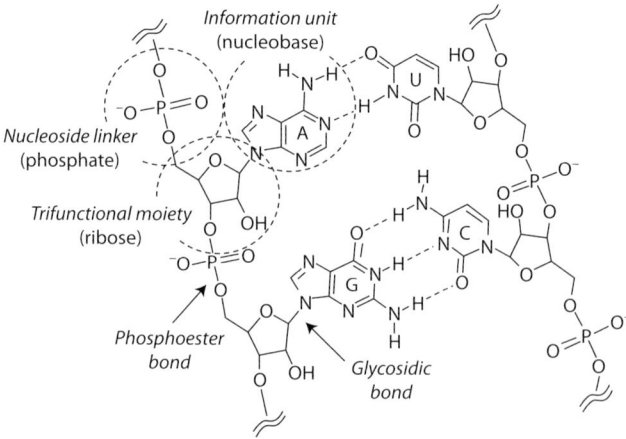

Figure 1. Two base-paired RNA dinucleotide steps with functional units discussed in the text annotated. In contemporary life, the nucleoside linker is phosphate, and the information unit is one of the canonical nucleobases (A, G, C, and U). The contemporary trifunctional moiety, ribose, is coupled via N,O-acetals to the informational unit and via phosphoesters to the nucleoside linker.

proto-RNAs as reason for pessimism. The possibility that RNA had one or more predecessors also implies the possibility that there was a polymer with more facile chemistry of assembly, which researchers could show to spontaneously assemble from simple precursors. Additionally, the chemical space within which proto-RNA was formed is constrained by the molecules that could have been present on the prebiotic Earth. Although we will never know the precise contents of the *prebiotic chemical inventory*, model prebiotic reactions continue to provide valuable information regarding plausible prebiotic molecules (Miyakawa et al. 2002; Plankensteiner et al. 2004; Saladino et al. 2004; Cleaves et al. 2006). Additionally, studies of our neighboring planets and their moons, meteorites, comets and even interstellar space (Schwartz and Chang 2002; Hollis et al. 2004; Pizzarello 2006; Martins et al. 2008) also provide clues to the types of molecules from which a proto-RNA might have emerged.

Sugars and nucleobases, two essential building blocks of nucleic acids, have been identified as products in a rather wide range of model prebiotic reactions. However, feasible prebiotic reactions have not been shown for the efficient coupling of ribose to all four nucleobases to create the canonical nucleosides, and only limited evidence has been presented that preformed nucleotides might be able to polymerize without resorting to nonprebiotic chemical activation. Thus, a conceptual gap still exists between our ideas for how the building blocks of RNA might have formed from the simple molecules that are ubiquitous in the universe versus the chemical processes that gave rise to the first RNA-like polymers.

All the present challenges to bridging this conceptual gap between the "small molecule world" and the "polymer world" can be classified as belonging to one of two major problems: molecular selection and polymerization reactions. There was likely a wide variety of molecules in the prebiotic chemical inventory, so the first formidable challenge is to find an abiotic mechanism by which "useful" building blocks were selected from a complex mixture. The second formidable challenge is to find reactions by which these building blocks could have been "correctly" joined to create "useful" (i.e., self-replicating, catalytic or both) proto-biopolymers. In the following sections, we will elaborate on the challenge of RNA building block selection and assembly, as well as possible solutions to "reinventing," to use Albert Eschenmoser's term (Eschenmoser 2007), a process for the de novo synthesis of an RNA-like polymer from plausible prebiotic molecules and reactions.

Some Possible Constraints to Guide the Search for Proto-RNA

Synthetic chemists have prepared numerous nucleic acid analogs, some as part of origin of life studies (Schneider and Benner 1990; Pitsch et al. 1993; Herdewijn 2001b; Benner 2004; Mittapalli et al. 2007a; Mittapalli et al. 2007b) and many as part of a widespread search for therapeutic agents (Kurreck 2003). These analogs have included changes to both the nucleobases and to the polymer backbone. With regard to origin of life investigations, some proposals for the difference between RNA and an earlier proto-RNA have been subtle, such as the proposal that hypoxanthine once functioned in the place of guanine (these nucleobases differ by only a single exocyclic amino group) (Crick 1968). Other proposals are much more radical, such as the suggestion that the nucleobases were once connected by a completely different and uncharged backbone, as in the synthetic polymer known as peptide nucleic acid (Nelson et al. 2000), and intermediate proposals, in which the backbone was a peptide, but still negatively charged (Mittapalli et al. 2007a; Mittapalli et al. 2007b) (Fig. 2). Physical and chemical studies of these and other nucleic acid analogs have provided invaluable information regarding which alternative polymers might have predated RNA. An exhaustive review of all proposed predecessors of RNA is beyond the scope of this article, and several excellent reviews have been published considering possible variations in sugars and bases, to which the reader is referred (Benner 2004; Benner et al. 2004; Eschenmoser 2007).

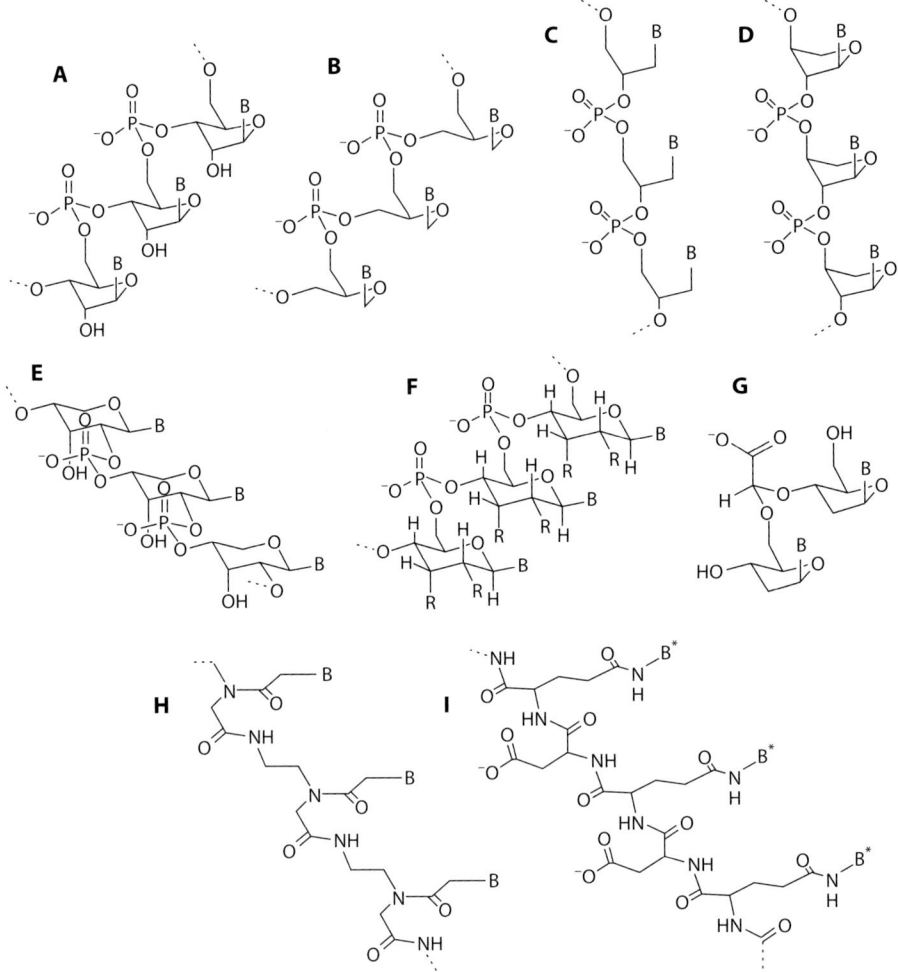

Figure 2. Structures of RNA and selected analogs discussed in the text. (*A*) RNA; (*B*) FNA or "flexible nucleic acid," originally named glycerol nucleic acid; (*C*) GNA, glycerol nucleic acid; (*D*) TNA, α-threofuranosyl nucleic acid; (*E*) pRNA, β-pyranosyl-RNA; (*F*) homo-DNA (R = H) and β-allopyranosyl (R = OH); (*G*) gaNA, glyoxylate-linked dinucleotide; (*H*) PNA, peptide nucleic acid; (*I*) poly-(L-Asp-L-Glu) with the γ-carboxyl function of Glu conjugated to a nucleobase by an isopeptide linkage. For all structures shown, B represents one of the canonical nucleobases, and B*, in structure I, represents a noncanonical base.

In this article, we focus on a subset of potential candidates for proto-RNA polymers that, in our opinion, would be closest to RNA in structure and physical properties but are composed of alternative building blocks that would have made prebiotic assembly more feasible. One could consider these structures as candidates for the later predecessors of RNA that resulted from proto-RNA evolution or, optimistically, as potential candidates for the first proto-RNA, if only a few changes were required to reach RNA. Our selection of proposed proto-RNA candidates on which to focus is guided by three hypotheses:

(1) *The original molecular components of proto-RNA may have been different from, but still similar to, those found in RNA.* Three distinct molecular components comprise RNA: the nucleobases, ribose and phosphate (Fig. 1). Each component has specific roles for which it appears to be very well (perhaps

optimally) suited. However, because of the limited availability of some components, or the difficulty in forming covalent bonds between others, there is good reason to entertain the possibility that alternative components were initially used by proto-RNA. Each of these three components could have been replaced on different timescales over the course of evolution. Although this criterion may seem obvious regarding what would be recognized as RNA-like polymers, it does exclude from our discussion the more radical models of "genetic takeover" (Cairns-Smith 1982), in which an inorganic or decidedly non-RNA-like system gave rise to RNA, models which come with the added burden of deciphering how a physicochemical transition could have transpired between completely different molecular systems.

(2) *The covalent bonds that connected the three molecular components of proto-RNA were (periodically) reversible.* In the early stages of life, it would have been highly advantageous for chemical building blocks to be available for reuse. Mechanisms for recycling are easy to imagine, if the covalent bonds that joined the molecular components of proto-biopolymers could have been formed, broken and re-formed repeatedly. Such a process would have allowed proto-biopolymers to be created that were thermodynamically favored structures (Li et al. 2002; Hud et al. 2007; Ura et al. 2009) and for "errors" in synthesis to be corrected (e.g., replacement of non- or mispaired nucleobases with pairing ones). Proto-RNA monomers might have even been repeatedly recycled into polymers with different nucleotide sequences (and corresponding functions) as survival pressures changed (Ura et al. 2009). Furthermore, without some regular cycle by which RNA or a proto-RNA could have been depolymerized, any monomer incorporated into a polymer that was not functionally useful would be wasted—a condition under which there simply may have not been enough material to get life started.[1]

(3) *Condensation-dehydration reactions were, from the beginning, integral to the formation of proto-RNA.* A wide range of biological bond-forming reactions involve the loss of water, including amide linkages in peptides, acetal linkages in polysaccharides and ester linkages in phosphoglycerols. Pertinent to the present discussion are the reactions required for nucleic acid formation—glycosylation of a nucleobase to form a nucleoside, phosphorylation of a nucleoside to form a nucleotide and condensation of nucleotides to form oligonucleotides—all of which are dehydration reactions. Such reactions are very appealing from a prebiotic standpoint, as their equilibria can be modulated via water activity (i.e., through drying-wetting cycles), suggesting that many polymerization reactions observed in contemporary life may have ancient roots in drying reactions driven by periodically fluctuating water activity—perhaps even dating back to the earliest stages of life.

We now will discuss the three molecular components of nucleic acid polymers and consider the implications of the three hypotheses presented earlier in considering alternative candidates for proto-RNA.

RECENT RESULTS

Selection of the Nucleobases: Was it C, U, A, and G from the Start?

Although nucleic acid chemists have provided some insights regarding what might have

---

[1] If the earliest effective ribozyme was 50nt in length (similar to the hammerhead ribozyme), and the earliest genetic code contained C, U, A, and G, this ribozyme would be one of $4^{50} \approx 10^{30}$ possible sequences. A collection of *one* molecule of every possible 50mer would require a carbon mass of around 10^7 kg. Assuming each polymer was produced in equal yield, and by an irreversible process, production of 1 picomole of the ribozyme would require 10^{19} kg of carbon (roughly the total weight of carbon in the Earth's crust).

predated RNA, as we hope we have conveyed herein, we concede that it is still not obvious by what mechanism the nucleobases were originally selected. The ability to form Watson-Crick base pairs should not be assumed to have been the sole criterion. One observation that runs counter to this assumption is that the free nucleobases and their nucleosides *do not* form Watson-Crick base pairs in aqueous solution (Ts'o et al. 1963), but rather, they form co-planar stacks, with their Watson-Crick edges interacting with the solvent. The hydrogen bonds formed with water are of comparable enthalpy to those that would be formed between paired bases. The local preorganization of nucleobases in oligonucleotides, on the other hand, promotes intra- and interstrand stacking of base pairs, which contributes substantial free energy to duplex stability, whereas Watson-Crick hydrogen bonds have been reported to contribute virtually no net free energy (De Voe and Tinoco 1962; Yakovchuk et al. 2006).

The observation that free nucleobases do not form Watson-Crick base pairs in aqueous solution inspired us to articulate a "paradox of base pairing" with regard to the emergence of RNA (Hud and Anet 2000; Hud et al. 2007). *Why would the nucleobases have been selected for inclusion in RNA or proto-RNA polymers for the purpose of Watson-Crick base pairing if they did not form base pairs before being linked by a common backbone?* To argue that the nucleobases formed nucleotides with ribose and phosphate *before* being involved in base pairing seems to imply that nature either predicted that the nucleobases would be useful for pairing polymers in the future (which would violate the nonpredictive property of evolution); that nature was simply lucky, and a series of chemical reactions coincidentally created rather complex molecules (i.e., nucleotides) that happened to form base pairs after polymerization; or that such molecules existed in a metabolism-first origin of life scenario, and these molecules were later incorporated into a genetic polymer. Serious conceptual challenges have been raised against metabolism-first scenarios (Anet 2004; Orgel 2008; Vasa et al. 2009). In any case, arguments for the prepolymer production and use of nucleotides by an abiotic metabolic cycle remain largely philosophical, as we simply do not know at what point in evolution (of both metabolism and genetic polymers) that the specific components of contemporary nucleotides were selected. It is feasible, for example, that phosphate was used first as a metabolic cofactor before being incorporated into genetic polymers.

As one possible solution to the paradox of base pairing, we have proposed that there was a template molecule, or a class of molecules, present in the prebiotic chemical inventory, that formed stacks with the nucleobases that were energetically favored over stacks with themselves. Each template, or "molecular midwife," would have been large enough to accommodate two or more nucleobases, if the bases were arranged in a particular hydrogen-bonded structure (e.g., a Watson-Crick base pair) (Hud and Anet 2000; Hud et al. 2007). The stacking of multiple midwife molecules, interleaved with nucleobases, would have locally concentrated and organized the bases, thereby aiding in the polymerization of the earliest proto-RNA. These midwife molecules would have no longer been necessary for polymer synthesis (and replication) once evolution had produced a superior means for carrying out these reactions (e.g., catalysis via protein or RNA enzymes). We envision that the midwife molecules would have been similar to the small planar molecules known to intercalate the bases of contemporary RNA and DNA (Ihmels and Otto 2005). Beyond a purely speculative model, dye molecules that intercalate DNA and RNA base pairs have been shown to promote the template-directed synthesis of nucleic acids. For example, intercalators can provide a >1000-fold enhancement to the ligation of short oligonucleotides (i.e., tri- and tetranucleotides) (Jain et al. 2004), and promote a specific base pairing (Hud, Jain et al. 2007; Horowitz et al. 2010).

Should we assume that Watson-Crick base pairs (i.e., adenine paired with uracil, guanine paired with cytosine) were used by the earliest proto-RNA? The successful incorporation of nonnatural base pairs into nucleic acid duplexes has given ample reason to seriously consider this question (Switzer et al. 1989; Piccirilli et al. 1990; Geyer et al. 2003). Most of the nonnatural base pairs studied thus far have maintained the form of a base comprised of a six-membered ring (like a pyrimidine) paired with a base comprised of fused five- and six-membered rings (like a purine), which can be accommodated in a helix also containing canonical Watson-Crick base pairs. However, duplexes that contain all noncanonical base pairs have also been shown, including duplexes with only purine–purine base pairs (Groebke et al. 1998; Battersby et al. 2007; Heuberger and Switzer 2008b; Engelhart et al. 2009), pyrimidine–pyrimidine base pairs (Mittapalli et al. 2007a; Mittapalli et al. 2007b), and even duplexes with tricyclic bases (Krueger et al. 2007).

If we restrict our consideration of potential proto-RNA base pairs to those with a Watson-Crick pairing geometry, there are still twelve possible types of nucleobases that can form six types of base pairs with distinct hydrogen bond donor–acceptor arrangements (Fig. 3) (Benner et al. 2004), of which only two types are commonly found in contemporary life. Some of these base pairs, at first glance, seem like reasonable candidates as proto-RNA alternatives to the canonical base pairs, such as isoC paired with isoG (Fig. 3, bottom left). However, even this close analog of the guanine-cytosine base pair comes with significant chemical challenges that could have prevented it from being a component of proto-RNA (Jaworski et al. 1985). In particular, isoguanine exists to about 10% as its enol tautomer in water (Sepiol et al. 1976; Voegel et al. 1993; Seela et al. 1995), which has a hydrogen-bonding pattern that is complementary to U/T. Moreover, some of the four noncanonical Watson-Crick-like base pairs could only be realized experimentally by using a nonpyrimidine or a nonpurine base, or by the connection of the base to the sugar through a carbon-carbon bond (C-nucleosides) (Fig. 3) (Benner 2004). The use of non- purine or pyrimidine heterocycles may not be so problematic from a prebiotic standpoint. Although

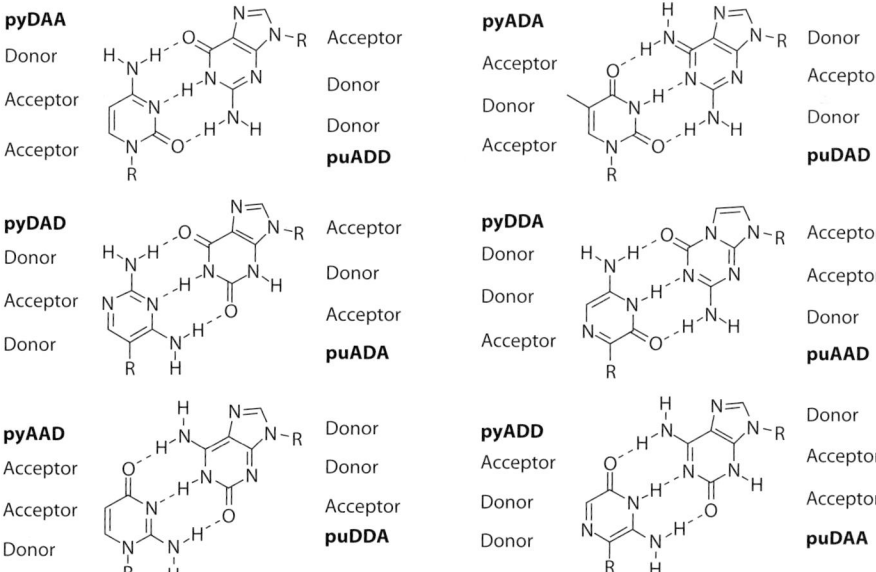

Figure 3. The six possible hydrogen-bonding patterns that can join 12 different nucleobases with Watson-Crick geometry. Nomenclature: pu, a [6,5] fused ring system; py, a six-membered ring. The hydrogen-bonding pattern of acceptor (A) and donor (D) groups from the major to the minor groove is indicated. For example, the standard nucleobase cytosine is pyDAA. R indicates the point of attachment of the backbone. Reprinted with permission from Benner et al., 2004. Copyright Elsevier Science Ltd.

purines are abundant in model reactions of the prebiotic chemical inventory using, for instance, formamide and HCN, other, helix-compatible parent heterocycles were surely present in the prebiotic chemical inventory (Mittapalli et al. 2007a; Mittapalli et al. 2007b), and these should be considered. With respect to C-nucleosides, although they are well-known, they have been less well-examined from a prebiotic standpoint than the canonical N,O-acetal linked nucleosides. Regardless of any potential difficulties in prebiotic synthesis, the demonstration of all six possible base pairs with a Watson-Crick paring geometry, made possible through the application of synthetic and physical organic chemistry, has provided invaluable constraints regarding what should be considered more or less likely as possible ancestral proto-RNA building blocks.

Among the proposals for proto-RNA non-canonical base pairs, that of purine-purine base pairs is one of the oldest (Crick 1968). The abiotic formation of proto-RNA containing only purine bases is appealing for several reasons. First, an all purine system requires only one type of heterocycle, and the variety of conditions under which purine production has been shown suggests that purines may be formed more easily in model prebiotic reactions than pyrimidines (Oro 1961; Sanchez et al. 1966; Saladino et al. 2004). Second, nucleoside formation, as discussed later, might have occured more efficiently with purine bases. Third, the purine bases have much more favorable stacking interactions in aqueous solution than the pyrimidine bases (Ts'o, Melvin et al. 1963; Inoue and Orgel 1983). The potential prebiotic obstacles presented by poor pyrimidine stacking in aqueous solution is illustrated by the inhibition of template-directed polymerization reactions that require the coupling of successive pyrimidine nucleotides (Joyce 1987). With respect to the molecular midwife hypothesis discussed earlier, purine–purine base pairs would also be expected to form assemblies with midwife molecules at lower concentrations and over a wider temperature range than Watson-Crick base pairs.

Crick was among the first to propose that life originally used purine-purine base pairs.

In particular, he proposed that A·I base pairs might have come first (Fig. 4) (Crick 1968). However, duplexes formed by homo-A and homo-I polymers proved to be relatively unstable, and these polymers were prone to forming triplexes (Howard and Miles 1977). Eschenmoser and coworkers have subsequently shown that homo-purine oligonucleotides (with G·isoG base pairs) can form very stable duplexes when the standard ribofuranose nucleosides are replaced by either a ribopyranose or a dideoxyhexopyranose sugar (Krishnamurthy et al. 1996; Groebke et al. 1998) (other sugar modifications are discussed later). More recently, evidence has been presented that standard oligodeoxynucleotides can form duplexes with mixed G·isoG and diaminopurine·xanthine base pairs (Fig. 4) that are of comparable stability to Watson-Crick duplexes (Heuberger and Switzer 2008a). In light of these recent advances, the

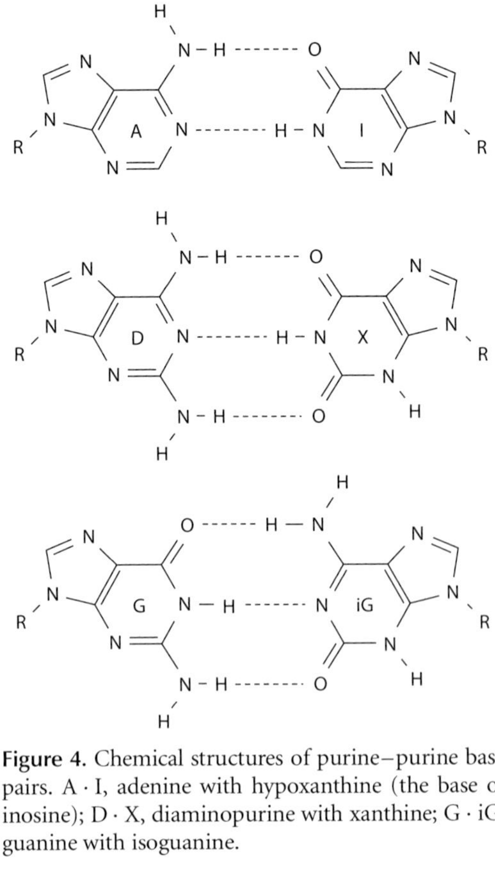

Figure 4. Chemical structures of purine–purine base pairs. A · I, adenine with hypoxanthine (the base of inosine); D · X, diaminopurine with xanthine; G · iG, guanine with isoguanine.

decades-old idea of purine-purine base pairs in proto-RNA now looks even more attractive.

The Trifunctional Moiety: Was It Ribose from the Start?

In RNA, ribose serves as the branch point between the polymer backbone and the side chains (i.e., the nucleobases). This role requires a chemical moiety that is, at least, trifunctional. Sugars with three or more carbons meet this minimum criterion and have been studied; among those investigated thus far, ribose appears optimal (Eschenmoser 1999). The conformational landscape of the β-furanose anomer of ribose is well accommodated in a helical arrangement when its nucleosides are linked by phosphate. Although some nucleic acids synthesized with alternative sugars form more stable Watson-Crick duplexes than RNA (Eschenmoser 1999), ribose appears to provide an optimal balance between duplex stability and the ability to adopt the more globular structures required for creating a catalytically active ribozyme, as exemplified by the variety of RNA structures observed in the ribosome (Ban et al. 2000; Wimberly et al. 2000).

The optimal functionality of ribose suggests that its inclusion in RNA is more the result of evolutionary refinement than prebiotic availability. The yield of ribose in the formose reaction, which is often cited as a prebiotically plausible synthetic route to sugars, is low with respect to the numerous other sugars produced in this reaction (Decker et al. 1982). Possible solutions to these challenges have included proposals that prebiotic ribose was preferentially stabilized by adduct formation with cyanamide (Springsteen and Joyce 2004), borate (Ricardo et al. 2004) or phosphate (Müller et al. 1990), each of these adducts being more stable than the unmodified sugar.

If ribose does not stand out for its ease of abiotic synthesis or chemical stability, then what sugars would have been more likely to have come before ribose in proto-RNA? Twenty years ago, Joyce et al. thoughtfully discussed this question (Joyce et al. 1987). With the goal of identifying a predecessor to ribose with enhanced prebiotic availability and tolerance for incorporation of mixed stereoisomers (i.e., D- and L-sugars), these authors proposed a proto-RNA with acyclic analogs of ribose, including an acyclic flexible nucleic acid (FNA) backbone that is tantamount to an RNA backbone with "deletion" of the 2′ carbon (Fig. 2). Subsequently, the Benner laboratory synthesized oligonucleotides containing these linkages and found that each substitution within a DNA oligonucleotide resulted in significantly depressed duplex stability, and complementary thirteen-base oligonucleotides containing eleven FNA thymine residues failed to hybridize to a DNA template (Schneider and Benner 1990). Later work by Merle and colleagues showed weak interaction of FNA oligomers with DNA—the strongest pairing being observed for a homo-FNA-adenine oligomer with the homo-dT complement, although this pairing was still depressed relative to the analogous DNA-DNA duplex (Merle et al. 1995). Thus, in addition to being well-accommodated in a Watson-Crick duplex, the conformational preorganization of ribose nucleosides is clearly an important component of duplex stability.

The Eschenmoser laboratory has performed an extensive and systematic investigation of how changes to the natural β-furanosyl ribonucleosides of RNA, to nucleosides with a sugar of different stereochemical configuration, phosphate connectivity, carbon chain length or ring form (i.e., pyranose or furanose), affect the ability of nucleic acids to base pair (Eschenmoser 2007). A few examples are shown as Figure 1 D–F. Among their findings was the observation that oligomers incorporating the pyranosyl form of nucleosides with a pentose sugar (including ribose) can form more stable duplexes than their furanosyl forms. This observation is intriguing, particularly taken in light of the observation that pyranosyl nucleosides are apparently formed along with furanosyl nucleosides in heating-drying reactions (Bean et al. 2007), suggesting that both forms could have been available for proto-RNA. Eschenmoser has interpreted the intermediate helical stability of the contemporary furanose nucleosides as an indication that the contemporary form of RNA prevailed due, in part, to an optimal, *intermediate* helical stability, not *maximal* helical stability. He suggests that the strong hybridization of pentopyranosyl backbones could have come at a price in a proto-RNA world—in the form of product inhibition and depressed sequence fidelity in replication.

The four carbon sugar threose is of particular interest as a potential ancestor of ribose (Schöning et al. 2000; Herdewijn 2001a; Wilds et al. 2002). The nucleosides of this threose can be connected by phosphate esters to create a backbone with a five-atom repeat (Fig. 2D), one atom shorter than the RNA backbone. Experiments with 3′,2′-linked threofuranose nucleic acid, or TNA, have generated significant excitement in the origin of life community. In addition to the potential relative ease of formation of threose in a prebiotic environment, duplexes of TNA oligonucleotides are of comparable stability to RNA and DNA duplexes, and TNA forms hybrid duplexes with RNA and DNA (Schöning et al. 2000; Wilds et al. 2002). It has often been speculated that cross-hybridization with RNA is an essential feature for an alternative nucleic acid to be considered a possible ancestor of RNA (Orgel 2000), as this ability would have been necessary for transfer of sequence information between the two polymers over the course of evolution.

A backbone containing even fewer atoms, but of the same polymer repeat length as TNA, has been termed GNA (Fig. 2C). The structure and stability of duplexes formed by oligonucleotides with this backbone have been examined in detail by Meggers and coworkers (Zhang and Meggers 2005; Schlegel et al. 2009). These investigators have shown that GNA oligonucleotides with Watson-Crick complementary sequences form antiparallel duplexes that are even more stable than DNA or RNA duplexes of the same nucleobase sequence (Zhang and Meggers 2005). In contrast, little or no hybridization is observed between complementary GNA and DNA oligonucleotides (relative to the DNA homoduplex) and base pairing with RNA is of roughly equal stability relative to homoduplex RNA. Meggers and coworkers have also provided solution-state evidence (by CD and NMR) for the preorganization of GNA (on the single-strand and single-residue levels) into a helix-compatible conformer. These data illustrate that even an acyclic backbone can be conformationally preorganized. A subsequent X-ray crystal structure revealed that the GNA duplex makes extensive interstrand stacking contacts and minimal intrastrand contacts (in contrast to RNA and DNA, for which the converse is true).

The Szostak laboratory recently showed that a GNA region of a DNA-GNA chimeric strand could act as a template for *Bst* polymerase (Tsai et al. 2007). Despite little or no hybridization between GNA and DNA, full-length polymerization of the dodecamer GNA template was achieved. Although it is unlikely that polymerase-catalyzed replication was contemporaneous with the first self-replicating polymers (which these authors do not suggest), their results show that, even if cross-hybridization is disfavored, the presence of a duplex-stabilizing molecule (a protein or RNA polymerase, or a small molecule) can relax this requirement for information transfer. A parallel line of evidence for this principle comes from the Switzer laboratory, who have shown that 2′,5′-linked DNA and FNA can also act as substrates for natural DNA polymerases, even though 2′,5′-linked DNA and FNA hybridize weakly or not at all to natural DNA (Sinha et al. 2004; Heuberger and Switzer 2008b).

Overall, synthetic organic chemists have shown that numerous alternative sugars from among the tetroses, pentoses, and hexoses, and alternative conformations of ribose (e.g., β-pyranosyl) can produce backbones that form double-stranded structures with Watson-Crick base pairs, with some of these being even more stable than duplex RNA (Bolli et al. 1997; Eschenmoser 1999; Schöning, Scholz et al. 2000; Eschenmoser 2004; Egli et al. 2006). These results support the proposal that proto-RNA backbones may have been constructed from a sugar (or a mixture of sugars) other than ribose. The search through sugar space is by no means complete. As one example, the ketose sugars have not been explored as possible ancestors to ribose, but they should also be considered as potentially simpler ancestors of ribose given the remarkably high yield with which the ketohexoses, such as fructose, are formed from DL-glyceraldehyde in a model prebiotic reaction (Weber 1992).

Finally, from a reactivity perspective, amino sugars are also an attractive possibility. As discussed later, the formation of phosphodiester bonds in aqueous solution is problematic. Recent work by Szostak and coworkers (Mansy et al. 2008; Chen et al. 2009; Schrum et al. 2009), as well as earlier work by Orgel and coworkers (Zielinski and Orgel 1985, 1987; Sievers and von Kiedrowski 1994), have shown the use of amino sugars in promoting polymerization because of the greater nucleophicity of amines compared with alcohols. The amino-substituted phosphoimidazole-activated nucleosides used in these studies were, obviously, prepared using classical synthetic organic techniques. Nevertheless, given the promising polymerization results that have been achieved with such nucleosides, as well as the presence of amino sugars in contemporary life (e.g., glucosamine), potential prebiotic reactions for the production of amino-substituted sugars are an intriguing area for future investigation.

Nucleoside Formation: How Did the Nucleobase and Trifunctional Moiety First Connect?

The prebiotic origin of the glycosidic bond (an N,O-acetal) that connects the nucleobase to ribose in a nucleoside has turned out to be one of the most vexing problems facing the origin of RNA. Given that the synthesis of adenine by HCN polymerization was one of the earliest successes of prebiotic chemistry (Oró 1960), and sugar formation from formaldehyde was known for a century prior (Butlerow 1861), many have speculated that prebiotic nucleoside formation resulted from the heating and drying of preexisting nucleobases and ribose. Orgel and coworkers first reported the formation of adenosine, inosine and guanosine when samples of free adenine, hypoxanthine and guanine, respectively, were dried and heated with ribose in the presence of magnesium salts (Fuller et al. 1972a, b). However, yields were low, and the synthesis of guanosine was especially problematic, because of the low solubility of guanine. More troublesome is the observation that the canonical pyrimidine bases of RNA, uracil and cytosine, do not form nucleosides when dried and heated with ribose (Orgel 2004). It is worth noting that even the previously reported formation of adenosine in heating-drying reactions has been described as difficult to reproduce (Zubay and Mui 2001).

Several alternative hypotheses for prebiotic nucleoside formation have emerged as a result of the difficulty of glycosidic bond formation. These hypotheses include the possibility that the nucleobases were formed on a preexisting sugar, or the converse, that the sugar was built off of the nucleobase. The earliest exploration of these alternative routes was again reported by the Orgel laboratory (Sanchez and Orgel 1970). Starting with 5'-phosphorylated ribose, followed by multiple steps, including chemical reagent addition, UV anomerization, and hydrolysis, Sanchez and Orgel showed the synthesis of cytidine monophosphate. The Sutherland laboratory has recently revisited this hypothesis with great vigor and has made impressive strides toward the presentation of a comprehensive pathway for the stepwise abiotic formation of β-DL-cytidine-2',3'-cyclic phosphate, using cyanamide, cyanoacetylene, glycolaldehyde, glyceraldehyde and inorganic phosphate as starting molecules (Ingar et al. 2003; Anastasi et al. 2006; Crowe and Sutherland 2006; Powner et al. 2009; Sutherland 2010).

The possibility that the nucleoside sugar was originally synthesized on the nucleobases, the converse of the previously described nucleosidation hypothesis, has also received some experimental support. Simply heating formamide in the presence of various mineral catalysts, as performed with a great variety of minerals by the Saladino and DiMauro laboratories (Saladino et al. 2007), have produced all four canonical RNA bases, as well as purine acyclonucleosides (albeit with rudimentary "sugar" moieties) (Saladino et al. 2003). Given the plausibly prebiotic nature of formamide as a starting material and solvent (Benner et al.

2004), the one-pot synthesis of a proto-nucleoside is a remarkable observation that certainly merits serious consideration as an alternative route to nucleosides.

The concept of nucleic acid evolution also opens the door to that possibility that proto-RNA contained nucleobases and/or a trifunctional moiety that more easily formed nucleosides. The merit of this hypothesis was first demonstrated by Miller and co-workers, who showed that urazole, a five-membered heterocycle with the same Watson-Crick hydrogen bonding pattern as uracil, reacts spontaneously with ribose to give α- and β- pyrano- and furanosides (Kolb et al. 1994; Dworkin and Miller 2000). More recently, Bean et al. investigated nucleoside formation with ribose and 2-pyrimidinone (Bean et al. 2007), a pyrimidine with the same 2-oxo functional group as uracil and cytosine, but with a proton instead of oxo- or amino-substitution at the 4 position. This base was shown to form nucleosides in heating-drying reactions, in which up to 12% of the remaining base was converted to the β-furanoside, possibly the highest conversion of a base to a nucleoside observed to date in a simple heating-drying reaction. Computational studies of this reaction have confirmed the importance of the in-plane lone pair of 2-pyrimidinone in nucleoside formation, which is also present in the imidazole ring of purines, and the importance of divalent cations in lowering the activation barrier to glycosidation (Bean, Sheng et al. 2007; Sheng et al. 2009). Together, the results obtained for urazole and 2-pyrimidinone illustrate that nucleoside formation from a preexisting base and sugar may not have been such a difficult step in proto-RNA formation, if the proto-nucleobases differed somewhat from those of contemporary RNA.

Selection of the Nucleoside Linker: When Did Nature Choose Phosphate?

The phosphate backbone is a chemical hallmark of contemporary nucleic acids. As a triacid, phosphate provides nucleic acids with a means to connect two alcohols, in the form of sugars, via ester linkages, while maintaining a negative charge (a key contributor to nucleic acid solubility in water). As Westheimer pointed out in his insightful paper entitled "Why Nature Chose Phosphates" (Westheimer 1987), it is difficult to conceive of a chemical arrangement that would provide better functional characteristics. The ester linkages provide connectivity that is sufficiently labile so that enzymes can hydrolyze the polymer (for proofreading or monomer salvage), whereas the negative charge provides sufficient screening of the phosphorus center from nucleophilic attack to stabilize the polymer against hydrolysis. Additionally, the intrastrand electrostatic repulsion afforded by the regularly-spaced negative charges may be crucial to an extended helical conformation and for information transfer that is rigorously dependent on Watson-Crick base pairing (contrast proteins, in which self-assembly via molecular recognition occurs, but not by the rule-based Watson-Crick system observed in nucleic acids). Westheimer noted that phosphate is unique here, as well: as a polyvalent acid with all acidic protons in close proximity, its negative charge will afford greater protection from nucleophilic attack and intraresidue repulsion than would a polyvalent acid with its ionizable moieties more distant from one another (e.g., citrate).

Benner and coworkers have experimentally explored the importance of backbone charge on the ability for RNA-like molecules to form duplexes with Watson-Crick base pairs. As a neutral structural analog of the phosphodiester-linkage, they studied dimethylenesulfone-linked nucleosides (in which the O3′ and O5′ sugar atoms are replaced by methyelene groups and phosphorus is replaced by sulfur) (Huang et al. 1991; Richert et al. 1996). The results from these studies were intriguing, but less than straightforward to interpret. Although short sulfone-linked DNA analogs (sNAs) were shown to support Watson-Crick base pairing (Roughton et al. 1995), longer oligosulfones appeared to have somewhat compromised pairing abilities (Huang et al. 1991; Richert et al. 1996) and small changes in nucleobase oligosulfone sequence resulted in appreciable changes in oligomer solubility, folding and aggregation (Eschgfaller et al. 2003; Schmidt et al. 2003). These authors interpreted the results of these studies as evidence that (1) charged linkages are important for molecular recognition by providing a repulsive energetic term between the two backbones of a duplex, ensuring molecular recognition is largely a function of the nucleobases, (2) that the regular repeating charge limits intramolecular folding of oligomers, thereby allowing these polymers to function well as linear duplexes and as templates for the same during replication, and (3) that, given the dominance of charge in governing the physical properties of these nucleic acids in aqueous solution, changes in DNA/RNA nucleobase sequence have only a second-order effect on polymer properties (i.e., sequence exerts more effect on the molecular recognition of a complementary strand and less effect on helical parameters), thereby allowing the use of virtually any possible nucleotide sequence (Benner et al. 2004).

Studies from the fields of biological and medicinal chemistry have also provided some insight regarding the importance of backbone charge on nucleic acid base pairing. For example, methylphosphonates, in which one phosphate oxygen is replaced by a methyl group, provide a neutral linkage that is among the closest possible structural analogs of the natural backbone (Miller et al. 1981). Oligonucleotides with methylphosphonate substitutions support Watson-Crick base pairing in water, as homoduplexes and as hybrid duplexes with natural nucleic acids (Miller et al. 1981; Kiblerherzog et al. 1991; Schweitzer and Engels 1999). Surprisingly, removal of negative charges does not result in a substantial increase in duplex stability, as might be expected because of the reduced Coulombic repulsion between backbones compared with DNA and RNA. In contrast, two neutral polymers with more radical structural changes, "morpholinos" and PNAs, have been shown to form hybrid duplexes with RNA and DNA that are of greater stability than homo-DNA or homo-RNA duplexes (Summerton and Weller 1989; Egholm et al. 1993; Brown et al. 1994; Summerton and Weller 1997), and PNA will form homoduplexes with Watson-Crick molecular recognition (Rasmussen et al. 1997; He et al. 2008). The hybridization of nucleic acids is the result of a complex interplay of numerous factors, including hydrogen bonds, nucleobase stacking interactions, backbone conformational landscapes, electrostatic interactions and hydration effects. Thus, the sensitivity of duplex stability to backbone charge is consistent with the predicted importance of the phosphate backbone in facilitating base pairing, but the observation of several successful neutral polymers suggest we presently lack sufficient information to understand the precise origin of the observed charge-dependent differential stabilities.

Although phosphate esters are *functionally* superior linkages, they show several *chemical* characteristics that may have made it difficult to incorporate them into the earliest biopolymers. The first is solubility. In the presence of divalent cations, free phosphate tends to form insoluble minerals; outside of living organisms, the phosphate resources on Earth even today are found as insoluble mineral deposits (Keefe and Miller 1995). Indeed, phosphate is a limiting reagent in much of contemporary life. Although recent hypotheses have been proposed for the availability of reduced forms of phosphorus on the earth which are more soluble (De Graaf and Schwartz 2000; Bryant and Kee 2006; Schwartz 2006; Pasek 2008), the low aqueous solubility of the contemporary (V) oxidation state of phosphorus, in the presence of divalent cations, and the positive ΔH associated with phosphoester formation present problems for prebiotic phosphate chemistry.

Another chemical characteristic of phosphate that would have made polymer formation challenging is the kinetic barrier to spontaneous phosphate ester formation. The negative charge of phosphate above pH $\approx$ 1 presents a barrier to ester formation, which, like hydrolysis, must proceed via nucleophilic attack on the phosphorus atom. It is possible to activate phosphate for ester formation (e.g., using methylimidazolides). However, such activated phosphates are necessarily high-energy compounds and, as would be expected, hydrolysis competes with polymerization.

Phosphate activation poses problems for both de novo production of oligomers from monomers and the incorporation of oligomers into higher polymers. In the case of monomer polymerization, Ferris and coworkers have shown that activated imidazolide derivatives of adenosine nucleotides will polymerize on mineral surfaces, but a large proportion of the monomer is incorporated into cyclic dimers (Miyakawa et al. 2006). In attempting to polymerize tiled half-complementary hexanucleotides into higher oligo- and polymers by carbodiimide activation, Kawamura and Okamoto also observed substantial amounts of cyclized starting material (Kawamura and Okamoto 2001). In both of these examples, a significant amount of starting material is irreversibly incorporated into an undesired side product. In addition to being a nuisance in the lab, these side products would have amounted to a fatal and committed step in the synthesis of a nascent proto-RNA. This problem illustrates a difficulty in nonenzymatic polymerization that must be taken into account when considering how the nature of the synthetic routes to and structural identities of early genetic polymers: irreversible linkages are adaptive for an informational polymer *only when mechanisms exist to make them conditionally reversible* (so as to allow proofreading).

The difficulties described earlier confound the efficient prebiotic chemical synthesis of phosphate esters—ligation of activated substrates yields mixed regioisomers and poor replication fidelity because of strand cyclization, base misincorporation, and premature product chain termination (Joyce and Orgel 1999; Kawamura and Okamoto 2001; Miyakawa al. 2006). Work by Usher and colleagues has provided encouraging evidence that the $2',5'$ linkage is more labile than the contemporary $3',5'$ regioisomer, suggesting that thermodynamic selection may be helpful in addressing a subset of the aforementioned problems with phosphodiester linkages (Usher and McHale 1976). However, all these difficulties can be circumvented by the use of a linker chemistry that initially forms a low-energy, reversible bond, allowing for selection of the thermodynamic product (i.e., in the case of RNA, the Watson-Crick base-paired product).

One remarkable example of the power of reversible linkages was provided by Lynn and coworkers, who showed that $5'$-deoxy,$5'$-amino, $3'$-deoxy, $3'$-formylmethyl-dT could polymerize via reductive amination in aqueous solution (Li et al. 2002). Polymerization occurred only in the presence of a d(A_8) template; without a template, no oligomer product was detected. Interestingly, only linear polymers were observed. Here, polymerization decreases the entropic cost of hybridization, driving base pairing (which is disfavored at the mononucleotide level in water) (Yakovchuk et al. 2006). In turn, hybridization increases the local concentration of amines and aldehydes, driving the formation of imine linkages (which are otherwise disfavored in 55 M water). Since the irreversible reduction step requires the imine linkage to be present, and imine formation was template dependent, only linear polymers were formed. In this system, the subtle interplay between base pairing and linkage formation drives the simultaneous formation of reversible covalent and noncovalent interactions which are otherwise disfavored. Several other intriguing examples of self-assembling reversible polymers have been shown, particularly by the Lehn laboratory (Sreenivasachary and Lehn 2005; Sreenivasachary et al. 2006; Sreenivasachary and Lehn 2008).

Acetal linkages have the potential to provide reversible bonds, thereby providing the benefits showed by the reversible imine chemistry of Lynn. For example, the $3',5'$-formacetal linkage (i.e., phosphate replaced by -O-CH_2-O-), has been examined by a number of laboratories (albeit not in fully modified oligonucleotides). Phosphorus (V) esters and acetals are both tetrahedral and approximately isosteric, and it is known that *a*NAs can base pair with RNAs (Matteucci and Bischofberger 1991). Specifically, point substitutions of this linkage impart a modest destabilization to DNA-DNA and DNA–RNA duplexes and show salt-dependent effects on RNA-RNA duplexes (stabilization at low salt, destabilization at high salt) (Jones et al. 1993; Rice and Gao 1997; Rozners et al. 2007; Kolarovic et al. 2009).

Investigators have invoked compatibility with the B-versus A-form helices, electrostatic, and hydration effects for these observed differential effects. Further, some evidence has been presented that the destabilization associated with this substitution in DNA–DNA duplexes is sequence-dependent (Pitulescu et al. 2008). Finally, the 3′-thioformacetal linkage has been shown to slightly destabilize DNA-DNA duplexes and slightly stabilize DNA-RNA duplexes (Jones et al. 1993). The latter stabilization was explained by a combination of sugar pucker, bond lengths, and accessible torsion angles resulting from the 3′-thio substitution. This sugar pucker effect was later confirmed by NMR spectroscopy (Rice and Gao 1997).

With the proper choice of aldehyde or ketone, these linkages can provide a negative charge that has proven critical to the phosphodiester backbone. For these reasons, acetal-linked nucleic acids (aNAs) are an intriguing potential ancestral polymer (Hud and Anet 2000). Acetal bond formation is favored over hydrolysis on water removal (Wiberg et al. 1994). The concentration of substrates that occurs during dry-down, driving bond formation, would also favor the hybridization of short, complementary sequences, providing an accessible means by which to drive polymerization.

The functionalization of an acetal with a negatively charged group, such as carboxylate group in the case of the aldehyde glyoxylate, would provide a localized charge near the backbone acetal linkage, thereby providing many of the favorable electrostatic properties afforded by phosphate. Glyoxylate has been shown to form in a prebiotically plausible reaction from glycolaldehyde (Weber 2001). We have speculated that glyoxylate could have preceded phosphate in an early proto-RNA, in the form of glyoxylate-acetal nucleic acids (gaNAs, Fig. 2G) (Bean et al. 2006). Based on energy minimized molecular models, it appears that a gaNA duplex would show helical properties very similar to RNA, suggesting that information transfer could occur between the two polymers. The formation of gaNA dinucleotides has been shown on heating and drying of a neutral solution containing nucleosides, glyoxylate and mono- or divalent cations (Bean et al. 2006). These acetal linkages formed by glyoxylate and nucleosides are surprisingly stable to hydrolysis—no detectable decomposition occurs after two weeks at room temperature and neutral pH. Importantly, gaNAs can be hydrolyzed at moderately elevated temperature in the presence of salt (i.e., 60 °C, 1 M $MgCl_2$). An accessible means to modulate polymer stability (in this case, temperature and salt concentration), which provides an attractive means by which to "turn on and off" thermodynamic control of polymerization—a crucial feature in the chemical synthesis of an early informational polymer.

It is possible that other, similar, molecules could form suitable linkages. For instance, pyruvate, like glyoxylate, provides a negative charge and tetrahedral geometry, and ketal bond formation is still expected to be more facile than phosphodiester bond formation. Further, pyruvate is ubiquitous in life and formed in excellent yield in model prebiotic reactions (Weber 2001).

It is important to note that additional promising models have presented for the prebiotic formation of oligonucleotides that begin with either a reduced form of phosphorus or mononucleotides activated as cyclic monophosphates. Reduced phosphorus species, such as phosphite, in addition to showing the aforementioned enhanced solubility relative to phosphates, enjoy enhanced reactivity. Schwartz and coworkers reported high yields of uridine 5′-H-phosphonate (as high as 44% in the presence of urea) in 60 °C drying reactions of aqueous ammonium phosphite, uridine and, optionally, urea. Additionally, when dried at 110 °C, these authors reported production of small amounts of putative higher-order products, the most abundant of which was produced in 4% yield, which they tentatively assigned as the 5′,5′ H-phosphonate-linked dinucleotide. This species could be oxidized with iodine to a second species, assigned as the phosphate-linked dinucleotide (De Graaf and Schwartz 2005). Like acetals, H-phosphonate diesters are a conditionally reversible, low-energy linkage.

Although they are uncharged and hydrolytically labile, upon oxidation, they become negatively charged, comparatively hydrolytically inert phosphate diesters, making phosphonates another attractive, conditionally reversible backbone linkage. Finally, DiMauro and coworkers have recently reported that 3′, 5′-cyclic AMP and GMP will spontaneously form homo-A and homo-G oligonucleotides, respectively, in aqueous solution (Costanzo et al. 2009). These authors propose that nucleotides activated as cyclic monophosphates are potentially prebiotic, based on their earlier demonstration of 3′, 5′-cyclic AMP formation in formamide solutions containing phosphate minerals (Costanzo et al. 2007).

As Westheimer, Benner, and others have elegantly discussed, phosphate affords RNA numerous positive characteristics (Westheimer 1987; Benner et al. 2004). When considering some of the adaptive traits of RNA as a polymer—hydrolytic resistance, water solubility, extended conformation, and heritability of an arbitrary sequence because of rule-based molecular recognition—it is tempting to speculate that *phosphate might be the optimal linker for all possible RNA-like polymers.* The roles of phosphate in contemporary life, however, are possible only because of a highly optimized suite of catalysts, which enable the transfer of energy between disparate substrates and the "phosphate economy" (i.e., via the agency of ATP and similar compounds). Absent such catalysis, the chemical synthesis (and subsequent proofreading) of phosphate esters and anhydrides is problematic. *Many of the same attributes of phosphate that are adaptive in contemporary life could have been maladaptive in the proto-biopolymer world.* We therefore suggest that other chemistries, including those discussed here, be considered as possible linkers for the nucleosides of proto-RNA. These linkers could have incorporated some of the favorable traits of phosphate (e.g., electrostatic benefits, significant hydrolytic stability) while addressing some of the problematic ones (e.g., disfavored bond formation, lack of a proofreading mechanism in chemical synthesis) (Westheimer 1987; Bean, Anet et al. 2006).

CHALLENGES AND FUTURE DIRECTIONS

Our primary objective in this article has been to present some of the known challenges for the prebiotic synthesis of RNA, as well as potential solutions to these challenges that become available if we accept the possibility that one or more proto-RNAs predated RNA. More than a mere academic exercise, the discovery (or "reinvention") of a self-assembling RNA-like polymer would have tremendous potential for applications in biotechnology, medicine, and materials science.

Although our current examples of self-assembling polymers with reversible linkages are limited, the properties of these polymers are impressive and indicative of the promise held by the development of this area of polymer science. For example, dynamic covalent assemblies have been examined by the Lehn laboratory in the form of guanosine hydrogels, and they have shown that these self-assembling materials can act as controlled release media for bioactive small molecules (Sreenivasachary and Lehn 2005, 2008). Similarly, these investigators, as well as Leibler and coworkers, have also shown that polymers with the capacity for constitutional reorganization show a number of intriguing material properties, including self-healing and chemoresponsive characteristics (Sreenivasachary et al. 2006; Cordier et al. 2008).

Most recently, Ghadiri and coworkers have showed the ability for peptide nucleic acid polymer with reversible linkages between a polycysteine backbone and thioester acyclonucleosides to undergo a form of evolution in response to the presence of different DNA templates (Ura et al. 2009). Almost simultaneously, the Liu group reported a similar system, involving amidation and reductive amination of a PNA backbone with acyclonucleosides containing carboxylate and aldehyde functional groups (Heemstra and Liu 2009). Acyclonucleosides formed in formamide mixtures recently reported by Di Mauro and coworkers share at least some structural similarity with those used by the Ghadiri and Liu laboratories, suggesting that prebiotic routes may exist to nucleosides that multiple investigators have now shown could couple to varied preformed backbones. One can now imagine the development of RNA-like polymers following a similar theme, creating functional polymers that are environmentally responsive, perhaps even evolving new functions in response to unanticipated environmental signals, which is completely feasible for polymers with reversible linkages that can assume the thermodynamically most favored state, as defined by their local environment.

Finally, the prospect of proto-RNA is reason for optimism that we will yet discover RNA-like polymers with self-assembling properties that are superior to any currently known. The chemical space that still must be searched is vast, but it is not infinite, and the studies discussed earlier have certainly provided clues to where we will most likely find such polymers, and perhaps even the origin of RNA.

ACKNOWLEDGMENTS

We thank R. Krishnamurthy, P. Herdewijn, A. Schwartz and C. Switzer for helpful comments on this article. Support from the NASA Exobiology Program and the National Science Foundation is gratefully acknowledged.

REFERENCES

Anastasi C, Crowe MA, Powner MW, Sutherland JD. 2006. Direct assembly of nucleoside precursors from two- and three-carbon units. *Angew Chem Int Ed Engl* **45:** 6176–6179.

Anet FAL. 2004. The place of metabolism in the origin of life. *Curr Opin Chem Biol* **8:** 654–659.

Ban N, Nissen P, Hansen J, Moore PB, Steitz TA. 2000. The complete atomic structure of the large ribosomal subunit at 2.4 Ångstrom resolution. *Science* **289:** 905–920.

Battersby TR, Albalos M, Friesenhahn MJ. 2007. An unusual mode of DNA duplex association: Watson-Crick interaction of all-purine deoxyribonucleic acids. *Chem Biol* **14:** 525–531.

Bean HD, Anet FAL, Gould IR, Hud NV. 2006. Glyoxylate as a backbone linkage for a prebiotic ancestor of RNA. *Orig Life Evol B* **36:** 39–63.

Bean HD, Sheng YH, Collins JP, Anet FAL, Leszczynski J, Hud NV. 2007. Formation of a β-pyrimidine nucleoside by a free pyrimidine base and ribose in a plausible prebiotic reaction. *J Am Chem Soc* **129:** 9556–9557.

Benner SA. 2004. Understanding nucleic acids using synthetic chemistry. *Acc Chem Res* **37:** 784–797.

Benner SA, Ricardo A, Carrigan MA. 2004. Is there a common chemical model for life in the universe? *Curr Opin Chem Biol* **8:** 672–689.

Bolli M, Micura R, Eschenmoser A. 1997. Pyranosyl-RNA: chiroselective self-assembly of base sequences by ligative oligomerization of tetra nucleotide-2′,3′-cyclophosphates (with a commentary concerning the origin of biomolecular homochirality). *Chem Biol* **4:** 309–320.

Brown S, Thomson S, Veal J, Davis D. 1994. NMR solution structure of a peptide nucleic acid complexed with RNA *Science* **265:** 777–780.

Bryant DE, Kee TP. 2006. Direct evidence for the availability of reactive, water soluble phosphorus on the early Earth. H-Phosphinic acid from the Nantan meteorite. *Chem Commun*: 2344–2346.

Butlerow A. 1861. Bildung einer zuckerartigen substanz durch synthese. *Liebigs Ann Chem* **120:** 295–298.

Cairns-Smith AG. 1982. *Genetic takeover and the mineral origins of life.* Cambridge University Press, Cambridge.

Chen JJ, Cai X, Szostak JW. 2009. N2′→P3′ phosphoramidate glycerol nucleic acid as a potential alternative genetic system. *J Am Chem Soc* **131:** 2119–2121.

Cleaves HJ, Nelson KE, Miller SL. 2006. The prebiotic synthesis of pyrimidines in frozen solution. *Naturwissenschaften* **93:** 228–231.

Cordier P, Tournilhac F, Soulié-Ziakovic C, Leibler L. 2008. Self-healing and thermoreversible rubber from supramolecular assembly. *Nature* **451:** 977–980.

Costanzo G, Saladino R, Crestini C, Ciciriello F, Di Mauro E. 2007. Nucleoside phosphorylation by phosphate minerals. *J Biol Chem* **282:** 16729–16735.

Crick FHC. 1968. The origin of the genetic code. *J Mol Biol* **38:** 367–379.

Crowe MA, Sutherland JD. 2006. Reaction of cytidine nucleotides with cyanoacetylene: support for the intermediacy of nucleoside-2′,3′-cyclic phosphates in the prebiotic synthesis of RNA. *ChemBioChem* **7:** 951–956.

De Graaf RM, Schwartz AW. 2000. Reduction and activation of phosphate on the primitive earth. *Orig Life Evol B* **30:** 405–410.

De Graaf RM, Schwartz AW. 2005. Thermal synthesis of nucleoside H-phosphonates under mild conditions. *Orig Life Evol B* **35:** 1–10.

De Voe H, Tinoco I. 1962. The stability of helical polynucleotides: base contributions. *J Mol Biol* **4:** 500–517.

Decker P, Schweer P, Pohlmann R. 1982. Identification of formose sugars, presumable prebiotic metabolites, using capillary gas chromatography/gas chromatography-mass spectroscopy of n-butoxime trifluoroacetates on OV-225. *J Chromatogr* **244:** 281–291.

Dworkin JP, Miller SL. 2000. A kinetic estimate of the free aldehyde content of aldoses. *Carbohydr Res* **39:** 359–365.

Egholm M, Buchardt O, Christensen L, Behrens C, Freier S, Driver D, Berg R, Kim S, Norden B, Nielsen P. 1993. PNA hybridizes to complementary oligonucleotides obeying the Watson–Crick hydrogen-bonding rules. *Nature* **365:** 566–568.

Egli M, Pallan PS, Pattanayek R, Wilds CJ, Lubini P, Minasov G, Dobler M, Leumann CJ, Eschenmoser A. 2006. Crystal structure of homo-DNA and Nature's choice of pentose over hexose in the genetic system. *J Am Chem Soc* **128**: 10847–10856.

Engelhart AE, Morton TH, Hud NV. 2009. Evidence of strong hydrogen bonding by 8-amino-guanine. *Chem Commun*: 647–649.

Eschenmoser A. 1999. Chemical etiology of nucleic acid structure. *Science* **284**: 2118–2124.

Eschenmoser A. 2004. The TNA-family of nucleic acid systems: Properties and prospects. *Orig Life Evol B* **34**: 277–306.

Eschenmoser A. 2007. The search for the chemistry of life's origin. *Tetrahedron* **63**: 12821–12844.

Eschgfaller B, Schmidt JG, Konig M, Benner SA. 2003. Synthesis and properties of oligodeoxynucleotide analogs with bis(methylene) sulfone bridges. *Helv Chim Acta* **86**: 2959–2997.

Fuller WD, Sanchez RA, Orgel LE. 1972a. Studies in prebiotic synthesis. VI. Synthesis of purine nucleosides. *J Mol Biol* **67**: 25–33.

Fuller WD, Sanchez RA, Orgel LE. 1972b. Studies in prebiotic synthesis: VII. Solid-state synthesis of purine nucleosides. *J Mol Evol* **1**: 249–257.

Gesteland R, Atkins JF. 2006. *The RNA world: The nature of modern RNA suggests a prebiotic RNA world* (3rd ed.) Cold Spring Harbor Laboratory Press, Cold Spring Harbor, NY.

Geyer CR, Battersby TR, Benner SA. 2003. Nucleobase pairing in Watson-Crick-like genetic expanded information systems. *Structure* **11**: 1485–1498.

Groebke K, Hunziker J, Fraser W, Peng L, Diederichsen U, Zimmermann K, Holzner A, Leumann C, Eschenmoser A. 1998. Why pentose- and not hexose-nucleic acids? Purine-purine pairing in homo-DNA: guanine, isoguanine, 2,6-diaminopurine, and xanthine. *Helv Chim Acta* **81**: 375–474.

Guerrier-Takada C, Gardiner K, Marsh T, Pace N, Altman S. 1983. The RNA moiety of ribonuclease P is the catalytic subunit of the enzyme. *Cell* **35**: 849–857.

He W, Hatcher E, Balaeff A, Beratan D, Gil R, Madrid M, Achim C. 2008. Solution structure of a peptide nucleic acid duplex from NMR data: Features and limitations. *J Am Chem Soc* **130**: 13264–13273.

Heemstra JM, Liu DR. 2009. Templated synthesis of peptide nucleic acids via sequence-selective base-filling reactions. *J Am Chem Soc* **131**: 11347–11349.

Herdewijn P. 2001a. TNA as a potential alternative to natural nucleic acids. *Angew Chem Int Ed Engl* **40**: 2249–2251.

Herdewijn P. 2001b. TNA as a potential alternative to natural nucleic acids. *Angew Chem Int Ed Eng* **40**: 2249–2251.

Heuberger BD, Switzer C. 2008a. An alternative nucleobase code: Characterization of purine-purine DNA double helices bearing guanine-isoguanine and diaminopurine-7-deaza-xanthine base pairs. *ChemBioChem* **9**: 2779–2783.

Heuberger BD, Switzer C. 2008b. A nonRNA candidate revisited: Both enantiomers of flexible nucleoside triphosphates are DNA polymerase substrates. *J Am Chem Soc* **130**: 412–413.

Hollis JM, Jewell PR, Lovas FJ, Remijan A. 2004. Green bank telescope obserations of interstellar glycolaldehyde: Low-temperature sugar. *Astrophys J* **613**: L45–L48.

Horowitz ED, Engelhart AE, Chen MC, Quarles KA, Smith MW, Lynn DG, Hud NV. 2010. Intercalation as a means to suppress cyclization and promote polymerization of base-pairing oligonucleotides in a prebiotic world. *Proc Natl Acad Sci* **107**: 5288–5293.

Howard FB, Miles HT. 1977. Interaction of poly(A) and poly(I), a reinvestigation. *Biochemistry* **16**: 4647–4650.

Huang Z, Schneider KC, Benner SA. 1991. Building-blocks for oligonucleotide analogs with dimethylene sulfide, sulfoxide, and sulfone groups replacing phosphodiester linkages. *J Org Chem* **56**: 3869–3882.

Hud NV, Anet FAL. 2000. Intercalation-mediated synthesis and replication: A new approach to the origin of life. *J Theor Biol* **205**: 543–562.

Hud NV, Jain SS, Li X, Lynn DG. 2007. Addressing the problems of base pairing and strand cyclization in template-directed synthesis – A case for the utility and necessity of 'molecular midwives' and reversible backbone linkages for the origin of proto-RNA. *Chem Biodiver* **4**: 768–783.

Ihmels H, Otto D. 2005. Intercalation of organic dye molecules into double-stranded DNA: General principles and recent developments. *Top Curr Chem* **258**: 161–204.

Ingar A-A, Luke RWA, Hayter BR, Sutherland JD. 2003. Synthesis of cytidine ribonucleotides by stepwise assembly of the heterocycle on a sugar phosphate. *ChemBioChem* **4**: 504–507.

Inoue T, Orgel LE. 1983. A non-enzymatic RNA polymerase model. *Science* **219**: 859–862.

Jain SS, Anet FAL, Stahle CJ, Hud NV. 2004. Enzymatic behavior by intercalating molecules in a template-directed ligation reaction. *Angew Chem Int Ed Engl* **43**: 2004–2008.

Jaworski A, Kwiatkowski JS, Lesyng B. 1985. Why isoguanine and isocytosine are not the components of the genetic code. *Int J Quantum Chem* **28, S12**: 209–216.

Jones RJ, Lin K-Y, Milligan JF, Wadwani S, Matteucci MD. 1993. Synthesis and binding properties of pyrimidine oligonucleotide analogs containing neutral phosphodiester replacements: The formacetal and 3′-thioformacetal internucleoside linkages. *J Org Chem* **58**: 2983–2991.

Joyce GF. 1987. Nonenzymatic template-directed synthesis of informational macromolecules. *Cold Spring Harbor Symp Quant Biol* **52**: 41–51.

Joyce GF, Orgel LE. 1999. Prospects for understanding the origin of the RNA world. In The RNA World, Second Edition: The Nature of Modern RNA Suggests a Prebiotic RNA World (Eds. J.F. Atkins, R.F. Gesteland), 49–77. Cold Spring Harbor Laboratory Press, Cold Spring Harbor, NY.

Joyce GF, Schwartz AW, Miller SL, Orgel LE. 1987. The case for an ancestral genetic system involving simple analogs of the nucleotides. *Proc Natl Acad Sci* **84**: 4398–4402.

Kawamura K, Okamoto F. 2001. Cyclization and dimerization of hexanucleotides containing guanine and cytosine with water-soluble carbodiimide. *Viva Origino* **29**: 162–167.

Keefe AD, Miller SL. 1995. Are polyphosphates or phosphate esters prebiotic reagents? *J Mol Evol* **41**: 693–702.

Kiblerherzog L, Zon G, Uznanski B, Whittier G, Wilson WD. 1991. Duplex stabilities of phosphorothioate, methylphosphonate, and RNA analogs of 2 DNA 14-mers. *Nucleic Acids Res* **19**: 2979–2986.

Kolarovic A, Schweizer E, Greene E, Gironda M, Pallan P, Egli M, Rozners E. 2009. Interplay of structure, hydration and thermal stability in formacetal modified oligonucleotides: RNA may tolerate nonionic modifications better than DNA. *J Am Chem Soc* **131**: 14932–14937.

Kolb VM, Dworkin JP, Miller SL. 1994. Alternative bases in the RNA world: the prebiotic synthesis of urazole and its riboside. *J Mol Evol* **38**: 549–557.

Krishnamurthy R, Pitsch S, Minton M, Miculka C, Windhab N, Eschenmoser A. 1996. Pyranosyl-RNA: base pairing between homochiral oligonucleotide strands of opposite sense of chirality. *Angew Chem Int Ed Engl* **35**: 1537–1541.

Krueger AT, Lu HG, Lee AHF, Kool ET. 2007. Synthesis and properties of size-expanded DNAs: Toward designed, functional genetic systems. *Acc Chem Res* **40**: 141–150.

Kruger K, Grabowski PJ, Zaug AJ, Sands J, Gottschling DE, Cech TR. 1982. Self-splicing RNA: Autoexcision and autocyclization of the ribosomal RNA intervening sequence of *Tetrahymena*. *Cell* **31**: 147–157.

Kurreck J. 2003. Antisense technologies—Improvement through novel chemical modifications. *Eur J Biochem* **270**: 1628–1644.

Li X, Zhan Z-YJ, Knipe R, Lynn DG. 2002. DNA-catalyzed polymerization. *J Am Chem Soc* **124**: 746–747.

Mandal M, Breaker RR. 2004. Gene regulation by riboswitches. *Nat Rev Mol Cell Biol* **5**: 451–463.

Mansy SS, Schrum JP, Krishnamurthy M, Tobe S, Treco DA, Szostak JW. 2008. Template-directed synthesis of a genetic polymer in a model protocell. *Nature* **454**: 122–125.

Martins Z, Botta O, Fogel ML, Sephton MA, Glavin DP, Watson JS, Dworkin JP, Schwartz AW, Ehrenfreund P. 2008. Extraterrestrial nucleobases in the Murchison meteorite. *Earth Planet Sci Lett* **270**: 130–136.

Matteucci MD, Bischofberger N. 1991. Sequence-defined oligonucleotides as potential therapeutics. *Annu Rep Med Chem* **26**: 287–296.

Merle Y, Bonneil E, Merle L, Sági J, Szemző A. 1995. Acyclic oligonucleotide analogues. *Int J Biol Macromol* **17**: 239–246.

Miller PS, McParland KB, Jayaraman K, Tso POP. 1981. Biochemical and biological effects of nonionic nucleic acid methylphosphonates. *Biochem Cell Biol* **20**: 1874–1880.

Mittapalli GK, Osornio YM, Guerrero MA, Reddy KR, Krishnamurthy R, Eschenmoser A. 2007a. Mapping the landscape of potentially primordial informational oligomers: oligodipeptides tagged with 2,4-disubstituted 5-aminopyrimidines as recognition elements. *Angew Chem Int Ed Engl* **46**: 2478–2484.

Mittapalli GK, Reddy KR, Xiong H, Munoz O, Han B, De Riccardis F, Krishnamurthy R, Eschenmoser A. 2007b. Mapping the landscape of potentially primordial informational oligomers: Oligodipeptides and oligodipeptoids tagged with triazines as recognition elements. *Angew Chem Int Ed Engl* **46**: 2470–2477.

Miyakawa S, Joshi PC, Gaffey MJ, Gonzalez-Toril E, Hyland C, Ross T, Rybij K, Ferris JP. 2006. Studies in the mineral and salt-catalyzed formation of RNA oligomers. *Origins Life Evol B* **36**: 343–361.

Miyakawa S, Yamanashi H, Kobayashi K, Cleaves HJ, Miller SL. 2002. Prebiotic synthesis from CO atmospheres: Implications for the origins of life. *Proc Natl Acad Sci* **99**: 14628–14631.

Müller D, Pitsch S, Kittaka A, Wagner E, Wintner CE, Eschenmoser A. 1990. Chemie von α-Aminonitrilen. Aldomerisierung von Glycolaldehyd-phosphat zu racemischen Hexose-2,4,6-triphosphaten und (in Gegenwart von Formaldehyd) reacemischen Pentose-2,4-diphosphaten: *rac*-Allose-2,4,6-triphosphat und *rac*-Ribose-2,4-diphosphat sind die Reaktionshauptprodukte. *Helv Chim Acta* **73**: 1410–1468.

Nelson KE, Levy M, Miller SL. 2000. Peptide nucleic acids rather than RNA may have been the first genetic molecule. *Proc Natl Acad Sci* **97**: 3868–3871.

Orgel L. 2000. A simpler nucleic acid. *Science* **290**: 1306–1307.

Orgel LE. 1968. Evolution of the genetic apparatus. *J Mol Biol* **38**: 381–393.

Orgel LE. 1998. The origin of life – a review of facts and speculations. *TIBS* **23**: 491–495.

Orgel LE. 2004. Prebiotic chemistry and the origin of the RNA world. *Crit Rev Biochem Mol Biol* **39**: 99–123.

Orgel LE. 2008. The implausibility of metabolic cycles on the prebiotic Earth. *Plos Biology* **6**: 5–13.

Oro J. 1961. Mechanism of synthesis of adenine from hydrogen cyanide under possible primitive Earth conditions. *Nature* **191**: 1193–&.

Oró J. 1960. Synthesis of adenine from ammonium cyanide. *Biochem Biophys Res Comm* **2**: 407–412.

Pasek MA. 2008. Rethinking early Earth phosphorus geochemistry. *Proc Natl Acad Sci* **105**: 853–858.

Piccirilli JA, Krauch T, Moroney SE, Benner SA. 1990. Enzymatic incorporation of a new base pair into DNA and RNA extends the genetic alphabet. *Nature* **343**: 33–37.

Pitsch S, Wendeborn S, Jaun B, Eschenmoser A. 1993. Why pentose- and not hexose-nucleic acids? Part VII. Pyranosyl-RNA ('p-RNA'). *Helv Chim Acta* **76**: 2161–2183.

Pitulescu M, Grapp M, Krätzner R, Knepel W, Diederichsen U. 2008. Synthesis of formacetal-linked dinucleotides to facilitate dsDNA bending and binding to the homeodomain of PAX6. *Eur J Org Chem*: 2100–2106.

Pizzarello S. 2006. The chemistry of life's origin: A carbonaceous meteorite perspective. *Acc Chem Res* **39**: 231–237.

Plankensteiner K, Reiner H, Schranz B, Rode BM. 2004. Prebiotic formation of amino acids in a neutral atmosphere by electric discharge. *Angew Chem Int Ed Engl* **43**: 1886–1888.

Powner MW, Gerland B, Sutherland JD. 2009. Synthesis of activated pyrimidine ribonucleotides in prebiotically plausible conditions. *Nature* **459**: 239–242.

Rajamani S, Vlassov A, Benner S, Coombs A, Olasagasti F, Deamer D. 2008. Lipid-assisted synthesis of RNA-like polymers from mononucleotides. *Orig Life Evol B* **38**: 57–74.

Rasmussen H, Kastrup J, Nielsen J, Nielsen J, Nielsen P. 1997. Crystal structure of a peptide nucleic acid (PNA) duplex at 1.7 Å resolution. *Nat Struct Biol* **4:** 98–101.

Ricardo A, Carrigan MA, Olcott AN, Benner SA. 2004. Borate minerals stabilize ribose. *Science* **303:** 196–196.

Rice JS, Gao X. 1997. Conformation of formacetal and 3′-thioformacetal nucleotide linkers and stability of their antisense RNA·DNA hybrid duplexes. *Biochemistry* **36:** 399–411.

Richert C, Roughton AL, Benner SA. 1996. Nonionic analogs of RNA with dimethylene sulfone bridges. *J Am Chem Soc* **118:** 4518–4531.

Roughton AL, Portmann S, Benner SA, Egli M. 1995. Crystal-structure of a dimethylene sulfone-linked ribodinucleotide analog. *J Am Chem Soc* **117:** 7249–7250.

Rozners E, Katkevica D, Strömberg R. 2007. Oligoribonucleotide analogues containing a mixed backbone of phosphodiester and formacetal internucleoside linkages, together with vicinal 2′-O-methyl groups. *Chembiochem* **8:** 537–545.

Saladino R, Ciambecchini U, Crestini C, Costanzo G, Negri R, Di Mauro E. 2003. One-pot TiO_2-catalyzed synthesis of nucleic bases and acyclonucleosides from formamide: Implications for the origin of life. *ChemBioChem* **4:** 514–521.

Saladino R, Crestini C, Ciciriellloc F, Costanzo G, Di Mauro E. 2007. Formamide chemistry and the origin of informational polymers. *Chem Biodiv* **4:** 694–720.

Saladino R, Crestini C, Costanzo G, DiMauro E. 2004. Advances in the prebiotic synthesis of nucleic acids bases: implications for the origin of life. *Curr Org Chem* **8:** 1425–1443.

Sanchez R, Ferris J, Orgel LE. 1966. Conditions for purine synthesis—Did prebiotic synthesis occur at low temperatures? *Science* **153:** 72–73.

Sanchez RA, Orgel LE. 1970. Studies in prebiotic synthesis. V. Synthesis and photoanomerization of pyrimidine nucleosides. *J Mol Biol* **47:** 531–543.

Schlegel MK, Xie XL, Zhang LL, Meggers E. 2009. Insight into the high duplex stability of the simplified nucleic acid GNA. *Angew Chem Int Ed Engl* **48:** 960–963.

Schmidt J, Eschgfaller B, Benner SA. 2003. A direct synthesis of nucleoside analogs homologated at the 3′- and 5′-positions. *Helv Chim Acta* **86:** 2937–2958.

Schneider KC, Benner SA. 1990. Oligonucleotides containing flexible nucleoside analogs. *J Am Chem Soc* **112:** 453–455.

Schöning KU, Scholz P, Guntha S, Wu X, Krishnamurthy R, Eschenmoser A. 2000. Chemical etiology of nucleic acid structure: The α-threofuranosyl-(3′→2′) oligonucleotide system. *Science* **290:** 1347–1351.

Schrum JP, Ricardo A, Krishnamurthy M, Blain JC, Szostak JW. 2009. Efficient and rapid template-directed nucleic acid copying using 2′-amino-2′,3′-dideoxyribonucleoside–5′-phosphorimidazolide monomers. *J Am Chem Soc* **131:** 14560–14570.

Schwartz AW. 2006. Phosphorus in prebiotic chemistry. *Philos Trans R Soc London B* **361:** 1743–1749.

Schwartz AW, Chang S. 2002. From Big Bang to primordial planet: Setting the stage for the origin of life. In Life's Origin: The Beginnings of Biological Evolution (Eds.) W. Schopf, 46–77. University of California, Berkeley.

Schweitzer M, Engels JW. 1999. Sequence specific hybridization properties of methylphosphonate oligodeoxynucleotides. *J Biomol Struct Dyn* **16:** 1177–1188.

Seela F, Wei C, Kazmierczuk Z. 1995. Substituent reactivity and tautomerism of isoguanosine and related nucleosides. *Helv Chim Acta* **78:** 1843–1854.

Sepiol J, Kazmierczuk Z, Shugar D. 1976. Tautomerism of iso-guanosine and solvent-induced keto-enol equilibrium. Zeitschrift für Naturforschung C-A. *J Biosci* **31:** 361–370.

Sheng Y, Bean HD, Mamajanova I, Hud NV, Leszczynski J. 2009. A comprehensive investigation of the energetics of pyrimidine nucleoside formation in a model prebiotic reaction. *J Am Chem Soc:* in press.

Sievers D, von Kiedrowski G. 1994. Self replication of complementary nucleotide-based oligomers. *Nature* **369:** 221–224.

Sinha S, Kim PH, Switzer C. 2004. 2′,5′-linked DNA is a template for polymerase-directed DNA synthesis. *J Am Chem Soc* **126:** 40–41.

Springsteen G, Joyce GF. 2004. Selective derivatization and sequestration of ribose from a prebiotic mix. *J Am Chem Soc* **126:** 9578–9583.

Sreenivasachary N, Hickman DT, Sarazin D, Lehn JM. 2006. DyNAs: Constitutional dynamic nucleic acid analogues. *Chem-Eur J* **12:** 8581–8588.

Sreenivasachary N, Lehn JM. 2005. Gelation-driven component selection in the generation of constitutional dynamic hydrogels based on guanine-quartet formation. *Proc Natl Acad Sci* **102:** 5938–5943.

Sreenivasachary N, Lehn JM. 2008. Structural selection in G-quartet-based hydrogels and controlled release of bioactive molecules. *Chem-Asian J* **3:** 134–139.

Summerton J, Weller D. 1989. Uncharged morpholino-based polymers having achiral intersubunit linkages. US Patent #5,034,506

Summerton J, Weller D. 1997. Morpholino antisense oligomers: design, preparation, and properties. *Antisense Nucleic Acid Drug Dev* **7:** 187–195.

Sutherland JD. 2010. Ribonucleotides. *Cold Spring Harb Perspect Biol* **2:** a005439.

Switzer C, Moroney SE, Benner SA. 1989. Enzymatic incorporation of a new base pair into DNA and RNA. *J Am Chem Soc* **111:** 8322–8323.

Ts'o P, Melvin I, Olson A. 1963. Interaction and association of bases and nucleosides in aqueous solutions. *J Am Chem Soc* **85:** 1289–1296.

Tsai CH, Chen JY, Szostak JW. 2007. Enzymatic synthesis of DNA on glycerol nucleic acid templates without stable duplex formation between product and template. *Proc Natl Acad Sci* **104:** 14598–14603.

Ura Y, Beierle JM, Leman LJ, Orgel LE, Ghadiri MR. 2009. Self-assembling sequence-adaptive peptide nucleic acids. *Science* **325:** 73–77.

Usher DA, McHale AH. 1976. Hydrolytic stability of helical RNA—Selective advantage for natural 3′,5′-bond. *Proc Natl Acad Sci* **73:** 1149–1153.

Vasas V, Szathmáry E, Santos M. 2010. Lack of evolvability in self-sustaining autocatalytic networks constrains metabolism-first scenarios for the origin of life. *Proc Natl Acad Sci* **107:** 1470–1475.

Voegel JJ, Altorfer MM, Benner SA. 1993. The donor-acceptor-acceptor purine analog-transformation of 5-aza-7-deaza-1H-isoguanine (= 4-aminoimidazo-[1,2-α]-1,3,5-triazin-2(1H)-one) to 2′-deoxy-5-aza-7-deaza-isoguanosine using purine nucleoside phosphorylase. *Helv Chim Acta* **76:** 2061–2069.

Weber AL. 1992. Prebiotic sugar synthesis: hexose and hydroxy acid synthesis from glyceraldehyde catalyzed by iron(III) hydroxide oxide. *J Mol Evol* **35:** 1–6.

Weber AL. 2001. The sugar model: Catalysis by amines and amino acid products. *Orig Life Evol B* **31:** 71–86.

Westheimer FH. 1987. Why Nature chose phosphates. *Science* **235:** 1173–1178.

Wiberg KB, Morgan KM, Maltz H. 1994. Thermochemistry of carbonyl reactions. 6. A study of hydration equilibria. *J Am Chem Soc* **116:** 11067–11077.

Wilds CJ, Wawrzak Z, Krishnamurthy R, Eschenmoser A, Egli M. 2002. Crystal structure of a B-form DNA duplex containing (L)-α- threofuranosyl (3′→ 2′) nucleosides: A four-carbon sugar is easily accommodated into the backbone of DNA. *J Am Chem Soc* **124:** 13716–13721.

Wimberly BT, Brodersen DE, Clemons WM, Morgan-Warren RJ, Carter AP, Vonrhein C, Hartsch T, Ramakrishnan V. 2000. Structure of the 30S ribosomal subunit. *Nature* **407:** 327–339.

Woese C. 1967. The evolution of the genetic code. In *The genetic code* (Eds.), 179–195. Harper & Row, New York.

Yakovchuk P, Protozanova E, Frank-Kamenetskii MD. 2006. Base-stacking and base-pairing contributions into thermal stability of the DNA double helix. *Nucleic Acids Res* **34:** 564–574.

Zhang LL, Meggers E. 2005. An extremely stable and orthogonal DNA base pair with a simplified three-carbon backbone. *J Am Chem Soc* **127:** 74–75.

Zielinski WS, Orgel LE. 1985. Oligomerization of activated derivatives of 3′-amino-3′-deoxyguanosine on poly(C) and poly(dG) templates. *Nucleic Acids Res* **13:** 2469–2484.

Zielinski WS, Orgel LE. 1987. Oligoaminonucleoside phosphoramidates. Oligomerization of dimers of 3′-amino-3′-deoxy-nucleotides (GC and CG) in aqueous solution. *Nucleic Acids Res* **15:** 1699–1715.

Zubay G, Mui T. 2001. Prebiotic synthesis of nucleosides. *Origins Life Evol B* **31:** 87–102.

Closing the Circle: Replicating RNA with RNA

Leslie K.L. Cheng and Peter J. Unrau

Simon Fraser University, 8888 University Drive, Burnaby, BC. V5A 1S6, Canada
Correspondence: punrau@sfu.ca

How life emerged on this planet is one of the most important and fundamental questions of science. Although nearly all details concerning our origins have been lost in the depths of time, there is compelling evidence to suggest that the earliest life might have exploited the catalytic and self-recognition properties of RNA to survive. If an RNA based replicating system could be constructed in the laboratory, it would be much easier to understand the challenges associated with the very earliest steps in evolution and provide important insight into the establishment of the complex metabolic systems that now dominate this planet. Recent progress into the selection and characterization of ribozymes that promote nucleotide synthesis and RNA polymerization are discussed and outstanding problems in the field of RNA-mediated RNA replication are summarized.

Cell division is a fundamental biological process in which genetic information is duplicated and shared between daughter cells. In extant cellular life, DNA serves as the repository of genetic information, but its replication is complicated by the daunting size and complex structural organization of modern genomes. For this reason, multiple enzymes are required to ensure faithful genomic replication in all higher life forms. Notably, simpler replicating systems such as viruses, have smaller genomes and tend to use correspondingly more error-prone replicative machinery (Kunkel and Bebenek 2000; Gago, Elena et al. 2009). Presumably, if the initial organisms on this planet also had small genomes, then the earliest genomic replication could have been a relatively simple and error-prone process compared with the complex replicative strategies of modern life.

ABC of Life: Abiotic to Biotic Chemistry, the Emergence of the RNA World

Modern biology is built up from a dense web of chemical reactions that are maintained by the catalytic reactions of hundreds of enzymes. These catalysts are all one dimensional, aperiodic polymers, built from protein or nucleic acids. Such polymers are able to adopt a diverse range of complex three-dimensional folds and by virtue of their linear sequence are simple to encode. A particular advantage of these biological polymers is that each polymer type is defined by a small set of monomers that can be polymerized by a single self-consistent chemistry. This type of polymer construction empowers evolution with a mechanism to rapidly adapt existing polymer folds to new functions by the mutation/recombination of polymer sequence. Based on the intrinsic simplicity of modern biological polymers, the earliest biological systems appear overwhelmingly likely to have been polymer based as well (Joyce 2002).

The emergence of the earliest biological replicating systems must have required considerable abiotic chemical organization. Such abiotic chemistry would not have immediately disappeared upon the emergence of life and would have provided chemical "sustenance" to nurture the increasing metabolic complexity of emergent life as it evolved. As a result if the first biological polymer can ever be reliably identified it would place an important constraint on the abiotic environments possible on the early Earth. This polymer would be the bridge between the world of abiotic chemistry and the biotic world of enzymatic reactions.

Based on this logic, the earliest replicative polymers should satisfy three primary conditions: (1) The initial polymers should have an intrinsic mechanism to facilitate their replication either abiotically or biotically. Ideally the polymer's monomer subunits should be able to be polymerized by a single uniform chemistry. (2) Abiotically, monomers should be easily synthesized and this synthesis should be compatible with abiotic polymerization. Abiotic polymerization from abiotically synthesized monomers would have presumably resulted in the synthesis of more monomer than polymer and would provide the earliest replicating systems with a source of monomers before the evolution of a biological metabolism. (3) The resulting polymers should intrinsically be endowed with the ability to promote a broad range of chemistry. In particular the metabolic synthesis of monomer units from abiotic sources should be tenable using the polymers themselves as catalysts. This monomer synthesis would serve as the basis for one of the earliest metabolic reactions associated with replication.

RNA is the simplest aperiodic polymer that biologically and chemically has been shown to satisfy the first and last of these three conditions. RNA is comprised of four distinct monomers, which can form a double-stranded RNA helix by simple and predictable pairing rules. The ability to form a regular homoduplex provides a fundamental mechanism for the templated replication of RNA strands and is used by simple RNA systems such as viruses to replicate. Equally important, RNA can fold into a variety of complex three-dimensional shapes that promote a broad range of chemical reactions (Wilson and Szostak 1999; Ellington, Chen et al. 2009). Biologically, RNA plays a fundamental role in modern life being responsible for ribosomal translation of mRNA into protein (Nissen, Hansen et al. 2000), RNase P and tRNA maturation (Pannucci, Haas et al. 1999; Marquez, Chen et al. 2006), riboswitches and gene regulation (Tucker and Breaker 2005), and as a template for the extension of telomeres by telomerase (Qiao and Cech 2008). The ribosome together with other critical biological RNA catalysts, such as RNase P, has been interpreted as a relic of an "RNA World," where RNA and not protein was the dominant catalyst (Crick 1958; Gilbert 1986; Orgel 2004). This evidence together with the metabolic importance of RNA and the nucleotide cofactors across biology (White 1976; Benner et al. 1989), strongly suggests that RNA played an important role early in evolution and that it might have been involved in the very first autocatalytic reactions.

Abiotic Nucleotide Synthesis

Until recently a primary difficulty in declaring RNA the earliest biological polymer has been the absence of convincing evidence that activated RNA monomers can be produced by abiotic processes. This has caused some to speculate that another even simpler polymer preceded RNA in the early evolution of life (Orgel 2004; Joyce and Orgel 2006). Although it has been known for some time that the purine and pyrimidine nitrogenous bases (Robertson and Miller 1995; Zubay and Mui 2001; Hill and Orgel 2002) and ribose sugar (Ricardo et al. 2004; Gesteland et al. 2006) required to construct ribonucleosides can be synthesized separately from plausible prebiotic compounds, the creation of a glycosidic linkage has been problematic. Heating purine bases together with ribose in dehydrating conditions leads to the production of nucleosides in low yield (Fuller et al. 1972) but the synthesis of pyrimidine nucleotides has proven difficult (Orgel 2004).

In 2009, Powner et al. elegantly demonstrated that pyrimidine ribonucleotides could be formed from the same precursor molecules used to make a pyrimidine and ribose, with the exception that these molecules first be reacted to generate an intermediate, 2-aminooxazole. This clever approach bypasses the need to synthesize a glycosidic linkage and produces a reasonable yield of a pyrimidine $2',3'$-cyclic phosphate nucleotide monomer (Powner et al. 2009). It is currently unknown how purine monomers could be efficiently synthesized by abiotic mechanisms. Nevertheless, if prebiotic routes for both pyrimidine and purine nucleotides can be found, then the abiotic synthesis of RNA polymers via their condensation (Ferris et al. 1996; Monnard et al. 2003) or via the efficient polymerization of $3'$-$5'$ cyclic nucleotides (Costanzo et al. 2009) might be sufficient to trigger the emergence of an RNA based replicating system. Such additional experimental evidence would remove the primary objections to an RNA early hypothesis and is an exciting area of current research (Ricardo and Szostak 2009).

TOWARD A REPLICATING SYSTEM: NUCLEOTIDE SYNTHESIS BY RNA

Prebiotic sources of nucleotides would have been quickly used by a rapidly growing population of ribo-organisms and would have presented a bottleneck to early evolution. To flourish, early RNA based life would have needed to develop the ability to synthesize nucleotide monomers from some more abundant supply of prebiotic material. Based on the chemistry of nucleotide synthesis in modern metabolism it would be satisfying if these abundant prebiotic compounds included ribose and the nucleotide bases (Joyce 1989; Robertson and Miller 1995; Orgel 1998). If so a solid abiotic chemical framework would exist for the evolution of RNA-mediated nucleotide synthesis reactions that resemble those found in modern metabolism.

Pyrimidine nucleotides and purine nucleotides (via salvage pathways) are synthesized by the formation of a glycosidic linkage using phosphoribosyl 1-pyrophosphate (PRPP) and the appropriate nucleobase. Given the fundamental importance of this chemistry to modern metabolism and its potential compatibility with abiotic supplies of ribose and nucleobases (Lau and Unrau 2009), the ability of RNA to mediate such chemistry serves as a potential bridge between the abiotic and biotic RNA World of nucleotide synthesis. However, the chemistry of glycosidic bond formation presents a number of hurdles for RNA catalysts. Relative to protein, RNA has relatively few functional groups and the nucleotide monomers from which an RNA catalyst must be built are considerably bulkier. Further, the chemistry of nucleotide synthesis is strongly influenced by nucleobase composition. Pyrimidine nucleotides are thermodynamically and kinetically much more difficult to form than purine nucleotides (Lau et al. 2004) making nucleotide synthesis by RNA an interesting enzymatic challenge.

One way to evaluate the ability of RNA to mediate such chemistries is to artificially select and evolve ribozymes in the laboratory using in vitro selection (Ellington and Szostak 1990; Robertson and Joyce 1990; Tuerk and Gold 1990). We have used this approach to isolate both pyrimidine nucleotide and purine nucleotide synthase ribozymes that are able to promote the formation of a glycosidic linkage between a base (4-thiouracil and 6-thioguanine, respectively) and tethered PRPP (Unrau and Bartel 1998; Lau et al. 2004). As expected based on the difficulty of pyrimidine glycosidic bond formation, purine nucleotide synthases were much more abundant in sequence space and were 50–100 times more efficient than their pyrimidine synthase counterparts. Interestingly pyrimidine nucleotide synthase ribozymes use charge stabilization to promote glycosidic bond formation (Unrau and Bartel 2003) presumably by the precise positioning of the ribozyme's phosphodiester backbone (Dinner et al. 2001). Recently we made the interesting discovery that a ribozyme selected for its ability to mediate chemistry between ribose and 6-thioguanosine could also promote purine nucleotide synthesis when the ribose substrate was substituted with PRPP (Lau and Unrau

2009). This promiscuous ribozyme suggests that metabolically relevant ribozymes making use of a small metabolite such as ribose could have easily evolved to promote nucleotide synthesis with PRPP. Together with our purine and pyrimidine nucleotide synthases, this in vitro evidence strongly suggests that RNA folds able to promote nucleotide synthesis would not have been difficult to discover early in the evolution of a RNA World.

Satisfyingly, both purine and pyrimidine nucleotide synthase ribozymes were able to robustly discriminate between even quite closely related nucleobase substrates. The family A pyrimidine nucleotide synthase showed a marked preference for 4-thiouracil over any other pyrimidine tested (including uracil), whereas the purine nucleotide synthases reacted two to three orders of magnitude slower when 6-thiopurine was substituted for 6-thioguanine (Unrau and Bartel 1998; Lau et al. 2004). This ability to accurately distinguish small substrates is a hallmark of the metabolic enzymes and provides encouraging evidence that in an early RNA World metabolically relevant ribozymes would have been able to efficiently distinguish and hence regulate the synthesis of important small molecule metabolites essential to life.

RNA REPLICATION: RNA POLYMERASES AND LIGASES

A RNA-mediated metabolism could only have been sustained if the RNA catalysts required to sustain metabolism were produced or repaired faster than they degraded. Although RNA repair is relatively uncommon in modern metabolism (Chan, Zhou et al. 2009), there exist numerous examples of RNA modification and editing that make use of short guide RNAs to promote chemistry at specific sites within a target RNA (Kiss 2002; Madison-Antenucci, Grams et al. 2002; Matera, Terns et al. 2007). Guide RNAs, which generally speaking are RNAs having the reverse complement of some target sequence, could have been very useful early in evolution and might have directed RNA ligase ribozymes to specifically ligate together short RNA fragments in a sequence specific fashion. This would allow the construction of larger and more complex metabolically useful ribozymes from simpler RNA elements. If in turn these RNA fragments and guide sequences were transcribed from short RNA "genomic elements" this would not only allow the replication of a relatively complex, but fragmented RNA genome, it would only require two replicative activities: RNA ligation and a RNA polymerase capable of transcribing and replicating the short RNA genomic elements. As the phosphodiester chemistry of RNA ligation and polymerization are related (Fig. 1) their emergence from an abiotic system could be evolutionarily linked and provide a simple route to biological replication.

Cross Catalytic Ligation Strategies

An ingenious continuous molecular evolution approach, developed by Kim and Joyce (Kim and Joyce 2004) and further optimized by Lincoln and Joyce (Lincoln and Joyce 2009) has made possible the study of cross-catalytic systems built entirely of RNA. In this ribozyme-based system an RNA ligase ribozyme hybridizes to two oligonucleotide substrates and specifically catalyzes their ligation (Fig. 1A). The ligated RNA results in a second ligase ribozyme that in turn can hybridize to two other oligonucleotide substrates catalyzing in turn the production of the first ribozyme. This cross-catalytic system was found to promote the continuous exponential doubling of RNA. A similar system was developed and optimized by the Lehman laboratory, which divided the Azoarcus Group I ribozyme into four parts and made use of its intrinsic ligation capability to autocatalytically reassemble the ribozyme (Hayden and Lehman 2006; Draper et al. 2008). These fascinating systems all make use of guide sequences built into the RNA catalyzing ligation. If in the future these systems can be engineered to use generic guide RNAs supplied in trans these systems would have tremendous in vitro evolutionary potential and make possible the synthesis of arbitrary RNA components from a set of short RNAs and their corresponding guide sequences.

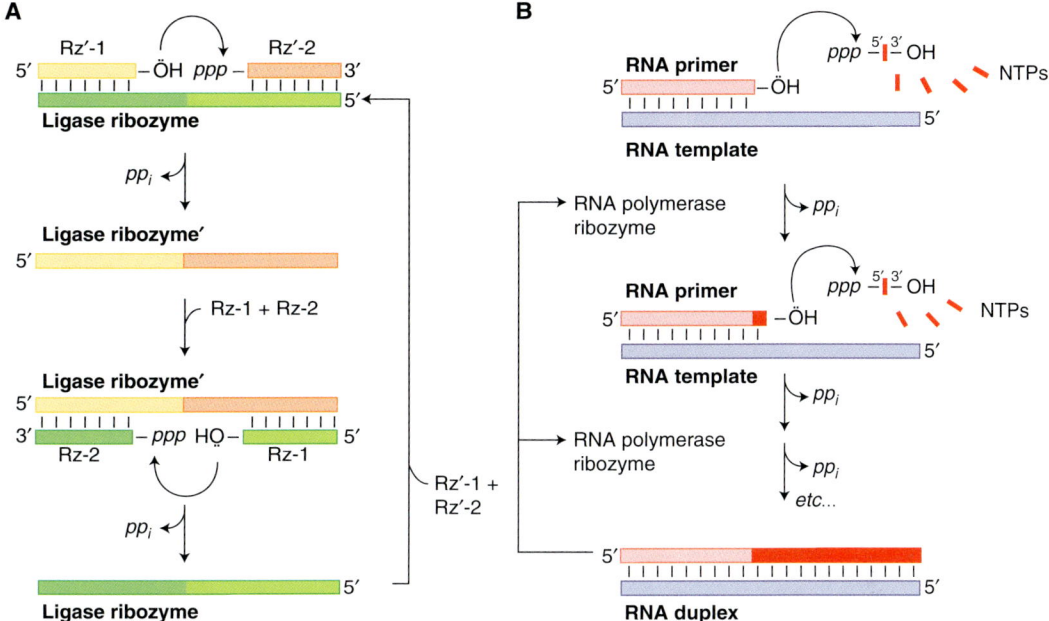

Figure 1. RNA-catalyzed polymerization and cross-replication by ligation. (*A*) Cartoon schematic of cross-replication of RNA ligase ribozymes: A ligase ribozyme (colored in two shades of green) catalyzes the ligation of two orange oligonucleotides (Rz′-1 and Rz′-2) to generate a ligase ribozyme that catalyzes the ligation of two green oligonucleotides (Rz-1 and Rz-2) to regenerate the first ligase ribozyme. (Adapted from Lincoln and Joyce 2009.) (*B*) Cartoon schematic scheme of ribozyme-catalyzed RNA polymerization. The complete extension of an RNA primer (pink) according to the sequence of a template (blue) by an RNA polymerase produces an RNA duplex.

Although these systems have elegantly shown the potential for RNA to replicate without the addition of other components and provide insight into how a complex multi-component RNA world might function, these experiments do not address how the RNA fragments being ligated together are themselves produced. Returning to our initial argument, if the emergence of life rapidly consumed abiotically generated RNA fragments it might be expected that early in evolution ligation strategies could have evolved hand in hand with ribozyme polymerases able to transcribe short nucleotide sequences that ultimately become substrates for ligation. In this model, as the RNA World became more established and complex, improvements in RNA polymerization would make possible the gradual concatenation of RNA genomic elements into longer and longer elements making the importance of RNA ligation less and less important as overall genome size grew.

Polymerization from Short RNA Genomic Elements

Although nature abounds with replicative strategies, nucleic acid based replication always involves the synthesis of sense and antisense strands following the canonical base pairing rules of nucleic acid. The fundamental symmetry of this copying process—where sense stand is copied to antisense to sense *ad infinitum*—is always broken in biological systems with the synthesis of one stand being elevated over the other so as to allow gene expression. This is most clear in the highly evolved organisms where dsDNA is the genetic material and mRNA is transcribed, but it is also true for the simplest homopolymeric systems, such as the

RNA based plant viroids, where more plus strand is expressed during rolling circle replication than the minus. Notably in the case of the plant viroids this asymmetry does not depend on translation but only on the functional aspects of the expressed RNAs (Soll et al. 2001). In an early RNA world such asymmetry would have been equally important for the stable expression of RNA based metabolic enzymes. If such "transcriptional" asymmetry could be combined with the replication of short RNA genomic elements, then together with the ligation strategies just discussed a self-consistent system of RNA replication could be constructed that simultaneously allows genomic replication and expression of metabolic ribozymes.

We favor a replicative model whereby the earliest genetic components were short dsRNAs. Although not fundamental for replication, dsRNA has a number of interesting virtues. First and in contrast to single-strand or folded RNAs the RNA duplex has a uniform and predictable double helix that makes it easily recognizable as a genetic component by replicative enzymes. Second, it is intrinsically difficult to copy single-stranded RNA without creating a RNA duplex. Thus generation of a reverse complement strand from a single-stranded RNA genome requires some structural mechanism to avoid the genome becoming double-stranded by default. Although such strategies are possible and are used by degenerate RNA systems such as RNA-X (Konarska and Sharp 1989) it is unclear how such strategies could be generalized to arbitrary sequence (Bartel 1999). A RNA polymerase ribozyme able to differentially transcribe from both strands of a short RNA duplex could however produce double-stranded genomic copies and an excess of single-stranded RNA that can be ligated by subsequent steps into functional RNA components (Fig. 2).

RNA POLYMERIZATION: EVOLVING RNA LIGASES

The advent of in vitro selection provided a mechanism to directly select template directed

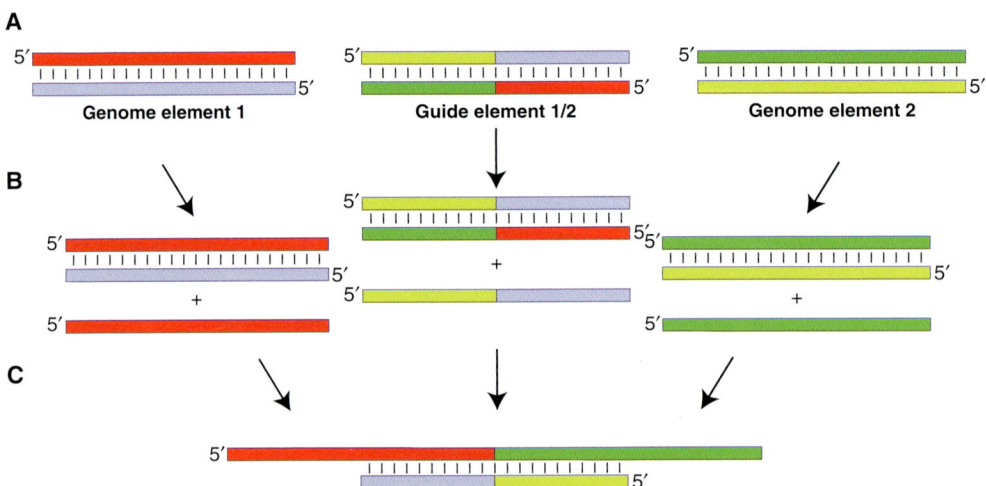

Figure 2. Genomic elements, their asymmetrical transcription resulting in genomic replication and synthesis of functional RNAs by ligation. (A) Short double-stranded elements define the total RNA genome together with guide sequences important for the later synthesis of full length functional RNAs. (B) Asymmetrical transcription from either strand of the duplex genomic element results in the synthesis of multiple genomic element copies together with an excess of one strand. This asymmetry could be generated by having one end of the genomic element fray with a higher propensity than the other. (C) Excess single-stranded transcripts can in turn hybridize to guide element sequences making them substrates for ligase enzymes and allowing the synthesis of long highly functional RNAs.

RNA ligase ribozymes. RNA ligation, which involves the nucleophilic attack of a 3′-hydroxyl from one substrate onto the 5′-triphosphate of a second, is identical to the chemistry required for RNA polymerization (Fig. 1). Initial efforts in the 1990s led to the isolation and structural characterization of the highly efficient Class I ligase ribozyme (Bartel and Szostak 1993; Ekland and Bartel 1995; Ekland et al. 1995). Different classes of ligase ribozymes were subsequently isolated by in vitro selection, including the hc ribozyme and its variants isolated from the *Tetrahymena* group 1 intron (Jaeger, Wright et al. 1999; Yoshioka, Ikawa et al. 2004) and smaller ligases such as the L1 ligase (Robertson and Ellington 1999; Robertson and Scott 2007). Unfortunately, these ribozymes were all much slower than the class I ligase implying that highly efficient ligases might be rare in sequence space. Recently however, the DSL ligase (Ikawa, Tsuda et al. 2004; Voytek and Joyce 2007) was evolved to rival the catalytic efficiency of the class I ligase. This second example of an efficient RNA ligase ribozyme strongly suggests that a diverse range of ligase ribozymes could have been available early in evolution. Further characterization of these new ligases should generalize our understanding of RNA mediated phosphodiester bond formation and by virtue of their identical chemistry provide insight into the function of RNA polymerases.

Template-Directed RNA Polymerization: Evolving the Class I Ligase

Surprisingly, Ekland and Bartel showed that a variant of the class I ligase ribozyme could be simply engineered to extend a primer by six nucleotide incorporations after a four day incubation (Ekland and Bartel 1996). This template-mediated extension required that the primer itself be hybridized to the ligase ribozyme. In vitro selection was used to overcome this problem and led to the isolation of the Round 18 ribozyme (Johnston, Unrau et al. 2001). This RNA polymerase ribozyme contains a 76-nucleotide accessory domain selected from a random sequence pool that was appended onto the 3′ end of the class I ligase ribozyme (now called the ligase core). The Round 18 ribozyme was capable of using nucleotide triphosphates in a template-dependent manner to extend 14 nucleotides of a *trans* RNA primer-template in 24 hours (i.e., see Fig. 1B). The engineering of the Round 18 ribozyme from the class I ligase directly demonstrates that polymerization can be evolved from RNA ligation and has yet to be shown for any other RNA ligase family.

The selection of the Round 18 ribozyme required that the RNA primer being extended by the polymerase be covalently attached to the ribozyme so as to maintain a correlation between RNA function (polymerization) and phenotype (RNA sequence). To overcome this limitation and allow a true *trans* selection for RNA polymerization, we used a compartmentalized approach (Fig. 3) to select for an improved variant of the Round 18 ribozyme called B6.61. This polymerase is able to extend a *trans* primer-template duplex by 20 nucleotides and has improved fidelity relative to the Round 18 ribozyme (Zaher and Unrau 2007). In the selection for B6.61, a diverse DNA pool containing approximately 9×10^{14} variants of the Round 18 ribozyme was ligated to an RNA primer-template complex. These tagged DNA genomes were encapsulated into water-in-oil vesicles (Tawfik and Griffiths 1998; Miller et al. 2006) where they were transcribed by T7 RNA polymerase. By this strategy RNA phenotypes were confined together with their DNA genotype. If an expressed RNA could extend the primer-template, enrichment of the corresponding DNA genome was made possible. By mimicking natural selection, the in vitro encapsulated selection of B6.61 allowed for a true *trans* correlation between the genotype and phenotype. Similar approaches might in future allow the isolation of RNA systems that work cooperatively in order to survive within a common compartment and allow the development of artificially evolving RNA systems (Bartel and Unrau 1999).

The Evolutionary Power of Constructing Modular RNA Polymerases

Protein polymerases are modular enzymes that have clearly delimited functional folds (Werner

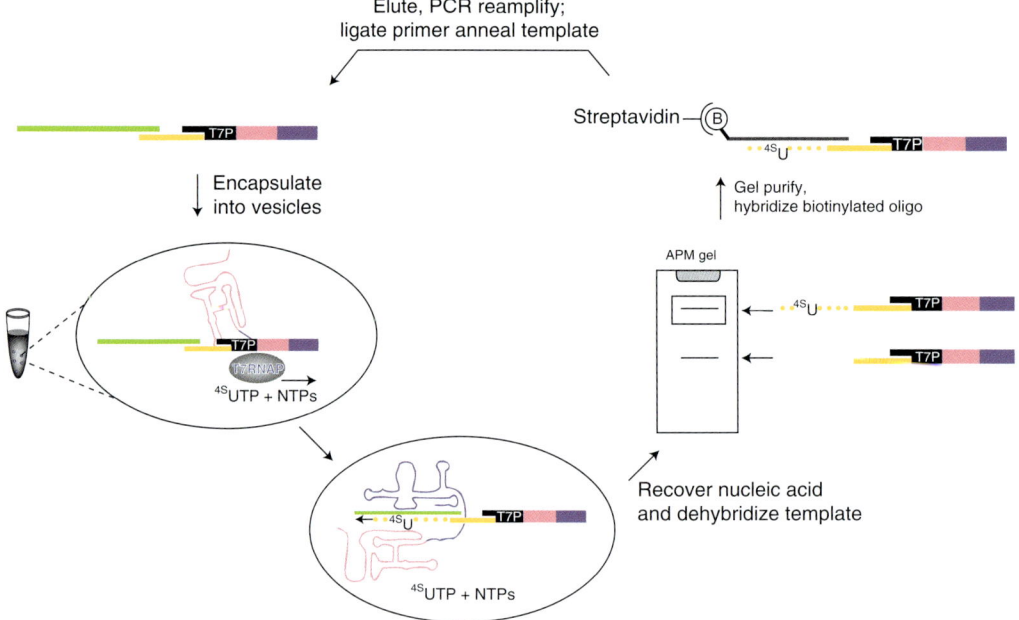

Figure 3. In vitro encapsulated selection scheme for *trans* acting RNA polymerase ribozymes. A DNA pool was generated that contained a T7 promoter allowing transcription of mutagenized Round 18 ribozyme sequence library. After ligating a RNA primer (orange)–template (green) complex to the DNA pool, genomes were encapsulated and RNA transcribed by T7 RNAP. Active RNA polymerase ribozymes extend the RNA primer tethered to their DNA genome and in the process incorporate 4-thiouridine residues in the growing strand. This allows selection of functional genomes using thiol-sensitive mercury gels and hybridization-based capture using biotinylated oligonucleotides. The captured DNA was then PCR amplified and used in a further round of selection (Zaher and Unrau 2007 and reprinted here with permission by the author).

2007). This is seen most clearly in the well studied DNA dependent RNA polymerases where initiation requires the static recognition of a dsDNA promoter element (by the σ-factors in bacterial RNAPs) and where elongation requires a structural rearrangement of the polymerase so as to produce a DNA–RNA heteroduplex that moves within the DNA transcription bubble to ensure processive elongation (Yin and Steitz 2002; Mooney et al. 2005). Although the B6.61 polymerase cannot rival the complexity of such beautiful machines, it does compare with the much less processive, primer dependent DNA repair enzymes such as *Taq* polymerase. Interestingly, experiments with this commonly used polymerase have found that appending a nonspecific DNA binding protein to the enzyme can dramatically increase its processivity (Wang et al. 2004). It might therefore be possible to enhance the ability of the B6.61 polymerase to extend long templates by adding additional RNA domains that enhance the ability of the polymerase to nonspecifically bind RNA duplex. Such modularity is extensively found in large biological RNAs such as the ribosome where the functions of peptide bond formation and mRNA decoding are apportioned between the large and small subunits. Similarly, the RNase P RNA can be dissected into two major submotifs, one responsible for substrate recognition and the other catalysis (Krasilnikov et al. 2004). Presumably the evolution of such modularity would have driven the early emergence of new function and is an important element in understanding complex molecular machines.

The laboratory evolution of the B6.61 ribozyme models this incremental expansion in enzyme functionality and was recently explored by characterizing the two domains of the polymerase. The ligase core and accessory domain of B6.61 are modular domains that fold independently, yet act cooperatively, to extend a *trans* primer-template substrate (Cheng et al., unpubl). Both domains fold well in *trans* and do not, based on chemical probing, appear to change their folds when transcribed separately. Notably nucleotide incorporation was completely abolished by removal of the accessory domain, but as shown in Figure 4A, adding the ligase and accessory domain in *trans* resulted in polymerization

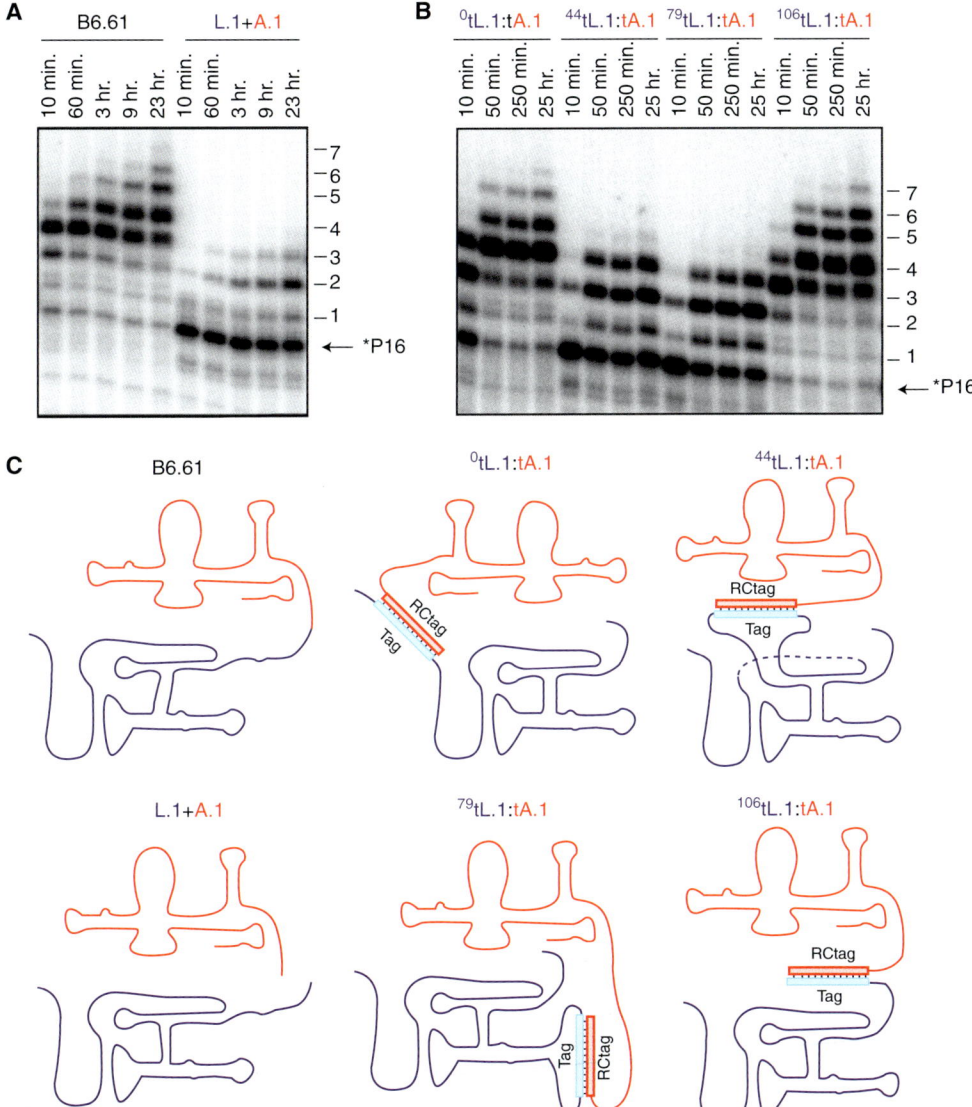

Figure 4. RNA polymerase ribozyme modularity: The B6.61 Ligase core and accessory domain in a range of equimolar (0.5 μm) contexts. (*A*) Comparison of polymerization activity of *trans* bimolecular constructs (L.1 + A.1) with that of B6.61. (*B*) Polymerization activity assay of different assemblies of the two hybridized *trans* bimolecular constructs as shown in panel C. (*C*) Cartoon schematic of the four bimolecular assemblies.

(last 5 lanes), albeit ~100-fold slower than unimolecular construct (first 5 lanes). Remarkably the two domains can be joined together by a range of hybridization sites some of which retain nearly wild-type levels of activity (Fig. 4 B,C). Orientation appears crucial based on the existence of a specific cross-linking pattern between the two domains that suggest that weak tertiary interactions are responsible for the correct positioning of the two domains with respect to each other. Consistent with the three-dimensional structure of the class I ligase ribozyme (Bergman et al. 2004; Shechner et al. 2009), the 5' and 3' tethering sites are close together and on the same side of the ligase core as the active site, whereas tethering the accessory domain to the two internal loops of the ligase core that are distal to the active site lowers polymerization activity. This is consistent with the accessory domain being positioned over the class I ligase active site so at to constrain the primer–template and presumably nucleotide triphosphates in the vicinity of the enzyme active site. Most interestingly, a ten-fold increase in polymerization rate was observed by tethering the primer-template to the polymerase. As this increase was dependent on the accessory domain being present, improvements in the accessory domain or the selection of new complementary functional modules might be expected to further improve RNA polymerase.

FUTURE RESEARCH DIRECTIONS: STRAND DISPLACEMENT AND INITIATION

The B6.61 polymerase fills in a primer-template to produce a RNA duplex. Currently, this dsRNA product is not a substrate for the polymerase, making it impossible to construct a replicating system of the sort sketched in Figure 2. If, however, this polymerase could be evolved to recognize and transcribe a RNA duplex, then a self-consistent RNA system could be constructed that requires only nucleotide triphosphates to replicate. Two critical challenges remain to be achieved: First and most problematic, a mechanism must be developed to allow the initiation of polymerization from a duplex RNA. Second a mechanism to expose the template strand of the duplex must be engineered. These two functionalities would allow the polymerase ribozyme to transcribe one strand of the duplex, resulting in a total of three RNA strands, one of which is by necessity must be single-stranded (Fig. 2). Because this ssRNA can hybridize to an antisense strand produced by transcription in the opposite direction a simple and effective mechanism for copying genomic elements would be built into this form of transcription. Asymmetric transcription from any particular genomic element would therefore not only provide a mechanism to copy the RNA genome but would naturally provide a source of ssRNA that could be ligated together to form larger more complex functional RNAs (Fig. 2).

RdRP Initiation: A Protein Point of View

Biologically, there are two main types (Ng et al. 2008) of initiation: primer-independent (de novo synthesis) and primer-dependent. As shown in Figure 5A, de novo synthesis begins with the base pairing of an initiation nucleotide triphosphate (often this is GTP, Ng et al. 2008) to the 3' end of the RNA template. Because this interaction is not sufficient to stabilize the single base pair thus formed, other interactions such as base stacking of aromatic protein residues with the initiation nucleotide are typically employed (Butcher et al. 2001). The extension of this first nucleotide by additional templated nucleotides can then trigger a rearrangement of protein structure that allows the formation of a stable elongation complex via an abortive cycling process (Ng et al. 2008). Primer-dependent mechanisms of initiation use a hydroxyl from a nearby source for nucleophilic attack on the triphosphate of the incoming nucleotide. Figure 5B–D depict the mechanism employed by three viral RdRPs in which a free hydroxyl is derived from: (1) a nearby protein residue (Paul et al. 1998), (2) a short oligonucleotide originating either from abortive cycling (McClure 1985) or from a cleaved mRNA (Hagen et al. 1995), or (3) by folding back a RNA template so as to produce a hairpin and a free 3'-hydroxyl that can be extended by a

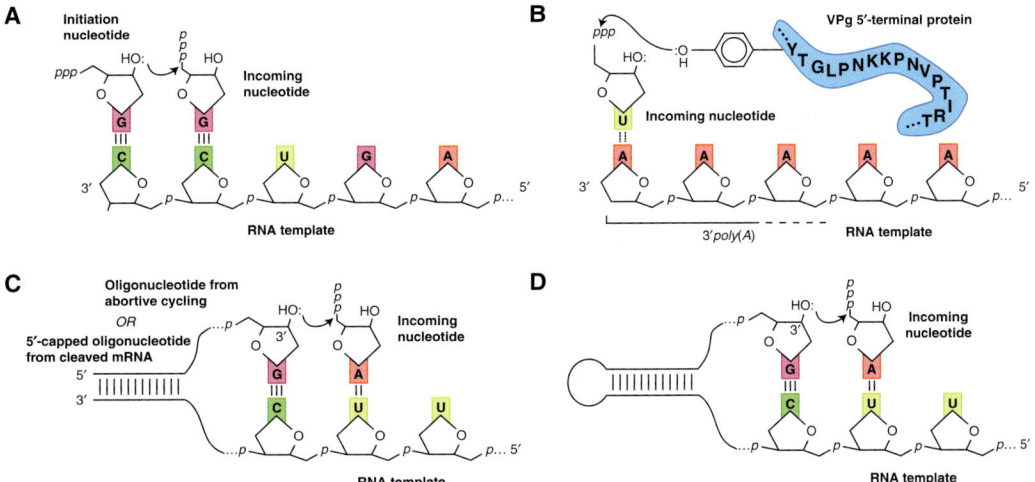

Figure 5. Comparison of initiation mechanisms. (A) Primer-independent (de novo) initiation B, C, and D—Primer-dependent initiation strategies: (B) "Borrowing" a hydroxyl from a nearby protein residue (C) Use of a short oligonucleotide from abortive cycling in de novo initiation or from a cleaved mRNA; (D) Template folds back to form a stable hairpin that is then extended. Adapted from (Paul et al. 1998; van Dijk et al. 2004 and reprinted with permission from The Journal of General Virology ©2004; Ng et al. 2008 and reprinted with express permission from the authors).

suitable polymerase (Laurila et al. 2002; Laurila et al. 2005).

In all of these examples the template strand must first be exposed so that a short primer can either be synthesized (primer-independent) or provided by an external source (primer dependent). While transcribing from a RNA duplex it would be essential that the duplex be first melted by some form of helicase activity. There are several ways to design a solution to this problem using RNA, although none appear particularly straightforward. The simplest approach might be to make use of the intrinsic properties of RNA to pry open the end of the duplex. This could be achieved by first making a temporary and reversible anchor to the 3′ terminus of the template strand that would ultimately orient the template strand correctly with respect to the polymerase's active site. This ligation might be expected to mildly destabilize the terminal duplex and could be dramatically enhanced by appending a binding motif that specifically recognizes either the 5′ triphosphate on the RNA duplex or equally a 5′ cap (Huang and Yarus 1997; Zaher et al. 2006). Correctly positioned this combination of a covalent anchor on the 3′ terminus of the template strand together with tight binding to the 5′ terminus would serve to unwind the RNA duplex in a fashion sufficient to allow either primer independent or dependent initiation. After several nucleotide incorporations the polymerase structure would change, induced by forces acting through the site of RNA ligation. This change in geometry would then mediate changes in the ligation site causing RNA cleavage to be favored instead of ligation and triggering the release of the template strand from the polymerase (potentially the polymerase could use a cyclic phosphate to mediate this interaction in analogy with the reversible ligation and cleavage mediated by the hammerhead ribozyme). This same structural change would hopefully distort the geometry of the 5′ binding motif, releasing the emerging single strand and allowing a functional elongation complex to form.

Solving the Strand Displacement Problem

The initiation just described would not lead to a transcriptional complex that is directly analogous to a biological transcriptional complex.

DNA dependent RNAPs form a stable transcription bubble complex (Yin and Steitz 2004; Mooney et al. 2005; Borukhov and Nudler 2008) that makes use of a RNA exit tunnel to precisely position a RNA–DNA heteroduplex with respect to the polymerase active site. A transcription bubble where the two downstream strands of DNA are reunited upstream of the RNA–DNA heteroduplex is only required for transcription of mixed polymer types and makes DNA dependent RNA transcription more complex. If RNA is being transcribed from a RNA duplex simpler topological nucleic acid structures can be used and the freshly synthesized strand need not be displaced, but instead can remain hybridized to the template strand. This change results in the original non-template strand being displaced and makes elongation simpler as a consequence.

Thermodynamically the free energy of the states sampled by polymerization should be nearly equivalent at any particular extension:

$$\Delta G°(n,m)_{complex} = \Delta G°(n,m)_{RNA\text{-}RNA} + \Delta G°(n,m)_{RNA\text{-}polymerase} = 0$$

where n and m are the nth and mth nucleotide extension, RNA–RNA represents the interactions formed in the substrate strands, and RNA-polymerase the interactions made specifically by the polymerase with the substrate. As the formation of a base pair in the downstream product helix is on average compensated by the disruption of an upstream base pair, $\Delta G°_{RNA\text{-}RNA}$ should remain small except for pathological sequences. Ideally $\Delta G°_{RNA\text{-}polymerase}$ can be made small as well by ensuring that the polymerase is only topologically associated with substrate RNA strands and does not make any specific contacts with the replicating nucleic acid. This can be achieved by building RNA domains onto the polymerase that are able to surround but not bind to either the upstream or downstream helixes. This clamp would provide a mechanism to ensure that the polymerase always stays associated with its template making it much more processive (Fig. 6, Accessory domain). Equally important the emerging

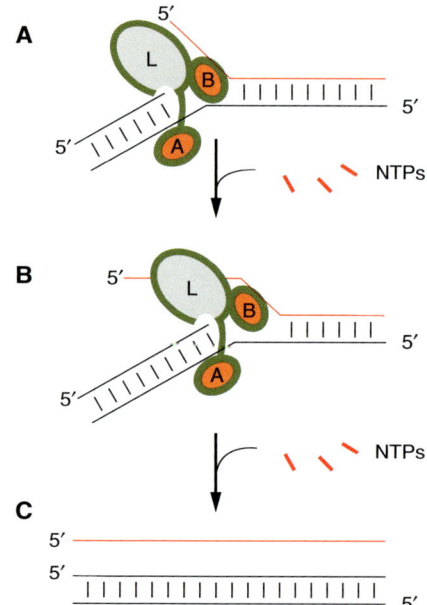

Figure 6. Evolving a processive strand-displacing RNA polymerase ribozyme from the B6.61 RNA polymerase. (A) After formation of an initiation complex, which exposes the template strand, a short RNA primer sequence is either synthesized de novo or allowed to hybridize. (B) Elongation requires that nucleotides immediately upstream of the site of nucleotide incorporation are single-stranded. If interactions between the ligase core (L) and the accessory domain (Circled A) can be improved so that the freshly synthesized strand together with template are held robustly and nonspecifically the polymerase should become much more processive. If in addition these two domains can be rigidly connected to a third domain (Circled B) that ensures dehybridization of the incoming duplex, transcription from dsRNA as well as ssRNA templates would be possible. This elongation complex should remain invariant in shape as it slides along the template. (C) Transcription ends with the release of one strand from the original duplex RNA and the duplex RNA being regenerated with a freshly synthesized RNA strand.

single-strand must be prevented from hybridizing to the template strand of the duplex immediately upstream of the site of nucleotide incorporation. A helix placed transversely between the template strand and the strand being displaced by polymerization might achieve this aim provided it was rigidly anchored to the duplex sliding clamp (Fig. 6, third domain). Failure to achieve strand displacement would prevent polymerization and is essential to success. Similarly, improving the stability of the B6.61 accessory domain and ligase core by further engineering and in vitro selection could well result in a suitable duplex sliding clamp. This clamp could then be selected to function with the new strand displacement motif by selecting for polymerization from artificially prepared mimics of a replication complex (Fig. 7).

Selecting polymerases for strand displacement should be the first priority for improving RNA dependent polymerase ribozymes. Such an enzyme even if it lacks the ability to initiate could still allow a number of very interesting experiments. For example, with the appropriate primers such a polymerase could mediate rolling circle replication of an RNA genome. If this circular genome contained a hammerhead ribozyme motif together with the polymerase ribozyme sequence, then primed rolling circle transcription of both plus and minus strands would allow the creation of a simple autocatalytic system requiring only primer and nucleotides to replicate. This would make possible the artificial continuous evolution of the RNA polymerase itself and with suitable selection

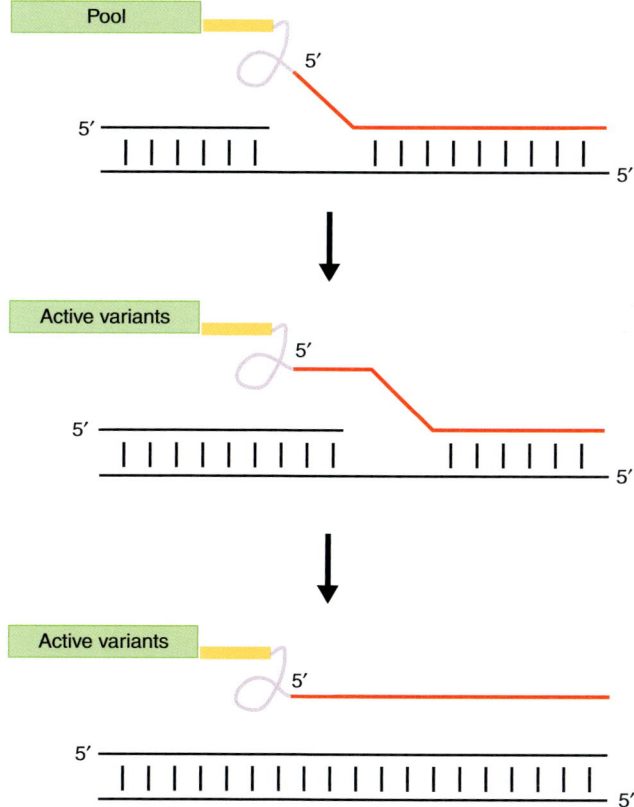

Figure 7. Selecting for strand displacement in an RNA polymerase ribozyme. A pool of potential strand displacing RNA polymerase ribozymes is tethered by a long flexible linker to a RNA (red strand) that is partially complementary to the template strand (bottom black strand). The unhybridized nucleotides in the red strand are not able to hybridize to the template strand resulting in the formation of an elongation complex mimic. Transcription by the tethered polymerase that is able to displace the red strand results in the polymerase disassociating from the extending primer template duplex (black strands). Freed polymerases can then be differentially recovered and amplified.

pressure might rapidly lead to the evolution of initiation strategies that allow the controlled expression of other potentially metabolically relevant RNAs.

CONCLUSIONS

Based on a number of independent lines of laboratory research RNA appears well suited to have served as the first replicative polymer on this planet. More importantly the catalytic and genetic properties intrinsic to RNA might in the near future make possible the first artificial forms of life. Such systems would allow the exploration of many important and outstanding questions concerning the robustness and early evolution of life and provide critical insight into the difficulty of evolving life elsewhere in the universe.

REFERENCES

Bartel DP. 1999. Re-creating an RNA replicase. In *RNA World*, 2nd ed. Cold Spring Harber Laboratory Press, Cold Spring Harber, NY.

Bartel DP, Szostak JW. 1993. Isolation of new ribozymes from a large pool of random sequences. *Science* **261**: 1411–1418.

Bartel DP, Unrau PJ. 1999. Constructing an RNA world. *Trends Cell Biol* **9**: M9–M13.

Benner SA, Ellington AD, Tauer A. 1989. Modern metabolism as a palimpsest of the RNA world. *Proc Natl Acad Sci* **86**: 7054–7058.

Bergman NH, Lau NC et al. 2004. The three-dimensional architecture of the class I ligase ribozyme. *RNA* **10**: 176–184.

Borukhov S, Nudler EC. 2008. RNA polymerase: The vehicle of transcription. *Trends Microbiol* **16**: 126–134.

Butcher SJ, Grimes JM, Makeyev EV, Bamford DH, Stuart DI. 2001. A mechanism for initiating RNA-dependent RNA polymerization. *Nature* **410**: 235–240.

Chan CM, Zhou C, Huang RH. 2009. Reconstituting bacterial RNA repair and modification in vitro. *Science* **326**: 247.

Costanzo G, Pino S, Ciciriellli F, Di Mauro E. 2009. Generation of long RNA chains in water. *J Biol Chem* **284**: 33206–33216.

Crick FH. 1958. On protein synthesis. *Symp Soc Exp Biol* **12**: 138–163.

Dinner AR, Blackburn GM, Karplus M. 2001. Uracil-DNA glycosylase acts by substrate autocatalysis. *Nature* **413**: 752–755.

Draper WE, Hayden EJ, Lehman N. 2008. Mechanisms of covalent self-assembly of the Azoarcus ribozyme from four fragment oligonucleotides. *Nucleic Acids Res* **36**: 520–531.

Ekland EH, Bartel DP. 1995. The secondary structure and sequence optimization of an RNA ligase ribozyme. *Nucleic Acids Res* **23**: 3231–3238.

Ekland EH, Bartel DP. 1996. RNA-catalysed RNA polymerization using nucleoside triphosphates. *Nature* **382**: 373–376.

Ekland EH, Szostak JW, Bartel DP. 1995. Structurally complex and highly active RNA ligases derived from random RNA sequences. *Science* **269**: 364–370.

Ellington AD, Szostak JW. 1990. In vitro selection of RNA molecules that bind specific ligands. *Nature* **346**: 818–822.

Ellington AD, Chen X, Robertson M, Syrett A. 2009. Evolutionary origins and directed evolution of RNA. *Int J Biochem Cell Biol* **41**: 254–265.

Ferris JP, Hill AR Jr, Lui R, Orgel LE. 1996. Synthesis of long prebiotic oligomers on mineral surfaces. *Nature* **381**: 59–61.

Fuller WD, Sanchez RA, Orgel LE. 1972. Studies in prebiotic synthesis. VI. Synthesis of purine nucleosides. *J Mol Biol* **67**: 25–33.

Gago S, Elena SF, Flores R, Sanjuan R. 2009. Extremely high mutation rate of a hammerhead viroid. *Science* **323**: 1308.

Gesteland RF, Cech T, Atkins JF. 2006. The RNA World: The nature of modern RNA suggests a prebiotic RNA world. Cold Spring Harbor Laboratory Press, NY.

Gilbert SD. 1986. The RNA world. *Nature* **319**: 618.

Hagen M, Tiley L, Chung TD, Krystal M. 1995. The role of template-primer interactions in cleavage and initiation by the influenza virus polymerase. *J Gen Virol* **76**: 603–611.

Hayden EJ, Lehman N. 2006. Self-assembly of a group I intron from inactive oligonucleotide fragments. *Chem Biol* **13**: 909–918.

Hayden EJ, von Kiedrowski G, Lehman N. 2008. Systems chemistry on ribozyme self-construction: Evidence for anabolic autocatalysis in a recombination network. *Angew Chem Int Ed Engl* **47**: 8424–8428.

Hill A, Orgel LE. 2002. Synthesis of adenine from HCN tetramer and ammonium formate. *Orig Life Evol Biosph* **32**: 99–102.

Huang F, Yarus M. 1997. Versatile 5′ phosphoryl coupling of small and large molecules to an RNA. *Proc Natl Acad Sci* **94**: 8965–8969.

Ikawa Y, Tsuda K, Matsumura S, Inoue T. 2004. De novo synthesis and development of an RNA enzyme. *Proc Natl Acad Sci* **101**: 13750–13755.

Jaeger L, Wright MC, Joyce GF. 1999. A complex ligase ribozyme evolved in vitro from a group I ribozyme domain. *Proc Natl Acad Sci* **96**: 14712–14717.

Johnston WK, Unrau PJ, Lawrence MS, Glasner ME, Bartel DP. 2001. RNA-catalyzed RNA polymerization: Accurate and general RNA-templated primer extension. *Science* **292**: 1319–1325.

Joyce GF. 1989. RNA evolution and the origins of life. *Nature* **338**: 217–224.

Joyce GF. 2002. The antiquity of RNA-based evolution. *Nature* **418:** 214–221.

Joyce GF, Orgel LE. 2006. Progress towards understanding the origin of the RNA world. In *RNA World*, 3rd ed. Cold Spring Harbor Laboratory Press, Cold Spring Harber, NY.

Kim DE, Joyce GF. 2004. Cross-catalytic replication of an RNA ligase ribozyme. *Chem Biol* **11:** 1505–1512.

Kiss T. 2002. Small nucleolar RNAs: An abundant group of noncoding RNAs with diverse cellular functions. *Cell* **109:** 145–148.

Konarska MM, Sharp PA. 1989. Replication of RNA by the DNA-dependent RNA polymerase of phage T7. *Cell* **57:** 423–431.

Krasilnikov AS, Xiao Y, Pan T, Mondragon A. 2004. Basis for structural diversity in homologous RNAs. *Science* **306:** 104–107.

Kunkel TA, Bebenek K. 2000. DNA replication fidelity. *Annu Rev Biochem* **69:** 497–529.

Lau MW, Unrau PJ. 2009. A promiscuous ribozyme promotes nucleotide synthesis in addition to ribose chemistry. *Chem Biol* **16:** 815–825.

Lau MW, Cadieux KE, Unrau PJ. 2004. Isolation of fast purine nucleotide synthase ribozymes. *J Am Chem Soc* **126:** 15686–15693.

Laurila MR, Makeyev EV, Bamford DH. 2002. Bacteriophage φ6 RNA-dependent RNA polymerase: Molecular details of initiating nucleic acid synthesis without primer. *J Biol Chem* **277:** 17117–17124.

Laurila MR, Salgado PS, Stuart DI, Grimes JM, Bamford DH. 2005. Back-priming mode of φ6 RNA-dependent RNA polymerase. *J Gen Virol* **86:** 521–526.

Lincoln TA, Joyce GF. 2009. Self-sustained replication of an RNA enzyme. *Science* **323:** 1229–1232.

Madison-Antenucci S, Grams J, Hajduk SL. 2002. Editing machines: the complexities of trypanosome RNA editing. *Cell* **108:** 435–438.

Marquez SM, Chen JL, Evans D, Pace NR. 2006. Structure and function of eukaryotic Ribonuclease P RNA. *Mol Cell* **24:** 445–456.

Matera AG, Terns RM, Terns MP. 2007. Non-coding RNAs: Lessons from the small nuclear and small nucleolar RNAs. *Nat Rev Mol Cell Biol* **8:** 209–220.

McClure WR. 1985. Mechanism and control of transcription initiation in prokaryotes. *Annu Rev Biochem* **54:** 171–204.

Miller OJ, Bernath K, Agresti JJ, Amitai G, Kelly BT, Mastrobattista E, Taly V, Magdassi S, Tawfik DS, Griffiths AD. 2006. Directed evolution by in vitro compartmentalization. *Nat Methods* **3:** 561–570.

Monnard PA, Kanavarioti A, Deamer DW. 2003. Eutectic phase polymerization of activated ribonucleotide mixtures yields quasi-equimolar incorporation of purine and pyrimidine nucleobases. *J Am Chem Soc* **125:** 13734–13740.

Mooney RA, Darst SA, Landick R. 2005. Sigma and RNA polymerase: an on-again, off-again relationship? *Mol Cell* **20:** 335–345.

Ng KK, Arnold JJ, Cameron CE. 2008. Structure-function relationships among RNA-dependent RNA polymerases. *Curr Top Microbiol Immunol* **320:** 137–156.

Nissen P, Hansen J, Ban N, Moore PB, Steitz TA. 2000. The structural basis of ribosome activity in peptide bond synthesis. *Science* **289:** 920–930.

Orgel LE. 1998. The origin of life—a review of facts and speculations. *Trends Biochem Sci* **23:** 491–495.

Orgel LE. 2004. Prebiotic chemistry and the origin of the RNA world. *Crit Rev Biochem Mol Biol* **39:** 99–123.

Pannucci JA, Haas ES, Hall TA, Harris JK, Brown JW. 1999. RNase P RNAs from some Archaea are catalytically active. *Proc Natl Acad Sci* **96:** 7803–7808.

Paul AV, van Boom JH et al. 1998. Protein-primed RNA synthesis by purified poliovirus RNA polymerase. *Nature* **393:** 280–284.

Powner MW, Gerland B et al. 2009. Synthesis of activated pyrimidine ribonucleotides in prebiotically plausible conditions. *Nature* **459:** 239–242.

Qiao F, Cech TR. 2008. Triple-helix structure in telomerase RNA contributes to catalysis. *Nat Struct Mol Biol* **15:** 634–640.

Ricardo A, Szostak JW. 2009. Origin of life on earth. *Sci Am* **301:** 54–61.

Ricardo A, Carrigan MA, Olcott AN, Benner SA. 2004. Borate minerals stabilize ribose. *Science* **303:** 196.

Robertson MP, Ellington AD. 1999. In vitro selection of an allosteric ribozyme that transduces analytes to amplicons. *Nat Biotechnol* **17:** 62–66.

Robertson DL, Joyce GF. 1990. Selection in vitro of an RNA enzyme that specifically cleaves single-stranded DNA. *Nature* **344:** 467–468.

Robertson MP, Miller SL. 1995. An efficient prebiotic synthesis of cytosine and uracil. *Nature* **375:** 772–774.

Robertson MP, Scott WG. 2007. The structural basis of ribozyme-catalyzed RNA assembly. *Science* **315:** 1549–1553.

Shechner DM, Grant RA et al. 2009. Crystal structure of the catalytic core of an RNA-polymerase ribozyme. *Science* **326:** 1271–1275.

Soll D, Nishimura S, Moore PB. 2001. *RNA*. Pergamon Press.

Tawfik DS, Griffiths AD. 1998. Man-made cell-like compartments for molecular evolution. *Nat Biotechnol* **16:** 652–656.

Tucker BJ, Breaker RR. 2005. Riboswitches as versatile gene control elements. *Curr Opin Struct Biol* **15:** 342–348.

Tuerk C, Gold LC. 1990. Systematic evolution of ligands by exponential enrichment: RNA ligands to bacteriophage T4 DNA polymerase. *Science* **249:** 505–510.

Unrau PJ, Bartel DP. 1998. RNA-catalysed nucleotide synthesis. *Nature* **395:** 260–263.

Unrau PJ, Bartel DP. 2003. An oxocarbenium-ion intermediate of a ribozyme reaction indicated by kinetic isotope effects. *Proc Natl Acad Sci* **100:** 15393–15397.

van Dijk AA, Makeyev EV, Bamford DH. 2004. Initiation of viral RNA-dependent RNA polymerization. *J Gen Virol* **85:** 1077–1093.

Voytek SB, Joyce GF. 2007. Emergence of a fast-reacting ribozyme that is capable of undergoing continuous evolution. *Proc Natl Acad Sci* **104:** 15288–15293.

Wang Y, Prosen DE, Mei L, Sullivan JC, Finney M, Vander Horn PB. 2004. A novel strategy to engineer DNA polymerases for enhanced processivity and improved performance in vitro. *Nucleic Acids Res* **32:** 1197–1207.

Werner F. 2007. Structure and function of archaeal RNA polymerases. *Mol Microbiol* **65:** 1395–1404.

White HBIII. 1976. Coenzymes as fossils of an earlier metabolic state. *J Mol Evol* **7:** 101–104.

Wilson DS, Szostak JW. 1999. In vitro selection of functional nucleic acids. *Annu Rev Biochem* **68:** 611–647.

Yin YW, Steitz TA. 2002. Structural basis for the transition from initiation to elongation transcription in T7 RNA polymerase. *Science* **298:** 1387–1395.

Yin YW, Steitz TA. 2004. The structural mechanism of translocation and helicase activity in T7 RNA polymerase. *Cell* **116:** 393–404.

Yoshioka W, Ikawa Y, Jaeger L, Inoue T. 2004. A ligase ribozyme obtained from a structured pool. *Nucleic Acids Symp Ser (Oxf)* **48:** 209–210.

Zaher HS, Unrau PJ. 2007. Selection of an improved RNA polymerase ribozyme with superior extension and fidelity. *RNA* **13:** 1017–1026.

Zaher HS, Watkins RA, Unrau PJ. 2006. Two independently selected capping ribozymes share similar substrate requirements. *RNA* **12:** 1949–1958.

Zubay G, Mui T. 2001. Prebiotic synthesis of nucleotides. *Orig Life Evol Biosph* **31:** 87–102.

The Origins of Cellular Life

Jason P. Schrum, Ting F. Zhu, and Jack W. Szostak

Howard Hughes Medical Institute, Department of Molecular Biology and the Center for Computational and Integrative Biology, Massachusetts General Hospital, Boston, Massachusetts 02114

Correspondence: szostak@molbio.mgh.harvard.edu

Understanding the origin of cellular life on Earth requires the discovery of plausible pathways for the transition from complex prebiotic chemistry to simple biology, defined as the emergence of chemical assemblies capable of Darwinian evolution. We have proposed that a simple primitive cell, or protocell, would consist of two key components: a protocell membrane that defines a spatially localized compartment, and an informational polymer that allows for the replication and inheritance of functional information. Recent studies of vesicles composed of fatty-acid membranes have shed considerable light on pathways for protocell growth and division, as well as means by which protocells could take up nutrients from their environment. Additional work with genetic polymers has provided insight into the potential for chemical genome replication and compatibility with membrane encapsulation. The integration of a dynamic fatty-acid compartment with robust, generalized genetic polymer replication would yield a laboratory model of a protocell with the potential for classical Darwinian biological evolution, and may help to evaluate potential pathways for the emergence of life on the early Earth. Here we discuss efforts to devise such an integrated protocell model.

The emergence of the first cells on the early Earth was the culmination of a long history of prior chemical and geophysical processes. Although recognizing the many gaps in our knowledge of prebiotic chemistry and the early planetary setting in which life emerged, we will assume for the purpose of this review that the requisite chemical building blocks were available, in appropriate environmental settings. This assumption allows us to focus on the various spontaneous and catalyzed assembly processes that could have led to the formation of primitive membranes and early genetic polymers, their coassembly into membrane-encapsulated nucleic acids, and the chemical and physical processes that allowed for their replication. We will discuss recent progress toward the construction of laboratory models of a protocell (Fig. 1), evaluate the remaining steps that must be achieved before a complete protocell model can be constructed, and consider the prospects for the observation of spontaneous Darwinian evolution in laboratory protocells. Although such laboratory studies may not reflect the specific pathways that led to the origin of life on Earth, they are proving to be invaluable in uncovering surprising and unanticipated physical processes that help us to reconstruct plausible pathways and scenarios for the origin of life.

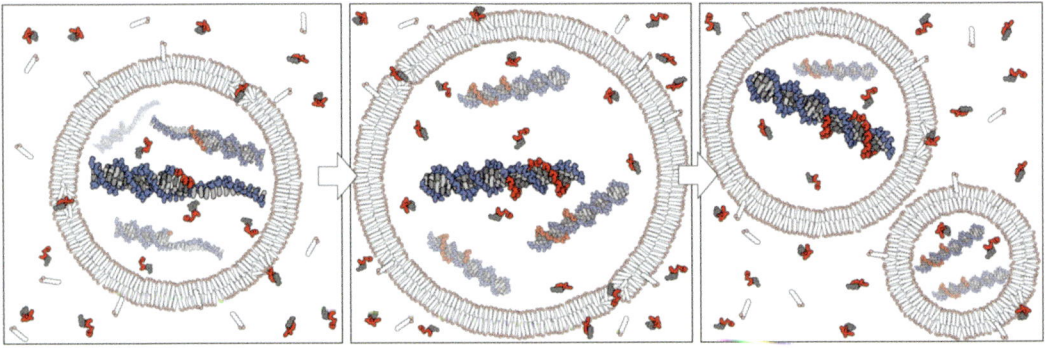

Figure 1. A simple protocell model based on a replicating vesicle for compartmentalization, and a replicating genome to encode heritable information. A complex environment provides lipids, nucleotides capable of equilibrating across the membrane bilayer, and sources of energy (*left*), which leads to subsequent replication of the genetic material and growth of the protocell (*middle*), and finally protocellular division through physical and chemical processes (*right*). (Reproduced from Mansy et al. 2008 and reprinted with permission from Nature Publishing ©2008.)

The term protocell has been used loosely to refer to primitive cells or to the first cells. Here we will use the term protocell to refer specifically to cell-like structures that are spatially delimited by a growing membrane boundary, and that contain replicating genetic information. A protocell differs from a true cell in that the evolution of genomically encoded advantageous functions has not yet occurred. With a genetic material such as RNA (or perhaps one of many other heteropolymers that could provide both heredity and function) and an appropriate environment, the continued replication of a population of protocells will lead inevitably to the spontaneous emergence of new coded functions by the classical mechanism of evolution through variation and natural selection. Once such genomically encoded and therefore heritable functions have evolved, we would consider the system to be a complete, living biological cell, albeit one much simpler than any modern cell (Szostak et al. 2001).

BACKGROUND

Membranes as compartment boundaries

All biological cells are membrane-bound compartments. The cell membrane fulfills the essential function of creating an internal environment within which genetic materials can reside and metabolic activities can take place without being lost to the environment. Modern cell membranes are composed of complex mixtures of amphiphilic molecules such as phospholipids, sterols, and many other lipids as well as diverse proteins that perform transport and enzymatic functions. Phospholipid membranes are stable under a wide range of temperature, pH, and salt concentration conditions. Such membranes are extremely good permeability barriers, so that modern cells have complete control over the uptake of nutrients and the export of wastes through the specialized channel, pump and pore proteins embedded in their membranes. A great deal of complex biochemical machinery is also required to mediate the growth and division of the cell membrane during the cell cycle. The question of how a structurally simple protocell could accomplish these essential membrane functions is a critical aspect of understanding the origin of cellular life.

Vesicles formed by fatty acids have long been studied as models of protocell membranes (Gebicki and Hicks 1973; Hargreaves and Deamer 1978; Walde et al. 1994a). Fatty acids are attractive as the fundamental building block of prebiotic membranes in that they are chemically simpler than phospholipids. Fatty acids with a saturated acyl chain are extremely stable compounds and therefore might have accumulated to significant levels, even given a relatively slow or episodic synthesis. Moreover, the condensation of fatty acids with glycerol to yield the corresponding glycerol esters provides a highly stabilizing membrane component (Monnard et al., 2002). Finally, phosphorylation and the addition of a second acyl chain yields phosphatidic acid, the simplest phospholipid, thus providing a conceptually simple pathway for the transition from primitive to more modern membranes. The prebiotic chemistry leading to the synthesis of fatty acids and other amphiphilic compounds is treated in more detail in Mansy (2010).

The best reason for considering fatty acids as fundamental to the nature of primitive cell membranes is not, however, their chemical simplicity. Rather, fatty-acid molecules in membranes have dynamic properties that are essential for both membrane growth and permeability. Because fatty acids are single chain amphiphiles with less hydrophobic surface area than phospholipids, they assemble into membranes only at much higher concentrations. This equilibrium property is mirrored in their kinetics: Fatty acids are not as firmly anchored within the membrane as phospholipids; they enter and leave the membrane on a time scale of seconds to minutes (Chen and Szostak 2004). Fatty acids can also exchange between the two leaflets of a bilayer membrane on a subsecond time scale. Rapid flip-flop is essential for membrane growth when new amphiphilic molecules are supplied from the environment. New molecules enter the membrane primarily from the outside leaflet, and flip-flop allows the inner and outer leaflet areas to equilibrate, leading to uniform growth.

Considering that protocells on the early Earth did not, by definition, contain any complex biological machinery, they must have relied on the intrinsic permeability properties of their membranes. Membranes composed of fatty acids are in fact reasonably permeable to small polar molecules and even to charged species such as ions and nucleotides (Mansy et al. 2008). This appears to be largely a result of the ability of fatty acids to form transient defect structures and/or transient complexes with charged solutes, which facilitate transport across the membrane. The subject of the permeability of fatty-acid based membranes is dealt with in greater detail by Mansy (2010).

Prebiotic vesicles were almost certainly composed of complex mixtures of amphiphiles. Amphiphilic molecules isolated from meteorites (Deamer 1985; Deamer and Pashley 1989) as well as those synthesized under simulated prebiotic conditions (McCollum et al. 1999; Dworkin et al. 2001; Rushdi and Simoneit 2001) are highly heterogeneous, both in terms of acyl chain length and head group chemistry. Membranes composed of mixtures of amphiphiles often have superior properties to those composed of single pure species. For example, mixtures of fatty acids together with the corresponding alcohols and/or glycerol esters generate vesicles that are stable over a wider range of pH and ionic conditions (Monnard et al. 2002), and are more permeable to nutrient molecules including ions, sugars and nucleotides (Chen et al. 2004; Sacerdote and Szostak 2005; Mansy et al. 2008). This is in striking contrast to the apparent requirement for homogeneity in the nucleic acids, where even low levels of modified nucleotides can be destabilizing or can block replication.

RECENT RESULTS

Pathways for Vesicle Growth

Fatty-acid vesicle growth has been shown to occur through at least two distinct pathways: growth through the incorporation of fatty acids from added micelles, and growth through fatty-acid exchange between vesicles. The growth of membrane vesicles from micelles has been observed following the addition of micelles or fatty-acid precursors to pre-formed vesicles (Walde et al. 1994a, 1994b; Berclaz et al. 2001). When initially alkaline fatty-acid micelles are mixed with a buffered solution at a lower pH, the micelles become thermodynamically unstable. As a consequence, the fatty-acid molecules can either be incorporated into pre-existing membranes, leading to growth (Berclaz et al. 2001), or can self-assemble into new vesicles

(Blochliger et al. 1998; Luisi et al. 2004). These pioneering studies were done by cryo-TEM, which does not allow growth to be followed in real time, and by light scattering, which is difficult to interpret in the case of samples with heterogeneous size distributions. We therefore adapted a fluorescence assay based on FRET (Förster resonance energy transfer) to measure changes in membrane area in real time. This assay is based on the distance dependence of energy transfer between donor and acceptor fluorescent dyes; thus when a membrane grows in area by incorporating additional fatty-acid molecules, the dyes are diluted and the efficiency of FRET decreases. Studies on small (typically 100 nm in diameter) unilamellar fatty-acid vesicles using this assay showed that the slow addition of fatty-acid micelles led to vesicle growth with an efficiency of $\sim$90% (Hanczyc et al. 2003).

The real-time FRET assay allowed for a kinetic dissection of the growth process, revealing a surprisingly complex series of events after the rapid addition of micelles (Chen and Szostak 2004). Two major processes were observed. The first fast phase resulted in membrane area growth that was limited to $\sim$40% increase in area, independent of the amount of added micelles. A second much slower phase led to a further increase in membrane area that varied with the amount of added micelles. We interpreted the fast phase as reflecting the rapid assembly of a layer of adhering micelles around the pre-formed vesicles, with rapid monomer exchange resulting in the efficient incorporation of this material into the pre-formed membrane. We interpreted the slow phase as the consequence of micelle–micelle interactions leading to the assembly of intermediate structures that could partition between two pathways—with some monomers dissociating and contributing to membrane growth and the remainder ultimately assembling into new membrane vesicles. Although these interpretations are consistent with our data, the experiments are rather indirect, and further exploration of the mechanism of membrane growth is certainly desirable.

A second, distinct pathway for vesicle growth involves fatty-acid exchange between vesicles. Under certain conditions this exchange can lead to growth of a subpopulation of vesicles at the expense of their surrounding neighbors. Within populations of osmotically relaxed vesicles, such exchange processes do not result in significant changes in size distribution with time. Similarly, a population of uniformly osmotically swollen vesicles does not change in size distribution, but such vesicles are in equilibrium with a lower solution concentration of fatty acids because the tension in the membrane of the swollen vesicles makes it more energetically favorable for fatty-acid molecules to reside in membrane. When osmotically swollen vesicles are mixed with osmotically relaxed (isotonic) vesicles, rapid fatty-acid exchange processes result in growth of the swollen vesicles and corresponding shrinkage of the relaxed vesicles (Chen et al. 2004). Because vesicles can be osmotically swollen as a result of the encapsulation of high concentrations of nucleic acids such as RNA, this process allows for the growth of vesicles containing genetic polymers at the expense of empty vesicles (or vesicles that contain less internal nucleic acid). Because faster replication would increase the internal nucleic acid concentration, this pathway of competitive vesicle growth provides the potential for a direct physical link between the rate of replication of an encapsulated genetic polymer and the rate of growth of the protocell as a whole.

Assuming that the division of osmotically swollen vesicles could occur either stochastically or at some threshold size, protocells that developed some heritable means of faster replication and growth would have a shorter cell cycle, on average, and would therefore gradually take over the population. This simple physical mechanism might therefore lead to the emergence of Darwinian evolution by competition at the cellular level. However, if replication is limited by the rapid reannealing of complementary strands (see later discussion), it may be difficult to reach osmotically significant concentrations. Furthermore, osmotically swollen vesicles are intrinsically difficult to divide owing to the energetic cost of reducing the volume of a spherical vesicle to that of two daughter vesicles of the same total surface area. One possibility is that osmotically driven competitive growth might alternate with the faster membrane growth that follows micelle addition. If new fatty-acid material was only available sporadically, rapid membrane growth might follow an influx of fresh fatty acids, facilitating division (see following discussion).

All of the experiments discussed earlier were done with small unilamellar vesicles prepared by extrusion through 100 nm pores in filters. In contrast, fatty-acid vesicles that form spontaneously by rehydrating dry fatty-acid films tend to be several microns in diameter and multilamellar (Hargreaves and Deamer 1978; Hanczyc et al. 2003). Such large multilamellar fatty-acid vesicles are so heterogeneous that quantitative studies of growth and division are difficult. We have recently developed a simple procedure for the preparation of micron-sized, monodisperse (homogeneous in size) multilamellar vesicles by large-pore dialysis (Zhu and Szostak 2009b). The preparation of large monodisperse multilamellar vesicles has allowed us to directly observe an unusual mode of vesicle growth (Zhu and Szostak 2009a). We showed that feeding a micron-sized multilamellar fatty-acid vesicle with fatty-acid micelles results in the formation of a thin membranous protrusion which extends from the side of the initially spherical parental vesicle. Over time, this thin membrane tubule elongates and thickens, gradually incorporating more and more of the parental vesicle, until eventually the entire vesicle is transformed into a long, hollow threadlike vesicle (Fig. 2). This pathway occurs with vesicles ranging in size from 1 to at least 10 μm in diameter, composed of a variety of different fatty acids and related amphiphiles. Only multilamellar vesicles grow in this manner and only when vesicle volume increases slowly (relative to surface area growth) because of a relatively impermeable buffer solute. Confocal microscopy has provided insight into the mechanism of this mode of growth: The outermost membrane layer grows first, and because there is little volume between it and the next membrane layer, and that volume cannot increase on the same time scale, the extra membrane area is forced into the form of a thin tubule. Over time, this tubule grows, and as a result of poorly understood exchange processes, the entire original vesicle is ultimately transformed into a long thread-like hollow vesicle (Fig. 3).

Pathways for Vesicle Division

Vesicle division by the extrusion of large vesicles through small pores is a way in which mechanical energy can be used to drive division (Hanczyc et al. 2003). Vesicle growth by micelle feeding followed by division by extrusion can

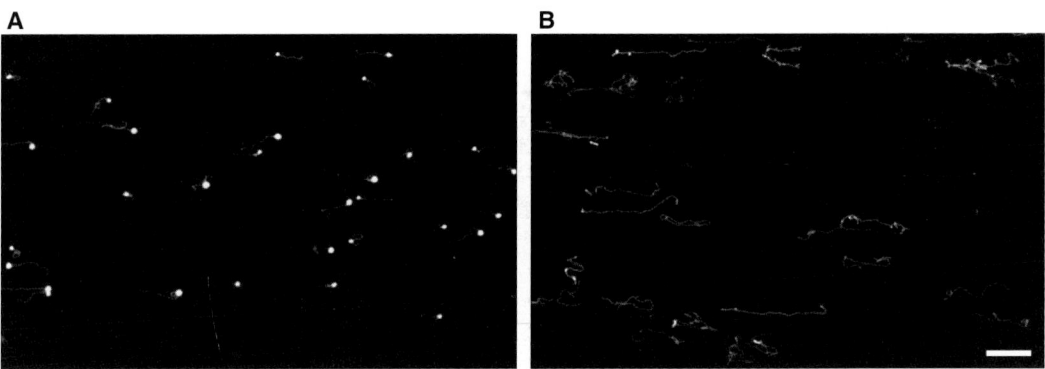

Figure 2. Vesicle shape transformations during growth. All vesicles are labeled with 2 mM encapsulated HPTS, a water-soluble fluorescent dye, in their internal aqueous space. (A) 10 min and (B) 30 min after the addition of five equivalents of oleate micelles to oleate vesicles (in 0.2 M Na-bicine, pH 8.5). Scale bar: 50 μm. (Reproduced from Zhu and Szostak 2009a and reprinted with permission from ACS Publications ©2009.)

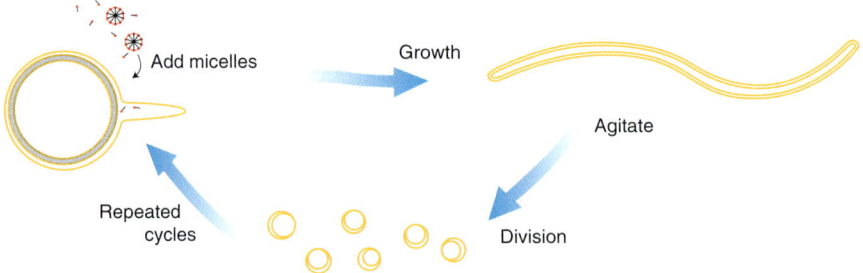

Figure 3. Schematic diagram of coupled vesicle growth and division. (Reproduced from Zhu and Szostak 2009a and reprinted with permission from the Journal of the American Chemical Society ©2009.)

be performed repetitively, resulting in cycles of growth and division in which both membrane material and vesicle contents are distributed to daughter vesicles in each cycle. However, division by extrusion results in the loss of 30%–40% of the encapsulated vesicle contents to the environment during each cycle (Hanczyc et al. 2003; Hanczyc and Szostak 2004). Most of this loss is a result of the unavoidable geometric constraint of dividing a spherical vesicle into two spherical (or subspherical) daughter vesicles with conservation of surface area; some additional loss may occur as a result of pressure-induced membrane rupture. Although extrusion is a useful laboratory model for vesicle division, an analogous extrusion process appears unlikely to occur in a prebiotic scenario on the early Earth because vesicle extrusion from the flow of suspended vesicles through a porous rock would require both the absence of any large pores or channels and a very high pressure gradient (Zhu and Szostak 2009a).

The above problems stimulated a search for a more realistic pathway for vesicle division. The possible spontaneous division of small unilamellar vesicles after micelle addition has been reported (Luisi et al. 2004; Luisi 2006), and electron microscopy has revealed structures that are possible intermediates in growth and division, notably pairs of vesicles joined by a shared wall (Stano et al. 2006). However, the mechanism of the proposed division as well as the nature of the energetic driving force remain unclear. Additional studies are required to clearly distinguish between the vesicle-stimulated assembly of new vesicles, and the more biologically relevant processes of growth and division.

We have recently found that the growth of large multilamellar vesicles into long threadlike vesicles, described above, provides a pathway for coupled vesicle growth and division (Zhu and Szostak 2009a). The long threadlike vesicles are extremely fragile, and divide spontaneously into multiple daughter vesicles in response to modest shear forces. In an environment of gentle shear, growth and division become coupled processes because only the filamentous vesicles can divide (Fig. 3). If the initial parental vesicle contains encapsulated genetic polymers such as RNA, these molecules are distributed randomly to the daughter vesicles and are thus inherited. The robustness and simplicity of this pathway suggests that similar processes might have occurred under prebiotic conditions. The mechanistic details of this mode of division remain unclear. One possibility, supported by some microscopic observations, is that the long thin membrane tubules are subject to the "pearling instability" (Bar-Ziv and Moses 1994), and minimize their surface energy by spontaneously transforming from a cylindrical shape to a string of beads morphology. The very thin tether joining adjacent spherical beads may be a weak point that can be easily disrupted by shear forces.

RNA-Catalyzed RNA Replication on the Early Earth and the Modern Laboratory

A core assumption of the RNA world hypothesis is that the RNA genomes of primitive cells were replicated by a ribozyme RNA polymerase

(Gilbert 1986). The idea of RNA-catalyzed RNA replication provides a solution to the apparent paradox of DNA replication catalyzed by proteins that are encoded by DNA. This simplification of early biochemistry gained instant plausibility from the discovery of catalytic RNAs almost 30 years ago (Kruger et al. 1982; Guerrier-Takada et al. 1983). In the time after the discovery of the first ribozymes, the RNA World hypothesis has continued to gain support, most dramatically from the discovery that the ribosome is a ribozyme (Nissen et al. 2000), and that all proteins are assembled through the catalytic activity of the ribosomal RNA. Support has also come from the in vitro evolution of a wide range of new ribozymes, including bona fide RNA polymerases made of RNA (Johnston et al. 2001). On the other hand, no ribozyme polymerase yet comes close to being a self-replicating RNA, like the replicase envisaged as the core of the RNA World biochemistry.

Why has the in vitro evolution of an RNA replicase been so much more difficult than originally expected? It is clear that the problem is not with catalysis of the chemical step, even with catalytically demanding triphosphate substrates. Evolutionarily optimized versions of the Class I ligase carry out multiple-turnover ligation reactions at $>1\text{ s}^{-1}$, with over 50,000 turnovers overnight (Ekland et al. 1995), and optimized versions of the smaller, simpler DSL ligase carry out sustained multiple turn-over ligation reactions at rates $>1/\text{min}$ (Voytek and Joyce 2007). These catalytic rates are more than sufficient to carry out the replication of a 100–200 nt ribozyme in minutes to hours, if these rates could be maintained in the context of a polymerase reaction using monomer substrates. However, even the best available polymerase ribozyme requires 1–2 days to copy 10–20 nucleotides of a template strand, apparently as a consequence of poor binding to both the ribonucleoside triphosphate (NTP) monomers and the primer-template substrate. The need to overcome the electrostatic repulsion between negatively charged NTP and RNA substrates and ribozyme is thought to contribute to the very high Mg^{++} requirement for the polymerase ribozyme (Glasner et al. 2000). Such high levels of Mg^{++} lead to hydrolytic degradation of the ribozyme, and are also not compatible with known fatty-acid based membranes because of crystallization of the fatty acid-magnesium salt. In addition, fatty-acid membranes are almost impermeable to NTPs (Mansy et al. 2008).

The incompatibility between currently available ribozyme polymerases and fatty-acid based vesicles suggests either that early replicases were quite different, or that RNA replication in early cells proceeded in a very different manner. For example, less charged and more activated nucleotides might be easier for a ribozyme polymerase to bind, with little or no Mg^{++}. Many potential leaving groups, such as imidazole, adenine or 1-Me-adenine have been examined in template-directed and nontemplated polymerization reactions, but have not yet been tested as substrates for ribozyme polymerases (Prabahar and Ferris 1997; Huang and Ferris 2006). The tethering of ribozyme and primer-template to hydrophobic aggregates has been examined as a way of increasing local substrate concentration (Müller and Bartel 2008). However, this approach did not lead to a dramatic improvement in the extent of template copying, apparently as a result of ribozyme inhibition at high effective RNA concentrations. It may be possible to evolve polymerases that operate well at high RNA concentrations, but membrane localization by chemical derivatization adds further complexity to a replication pathway because of the need for a specific catalyst for the derivatization step. Alternative means of facilitating the interaction of a ribozyme with a primer-template substrate, such as the presence of basic peptides or other cofactors, might overcome this problem. Finally it is noteworthy that very small self-aminoacylating ribozymes have been obtained using RNA libraries that have an unconstrained sequence at their 3′-end (Chumachenko et al. 2009). This strategy may also be fruitful in selections for ribozyme polymerases.

Given the above constraints and uncertainties, what can we say about the emergence of RNA-catalyzed RNA replication in the origin of life? It is still possible that under the proper conditions, and using the right substrates, a small simple ribozyme could effectively catalyze RNA replication. However, a replicase must do more than catalyze a simple phospho-transfer reaction. Binding in a nonsequence-specific manner to a primer-template complex, facilitating binding of the proper incoming monomer, catalyzing primer extension, and repeating this process until the end of the template (or set of templates for a multi-component replicase) is reached might require a complex replicase structure. Such a replicase would presumably be rare in collections of random RNA sequences. If life required a very special sequence to get started, then the origin of life on earth could have been a low probability event and life on other earth-like planets might be very rare. If, on the other hand, RNA-catalyzed RNA replication could have emerged gradually in a series of simpler steps, it might have been easier and thus more likely for life to begin, and life elsewhere might be common. For this reason, we now turn to a consideration of nonenzymatic template-directed replication chemistry.

Chemical Template Replication Revisited

The nonenzymatic template-directed polymerization of activated ribonucleotides was studied in depth by Leslie Orgel, together with his students and colleagues, over a period of several decades (Orgel 2004). Here, the template itself acts as a catalyst by helping to align and orient the monomers so that they are pre-organized for polymerization. The main lesson from this work is that spontaneous chemical copying of RNA sequences is indeed possible, but is subject to several important constraints and limitations. The constraints make template-directed RNA copying incompatible with currently available membrane vesicle systems, and the limitations have, so far at least, made it impossible to obtain repeated cycles of RNA replication through chemical copying.

To obtain reasonable reaction rates, Orgel made use of nucleoside monophosphates activated with a good leaving group such as imidazole. The ribonucleotide 5′-phosphorimidazolides spontaneously assemble on a template oligonucleotide and polymerize over several days, generating a complementary strand. However, monomer binding to RNA templates is weak and concentrations on the order of 0.1 M were required for optimal copying. In addition, a Mg^{++} concentration of ~ 0.1 M, which as noted above is incompatible with the presence of fatty-acid based membranes, is required for optimal polymerization. Even under these rather extreme conditions, polymerization proceeds at only 1–2 nucleotides per day, and is therefore limited by monomer hydrolysis. Beyond these constraints, three aspects of the copying reaction present major hurdles to multiple rounds of replication. First, the chemical structure of RNA results in a problem of regiospecificity, because new linkages can be either 3′–5′ or 2′–5′ phosphodiester bonds. Surprisingly, under most conditions, it is the 2′–5′ phosphodiester bonds that are most common in polymerization products. This problem can be ameliorated by the choice of ions and leaving groups; for unknown reasons, Zn^{2+} ions and the 2-methylimidazole leaving group favor the synthesis of 3′–5′ linkages (Lohrmann et al. 1980; Inoue and Orgel 1982). Studies of oligonucleotide ligation showed that the helical context of an extended RNA duplex also favors the formation of 3′–5′ linkages (Rohatgi et al. 1996). However, it remains difficult to obtain a homogeneous RNA backbone without enzymatic catalysis. Second, adenine (A) residues in the template are difficult to copy, and two or more As in succession block chain growth (Inoue and Orgel 1983), presumably as a result of the poor base-stacking propensity of the incoming U monomers. It has recently been found that template copying at subzero temperatures can proceed past multiple A residues in the template (Vogel and Richert 2007), but this required sequential additions of oxyazabenzotriazole-activated monomer together with a series of helper oligos. Third, there is the issue of fidelity. The copying of G and C residues is remarkably accurate, with error rates estimated at 0.5% or less. However, the addition of A and U residues causes problems, most significantly the formation of G:U wobble base-pairs

(Wu and Orgel 1992), which would lead to significant error rates in a four base system. Might an alternative genetic polymer, perhaps even a close relative of RNA, overcome these problems and enable chemical replication? The identification of such a system might ultimately lead to the discovery of plausible progenitors of RNA, or, alternatively, to the discovery of new replication strategies that allow for the chemical replication of RNA itself.

Phosphoramidate Nucleic Acids

Early studies of template copying using more reactive nucleotide derivatives were performed by Orgel et al., who examined both 2′- and 3′-amino ribonucleotide 5′-phosphorimidazolides (Lohrmann and Orgel 1976; Zielinksi and Orgel 1985; Tohidi et al. 1987). Polymerization of these nucleotides yields phosphoramidate nucleic acids, which are generally similar to standard phosphodiester linked nucleic acids except that the phosphoramidate linkage is more acid labile. As expected, replacing a sugar hydroxyl in the monomer with a more nucleophilic amino group resulted in a large increase in monomer reactivity. The activated 3′-amino ribonucleotides participated in rapid copying of short oligonucleotide templates 5-13 nucleotides in length, yielding $N3' \rightarrow P5'$ linked complementary oligonucleotides (Tohidi et al. 1987). The increased reactivity also led to faster intramolecular monomer cyclization, which depleted the template copying reactions of activated substrate molecules (Hill et al. 1988).

More recently, the Richert group has begun to explore the potential of 3′-amino- nucleotide analogs in template-directed condensation reactions. Deoxyribonucleotide monomers, activated with an oxyazabenzotriazole leaving group, completed a template-directed reaction with a 3′-amino-terminated primer in seconds (Rothlingshofer et al. 2008). The fidelity of this reaction is sufficient to allow for sequencing by nonenzymatic primer extension. However, the monomer is rapidly consumed by internal cyclization.

The high intrinsic reactivity of the amino-sugar modified nucleotides suggested to us that an alternative phosphoramidate nucleic acid might act as a good platform for chemical self-replication. Our group has therefore started to study a series of phosphoramidate nucleic acids with sugar phosphate backbones that vary in their degree of conformational flexibility or constraint. We are currently focusing on the 2′-5′ linked phosphoramidate analog of DNA, and the corresponding monomers, the 2′-amino dideoxyribonucleotide 5′-phosphorimidazolides (Fig. 4). The 2′-amino ImpddNs are advantageous as monomers because they cannot undergo intramolecular cyclization because of the steric constraint of the ribose ring; they are only depleted during polymerization reactions by competing hydrolysis.

Our first experiments with 2′-amino ImpddG led to rapid and efficient primer-extension across a dC_{15} template, generating full-length product in ∼6 hour (Mansy et al. 2008). Encouraged by the rapid copying of oligo-dC templates by 2′-amino ImpddG, we performed a more extensive study of the copying of templates with differing sugar-phosphate backbones, lengths and sequences (Schrum et al. 2009). The most important property contributing to good template activity appears to be preorganization in the form of an A-type helix. Thus, RNA templates were uniformly superior

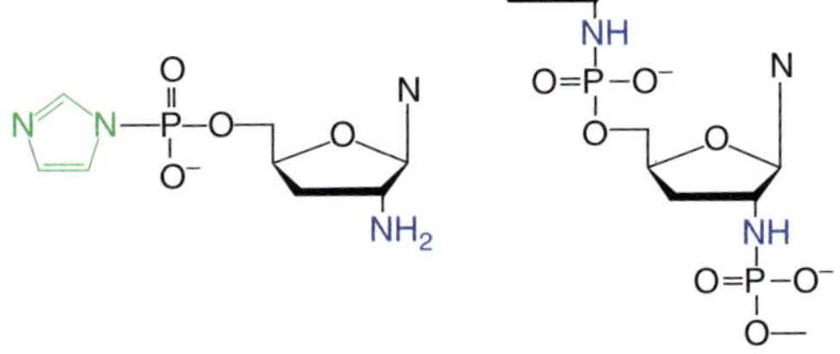

Figure 4. Structures of 2′-5′ phosphoramidate DNA, and the corresponding activated monomers.

to DNA templates of the same sequence, and LNA (locked nucleic acid) templates, which are chemically locked in a C3′-endo sugar conformation, were superior to RNA templates. This result is consistent with previous observations that under most conditions RNA-template directed polymerization of ribonucleotides leads to a majority of 2′-5′ linkages; it appears that an A-type helical geometry generally favors polymerization through attack by a 2′ nucleophile. A second key factor is that enhanced monomer affinity for the template increases the reaction efficiency. Thus G:C base-pairs lead to efficient copying, whereas A:U base-pairs were very poorly copied. Replacing A with D (diaminopurine) results in a D:U base-pair with three hydrogen bonds, and slightly improved primer-extension. However, the poor base stacking of U residues must also be improved to obtain efficient template copying. When we replaced U with C5-propynyl-U, the resulting A:U^P base-pairs led to improved copying, but the D:U^P combination had a clearly synergistic effect (Fig. 5). In the context of a DNA duplex, the D:U^P base-pair has previously been shown to be energetically almost equivalent to a C:G base-pair (Chaput et al. 2002). When we used G, C, D, and U^P as the four monomer and template bases, we were able to copy mixed-sequence RNA templates over 15 nts in length in about a day. This represents a significant step toward developing a robust, generalized chemical replication system.

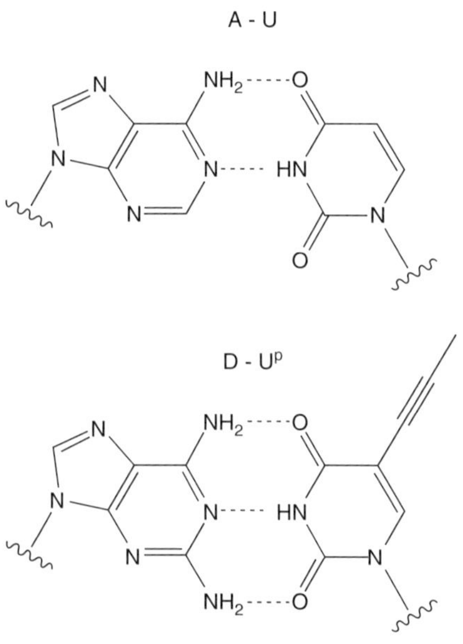

This system is, however, far from ideal, and there are strong indications that the fidelity of template copying becomes an issue when all four nucleotides are present, largely because of the formation of G:U wobble base-pairs. Given that primer-extension on 2′–5′ linked DNA templates is approximately similar to that on the corresponding RNA templates, we are currently synthesizing 2′–5′ linked phosphoramidate DNA templates to assess self-replication in this system. In light of the efficient templating of LNA oligonucleotides, we are also interested in exploring prebiotically plausible nucleic acids that are more conformationally constrained than RNA. A particularly interesting candidate is TNA (threose nucleic acid) (Schöning et al. 2000) and its 2′-amino substituted phosphoramidate version, NP-TNA (Wu et al. 2002). These are both base-pairing systems that form standard Watson-Crick duplexes, despite having only five atoms per backbone repeat unit (vs. six for RNA and DNA). It will be of great interest to see if the resulting decrease in flexibility leads to increased fidelity in template copying reactions.

Protocell Assembly

In principle, protocell-like objects could form spontaneously as new membranes self-assemble and encapsulate genetic molecules in solution. Recently, simple physical processes that would enhance the efficiency of the coassembly of nucleic acids and membrane vesicles have been proposed. One such alternative scenario is based on the fact that the clay mineral montmorillonite is not only a catalyst of RNA polymerization (Ferris et al. 1996; Huang and Ferris 2006) but also catalyzes membrane assembly (Hanczyc et al. 2003). Experiments with clay particles containing surface-adsorbed

Figure 5. Watson Crick base-pairs. *Top*: Standard A:U base-pair. *Bottom*: Alternative diaminopurine (D): C5-propynyl-uracil (U^P) base-pair.

RNA showed that such particles stimulated vesicle assembly, and frequently became trapped inside the vesicles whose assembly they had catalyzed. Thus, a common mineral can catalyze both the assembly of a genetic polymer and the assembly of a membrane vesicle, and bring these two components together to generate a protocell-like structure consisting of a genetic polymer trapped within a membrane compartment. Although the effectiveness of this process is attractive, some means of releasing at least some of the bound nucleic acid from the mineral surface, and/or replicating it on the surface, would be necessary for subsequent replication to occur.

More recent experiments suggest a very different geochemical scenario leading to the assembly of similar protocell-like structures. The hollow channels within the rocks of the alkaline off-axis hydrothermal vents provide a protected compartmentalized environment where it has been suggested that primitive metabolic activities might have originated (Martin and Russell 2003). Recent theoretical studies suggested that the strong thermal gradients present in hydrothermal vents, together with the thin channels produced by mineral precipitation, could greatly concentrate small organic molecules such as nucleotides as well as larger nucleic acids from a very dilute external reservoir (Baaske et al. 2007). Work from our laboratory (Budin et al. 2009) has confirmed the predicted concentration effect, and has also shown that subcritical concentrations of fatty acids can be concentrated to the extent that they self-assemble into vesicles at the bottom of the capillary channels. Moreover, DNA oligonucleotides can also be greatly concentrated and can become encapsulated within the vesicles, resulting in the spontaneous assembly of protocell-like structures.

Encapsulated Template Replication: Emergence of a Protocell

The experiments discussed earlier suggest that the assembly of protocell-like structures is not that difficult, because it appears to be possible through multiple distinct mechanisms. The more challenging question is, how could such a structure replicate? We have already considered the replication of the protocell membrane and the genetic material as separate entities. To address the question of their replication as a combined structure we must consider in more detail the molecular constituents of the protocell membrane and the molecular nature of the encapsulated genetic material.

Genome replication within a protocell can only occur if the building blocks used to copy template strands are able to enter the fatty-acid vesicle compartment. Early work using phospholipid-based vesicles and protein enzymes showed the feasibility of constructing primitive cell-like compartments (Chakrabarti et al. 1994; Luisi et al 1994). More recent permeability studies (Mansy et al. 2008; see also Mansy 2010) showed that nucleotides could spontaneously diffuse across simple fatty-acid membranes, but that net negative charge is a critical determinant of permeability. Thus, nucleotides that are chemically activated, e.g., by conversion of the $5'$-phosphate to a $5'$-phosphorimidazolide, equilibrate across vesicle membranes much more rapidly on account of the reduction of the net negative charge. In addition, mixtures of fatty acids with their glycerol esters generate membranes that are more permeable to polar and charged molecules. By combining these observations, we were able to show that activated $2'$-amino-$2'$,$3'$-dideoxyguanosine-$5'$-phosphorimidazolide, the same nucleotide previously shown to rapidly copy oligo-dC templates, could be added to the outside of fatty-acid vesicles containing an encapsulated primer-template complex, and copy the internal dC_{15} template. Copying of the encapsulated dC_{15} template by primer-extension reached $>95\%$ completion in 12–24 hour (Mansy et al. 2008), compared with 6–12 hour in free solution. The longer time required for copying encapsulated templates reflects the time required for entry of external nucleotides to the interior of the vesicles. Importantly, the presence of high concentrations (5–10 mM) of highly reactive activated nucleotides did not have any disruptive effects on the integrity of the vesicle membrane, as no leakage of encapsulated primer-template complexes was observed. It is also important to note that control experiments with phospholipid membranes showed no copying of internal template, because the activated nucleotides could not enter the vesicle; similarly, "modern" activated nucleotides such as nucleoside triphosphates cannot cross fatty-acid based membranes. Successful copying of encapsulated templates therefore requires both "primitive" nucleotides with reduced charge, and "primitive" membranes composed of single chain amphiphiles.

The copying of a genetic polymer inside a membrane compartment is an important step toward the realization of a self-replicating system capable of Darwinian evolution. What, then, are the remaining barriers to the assembly of such a system? The copying of a single-stranded template produces a double-stranded product; these strands would have to separate before a second cycle of genome replication could begin. Separate follow-up experiments by our group showed that some fatty-acid based vesicles are able to retain encapsulated DNA and RNA oligonucleotides over a temperature range of 0 °C to 100 °C (Mansy and Szostak 2008). As with permeability, mixtures of amphiphiles lead to improved thermostability, with glycerol esters being particularly stabilizing, possibly because of the additional hydrogen bond donors and acceptors provided by the glycerol head group. Furthermore, we found that encapsulated double-stranded DNA could be denatured at elevated temperatures, with the strands reannealing once the temperature was lowered (Mansy and Szostak 2008). This implies that thermal fluctuations could provide a mechanism for strand separation that is compatible with the integrity of fatty-acid vesicles, potentially allowing for complete cycles of replication of encapsulated genetic polymers. The mutual compatibility of nucleic acid replication and fatty-acid compartment growth is very encouraging because it alleviates concerns related to the permeability and stability of membrane vesicles. Vesicles therefore do seem to be a physically plausible way to segregate and spatially localize genomes, keep emergent catalytic polynucleotides physically close to their encoding genome, and protect the nascent evolving system from parasitic polymers.

CHALLENGES AND FUTURE RESEARCH DIRECTIONS

Prospects for a Complete Protocell Model

Although considerable progress has been made toward the assembly of model protocells, several remaining issues must be solved before multiple cycles of protocell replication can be achieved in the laboratory. These factors are also relevant to protocell replication on the early Earth. The most important factor at this time appears to be the competition between strand reannealing and strand copying, after thermal strand separation. PCR reactions generally plateau at about 1-μM DNA strand concentration, which is the concentration at which strand reannealing and strand copying occur on a similar time scale of about 1 minute However, nonenzymatic template copying requires on the order of a day for completion, which implies that either template copying must be much faster, or reannealing must be much slower. One way to make reannealing sufficiently slow is to keep strand concentrations subnanomolar. Low strand concentrations are possible in large vesicles, but it is hard to see how a few molecules of a genetic polymer could have any significant phenotypic effect on a large vesicle composed of millions of amphiphilic molecules. The emergence of metabolic ribozymes would be more plausible if nucleic acid strand concentrations were much higher, so that a catalyst of modest efficiency could generate enough product to influence cell properties.

What other factors might affect the rate of strand reannealing? Perhaps the most obvious possibility is that secondary structure, which can form extremely rapidly because the interactions are intramolecular, could greatly slow down strand annealing. This phenomenon is essential to the viability of single-stranded RNA phage such as Qβ° (Axelrod et al. 1991). Significant intrastrand secondary structure would also be an expected consequence of selection for sequences that fold into functional shapes with catalytic activity. On the other hand, chemical replication through dense secondary structure would probably be much slower than replication of an unfolded, open template. The outcome of simultaneous selection for an open, accessible template sequence, and a folded functional structure remains unclear. An alternative but even more speculative solution might result from the rapid binding by base-pairing of short oligonucleotides or even monomers to freshly separated strands. If a template strand was largely occupied by monomers or short oligomers, even if these were in rapid exchange, the strand might be prevented from annealing to a complementary strand. This possibility has the advantage that it need not block or slow the copying reaction, however its effectiveness remains to be tested experimentally.

Another challenge faced by replicating protocells, whether on the early earth or in the modern laboratory, is the continuous dilution of protocells through the competing formation of new empty vesicles. When new fatty acids are supplied as micelles, the efficiency of incorporation into pre-formed vesicles (or protocells) can be quite high, but some new vesicles are always formed. Thus over time, the descendants of a given protocell will gradually be diluted out by the continuous formation of these new vesicles. To avoid extinction by dilution, the protocells must out-compete other vesicles either by having a more rapid cell cycle, thereby generating more progeny during division, or by surviving destructive processes more efficiently. We have previously proposed (Chen et al. 2004) that faster growth, driven by the osmotic pressure of encapsulated nucleic acid, could lead to an effectively shorter cell cycle for protocells that contain high copy numbers of their replicating genome. However, in light of the problems associated with this approach, it is of considerable interest to explore new ways in which a protocell genome could lead to faster growth or growth that occurs at the expense of empty vesicles. An alternative strategy for surviving dilution would be for a protocell genome to colonize empty vesicles. This could occur through a low level of stochastic vesicle–vesicle fusion events, possibly catalyzed by low levels of divalent cations such as Ca^{2+}. Systematic efforts to measure vesicle fusion frequencies under different environmental conditions could therefore be quite useful. Finally, it is possible that this problem could be circumvented entirely if early life was discontinuous (Budin et al. 2009). For example, protocells could be occasionally disrupted by drying, or simply dissolve as a result of dilution with water to a level below the critical aggregate concentration; subsequent re-hydration or concentration would result in reformation of vesicles encapsulating genomic nucleic acids, thus generating a new "randomized" set of protocells. As long as such events were fairly uncommon, and assuming that genomic replication had kept ahead of vesicle replication so that each vesicle contained multiple genome copies prior to disruption, this process of disruption and reformation would lead to the spread of evolving genomic sequences through the "new" vesicles.

Laboratory models of protocell systems should be helpful in modeling many of the above scenarios. Assuming that protocell reproduction can be achieved, and made efficient enough to continue through many generations, it should then be possible to observe the spontaneous evolution of adaptive innovations in this relatively simple chemical system. The nature of such adaptations may provide clues as to how modern cells evolved from their earliest ancestors. Ultimately this line of research may also tell us whether the conserved biochemistry of life is driven by chemical necessity, or whether biochemically very different forms of life are also possible.

ACKNOWLEDGMENTS

We thank Itay Budin, Matt Powner, and other members of the Szostak lab for helpful discussions.

REFERENCES

Axelrod VD, Brown E, Priano C, Mills DR. 1991. Coliphage Q β RNA replication: RNA catalytic for single-strand release. *Virology* **184**: 595–608.

Baaske P, Weinert FM, Duhr S, Lemke KH, Russell MJ, Braun D. 2007. Extreme accumulation of nucleotides in simulated hydrothermal pore systems. *Proc Natl Acad Sci* **104:** 9346–9351.

Bar-Ziv R, Moses E. 1994. Instability and "pearling" states produced in tubular membranes by competition of curvature and tension. *Phys Rev Lett* **73:** 1392–1395.

Berclaz N, Muller M, Walde P, Luisi PL. 2001. Growth and transformation of vesicles studied by ferritin labeling and cryotransmission electron microscopy. *J Phys Chem B* **105:** 1056–1064.

Blochliger E, Blocher M, Walde P, Luisi PL. 1998. Matrix effect in the size distribution of fatty acid vesicles. *J Phys Chem B* **102:** 10383–10390.

Budin I, Bruckner R, Szostak JW. 2009. Formation of protocell-like vesicles in a thermal diffusion column. *J Am Chem Soc* **131:** 9628–9629.

Chakrabarti AC, Breaker RR, Joyce GF, Deamer DW. 1994. Production of RNA by a polymerase protein encapsulated within phospholipid vesicles. *J Mol Evol* **39:** 555–9.

Chaput JC, Sinha S, Switzer C. 2002. 5-propynyluracil.diaminopurine: An efficient base-pair for non-enzymatic transcription of DNA. *Chem Commun* **15:** 1568–9.

Chen IA, Roberts RW, Szostak JW. 2004. The emergence of competition between model protocells. *Science* **305:** 1474–1476.

Chen IA, Szostak JW. 2004. A kinetic study of the growth of fatty acid vesicles. *Biophys J* **87:** 988–998.

Chumachenko NV, Novikov Y, Yarus M. 2009. Rapid and simple ribozymic aminoacylation using three conserved nucleotides. *J Am Chem Soc* **131:** 5257–63.

Deamer DW. 1985. Boundary structures are formed by organic-components of the Murchison carbonaceous chondrite. *Nature* **317:** 792–794.

Deamer DW, Pashley RM. 1989. Amphiphilic components of the Murchison carbonaceous chondrite: Surface properties and membrane formation. *Orig Life Evol Biosph* **19:** 21–38.

Dworkin J, Deamer D, Sandford S, Allamandola L. 2001. Self-assembling amphiphilic molecules: Synthesis in simulated interstellar/precometary ices. *Proc Natl Acad Sci* **98:** 815–819.

Ekland EH, Szostak JW, Bartel DP. 1995. Structurally complex and highly active RNA ligases derived from random RNA sequences. *Science* **269:** 1319–25.

Ferris JP, Hill AR Jr, Liu R, Orgel LE. 1996. Synthesis of long prebiotic oligomers on mineral surfaces. *Nature* **381:** 59–61.

Gebicki JM, Hicks M. 1973. Ufasomes are stable particles surrounded by unsaturated fatty acid membranes. *Nature* **243:** 232–234.

Gilbert W. 1986. The RNA World. *Nature* **319:** 618.

Glasner ME, Yen CC, Ekland EH, Bartel DP. 2000. Recognition of nucleoside triphosphates during RNA-catalyzed primer extension. *Biochemistry* **39:** 15556–62.

Guerrier-Takada C, Gardiner K, Marsh T, Pace N, Altman S. 1983. The RNA moiety of ribonuclease P is the catalytic subunit of the enzyme. *Cell* **35:** 849–57.

Hanczyc MM, Szostak JW. 2004. Replicating vesicles as models of primitive cell growth and division. *Curr Opin Chem Biol* **8:** 660–664.

Hanczyc MM, Fujikawa SM, Szostak JW. 2003. Experimental models of primitive cellular compartments: Encapsulation, growth, and division. *Science* **302:** 618–622.

Hargreaves WR, Deamer DW. 1978. Liposomes from ionic, single-chain amphiphiles. *Biochemistry* **17:** 3759–3768.

Hill AR Jr, Nord LD, Orgel LE, Robins RK. 1988. Cyclization of nucleotide analogues as an obstacle to polymerization. *J Mol Evol* **28:** 170–1.

Huang W, Ferris JP. 2006. One-step, regioselective synthesis of up to 50-mers of RNA oligomers by montmorillonite catalysis. *J Am Chem Soc* **128:** 8914–9.

Inoue T, Orgel LE. 1982. Oligomerization of (guanosine 5′-phosphor)-2-methylimidazolide on poly(C). An RNA polymerase model. *J Mol Biol* **162:** 201–17.

Inoue T, Orgel LE. 1983. A nonenzymatic RNA polymerase model. *Science* **219:** 859–62.

Johnston WK, Unrau PJ, Lawrence MS, Glasner ME, Bartel DP. 2001. RNA-catalyzed RNA polymerization: Accurate and general RNA-templated primer extension. *Science* **292:** 1319–25.

Kruger K, Grabowski PJ, Zaug AJ, Sands J, Gottschling DE, Cech TR. 1982. Self-splicing RNA: Autoexcision and autocyclization of the ribosomal RNA intervening sequence of Tetrahymena. *Cell* **31:** 147–57.

Lohrmann R, Orgel LE. 1976. Template-directed synthesis of high molecular weight polynucleotide analogues. *Nature* **261:** 342–344.

Lohrmann R, Bridson PK, Orgel LE. 1980. Efficient metal-ion catalyzed template-directed oligonucleotide synthesis. *Science* **208:** 1464–5.

Luisi PL. 2006. *The emergence of life: From chemical origins to synthetic biology*, Cambridge University Press, Cambridge.

Luisi PL, Walde P, Oberholzer T. 1994. Enzymatic RNA synthesis in self-reproducing vesicles: An approach to the construction of a minimal synthetic cell. *Ber Bunsenges Phys Chem* **98:** 1160–5.

Luisi PL, Stano P, Rasi S, Mavelli F. 2004. A possible route to prebiotic vesicle reproduction. *Artif Life* **10:** 297–308.

Mansy SS. 2010. Membrane transport in primitive cells. *Cold Spring Harb Perspect Biol* **2:** a002188.

Mansy SS, Szostak JW. 2008. Thermostability of model protocell membranes. *Proc Natl Acad Sci* **105:** 13351–13355.

Mansy SS, Schrum JP, Krishnamurthy M, Tobé S, Treco DA, Szostak JW. 2008. Template-directed synthesis of a genetic polymer in a model protocell. *Nature* **454:** 122–125.

Martin W, Russell MJ. 2003. On the origins of cells: A hypothesis for the evolutionary transitions from abiotic geochemistry to chemoautotrophic prokaryotes, and from prokaryotes to nucleated cells. *Philos Trans R Soc Lond B Biol Sci* **358:** 59–83.

McCollom TM, Ritter G, Simoneit BR. 1999. Lipid synthesis under hydrothermal conditions by Fischer-Tropsch-type reactions. *Orig Life Evol Biosph* **29:** 153–166.

Monnard PA, Apel CL, Kanavarioti A, Deamer DW. 2002. Influence of ionic inorganic solutes on self-assembly and polymerization processes related to early forms of life: Implications for a prebiotic aqueous medium. *Astrobiology* **2:** 139–52.

Müller UF, Bartel DP. 2008. Improved polymerase ribozyme efficiency on hydrophobic assemblies. *RNA* **14:** 552–62.

Nissen P, Hansen J, Ban N, Moore PB, Steitz TA. 2000. The structural basis of ribosome activity in peptide bond synthesis. *Science* **289:** 920–30.

Orgel LE. 2004. Prebiotic chemistry and the origin of the RNA world. *Crit Rev Biochem Mol Biol* **39:** 99–123.

Prabahar KJ, Ferris JP. 1997. Adenine derivatives as phosphate-activating groups for the regioselective formation of 3',5'-linked oligoadenylates on montmorillonite: Possible phosphate-activating groups for the prebiotic synthesis of RNA. *J Am Chem Soc* **119:** 4330–7.

Rohatgi R, Bartel DP, Szostak JW. 1996. Nonenzymatic, template-directed ligation of oligoribonucleotides is highly regioselective for the formation of 3'-5' phosphodiester bonds. *J Am Chem Soc* **118:** 3340–4.

Röthlingshöfer M, Kervio E, Lommel T, Plutowski U, Hochgesand A, Richert C. 2008. Chemical primer extension in seconds. *Angew Chem Int Ed Engl* **47:** 6065–8.

Rushdi AI, Simoneit BR. 2001. Lipid formation by aqueous Fischer-Tropsch-type synthesis over a temperature range of 100 to 400 degrees C. *Orig Life Evol Biosph* **31:** 103–118.

Sacerdote MG, Szostak JW. 2005. Semi-permeable lipid bilayers exhibit diastereoselectivity favoring ribose; implications for the origins of life. *Proc Natl Acad Sci* **102:** 6004–6008.

Schöning K, Scholz P, Guntha S, Wu X, Krishnamurthy R, Eschenmoser A. 2000. Chemical etiology of nucleic acid structure: The α-threofuranosyl-(3'→2') oligonucleotide system. *Science* **290:** 1347–51.

Schrum JP, Ricardo A, Krishnamurthy K, Blain JC, Szostak JW. 2009. Efficient and rapid template-directed nucleic acid copying using 2'-amino-2', 3'-dideoxyribonucleoside-5'-phosphorimidazolide monomers. *J Am Chem Soc* **31:** 14560–14570.

Stano P, Wehrli E, Luisi PL. 2006. Insights into the self-reproduction of oleate vesicles. *J Phys: Condens Matter* **18:** S2231–S2238.

Szostak JW, Bartel DP, Luisi PL. 2001. Synthesizing life. *Nature* **409:** 387–390.

Tohidi M, Zielinski WS, Chen CH, Orgel LE. 1987. Oligomerization of 3'-amino-3'deoxyguanosine-5'phosphorimidazolidate on a d(CpCpCpCpC) template. *J Mol Evol* **25:** 97–99.

Vogel SR, Richert C. 2007. Adenosine residues in the template do not block spontaneous replication steps of RNA. *Chem Commun* **19:** 1896–8.

Voytek SB, Joyce GF. 2007. Emergence of a fast-reacting ribozyme that is capable of undergoing continuous evolution. *Proc Natl Acad Sci* **104:** 15288–93.

Walde P, Wick R, Fresta M, Mangone A, Luisi PL. 1994a. Autopoietic self-reproduction of fatty acid vesicles. *J Am Chem Soc* **116:** 11649–11654.

Walde P, Goto A, Monnard P-A, Wessicken M, Luisi PL. 1994b. Oparin's reactions revisited: Enzymatic synthesis of poly(adenylic acid) in micelles and self-reproducing vesicles. *J Am Chem Soc* **116:** 7541–7547.

Wu T, Orgel LE. 1992. Nonenzymatic template-directed synthesis on hairpin oligonucleotides. 3. Incorporation of adenosine and uridine residues. *J Am Chem Soc* **114:** 7963–9.

Wu X, Guntha S, Ferencic M, Krishnamurthy R, Eschenmoser A. 2002. Base-pairing systems related to TNA: α-Threofuranosyl oligonucleotides containing phosphoramidate linkages. *Org Lett* **4:** 1279–82.

Zhu TF, Szostak JW. 2009a. Coupled growth and division of model protocell membranes. *J Am Chem Soc* **131:** 5705–5713.

Zhu TF, Szostak JW. 2009b. Preparation of large monodisperse vesicles. *PLoS ONE* **4:** e5009.

Zielinski WS, Orgel LE. 1985. Oligomerization of activated derivatives of 3'-amino-3'-deoxyguanosine on poly(C) and poly(dC) templates. *Nucleic Acids Res* **13:** 2469–2484.

Origin and Evolution of the Ribosome

George E. Fox

Department of Biology and Biochemistry, University of Houston, Houston, Texas 77204-5001

Correspondence: fox@uh.edu

The modern ribosome was largely formed at the time of the last common ancestor, LUCA. Hence its earliest origins likely lie in the RNA world. Central to its development were RNAs that spawned the modern tRNAs and a symmetrical region deep within the large ribosomal RNA, (rRNA), where the peptidyl transferase reaction occurs. To understand pre-LUCA developments, it is argued that events that are coupled in time are especially useful if one can infer a likely order in which they occurred. Using such timing events, the relative age of various proteins and individual regions within the large rRNA are inferred. An examination of the properties of modern ribosomes strongly suggests that the initial peptides made by the primitive ribosomes were likely enriched for L-amino acids, but did not completely exclude D-amino acids. This has implications for the nature of peptides made by the first ribosomes. From the perspective of ribosome origins, the immediate question regarding coding is when did it arise rather than how did the assignments evolve. The modern ribosome is very dynamic with tRNAs moving in and out and the mRNA moving relative to the ribosome. These movements may have become possible as a result of the addition of a template to hold the tRNAs. That template would subsequently become the mRNA, thereby allowing the evolution of the code and making an RNA genome useful. Finally, a highly speculative timeline of major events in ribosome history is presented and possible future directions discussed.

A major commonality of all cellular life is the coupling between translation and transcription mediated by the genetic code. Comparative genomics has further refined this by revealing the presence of an "RNA metabolism" (Anantharaman et al. 2002) or "Persistent proteome" (Danchin et al. 2007) that is basically a compendium of essentially universal genes involved in translation, transcription, RNA processing and degradation, intermediary and RNA metabolism, and compartmentalization. DNA replication likely arose later because the core enzymes involved in the process are not related (Bailey et al. 2006 and others). Together these universal genes comprise what is frequently referred to as LUCA, the last universal common ancestor (Benner et al. 1993; Lazcano 1994; Mushegian and Koonin 1996; Kyrpides et al. 1999). It is noteworthy that no matter how they are defined, by far the largest numbers of genes in LUCA are associated with translation. Indeed, the translation machinery as represented in LUCA is essentially complete indicating that major events in its origins occurred before LUCA. Thus, it might appear that the origins of the translation machinery would be hopelessly obscured by time. Nevertheless, as will be discussed herein, substantial although necessarily incomplete, evidence relating to the origins and early development of the translation machinery and its relation to other core cellular processes continues to exist in the primary sequences, three-dimensional folding, and functional interactions of the various macromolecules involved in the modern versions of the translation machinery.

The modern ribosome consists of small and large subunits (30S and 50S in Bacteria and Archaea) that come together during the initiation of protein synthesis remain together as individual amino acids are added to a growing peptide according to information encoded on the mRNA, and finally separate again in conjunction with the release of the finished protein. Each subunit is an RNA/protein complex. In Bacteria and Archaea, the 50S subunit typically contains a 23S rRNA and a 5S rRNA whereas the 30S subunit contains the 16S rRNA. Peptide bond synthesis occurs in the 50S subunit at the peptidyl transferase center, (PTC), and codon recognition occurs at the decoding site, which is in the small subunit. Transfer RNAs, (tRNA), bridge the two subunits occupying, at various times in the synthesis cycle, the A, P, or E (exit) sites of the 50S subunit and the decoding site in the 30S subunit. A universal CCA sequence at the 3′ end of the tRNA is the point of attachment of the amino acid and later the growing peptide chain to the tRNA. The A, P, and E sites are partly in the small subunit and partly in the large subunit such that a tRNA can be in a hybrid site (e.g., the A site in the 30S and P site in the 50S. The mRNA is exclusively found in the small subunit where it interacts with the anticodon loops of the tRNAs. As the nascent protein is synthesized it passes through an exit tunnel that begins at the PTC center and ultimately exits from the back of the 50S subunit. Synthesis is a dynamic cyclic process in which tRNAs enter the ribosome bringing amino acids as specified by the mRNA and move through the machinery, which undergoes a series of coordinated motions that drive the process (Steitz 2008). These include the movements of the tRNAs between sites, opening and closing of the L1 stalk on the 50S subunit and the ratcheting of the small subunit relative to the large subunit (Frank and Agrawal, 2000), which has recently been elucidated in structural detail (Zhang et al. 2009).

Diverse species (*Escherichia coli, Haloarcula marismortui, Thermus thermophiles,* and *Deinococcus radiodurans*) are represented among the various atomic resolution ribosome structures now available (Ban et al. 2000; Yusupov et al. 2000; Wimberly et al. 2000; Schuwirth et al. 2005; Selmer et al. 2006; and others). These structures encompass 30S and 50S subunits as well as the whole 70S ribosome. In addition, cryoelectron microscopy studies have revealed dynamic motions associated with the ribosome (Frank and Agrawal 2000; Connell et al. 2007; and others). These ongoing high resolution structural studies provide the opportunity to examine the relative age of features within the ribosome such as the A, P, and E sites, the exit tunnel, the L7/L12 region, and the L1 region that facilitate the entry and exit of tRNAs.

PEPTIDYL TRANSFERASE CENTER

It is believed that the peptidyl transferase center, (PTC), which encompasses the large subunit portions of the A and P sites of the ribosome, is structurally the same in both the 50S and 70S subunits (Steitz 2008). When comparing 50S subunit structures between Archaea and Bacteria one again finds that the structures are essentially the same. However, the E site structure is different. In Archaea L44e interacts with the E-site tRNA but this protein is missing in Bacteria with the result that the tRNA CCA end is positioned differently. Hence, the A and P sites likely predate the E site, which may have been added post-LUCA (Steitz 2008).

The portion of 23S rRNA comprising the PTC contains a region of approximately 165 bases that shows high twofold pseudo symmetry (Agmon et al. 2005; Zimmerman & Yonath 2009). The two 82 nucleotide halves of the symmetrical region correspond to the 50S portion of the A and P sites of the ribosome. In fact, the essence of this region is contained in a single contiguous self-folding RNA (Smith et al. 2008). The PTC is located in Domain 5 of the 23S rRNA structure.

Recently, Hsiao et al. (2009) superimposed the structure of the large subunit RNAs from two ribosome crystal structures and sectioned the resulting structure into concentric shells with the PTC at the center. They, like others (Ban et al. 2000; Wimberly et al. 2000), found that ribosomal proteins (r-proteins) are effectively absent from the PTC region, which is why the ribosome is regarded as fundamentally an RNA machine. To the extent that protein elements are in proximity to the PTC, they are short, largely unstructured peptides rather than globular elements. The globular regions are mainly on the surface of the ribosome (Ban et al. 2000; Wimberly et al. 2000). A major stabilizing element in the PTC region is instead Mg^{2+} interactions. In many cases, the phosphate oxygen atoms act as inner sphere Mg^{2+} ligands (Hsiao et al. 2009; Hsiao and Williams 2009). Thus, consistent with the notion of a preceding RNA world, the structure of the PTC seems to have evolved before the availability of proteins.

Although the modern translation machinery is very complex, two small RNAs, the PTC RNA fragment and tRNAs are at its core. Both of these are less than 100 nucleotides in length, and their importance supports the notion that the translation machinery was originally a discovery of the RNA world. In fact, the ability to synthesize coded peptides of increasing complexity would eventually terminate the RNA world and create the RNA/protein world. The seldom discussed issue is whether such a termination would have occurred before (e.g., brief RNA world) or after the discovery of an RNA replicase (extended RNA world). If peptide synthesis arises quickly, then their will neither be time nor need for extensive catalysis of biochemical reactions by RNA. If reasonable, the rapid appearance of a translation system may even eliminate the need to validate the RNA world by demonstrating the self-replicating RNA system that has proven experimentally difficult to achieve.

tRNA ORIGINS AND INCREASING RNA COMPLEXITY

Because of its obvious importance, considerable attention has been focused on the origins of the tRNA and numerous models have been proposed and recently reviewed (Di Giulio 2009). The most popular model (Noller 1993; Maizels and Weiner 1993 and 1994; Schimmel et al. 1993; Schimmel and Henderson 1994), envisions the tRNA as having two domains, each encompassing half the molecule. One domain contains the terminal CCA sequence to which the incoming amino acid or growing peptide is attached. The second domain contains the anticodon and associated loop that interact with the mRNA. The two domains are frequently envisioned as being of different age with the CCA domain being older. Support for this idea stems from the fact that the CCA domain alone forms a "minihelix" to which modern tRNA synthetases can readily attach specific amino acids. Such aminoacylation has also been shown with evolved ribozymes (Lee et al. 2000), which can be surprisingly small (Chumachenko et al. 2009). In fact, aminoacylation has been reported without any enzyme or ribozyme at all (Tamura and Schimmel 2004). Furthermore, it has also been reported that a minihelix when incorporated into the 50S subunit can participate in peptide bond formation (Sardesai et al. 1999). Indeed, even the addition of a single cytosine (equivalent to C75 of modern tRNAs) to puromycin is apparently sufficient to allow peptide bond formation (Brunelle et al. 2006). Thus, it may initially only be necessary to have the CCA segment alone (Nissen et al. 2000). The 5′ domain of the tRNA is not consequential to peptide bond formation and could have been added later. If the tRNAs evolved from the one domain structure or an even simpler structure, then protein synthesis would likely have begun as a noncoded process (Schimmel and Henderson 1994). Single domain or even smaller aminoacylated RNAs are especially attractive in an RNA world where synthesis of larger RNAs is likely to be difficult. Synthesis of random oligomers in the 20–40 size range has been shown (Joshi et al. 2009; Powner et al. 2009; Szostak, 2009; Ferris et al. 1996) but the path to prebiotic synthesis of large RNAs is not without difficulties (Orgel, 2004).

How does one obtain RNAs of increasing complexity, such as those of modern tRNAs or the PTC RNA, without a true RNA replicase? There are two core possibilities, ligation and hybridization. RNA ligation has been shown to be feasible in an RNA World (Hager et al. 1996; Hager and Szostak 1997; McGinness and Joyce 2002). Thus, it is of interest that the tRNA "cloverleaf" secondary structure can be formed by a direct duplication, e.g., ligation, of an appropriate stem loop structure (Di Guilio 2002). The possible relevance of this idea was enhanced further by the demonstration that it was possible to actually replicate all the major tertiary interactions seen in modern tRNAs when two appropriate stem loop structures were ligated together (Nagaswamy and Fox 2003).

An alternative method of readily obtaining more complex structures is to simply hybridize small fragments to one another such that a larger RNA with many "nicks" is assembled. These nicks might or might not be sealed at a later stage. In *Nanoarchaeum equitans*, several tRNAs are encoded as partially complementary half molecules, which are then ligated together to form a tRNA (Randau et al. 2005a and b). In *Euglena gracilis* the large subunit rRNA is comprised of 14 discrete RNA fragments held together by hybridization events that form various helical elements. Not only are the fragments not coded in the order they appear in the final RNA but they are actually intermingled in the genome with similar fragments of the small subunit RNA (Smallman et al. 1996).

Chirality and the Ribosome

In modern organisms, mechanisms for nonribosomal peptide synthesis exist for specialized purposes and can produce peptides with unusual structures and mixed chirality (Marahiel and Essen 2009). Modern rRNAs and tRNAs are chiral with D sugars and during translation they work together to make chiral proteins with exclusively L-amino acids. This is highly advantageous to modern organisms because mixed chirality is clearly undesirable for the synthesis of structural elements such as α-helices and β-sheets that characterize modern proteins (Bada 2001; Sandars 2005). It is generally assumed that the modern chiral preferences reflect chiral synthesis of D-ribose in the RNA world (Tamura and Schimmel 2006; Tamura 2008). It thus is likely widely believed that charging of the tRNA by aminoacyl tRNA synthetases and peptide bond formation from their beginnings were chiral. In fact, with atomic resolution structural data now available, a theoretical analysis of the PTC indicates that the natural chirality of the sugar ring in the RNA is well paired with the choice of L-amino acids (Thirumoorthy and Nandi 2008).

Nevertheless, it has been shown that D-amino acids can bind to both the A and P sites of the ribosome in competition with their L-isomers (Quiggle et al. 1981; Bhuta et al. 1981). When elongation tRNAs carrying D-amino acids are presented to the ribosome in vitro they are incorporated extremely poorly, but incorporated nevertheless (Yamane et al. 1981; Heckler et al. 1988). It has recently been found that peptide synthesis can be effectively initiated with D-amino acids (Goto et al. 2008). In addition, by introducing mutations in the PTC region and/or other nearby regions of the 23S rRNA, it was possible to obtain enhanced tolerance of D-amino acids in vitro (Starck et al. 2003; Tan et al. 2004; Dedkova et al. 2003 and 2006). These results suggest that even though the rRNA is itself chiral, the essentially exclusive chirality of the modern ribosome is likely the result of selection rather than being a fundamental property of the PTC.

Regardless, of the preference of the ribosome in the modern machinery, D-amino acids will typically not reach the modern ribosome because the charging reaction also shows a strong but again imperfect chiral preference. For example, tyrosyl-tRNA synthetase is able to transfer both D and L tyrosine to its cognate tRNA although the L form is significantly preferred (Sheoran et al. 2009). Consistent with the notion that charging with D-amino acids can occur in vivo, the modern cellular machinery has a variety of mechanisms in place to prevent it. These include deacylases that remove D-amino acids from incorrectly charged tRNAs before they reach the ribosome (Soutourina et al. 2000; Yang et al.

2003). In addition, many aminoacyl tRNA synthetases have an editing domain. Nevertheless, the charging reaction has a strong chiral preference even in minihelix reactions (Tamura and Schimmel 2006; Tamura 2008).

Given that the first tRNA charging process was unlikely to better than the modern version, the onset of the ribosome as the machine for making chiral proteins likely emerged not from the chiral exclusivity of its processes, but rather from the fact it is a two tiered process with the same preference at both steps. Thus, if at earlier times 80% of the tRNAs were charged with an L-amino acid and 80% of the tRNAs charged with a D-amino acid were subsequently excluded by the ribosome then 96% of the residues incorporated into the growing protein would be of the L type. Thus, a modest peptide of 50 amino acids would perhaps have only two D-amino acids and thus have a good chance of being functional. In fact, incorporation of a D-amino acid into a modern protein is not necessarily destructive to the protein, but instead depends on where it occurs (Dedkova et al. 2003, 2006).

In summary, given that the modern ribosome is not exclusively chiral the early peptide synthesis machinery likely had significantly less chiral specificity. This would be true even if its RNA components were exclusively chiral as the various editing mechanisms associated with the modern charging process would not have been available. Thus, the peptides made by early ribosomes likely included D-amino acids and hence would tend to be unstructured.

Would such peptides be useful? Probably yes. For example, small, largely unstructured peptide segments are found in what are likely the older (not oldest) areas of the modern ribosome. Once the two tier chiral selection procedure used in the modern ribosome was established, refinements in either aspect could quickly improve the likely use of the product peptide. Future studies of partially chiral peptides might provide better insight to the nature of the earliest peptides and clarify the extent of chiral preference in the ribosomal machinery that is needed to begin to produce reasonable numbers of peptides of the modern type.

RNA HISTORY—INFERRING EVENTS BEFORE LUCA

The PTC and tRNAs clearly existed before LUCA. The fact that we can infer likely if not proven aspects of their history suggests that we can learn about events that occurred before LUCA. But, can we do this in a more general way? A key step is recognition that there are many opportunities to gain insight into relative timing. For example, if a ribosomal protein, (r-protein), is modified after translation by an enzyme, then there is a timing association between the two proteins. The more difficult second step is to develop evidence pertaining to the relative age of the associated entities, e.g., which is older? In the example, one might initially speculate that the modifying enzyme is newer than its target protein because without a target what good would the modifying enzyme be? However, this might not be the case because the modifying enzyme may have been recruited from somewhere else after the emergence of the r-protein. Finally, the two proteins could have emerged essentially simultaneously.

Given such essentially opposite alternatives, how can they be resolved? The answer will typically be independent information or a likely assumption. If the r-protein is found in all organisms and the modifying enzyme is only in Gram-positive bacteria, then it is more likely, but not proven, that the r-protein is older. Not proven because the gene for the modifying protein might have been lost in other lineages. If the protein in question is associated with a duplication event, as is the case with ribosomal proteins L15 and L18e, it may be possible to use phylogenetic arguments to deduce relative age (Roberts et al. 2008). In the example used here, when one examines the structure of the two proteins, the r-protein is much smaller and far simpler, composed only of α helices. Thus, the growing evidence for one hypothesis makes it increasing likely to be correct. Finally, the third step is to find more extensive associations involving multiple timing events.

An examination of the evidence relating to elongation factor G (EF-G) illustrates how timing information might be combined into a more general hypothesis. It has been argued from atomic resolution structures that EF-G is a structural mimic of elongation factor EF-Tu when it is part of a ternary complex with GTP and an aminoacylated tRNA (Nissen et al, 1995; Moore, 1996). If true, this hypothesis raises the obvious question of who mimics whom? Because the tRNA is at the core of the translation machinery, it likely arose very early. From this, one can logically argue that EF-G is likely newer. Regardless of the status of molecular mimicry, EF-G is actually not completely essential for the translocation process (Gavrilova & Spirin, 1971; Gavrilova et al. 1976; Spirin, 2002). Is the timing relationship proven? No, but at this stage it appears more likely than the alternative. Other claims of mimicry in the ribosomal machinery have also been made (Selmer et al. 1999; Nakamura and Ito, 2003).

The nonobvious benefit of the mimicry argument is that one can combine it with the previously discussed argument that tRNAs began as one domain RNAs (Schimmel and Henderson, 1994; Di Guilio, 1994; Schimmel and Ribas de Pouplana 1995). Thus, one produces a combined core time-line for the development of the translation machinery from a one domain tRNA predecessor, to a two domain tRNA, followed by the addition of EF-Tu, and finally the addition of EF-G. However, EF-Tu has been shown in at least one case to bind to a model one domain (minihelix) tRNA (Rudinger et al. 1994) which supports a hypothetical time line that delays the onset of the second tRNA until after the emergence of EF-Tu. This is interesting, because it also delays the possibility of mRNA-tRNA interaction and thus the onset of coding. However, in the absence of coding it is not clear how one produces a sophisticated protein.

DEDUCING THE RELATIVE AGE OF r-PROTEINS

Given that the ribosome is quite ancient, one might have expected the early r-proteins to have diverged to spawn later ones and possibly even super families of proteins used elsewhere (Ohnishi 1984; Leijonmarck et al. 1987). In fact, there are clear examples of genetic events such as gene fusion, insertion and duplication (Ramakrishnan and White 1998), but in general most r-proteins are structurally distinct. Some likely evolved with the early ribosome and then in some cases were recruited to other functions, whereas others likely evolved elsewhere and got incorporated into the ribosome at later stages. Thus, there are likely to be historical relationships between some of the proteins that will provide timing insights to the development of the subunits. One such example, e.g., S6 and S10, has been uncovered (Jue et al. 1980) and verified by structural data (Brodersen et al. 2002).

Can one deduce the relative age of the various r-proteins? Phylogenetic distribution is an obvious initial indicator with the more widely distributed proteins likely being older. However, many r-proteins are universal in all three Domains of life (Lecompte et al. 2002; Hartman et al. 2006), and hence it is not immediately obvious how one might infer relative age among members of this rather large group. Also, one must be alert to the fact that like the RNAs, all parts of the r-proteins are probably not equally old (Vishwanath et al. 2004). Experimental studies have shown that ribosomal components are assembled in a reproducible manner, which might recapitulate to a significant extent the history of the ribosome and thereby provide timing information. For this reason, it was hypothesized that the oldest proteins would assemble first and be at the core of the process whereas newer proteins would be incorporated into later stages of assembly and the newest proteins would be last (Fox and Naik 2004). The process of in vivo assembly is currently being actively studied at a very detailed level (Klein et al. 2004, Nierhaus 2007). However, traditional maps (Rohland and Nierhaus 1982; Herold and Nierhaus 1987; Nierhaus 1991) that summarize in vitro assembly remain a reasonable approximation of what actually occurs. Consistent with the hypothesis regarding assembly, an initial inspection of the traditional maps shows that the nonuniversal and hence likely newer r-proteins are largely incorporated into the ribosome at the final stages of assembly.

To focus on what are likely the oldest proteins, all the nonuniversal proteins and their associated connections were removed from the 50S subunit assembly map (Fig. 1). The hypothesis here is that the assembly order of the remaining universal r-proteins speaks to their relative age. Thus, although L23 binds directly to 23S rRNA, its assembly is also facilitated by L3 and hence it is likely a newer addition than L3. Another aspect of Figure 1 is that essentially all the remaining r-proteins are still interconnected but some are more connected than others. In complex systems, greater interdependence is likely to be associated with longer association and hence suggests greater age. For example, L1 and L3 are both universal and directly interact with the RNA, however, L3 is more central to the process of assembly and hence likely older.

Genomic organization can also be considered. Universally conserved gene clusters (e.g., operons) are very rare. When genes are associated in conserved operons they are likely to share regulatory relationships. In the case of r-proteins, four clusters of r-proteins (the S10, Str, Spc and L13 operons) are preserved in the Archaea and Bacteria (Siefert et al. 1997). In general, the universal r-proteins are encoded in the universal operons. Thus, r-proteins L2, L3, and L4 are all encoded by the S10 operon. When the assembly information is considered in combination with the other criteria, the results suggest that L2, L3, and L4, are among the oldest r-proteins (Fox and Naik 2004; Tran et al. in preparation). Overall, the conserved large subunit proteins have initially been grouped into four clusters ranging from oldest to most recent. These four groups are (1): L2, L3, L4; (2): L22, L23, L24; (3): L5, L6, L10, L11, L13, L18, L29; and (4): L9, L31, L32-L34.

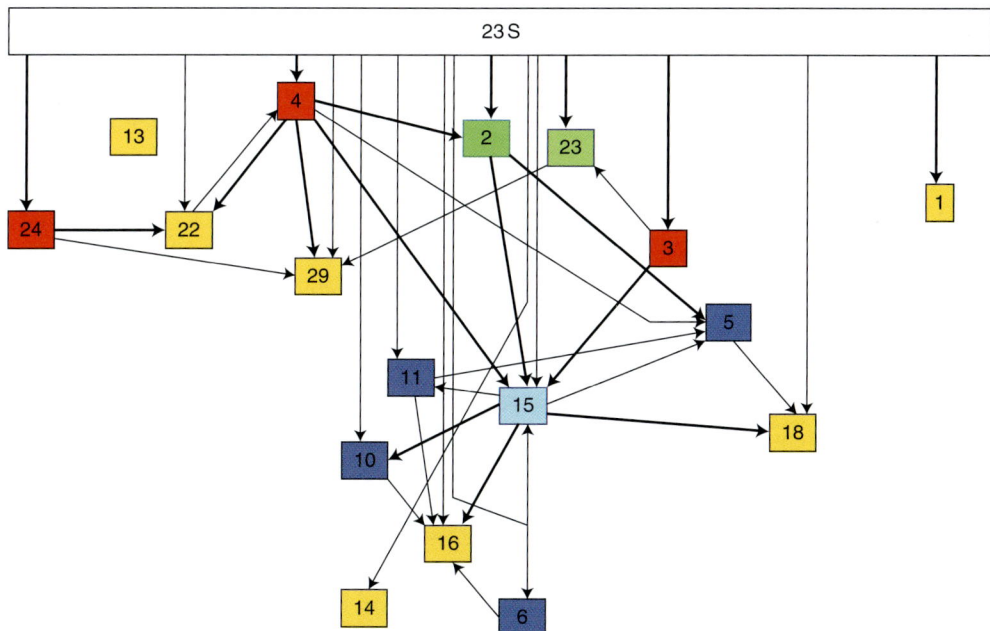

Figure 1. Assembly map of 50S ribosomal subunit with all nonuniversal protein omitted. The map was derived from Nierhaus (2001) and is a slightly modified version of that presented previously Fox and Naik (2004). Each protein is indicated by a numbered box with the 23S rRNA indicated at the top. Lines with arrows indicate order in assembly with darker lines representing stronger dependencies. Thus L4 and L24 bind directly to the RNA and work together to facilitate the incorporation of L22. Boxes are colored with regard to the similarity of their position in assembly. For example, yellow indicates terminal proteins, which are not required for addition of any other universal protein.

INFERENCES FROM r-PROTEIN STRUCTURE

Detailed examination of the structure of the older r-proteins and how they interact with the rRNAs is likely to provide insight to the development of the ribosome before LUCA. There are large amounts of information in this regard and to illustrate what might be learned two interesting examples, S1 and L2 will be discussed in some detail. Ribosomal protein L2 is universally distributed, plays a central role in ribosome assembly, is encoded in the universal L10 operon, and is near the PTC although not involved in peptide bond synthesis. Analysis of the assembly map discussed earlier suggests L2 is in fact one of the very oldest proteins. L2 has a RNA binding domain comprised of an OB-fold and an SH3 -like barrel.

The SH3 domain is homologous to similar domains found in the NusG protein (involved in Rho dependent termination of transcription), and two r-proteins, L24 (universal) and L21e (not universal). Although L24 is universal it is actually only required to initiate subunit assembly (Spillmann and Nierhaus 1978), In actuality this role can be assumed by L20 at low temperatures (Franceschi and Nierhaus 1988) and a mutant *E. coli* strain defective in L24 is viable (Herold et al. 1986). L21e and L24 are thus likely newer than L2 even though L24 binds directly to the rRNA.

The OB-fold is found in r-proteins S1, S12, S17, and S28e. The OB-fold is a small β-barrel formed from 5 strands connected by modulating loops; two or three loops on the same face of barrel are consistently observed acting as clamps to bind to their ligands (Agrawal and Kishan 2003). The SH3 domain has a characteristic fold with β-barrel architecture, which consists of five or six β-strands arranged as two tightly packed antiparallel β sheets. Two prominent loops, termed the RT and n-Src loops, are often seen in the fold (Boggon and Eck 2004). What is especially interesting from the perspective of ribosome origins is that these two folds are actually very similar. In particular, the insertion of strand β1 between β4 and β5 in the SH3-fold would actually create an OB-fold like topology (Agrawal and Kishan 2001). Thus, not only is L2 a possible progenitor of multiple r-proteins, its modern version may have arisen as a result of a very early (pre LUCA) duplication event creating two copies of one of the folds followed by a rearrangement in one of the domains to create one fold of each type. If this is correct, the obvious next question is which folding domain is older? In fact, the SH3 domain is encompassed entirely within a universal sequence block (Vishwanath et al. 2004) whereas the OB fold is partially in a block that distinguishes Bacteria and Archaea. This observation suggests that the SH3 domain may be older although position in the ribosome should also be considered.

Ribosomal protein S1 is substantially larger than all other r-proteins and in contrast with L2 is not integrally part of the ribosome. It is involved in initiation and has been associated with antitermination and trans-translation as well. It lacks an Archaeal homolog and is sometimes missing even in Bacteria suggesting it is post-LUCA addition to the ribosomal machinery. S1 contains six copies of an RNA binding domain (OB fold) that is known as the S1 domain. Many proteins in fact have one or more S1 domains. These include, but are not limited to: Polynucleotide phosphorylase, a bacterial exonuclease that degrades mRNA from 3' to 5' (Regnier et al. 1987); the α subunit of the eukaryotic initiation factor 2 (Gribskov, 1992); yeast PRP22, an RNA helicase like protein required for the release of the mRNA from the spliceosome (Company et al. 1991); and the amino-terminal end of ribonuclease E, which is involved in both 5S rRNA processing and the rapid degradation of mRNA in E. coli (Kaberdin et al. 1998). Perhaps, the most notable of the proteins that contain S1 domains for the present purposes are the translation initiation factor IF1 and its eukaryotic equivalent eIF1a, both of which also have the characteristic five stranded β barrel arrangement (Sette et al. 1997; Battiste et al. 2000). The proteins containing S1 domains can be broadly grouped into three main functional groups of RNA processing, involvement in transcription or translation and chromatin or septum regulation. The S1 motif is found in all three domains of life with the IF-1/eIF1A type are universally distributed suggesting this might be the original source of the fold. It seems likely that ribosomal protein S1 is a late addition to the ribosome, possibly derived from the initiation machinery.

EVOLUTION OF THE LARGE RIBOSOMAL RNA

The substantial structural and sequence conservation seen in comparisons of the rRNAs from all three Domains of life suggest that they reached their modern size early in the development of the ribosome. However, it is not necessarily true and in fact it is extremely unlikely that all parts of the rRNAs are of the same age. Instead, like some of the r-proteins, they have increased in size over time, perhaps beginning as an amalgamation of smaller fragments (Clark 1987; Gray and Schnare 1996). Indeed, the eukaryotic RNAs are tolerant of insertions in certain locations and have clearly grown larger since LUCA (Gray and Schnare 1996; Yokoyama and Suzuki 2008). Thus, cogent arguments have been presented that certain portions of the rRNAs are older than others (Gray and Schnare 1996; Wuyts et al. 2001; Mears et al. 2002; Caetano-Anolles 2002; Hury et al. 2006).

The secondary structure of the large rRNA reveals the presence of six domains in the RNA with the PTC being located in Domain V. Hury et al. (2006) argue that interconnectivity among distant regions can provide insight to historical timing. The argument is that older regions would have more time to be integrated into the structure and hence would show greater connectivity to other regions than newer additions. To implement this timing argument, all of the base–base interactions between regions that were not contiguous in the secondary sequence were counted. A number of regions were identified as being the most connected and hence likely to be the oldest (Fig. 2).

The oldest regions largely overlap with the minimal RNA previously deduced by comparative analysis (Mears et al. 2002) and the observed minimal core (Gutell 1992; Gray and Schnare 1996). In fact, rRNAs comprised of essentially only the regions highlighted in Figure 2 are found in various minimize mitochondrial rRNAs such as the large subunit RNA of *Trypanosoma brucei* (Sloof et al. 1985). The only exception is the GTPase region (above Region 2.5 on Fig. 2), which is involved in conformational changes during protein synthesis and hence not involved in interactions. It was argued from these results that in addition to Domain V, Domain IV, and a portion of Domain II (2.1 and 2.3 on Fig. 2) were also extremely old but no definitive decision was made regarding relative age. Domain IV is of special interest because it has major contacts with the 30S subunit (Yusupov et al. 2001) The addition of Domain IV to the structure likely corresponds with the beginning of the formation of the 30S subunit. Hsiao and Williams (2009) have observed that there are four magnesium microclusters that are shared in ribosome structures by the Bacteria (*Thermus thermophilus*) and the Archaea (*Haloarcula marismortui*). These complex clusters occur four times in the large subunit. One cluster is exclusively in the PTC whereas two others connect parts of Domain 2 (Regions 2.1 and 2.3) to Domain 4 and the PTC.

Recently, Bokov and Steinberg (2009) have improved on the connectivity argument by recognizing the potential of using the many A-minor motifs in the large subunit rRNA as timing events. The A-minor motif occurs when a stack of adenosines pack into the minor groove of a duplex region that can be some distance away in the primary sequence of the RNA (Nissen et al. 2001). Such a two-component interaction is inherently a timing event, if one component of the interaction is likely to predate the other. Bokov and Steinberg observed that in most A-minor interactions involving the PTC region (very old) the A stack was in the PTC and the helix was elsewhere (presumably newer). This implied that the A stack usually predated the helix it interacted with. From this assumption, they were able to deduce an order of addition of individual RNA regions as the RNA grew over evolutionary time, and created a hierarchical map of the RNA. The only other region where the A stack portion of the

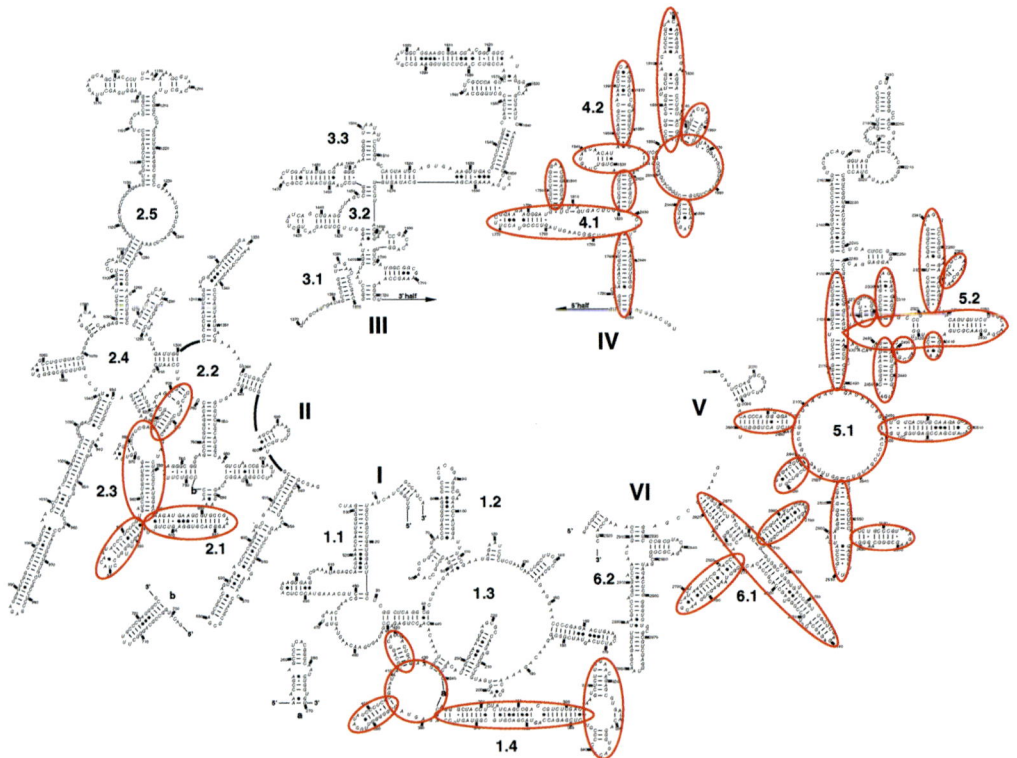

Figure 2. The secondary structure of *Haloarcula marismortui* 23S rRNA is broken into six major domains (I through VI) with subregions in the various domains denoted as 1.1, 1.2, etc. The highlighted regions are the most interconnected as measured by the numbers of base-base interactions between a residue in one domain and a residue in another. The figure is taken from Hury et al. 2006.

interaction is is concentrated is in Domain II (regions 2.1 and 2.3), which suggests that this region is also very old and in fact therefore likely older than Domain IV. Thus, the PTC region is envisioned as beginning its expansion before the small subunit RNA evolved. Because the decoding site is in the small subunit, this suggests that significant enhancement of the machinery occurred well before the beginning of coded peptide synthesis.

The immediate importance of the Bokov and Steinberg (2009) analysis is the ability to derive a hierarchical model of 23S rRNA. The core areas of the structure are again the PTC by assumption, the same portions of Domain II (region 2.1 in red on Fig. 2) and parts of Domain 4 as seen by Hury et al. (2006). However, in the context of studies of ribosome evolution, a hierarchical organization is useful in that it offers the potential for organizing diverse data into a single framework. Thus, one may be able to map the emergence of specific regions of various r-proteins to the emergence of particular rRNA segments and using other data possibly time adjust the events on different branches of the rRNA evolution map. Using this perspective, Bokov and Steinberg argue that the acquisition of the GTPase center and L1 protuberance and even the addition of the 30S subunit are relatively late additions to the ribosome. These suggestions are largely consistent with the timeline proposed earlier (Fox and Naik 2004).

RECENT RNA AND PROTEIN COEVOLUTION

An alternative approach to gain insight to how the RNA can change over time is to study changes that have occurred since LUCA. The availability of crystal structures of 50S ribosomal subunits from both Archaeal and Bacterial species has provided detailed information about the r-proteins that are unique to either the Archaea or Bacteria and how they interact with the RNA (Ban et al. 2000; Yusupov et al. 2001; Schuwirth et al. 2005). Frequently, a protein missing in one system is replaced by a different protein in the other with the implication that these diversifications developed after the divergence of the Bacteria and Achaea (Klein et al. 2004). By examining the relationship between these variable r-proteins and their associated RNA regions in detail, it is possible to document the nature of the coevolution that has occurred between the r-proteins and rRNA since LUCA. One might then, with some risk, infer that similar principles applied in the pre-LUCA era.

A detailed examination revealed that there are few completely unique proteins in either Domain of life. Instead, there are many examples of analogs. These have been analyzed in detail from a structural perspective (Klein et al. 2004). Table 1 summarizes the data for each of the nonuniversal proteins (Wang, 2006). In addition to three Archaeal proteins that have no homolog or analog in the Bacteria, there are three examples in which a single Archaeal protein has a clear analog in the Bacteria and five examples in which the Bacterial analog is comprised of two proteins. In each case, a

Table 1. Nonuniversal r-proteins in Archaea and Bacteria.

Arch.	Bact.	Conserved rRNA interaction region	Difference in 23S rRNA
L18e	X	23S rRNA domain II H28, H30; domain III H38, H42,	Extra H30 in Archaea
L19e	X	23S domain I H34; domain II H47, H53, H57, H58, H60; domain IV H62, H63; domain VI H96	H57, H63 are different for Bacteria and Archaea
L37ae	X	23S domain II H28, domain III H56, H58, domain IV H62, H67	H56 is different for Bacteria and Archaea
L39e	X	23S domain I H6, H10, domain III H49, H50, H51	No obvious difference
X	L25	5S rRNA	No obvious difference
X	L36	23S domain H42, domain V H89, domainVI H91, H97	No obvious difference
L21e	L27	23S domain II H34, domain V H81, H86; 5S rRNA	Slight difference at the end of H86
L24e	L19	23S domain VI H96, H101	No differences for RNA, but different L3 in Archaea and Bacteria next to L24e or L19
L31e	L17	H47, H61, H96, H100	Same secondary and tertiary structures for both Archaea & Bacteria
	L32 C-end	H100	
L37e	L34	23S domain I H5, H8, H23; domain II H32, H33; domain III H49	No differences for RNA; has interaction with proteins.
L15e	L9	Several residues connecting 23S domain IV H75, H76	Extra H15 in some prokaryotic species, different H10, H79 nearby between Archaea and Bacteria
	L31	23S domain I H11, H13, H15, H21; domain III H52; domain V H75	
L32e	L20	H2, H25, H40, H41, H45, H46	Different H25 between Archaea and Bacteria
	L21	H26, H40	
L44e	L33	H86, H88	H68 (only interacts with L44e) and H88 differ slightly between Arch and Bact.
	L35	H13, H86, H88	

The symbol X indicates no protein is present. Helix numbers are from Yusupov et al. 2001. L7ae is not considered because although it is not universal it is found in some Gram positive bacteria.

rather major change in the protein make up in the ribosome is associated with a very modest change in the RNA.

As an example, the region of the 23S rRNA that interacts with L17 and L19 in the *E. coli* 50S particle interacts instead with L24e and L31e in the *H. marismortui* 50S subunit. Primary sequence and structural comparisons of these proteins make it clear they are completely unrelated. Thus, the L17/L24e and L19/L24e pairs are clearly analogs created by convergent evolution. In contrast, the RNA structure is largely the same in this region in both the Archaea and Bacteria. Because the RNA structure is largely unchanged, it is likely that the protein components represent independent enhancements (probably mainly stabilization) of an even older RNA. The proteins, however, have likely been added in the Archaeal and Bacterial lineages since the common ancestor. In some cases, the extra protein is found in only one lineage and remains as a "hole" in the other, thus leaving us to speculate whether it has been lost or the use of a having protein at that location has only so far discovered in one lineage. In general, the RNA shows either no structural change or minimal change, whereas the proteins are dramatically different. In essence, we learn nothing about how the RNA grew, but it is very clear that the proteins at least in the post-LUCA era were not the driving force.

TIMELINE OF RIBOSOMAL EVOLUTION

To better organize the information regarding the origins and subsequent history of the ribosome discussed in the previous sections, it is perhaps useful to attempt to construct a time-line to outline a possible sequence of major events in the context of key historical events (Gray and Schnare 1996; Fox and Naik 2004; Wolf and Koonin 2007). Initial ribosomal development was likely fairly serial, but as its complexity increased it is probable that many developments began to occur in parallel, thus making a linear time line increasingly unrealistic as one approaches the post-LUCA age. In what follows, a scenario that attempts to incorporate the various insights discussed earlier is outlined.

The ribosome as envisioned here would have its earliest beginnings in an RNA world. Amino acids or similar molecules would be attached to very small RNA oligomers. When these RNAs encountered one another in the presence of a RNA ancestral to the PTC RNA, amide bond formation would occur with the result that larger peptide-like molecules would be created. Such a reaction has been shown to be in the realm of possibility in an RNA world (Zhang and Cech 1997). These earliest RNAs would be stabilized by Mg^{2+}. The peptides would be of mixed chirality but enriched for L-amino acids perhaps as a result of an excess of D-ribose in the RNAs of the RNA world. The early peptides might stabilize various RNAs in the RNA world and hence be advantageous. As complexity increased single domain tRNAs and the PTC region would emerge. The PTC region already encompassing the beginnings of the exit tunnel would grow, adding first the core region of Domain II and shortly thereafter portions of Domain IV. At some stage, the decoding domain of the tRNA will be added creating the modern two domain tRNA.

Although there is currently no evidence addressing this, the second domain of the tRNA may have offered the opportunity of anchoring the tRNA to an accessory RNA thereby increasing the amount of time the tRNA is associated with the PTC and hence perhaps increasing the probability of reaction (Wolf and Koonin 2007). The introduction of an anchoring RNA would have been a huge advance. By moving the anchoring RNA, one could move the primitive tRNAs and hence improve their orientations relative the PTC. The growing small ribosomal subunit likely soon took on the task of moving the template leading to the ability to eject used tRNAs and encourage arrival of new ones. Once such an anchoring RNA exists, the unexpected occurs. The anchoring RNA can serve as a template and later as a true mRNA, making it feasible to develop coded synthesis.

Alexander Mankin (see reviewer comments to Wolf and Koonin 2007) and perhaps others have raised the possibility that portions of the small ribosomal subunit RNA originated not in later times as an addition to the growing ribosome, but rather separately in the RNA world where it may have originally served as a replicase. Thus, when recruited to the emerging protein synthesis machinery, this RNA would be capable of traversing a template. Many find this model attractive as it preserves the notion of emergence of an RNA RNA replicase in the RNA world.

The key question regarding the genetic code is not the nature of the assignments, but rather when did a proto mRNA get added to the system? Decoding is inherent to the small subunit as are many of the movements associated with protein synthesis. In particular, the ratcheting motions of the small subunit are largely responsible for the movements of the tRNAs among the A, P, and E sites. Thus, we need to know the order of development of various regions in the small subunit and how their emergence tracks the development of the large subunit. At present this information is not readily available.

Once a true mRNA and core small subunit movements are in place, the ribosome would become increasingly complex by adding early conserved proteins such as L2, L3 and L4. Further expansion of the rRNA could occur by subsequent additions, for example the 5S rRNA and its associated proteins. With the onset of coding, it would be useful to store information, so an early RNA genome perhaps consisting of multiple 10KB or less RNA fragments would likely exist. What would that first genome encode? Clearly, one possibility is the conserved r-protein clusters, all of which are regulated at the RNA level (Siefert et al. 1997; Olsen and Woese 1997).

The next major step would be the addition of the modern versions of the GTPase center to the large ribosomal subunit with a resulting major increase in synthesis rates. This would allow a great radiation of cell types and likely end the age of progenotes (Woese and Fox 1977) while bringing on the post-LUCA age. Consistent with this late addition of the GTPase center is the recent argument (Frank and Gonzalez 2010) that the ribosome is essentially a Brownian motor and that EF-G is ancillary rather than instrumental in promoting movements.

Further refinements would be ongoing at this stage such as improvements in initiation, the addition of the exit site, the addition of L1, which facilitates entrance of tRNAs, introduction of posttranscriptionally modified nucleotides, and the enzymes that create them etc. The Archaeal and Bacterial RNAs would be largely fixed but newer nonuniversal proteins would be added and integration between protein synthesis and transcription increased. Ultimately limitations on genome size and stability would lead to early RNA genomes being replaced by DNA genomes.

CHALLENGES AND FUTURE DIRECTIONS

It is clear from what is presented here that much can already be inferred about the history of the ribosome in times that preceded LUCA. In the earliest stages of ribosome evolution, the cellular entities carrying "protoribosomes" would have lacked a genetic code and the complex dynamic systems of the modern ribosome. Such an entity would thus be in the "throes of evolving the genotype-phenotype relationship" and would be properly considered to be a progenote (Woese and Fox 1977). By the time of LUCA, the ribosome clearly exists in essentially its modern form. This strongly suggests that the ribosome reached a critical stage of development that facilitated the final transition from the RNA world to the RNA /protein world. What was the causative event in ribosome history? It might be argued that it was coding, but if this were the case the LUCA ribosome would likely be much more primitive. It should instead be a development that is taking place as the LUCA ribosome emerges. It is argued here and elsewhere (Hury et al. 2006; Grela et al. 2008) that this key event was the addition of the GTPase center to the ribosome. Although not essential to synthesis, the GTPase center dramatically increases the rate of peptide synthesis (Gavrilova and Spirin 1971; Gavrilova et al. 1976; Spirin 2002). Such an increase may have facilitated the transition from an RNA world to a RNA/protein world.

Looking toward future studies, the evolution of the small ribosomal subunit and its

RNA are starkly missing from what is presented here. There is an assembly map of the 30S subunit (Nomura et al. 1984) and efforts to refine it are being actively pursued (Sykes and Williamson 2009 and others). It is clear that the head region, which includes the decoding site is actively involved in the ratcheting motions (Frank and Agrawal 2000) and hence the universal proteins in the 3' domain of the 16S rRNA such as S7 are likely among the oldest. Clearly a major next step will be to examine the small subunit in detail with particular emphasis on the dynamic motions that occur during translation. A key to understanding small subunit history will be detailed knowledge of how and especially where these structural rearrangements occur. Such knowledge is just now reaching the literature (Bashan and Yonath 2008; Munro et al. 2009; Zhang et al. 2009) and has not yet been digested by the origins community. The small subunit is not the only missing piece. There are other aspects of the story that have not been addressed here. These include the evolutionary development of the aminoacyl tRNA synthetases, the initiation and termination aspects of translation, and the maturation and modification process that the RNAs and to a lesser extent the proteins undergo.

There are already substantial amounts of information in the literature regarding these and other issues which need to be brought together in the near future, perhaps as a community Wiki site on ribosome evolution similar to what is being performed for RNA families (Daub et al. 2008). This is especially true for aspects of translation that evolved entirely or in part after LUCA. For example, initiation differs significantly between Bacteria and the Archaea/Eucaryota, but nevertheless several key components are shared (Hernandez, 2008). Thus, IF-1 and eIF-2 in share an RNA-binding motif with r-protein S1 (Gribskov 1992). An examination of the Archaeal unique r-proteins (Wang et al. 2009) showed that many are genomically clustered with genes involved in transcription and initiation. In contrast, the older universal r-proteins are exclusively associated with one another with the single exception of integration with the core subunits of the RNA polymerase. Thus, there is some possibility that studies of ribosome origins may eventually expand to include other cellular processes.

In the end, no matter how complete a picture is developed of ribosomal development over time it will be hypothetical. The ultimate issue will be to prove at least the major parts of it. Thus, laboratory reconstructions will be needed. However, there would be limited value in resurrecting the complete ribosome of LUCA, because it was in effect a modern ribosome itself. An easier and likely equally informative task would be to obtain high resolution structural information on the minimalized ribosomes found in various mitochondria. Laboratory reconstructions may instead best focus on examining meaningful pieces.

For example, in the case of both major tRNA synthetase families, it is the catalytic subunit that is by far the most conserved (O'Donoghue et al. 2003). Other less conserved subunits provide the ability to recognize specific tRNAs and to edit charging errors. One can therefore infer a timeline for increased complexity of these multisubunit enzymes in which the ability to aminoacylate precedes these other features. That is to say, the ability to aminoacylate small RNAs may predate the ability to distinguish individual RNAs as being appropriate targets for the addition of particular amino acids. Thus, the first synthetases may have aminoacylated largely randomly. A relevant experiment then would be to reconstruct an ancestral synthetase catalytic subunit and see if it can charge a one domain tRNA and if so, with what amino acids. However, the critical first target for reconstruction will be the PTC and efforts in this direction have already begun (Davidovich et al. 2009). A full fledged experimental program will become possible if it can be shown that a PTC fragment can catalyze peptide bond formation when presented with CCA terminated RNAs carrying amino acids.

REFERENCES

Agmon I, Bashan A, Zarivach R, Yonath A. 2005. Symmetry at the active site of the ribosome: Structural and functional implications. *Biol Chem* **386:** 833–844.

Agrawal V, Kishan RK. 2001. Functional evolution of two subtly different (similar) folds. *BMC Struct Biol* **1**: 5.

Agrawal V, Kishan KV. 2003. OB-fold: Growing bigger with functional consistency. *Curr Protein Pept Sci* **4**: 195–206.

Anantharaman V, Koonin EV, Aravind L. 2002. Comparative genomics and evolution of proteins involved in RNA metabolism. *Nucleic Acids Res* **30**: 1427–1464.

Bada JL. 2001. State-of-the-art instruments for detecting extraterrestrial life. *Proc Natl Acad Sci* **98**: 797–800.

Bailey S, Wing RA, Steitz TA. 2006. The structure of *T. aquaticus* DNA polymerase III is distinct from eukaryotic replicative DNA polymerases. *Cell* **126**: 893–904.

Ban N, Nissen P, Hansen J, Moore PB, Steitz TA. 2000. The complete atomic structure of the large ribosomal subunit at 2.4 A resolution. *Science* **289**: 905–920.

Bashan A, Yonath A. 2008. Correlating ribosome function with high resolution structures. *Trends Microbiol* **16**: 326–335.

Battiste JL, Pestova TV, Hellen CU, Wagner G. 2000. The eIF1A solution structure reveals a large RNA–binding surface important for scanning function. *Mol Cell* **5**: 109–119.

Benner SA, Cohen MA, Gonnet GH, Berkowitz DB, Johnsson KP. 1993. Reading the palimpset: Contemporary biochemical data and the RNA world. In Gasteland R.F., Atkins J.F. eds. *The RNA world*. 1st ed. Cold Spring Harbor: Cold Spring Harbor Laboratory Press, pp 27–70.

Bhuta A, Quiggle K, Ott T, Ringer D, Chladek S. 1981. Stereochemical control of ribosomal peptidyltransferase reaction. Role of amino acid side chain orientation of acceptor substrate. *Biochemistry* **20**: 8–15.

Boggon TJ, Eck MJ. 2004. Structure and regulation of Src family kinases. *Oncogene* **23**: 7918–7927.

Bokov K, Steinberg SV. 2009. A hierarchical model for evolution of 23S ribosomal RNA. *Nature* **457**: 977–980.

Brimacombe R. 1991. RNA-protein interactions in the *Escherichia coli* ribosome. *Biochimie* **73**: 927–936.

Brodersen DE, Clemons WM Jr, Carter AP, Wimberly BT, Ramakrishnan V. 2002. Crystal structure of the 30 S ribosomal subunit from *Thermus thermophilus*: Structure of the proteins and their interactions with 16 S RNA. *J Mol Biol* **316**: 725–768.

Brunelle JL, Youngman EM, Sharma D, Green R. 2006. The interaction between C75 of tRNA and the A loop of the ribosome stimulates peptidyl transferase activity. *RNA* **12**: 33–39.

Caetano-Anolles G. 2002. Tracing the evolution of RNA structure in ribosomes. *Nucleic Acids Res* **30**: 2575–2587.

Calendar R, Berg P. 1967. D-Tyrosyl RNA: Formation, hydrolysis and utilization for protein synthesis. *J Mol Biol* **26**: 39–54.

Chumachenko NV, Novikov Y, Yarus M. 2009. Rapid and simple ribozymic aminoacylation using three conserved nucleotides. *J Am Chem Soc* **131**: 5257–5263.

Clark CG. 1987. On the evolution of ribosomal RNA. *J Mol Evol* **25**: 343–350.

Company M, Arenas J, Abelson J. 1991. Requirement of the RNA helicase-like protein PRP22 for release of messenger RNA from spliceosomes. *Nature* **349**: 487–493.

Connell SR, Takemoto C, Wilson DN, Wang H, Murayama K, Terada T, Shirouzu M, Rost M, Schüler M, Giesebrecht J, et al. 2007. Structural basis for interaction of the ribosome with the switch regions of GTP-bound elongation factors. *Mol Cell* **25**: 751–764.

Danchin A, Fang G, Noria S. 2007. The extant core bacterial proteome is an archive of the origin of life. *Proteomics* **7**: 875–889.

Daub J, Gardner PP, Tate J, Ramskold D, Manske M, Scott WG, Weinberg Z, Griffiths-Jones S, Bateman A. 2008. The RNA WikiProject: Community annotation of RNA families. *RNA* **14**: 2462–2464.

Davidovich C, Belousoff M, Bashan A, Yonath A. 2009. The evolving ribosome: From non-coded peptide bond formation to sophisticated translation machinery. *Res Microbiol. Jul 18 [Epub ahead of print]*

Dedkova LM, Fahmi NE, Golovine SY, Hecht SM. 2003. Enhanced D-amino acid incorporation into proteins by modified ribosomes. *J Am Chem Soc* **125**: 6616–6617.

Dedkova LM, Fahmi NE, Golovine SY, Hecht SM. 2006. Construction of modified ribosomes for incorporation of D-amino acids into proteins. *Biochemistry* **45**: 15541–15551.

Di Giulio M. 1992. On the origin of the transfer RNA molecule. *J Theor Biol* **159**: 199–214.

Di Giulio M. 1994. On the origin of protein synthesis: A speculative model based on hairpin RNA structures. *J Theor Biol* **171**: 303–308.

Di Giulio M. 2009. A comparison among the models proposed to explain the origin of the tRNA molecule: A synthesis. *J Mol Evol* **69**: 1–9.

Ferris JP, Hill AR Jr, Liu R, Orel LE. 1996. Synthesis of long prebiotic oligomers on mineral surfaces. *Nature* **381**: 59–61.

Fox GE, Naik AK. 2004. The evolutionary history of the ribosome, In *The genetic code and the origin of life* (Ribas de Pouplana L. ed), Landes Bioscience Chapter **6**, pp 92–105.

Franceschi FJ, Nierhaus KH. 1988. Ribosomal protein L20 can replace the assembly-initiator protein L24 at low temperatures. *Biochemistry* **27**: 7056–7059.

Frank J, Agrawal RK. 2000. A ratchet-like inter-subunit reorganization of the ribosome during translocation. *Nature* **406**: 318–322.

Frank J, Gonzalez RL Jr. 2010. Structure and dynamics of a processive Brownian motor: The translating ribosome. *Annu Rev Biochem 2010 Mar 17 [Epub ahead of print]*.

Gavrilova LP, Spirin AS. 1971. Stimulation of "non-enzymic" translocation in ribosomes by p-chloromercuribenzoate. *FEBS Lett* **17**: 324–326.

Gavrilova LP, Kostiashkina OE, Koteliansky VE, Rutkevich NM, Spirin AS. 1976. Factor-free ("Non-enzymic") and factor-dependent systems of translation of polyuridylic acid by *Escherichia coli* ribosomes. *J Mol Bio* **101**: 537–552.

Goto Y, Murakami H, Suga H. 2008. Initiating translation wih D-amino acids. *RNA* **14**: 1390–1398.

Gray MW, Schnare MN. 1996. Evolution of rRNA gene organization, in Ribosomal RNA Structure, Evolution, Processing, and Function in Protein Biosynthesis (eds

R.A. Zimmerman, and A.E. Dahlberg), CRC Press, Boca Raton FL. pp49–69.

Grela P, Bernado P, Svergun D, Kwiatowski J, Abramczyk D, Grankowski N, Tchorzewski M. 2008. Structural relationships among the ribosomal stalk proteins from the three Domains of life. *J Mol Evol* **67:** 154–167.

Gribskov M. 1992. Translational initiation factors IF-1 and eIF-2 α share an RNA-binding motif with prokaryotic ribosomal protein S1 and polynucleotide phosphorylase. *Gene* **119:** 107–111.

Gutell RR. 1992. Evolutionary characteristics of 16S and 23S rRNA structures, in *The Origin and Evolution of the Cell* (eds H. Hartman and K. Matsuno), World Scientific, pp. 243–309.

Hager AJ, Szostak JW. 1997. Isolation of novel ribozymes that ligate AMP-activated RNA substrates. *Chem Biol* **4:** 607–617.

Hager AJ, Pollard JD, Szostak JW. 1996. Ribozymes: aiming at RNA replication and protein synthesis. *Chem Biol* **3:** 717–725.

Heckler TG, Roesser JR, Xu C, Chang PI, Hecht SM. 1988. Ribosomal binding and dipeptide formation by misacylated tRNA[Phe's]. *Biochemistry* **27:** 7254–7262.

Hernandez G. 2008. Was the initiation of translation in early eukaryotes IRES-driven? *Trends Biochem Sci* **33:** 58–64.

Herold M, Nierhaus KH. 1987. Incorporation of six additional proteins to complete the assembly map of the 50 S subunit from *Escherichia coli* ribosomes. *J Biol Chem* **262:** 8826–8833.

Herold M, Nowotny V, Dabbs ER, Nierhaus KH. 1986. Assembly analysis of ribosomes from a mutant lacking the assembly-initiator protein L24: lack of L24 induces temperature sensitivity. *Mol Gen Genetics* **203:** 281–287.

Hsiao C, Williams LD. 2009. A recurrent magnesium-binding motif provides a framework for the ribosomal peptidyl transferase center. *Nucl Acids Res* **37:** 3134–3142.

Hsiao C, Mohan S, Kalahar BK, Williams LD. 2009. Peeling the onion: Ribosomes are ancient molecular fossils. *Mol Biol Evol* **26:** 2415–2425.

Hury J, Nagaswamy U, Larios-Sanz M, Fox GE. 2006. Ribosome origins: The relative age of 23S rRNA domains. *Orig Life Evol Biosphere* **36:** 421–429.

Joshi PC, Aldersley MF, Delano JW, Ferris JP. 2009. Mechanism of montmorillonite catalysis in the formation of RNA oligomers. *J Am Chem Soc* **131:** 13369–13374.

Jue RA, Woodbury NW, Doolittle RF. 1980. Sequence homologies among *E. coli* ribosomal proteins: Evidence for evolutionarily related groupings and internal duplications. *J Mol Evol* **15:** 129–148.

Kaberdin VR, Miczak A, Jakobsen JS, Lin-Chao S, McDowall KJ, von Gabain A. 1998. The endoribonucleolytic N-terminal half of *Escherichia coli* RNase E is evolutionarily conserved in *Synechocystis* sp. and other bacteria but not the C-terminal half, which is sufficient for degradosome assembly. *Proc Natl Acad Sci* **95:** 11637–11642.

Klein DJ, Moore PB, Steitz TA. 2004. The roles of ribosomal proteins in the structure assembly, and evolution of the large ribosomal subunit. *J Mol Biol* **340:** 141–177.

Kyrpides N, Overbeek R, Ouzounis C. 1999. Universal protein families and the functional content of the last universal common ancestor. *J Mol Evol* **49:** 413–423.

Lazcano A. 1994. Cellular evolution during the early Archaea: What happened between the progenote and the cenancestor? *Microbiologia SEM* **11:** 13–18.

Lecompte O, Ripp R, Thierry JC, Moras D, Poch O. 2002. Comparative analysis of ribosomal proteins in complete genomes: An example of reductive evolution at the domain scale, *Nucleic Acids Res* **30:** 5382–5390.

Lee N, Bessho Y, Wei K, Szostak JW, Suga H. 2000. Riboszyme-catalyzed tRNA aminoacylation. *Nat Struct Biol* **7:** 28–33.

Leijonmarck M, Liljas A. 1987. Structure of the C-terminal domain of the ribosomal protein L7/L12 from *Eschericia coli* at 1.7A. *J Mol Biol* **195:** 555–579.

McGinness KE, Joyce GF. 2002. RNA-catalyzed RNA ligation on an external RNA template. *Chem Biol* **9:** 585–596.

Maizels N, Weiner AM. 1993. The genomic tag hypothesis: modern viruses as molecular fossils of ancient strategies for genomic replication. In: Gesteland R.F., Atkins J.F. (eds) The RNA World, Cold Springs Harbor Laboratory Press, Plainview, NY, pp 577–602.

Maizels N, Weiner AM. 1994. Phylogeny from function: evidence from the molecular fossil record that tRNA originated in replication, not translation. *Proc Natl Acad Sci* **91:** 6729–6734.

Marahiel MA, Essen O. 2009. Chapter 13. Nonribosomal peptide synthetases mechanistic and structural aspects of essential domains. *Methds Enzymol* **458:** 337–351.

Mears JA, Cannone JJ, Stagg SM, Gutell RR, Agrawal RK, Harvey SC. 2002. Modeling a minimal ribosome based on comparative sequence analysis. *J Mol Biol* **321:** 215–234.

Moore PB. 1996. Molecular mimicry in protein synthesis. *Science* **270:** 1453–1454.

Munro JB, Sanbonmatsu KY, Spahn CM, Blanchard SC. 2009. Navigating the ribosome's metastable energy landscape. *Trends Biochem Sci* **34:** 390–400.

Mushegian AR, Koonin EV. 1996. A minimal gene set for cellular life derived by comparison of complete bacterial genomes. *Proc Natl Acad Sci* **93:** 10268–10273.

Nagaswamy U, Fox GE. 2003. RNA ligation and the origin of tRNA. *Orig Life Evol Biosph* **36:** 421–429.

Nakamura Y, Ito K. 2003. Making sense of mimic in translation termination. *Trends in Biochem Sci* **28:** 99–105.

Nierhaus KH. 1991. The assembly of prokaryotic ribosomes. *Biochimie* **73:** 739–755.

Nierhaus KH. 2007. Question 6: Early steps of evolution and some ideas about a simplified translational machinery. *Orig Life Evol Biosph* **37:** 391–398.

Nissen P, Hansen J, Ban H, Moore PB, Steitz TA. 2000. The structural basis of ribosome activity in peptide bond synthesis. *Science* **289:** 920–930.

Nissen P, Ippolito JA, Ban N, Moore PB, Steitz TA. 2001. RNA tertiary interactions in the large ribosomal subunit: The A-minor motif. *Proc Natl Acad Sci* **98:** 4899–4903.

Nissen P, Kjeldgaard M, Thirup S, Polekhina G, Reshetnikova L, Clark BF, Nyborg J. 1995. Crystal structure of the ternary complex of PhetRNA[Phe], EFTu and a GTP analog. *Science* **270**: 1464–1472.

Noller HF. 1993. On the origin of the ribosome: Co-evolution of sub-domains of tRNA and rRNA, In: Gesteland R.F., Atkins J.F. (eds) The RNA world, Cold Springs Harbor Laboratory Press, Plainview, NY, pp 137–156.

Nomura M, Gourse R, Baughman G. 1984. Regulation of the synthesis of ribosomes andribosomal components. *Annu Rev Biochem* **53**: 75–117.

O'Donoghue P, Luthey-Schulten Z. 2003. On the evolution of structure in aminoacyl-tRNA synthetases. *Microbiol Mol Biol Rev* **67**: 550–573.

Ohnishi K. 1984. Towards a classification of *E. coli* ribosomal proteins: a hypothetical 'small ribosome' as a primitive protein-synthesizing apparatus. *Orig Life* **14**: 717–724.

Olsen GJ, Woese CR. 1997. Archaeal genomics: an overview. *Cell* **89**: 991–994.

Orgel LE. 2004. Prebiotic chemistry and the origin of the RNA world. *Crit Rev Biochem Mol Biol* **39**: 99–123.

Powner MW, Gerland B, Sutherland JD. 2009. Synthesis of activated pyrimidines ribonucleotides in prebiotically plausible conditions. *Nature* **459**: 239–242.

Quiggle K, Kumar G, Ott TW, Ryu EK, Chladek S. 1981. Donor site of ribosomal peptidyltransferase: Investigationof substrate specificity using 2′(3′)-O-(N-acyaminoacyl)dinucleoside phosphates as models of the 3′ terminus of N-acylaminoacyl transfer ribonucleic acid. *Biochemistry* **20**: 3480–3485.

Ramakrishnan V, White SW. 1998. Ribosomal protein structures: insights into the architecture, machinery and evolution of the ribosome. *Trends Biochem Sci* **23**: 208–212.

Randau L, Calvin K, Hall M, Yuan J, Podar M, Li H, Söll D. 2005a. The heteromeric *Nanoarchaeum equitans* splicing endonuclease cleaves noncanonical bulge-helix-bulge motifs of joined tRNA halves. *Proc Natl Acad Sci* **102**: 17934–17939.

Randau L, Münch R, Hohn MJ, Jahn D, Söll D. 2005b. *Nanoarchaeum equitans* creates functional tRNAs from separate genes for their 5′- and 3′-halves. *Nature* **433**: 537–541.

Regnier P, Grunberg-Manago M, Portier C. 1987. Nucleotide sequence of the pnp gene of *Escherichia coli* encoding polynucleotide phosphorylase. Homology of the primary structure of the protein with the RNA-binding domain of ribosomal protein S1. *J Biol Chem* **262**: 63–68.

Roberts E, Montoya J, Sethi A, Woese CR, Luthey-Schulten Z. 2008. Molecular signatures of the past. *Proc Natl Acad Sci USA* **105**: 13953–13958.

Rohland R, Nierhaus KH. 1982. Assembly map of the large subunit (50S) of *Escherichia coli* ribosomes. *Proc Natl Acad Sci* **79**: 729–733.

Rudinger J, Blechschmitd B, Ribeiro S, Sprinzl M. 1994. Minimalist aminoacylated RNAs as efficient substrates for elongation factor Tu. *Biochemistry* **33**: 5682–5688.

Sandars PGH. 2005. Chirality in the RNA world and beyond. *Intn J Astrobiol* **4**: 49–61.

Sardesai NY, Green R, Schimmel P. 1999. Efficient 50S ribosome-catalyzed peptide bond synthesis with an aminoacyl minihelix. *Biochemistry* **38**: 12080–12088.

Schimmel P, Giege R, Moras D, Yokoyama S. 1993. An operational RNA code foramino acids and possible relationship to genetic code. *Proc Natl Acad Sci USA* **90**: 8763–8768.

Schimmel P, Henderson B. 1994. Possible role of aminoacyl-RNA complexes in noncoded peptide synthesis and origin of coded synthesis. *Proc Natl Acad Sci* **91**: 11283–11286.

Schimmel P, Ribas de Pouplana L. 1995. Transfer RNA: From minihelix to genetic code. *Cell* **81**: 983–986.

Schuwirth BS, Borovinskaya MA, Hau CW, Zhang W, Vila-Sanjurjo A, Holton JM, Cate JH. 2005. Structure of the bacterial ribosome at 3.5 A resolution. *Science* **310**: 827–834.

Selmer M, Al-Karadaghi S, Hirokawa G, Kaji A, Liljas A. 1999. Crystal structure of *Thermotoga maritima* ribosome recycling factor: a tRNA mimic. *Science* **286**: 2349–2352.

Selmer M, Dunham CM, Murphy FV 4th, Weixlbaumer A, Petry S, Kelley AC, Weir JR, Ramakrishnan V. 2006. Structure of the 70S ribosome complexed with mRNA and tRNA. *Science* **313**: 1935–1942.

Sette M, van Tilborg P, Spurio R, Kaptein R, Paci M, Gualerzi CO, Boelens R. 1997. The structure of the translational initiation factor IF1 from *E. coli* contains an oligomer -binding motif. *EMBO J* **16**: 1436–1443.

Sheoran A, Sharma G, First EA. 2008. Activation of D-tyrosine by *Bacillus stearothermophilus* tyrosyl-tRNA synthetase: 1. Pre-steady-state kinetic analysis reveals the mechanistic basis for the recognition of D-tyrosine. *J Biol Chem* **283**: 12971–12980.

Siefert JL, Martin KA, Abdi F, Widger WR, Fox GE. 1997. Conserved gene clusters in bacterial genomes provide further support for the primacy of RNA. *J Mol Evol* **45**: 467–472.

Simonović M, Steitz TA. 2008. Cross-crystal averaging reveals that the structure of the peptidyl-transferase center is the same in the 70S ribosome and the 50S subunit *Proc Natl Acad Sci USA* **105**: 500–505.

Sloof P, Van den Burg J, Voogd A, Benne R, Agostinelli M, Borst P, Gutell R, Noller H. 1985. Further characterization of the extremely small mitochondrial ribosomal RNAs from trypanosomes: a detailed comparison of the 9S and 12S RNAs from Crithidia fasciculate and *Trypanosoma brucei* with rRNAs from other organisms. *Nucleic Acids Res* **13**: 4171–4190.

Smallman DS, Schnare MN, Gray MW. 1996. RNA:RNA interactions in the large subunit ribosomal RNA of *Euglena gracilis*. *Biochim Biophys Acta* **1305**: 1–6.

Smith TF, Lee JC, Gutell RR, Hartman H. 2008. The origin and evolution of the ribosome. *Biol Direct* **3**: 16.

Soutourina J, Plateau P, Blanquet S. Metabolism of D-aminoacyl-tRNAs in *Escherichia coli* and *Saccharomyces cerevisiae* cells. *J Biol Chem* **275**: 32535–32542.

Spillmann S, Nierhaus KH. 1978. The ribosomal protein L24 of *Escherichia coli* is an assembly protein. *J Biol Chem* **253**: 7047–7050.

Spirin AS. 2002. Ribosome as a molecular machine. *FEBS Lett* **514**: 2–10.

Starck SR, Qi X, Olsen BN, Roberts RW. 2003. The puromycin route to assess stero- and regiochemical constraints on peptide bond formation in eukaryotic ribosomes. *J Am Chem Soc* **125:** 8090–8091.

Steitz TA. 2008. A structural understanding of the ribosome. *Nat Rev Mol Cell Biol* **9:** 242–253.

Sykes MT, Williamson JR. 2009. A complex assembly landscape for the 30S ribosomal subunit. *Annu Rev Biophys* **38:** 197–215.

Szostak JW. 2009. Systems chemistry on early earth. *Nature* **459:** 171–172.

Tamura K. 2008. Origin of amino acid homochirality: relationship with the RNA world and origin of tRNA aminoacylation. *Biosystems* **92:** 91–98.

Tamura K, Schimmel P. 2004. Non-enzymatic aminoacylation of an RNA minihelix with an aminoacyl phosphate oligonucleotide. *Nucleic Acids Symp Ser* **48:** 269–270.

Tamura K, Schimmel PR. 2006. Chiral-selective aminoacylation of an RNA minihelix: mechanistic features and chiral suppression. *Proc Natl Acad Sci* **103:** 13750–13752.

Tan Z, Forster AC, Blacklow SC, Cornish VW. 2004. Amino acid backbone specificity of the *Escherichia coli* translation machinery. *J Am Chem Soc* **126:** 12752–12753.

Thirumoorthy K, Nandi N. 2008. Role of chirality of the sugar ring in the ribosomal peptide synthesis. *J Phys Chem B* **112:** 9187–9195.

Vishwanath P, Favaretto P, Hartman H, Mohr SC, Smith TF. 2004. Ribosomal protein-sequence block structure suggests complex prokaryotic evolution with implications for the origin of eukaryotes. *Mol Phylo Genet Evol* **33:** 615–625.

Wang J. 2006. From genome to structure: comparative studies of archaeal unique ribosomal proteins. Ph. D. Dissertation, University of Houston, Houston, TX.

Wang J, Dasgupta I, Fox GE. 2009. Many non-universal archaeal ribosomal proteins are found in conserved gene clusters. *Archaea* **2:** 241–251.

Wimberly BT, Brodersen DE, Clemons WM Jr, Morgan-Warren RJ, Carter AP, Vonrhein C, Hartsch T, Ramakrishnan V. 2000. Structure of the 30S ribosomal subunit. *Nature* **407:** 327–339.

Woese CR, Fox GE. 1977. The concept of cellular evolution. *J Mol Evol* **10:** 1–6.

Wolf YI, Koonin EV. 2007. On the origin of the translation system and the genetic code in the RNA world by means of natural selection, exaptation, and subfunctionalization. *Biology Direct* **2:** 14.

Wuyts J, Van de Peer Y, De Wachter R. 2001. Distribution of substitution rates and locations of insertion sites in the tertiary structure of ribosomal RNA. *Nucleic Acids Res* **29:** 5017–5028.

Yamane T, Miller DL, Hopfield JJ. 1981. Discrimination between D and L-tyrosyl transfer ribonucleic acids in peptide chain elongation. *Biochemistry* **20:** 7059–7068.

Yang H, Zheng G, Peng X, Qiang B, Yuan J. 2003. D-Amino acids and D-Tyr-tRNA^{tyr} deacylase: Stereospecificity of the translation machine revisited. *FEBS Lett* **552:** 95–98.

Yokoyama T, Suzuki T. 2008. Ribosomal RNAs are tolerant towards genetic insertions: Evolutionary origin of expansion segments. *Nucleic Acids Res* **36:** 3539–3551.

Yusupov MM, Yusupova GZ, Baucom A, Lieberman K, Earnest TN, Cate JH, Noller HF. 2001. Crystal structure of the ribosome at 5.5 A resolution. *Science* **292:** 883–896.

Zhang B, Cech TR. 1997. Peptide bond formation by *in vitro* selected ribozymes. *Nature* **390:** 96–100.

Zhang W, Dunkle JA, Cate JHD. 2009. Structures of the ribosome in intermediate states of ratcheting. *Science* **325:** 12014–1017.

Zimmerman E, Yonath A. 2009. Biological implications of the ribosomes's stunning stereochemistry. *Chembiochem* **10:** 63–72.

Deep Phylogeny—How a Tree Can Help Characterize Early Life on Earth

Eric A. Gaucher, James T. Kratzer, and Ryan N. Randall

School of Biology, School of Chemistry, and Parker H. Petit Institute for Bioengineering and Biosciences, Georgia Institute of Technology, Atlanta, Georgia

Correspondence: eric.gaucher@biology.gatech.edu

The Darwinian concept of biological evolution assumes that life on Earth shares a common ancestor. The diversification of this common ancestor through speciation events and vertical transmission of genetic material implies that the classification of life can be illustrated in a tree-like manner, commonly referred to as the Tree of Life. This article describes features of the Tree of Life, such as how the tree has been both pruned and become bushier throughout the past century as our knowledge of biology has expanded. We present current views that the classification of life may be best illustrated as a ring or even a coral with tree-like characteristics. This article also discusses how the organization of the Tree of Life offers clues about ancient life on Earth. In particular, we focus on the environmental conditions and temperature history of Precambrian life and show how chemical, biological, and geological data can converge to better understand this history.

> *"You know, a tree is a tree. How many more do you need to look at?"*
> –Ronald Reagan (Governor of California), quoted in the *Sacramento Bee*, opposing expansion of Redwood National Park, March 3, 1966

The following article addresses a period in life most removed from life's origins compared with other articles in this collection. The article discusses an advanced form of life that seems to have lived on the order of 3.5–4.0 billion years ago, around the time when life as we know it began to diversify in a Darwinian sense. The life from this geological period is located deep within an illustrated taxonomic tree of life. The hope is that by understanding how early life evolved, we can better understand how life originated. In this sense, the article attempts to travel backwards in time, starting from modern organisms, to understand life's origin.

The Darwinian concept of evolution suggests that all modern life shares a single common ancestor, often referred to as the last universal common ancestor (LUCA). Throughout evolutionary history, this ancestor has for the most part generated descendants as successive bifurcations in a tree-like manner. This so called Tree of Life, and phylogenetics in general provides much of the framework for the field of molecular evolution. Taxonomic trees allow us to better understand relationships and commonalities shared by life. For instance, a tree may tell us whether a trait or phenotype shared between two organisms is the result of shared-common ancestry (termed homologous traits) or whether the trait has evolved multiple times independent of ancestry (analogous traits such as wings).

Taxonomic trees can be built using diverse sources of information. These can include morphological and phenotypic data at the macro-level down to DNA and protein sequence data at the micro-level. Ideally, trees built from multiple sources of input have identical taxonomic relationships and branching patterns, and such trees are said to be congruent. In practice, however, trees built from morphological data (say, presence or absence of wings) are often different than a tree built from molecular data (DNA or protein sequences). This requires the biologist to determine which of the two data sets is misleading and/or which taxonomic tree-building algorithm is most appropriate to use for a particular data set. Such an artform is common in the field of molecular evolution because rarely are trees congruent when built from two sources of input data.

In light of this fact, we have provided the quote at the beginning of this article as a reflection about the field of molecular evolution and its interpretations of taxonomic trees. Although Reagan was not speaking about taxonomic trees in his quote, the same sort of disconnect exists between evolutionary biologists and molecular biologists (Woese and Goldenfeld 2009), as it did between conservationists and Ronald Reagan. A molecular biologist may be inclined to say that once you have seen one phylogenetic tree, you have seen them all. And in fairness, there is some validity to such a notion because historically a phylogenetic tree could not help a molecular biologist to better describe their system. An evolutionary biologist, however, will argue that individual trees have nuances that can dramatically alter our interpretation of evolutionary processes.

We intend to show in this article that not all (taxonomic) trees look similar and describe identical evolutionary scenarios. We will discuss how our concept of the Tree of Life has changed over the past couple of decades, how trees can be interpreted, and what a tree can tell us about early life. In particular, the article will focus on the temperature conditions of early life because this topic has received much attention over the past few years as a direct result of improved DNA sequencing technology and a better understanding of molecular evolutionary processes. We will also describe how trees can be used to guide laboratory experiments in our attempt to understand ancient life. Lastly, we will discuss how phylogenetic trees will serve as the foundation for an "evolutionary synthetic biology" that should allow us to better understand the evolution of cellular pathways, macromolecular machines such as the ribosome, and other emergent properties of early life.

BACKGROUND

Prokaryotes and Eukaryotes

All natural and physical scientists have been taught that biological classification is the manner in which organisms are categorized according to common or shared traits. Two organisms will be located in close proximity within a classification system if those two organisms have similar characteristics. The greater the number of shared characteristics, the closer the two organisms will be grouped within the classification system.

The ability to classify organisms is probably a reflection of the notion that all living organisms share a common ancestor. In essence then, Darwinian evolution has already created a classification scheme, and it is our job to illustrate this scheme in a taxonomic context. Our ability to recapitulate life's phylogeny depends of course on our ability to identify all life forms and describe these life forms at a sufficiently detailed level, allowing us to identify shared characteristics resulting from common ancestry.

Biologists have made tremendous progress in their classification scheme during the past couple of centuries. For instance, Edouard Chatton outlined his classification scheme that divided life into two categories in the 1920s—he divided life into prokaryotes and eukaryotes (Fig. 1A) (Chatton 1925). Eukaryotic cells had a nucleus that encapsulates their genomic DNA, whereas prokaryotic cells had no nuclear organelle. From an evolutionary perspective, and by

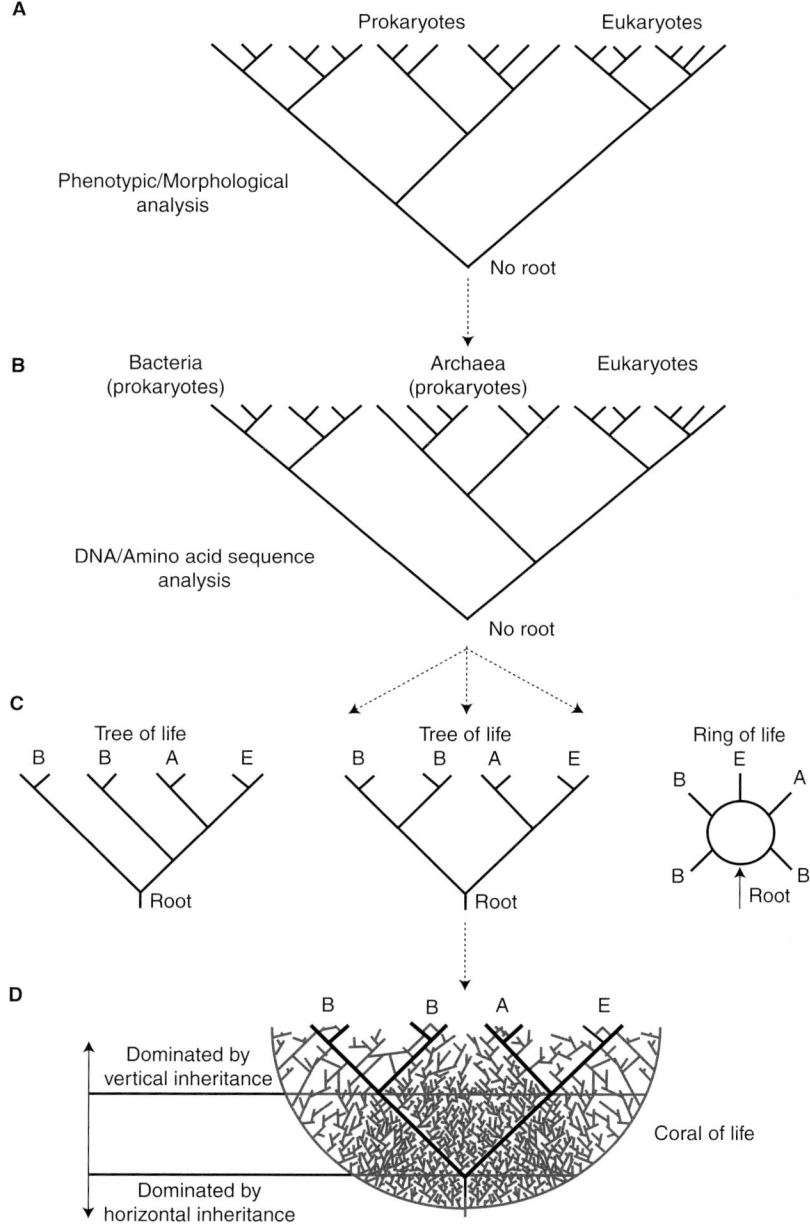

Figure 1. Tree of Life and its evolution over 100 years. (A) Taxonomy of life based on morphological characteristics (Chatton 1925). (B) Phylogenetic tree of life based on DNA sequence analysis (Woese and Fox 1977). (C) Competing views for the rooting of the phylogenetic tree of life. Initials A, B, and E represent Archaea, Bacteria, and Eukarotes, respectively. The tree on the *left* is based on membrane architecture and insertion/deletion events in gene (Cavalier-Smith 2002), the tree in the *center* is based on ancient gene duplication events (Gogarten et al. 1989; Iwabe et al. 1989; Brown and Doolittle 1995; Brown et al. 1997; Gribaldo and Cammarano 1998), and the tree on the *right* is based on phylogenetic analysis of hundreds of genes (Rivera and Lake 2004). (D) Most recent view about the tree of life in light of vertical and horizontal gene transmission (Fournier et al. 2009).

definition, prokaryotes (meaning before nucleus) were the progenitors to eukaryotes (dawn of the nucleus). Prokaryotic organisms were microscopic and morphologically similar but metabolically very diverse in their ability to inhabit "extreme" environments. Conversely, eukaryotic organisms were both microscopic and macroscopic and metabolically similar but morphologically they are very diverse because of their multicellularity. The prokaryotic/eukaryotic perspective seemed reasonable at the time because prokaryotes were for the most part morphologically "simple" single-celled organisms, whereas eukaryotes were for the most part "complex" multicellular organisms. The use of simple and complex were of course anthropomorphic. A flatworm would be considered complex because it morphologically looks more similar to humans than say to a simple bacterium, even if that bacterium can live in an anoxic, lightless, and boiling hot environment. The notion of similarity and complexity, however, would be uprooted nearly 50 years after Chatton's classification scheme as a result of a revolution in biochemistry and molecular biology.

Three Domains of Life

Although the field of biology is fundamentally concerned with the classification of living and extinct organisms, the field is highly dependent on technological advances that allow the biologist to gather ever more information that can in turn be used to classify organisms (e.g., the microscopic). One such technological advance took place in the 1970s, when chemists developed efficient methods to sequence DNA. The ability to extract information from the heritable material of life and use this information to classify life would revolutionize the classification system.

Carl Woese and George Fox would turn out to be the leaders of the revolution (Woese and Fox 1977). These microbiologists sequenced the DNA that encodes the RNA components of the ribosome from a diverse set of organisms. The DNA sequence information was then used to construct taxonomic trees (or so-called phylogenetic trees when they are generated from sequence data). Using the basic principles of taxonomy, a phylogenetic analysis attempts to group organisms, or gene sequences, based on similarity. The rRNA gene was an ideal gene to study because certain portions of the gene accumulate mutations very slowly so these portions could be used to elucidate ancient evolutionary relationships "deep" in the Tree of Life. The phylogenetic analysis of rRNA sequences by Woese and Fox would show that prokaryotes are divided into two groups, and that one of these groups shared a common ancestor with eukaryotes to the exclusion of the other prokaryotic group (Fig. 1B). This meant that prokaryotes should no longer be considered monophyletic (that all prokaryotes share a common ancestor to the exclusion of eukaryotes) and it meant that our Lemarckian notion of eukaryotes evolving from prokaryotes needed to be abandoned.

The use of DNA sequence information not only changed our view about the classification of life, but it also changed our confidence in taxonomy. Comparing sequences provides a discrete observation on a level at which development proceeds from and evolution acts on. Biologists identified the level at which natural selection and Darwinian evolution enabled life to diversify, thus comparing organisms at this level would allow biologists to accurately illustrate the natural classification scheme of life.

Root of the "Tree of Life"

The Darwinian notion of the Tree of Life implies that a trunk exists from which all branches extend if life does indeed share a common ancestor. The point where the all branches collapse and connect to the trunk is called the *root* in taxonomy and phylogenetics. Rooting (curiously not termed trunking) trees is theoretically possible if life shared a common ancestor and if a gene made a duplicate copy of itself (paralog) before the three domains of life diverged and both copies have since been retained in all three domains. A phylogenetic analysis of such anciently duplicated paralogs could then generate a tree consisting of two subtrees. Each subtree would be topologically identical to the

Tree of Life and the point or node where the two subtrees connected to one another would then represent the root of the phylogeny. Rooting a tree in this manner requires explicit models of sequence evolution because the analysis is attempting to extract a very ancient signal from the DNA/amino acid sequences.

The accumulation of DNA sequence data has allowed biologists to identify multiple paralogous gene families that appear to have undergone duplications before the three domains of life emerged and yet these genes have been evolving slowly enough to identify them clearly as paralogs. Some examples of gene families include ATPases, elongation factors, tRNA synthetases, signal recognition particles, and *inter alia* (Gogarten et al. 1989; Iwabe et al. 1989; Brown and Doolittle 1995; Brown et al. 1997; Gribaldo and Cammarano 1998). Initial phylogenetic analyses of these families place the root of the tree on the branch that separates bacteria from archaea/eukaryotes (Fig. 1C). This implies that the oldest separation or bifurcation on the Tree of Life was when LUCA split to give rise to the branch that would evolve into bacteria on one side of the tree and a branch that would later serve as the common ancestor of archaea and eukaryotes on the other side of the tree.

This is currently the prevailing view for the root of the Tree of Life. We must note, however, that other studies have criticized details of the approach discussed previously and reach different conclusions. In one alternative view, a ring has replaced the tree properties for illustrating the taxonomy of life. The so-called Ring of Life developed by Jim Lake suggests that genomes have been created by multiple fusion events during the evolution of early life and that these obviate a bifurcating branching pattern in the tree (Fig. 1C) (Rivera and Lake 2004). One advantage of this view is that it considers the likely widespread horizontal or lateral transfer of DNA during early life. Such transfer violates assumptions of the phylogenetic models used to analyze sequence data.

Another approach to root the Tree of Life is to use morphological or phenotypic observations in an attempt to define a clear splitting or bifurcation in the tree. For instance, membrane architecture and insertion/deletion events (indels) in gene sequences have been used to argue that the root of the Tree of Life exists *within* the bacterial domain, not on the branch that separates bacteria from archaea/eukaryotes (Fig. 1C) (Cavalier-Smith 2002; Cavalier-Smith 2006). Whereas a phylogenetic analysis of paralogous genes is heavily dependent on explicit models of sequence evolution (which can potentially and grossly mislead or bias results), the use of membrane architecture and indel events are conversely independent of models (but which cannot account for parallel or convergent evolution).

Biologists' ability to model evolution has improved substantially over the past decade. Biologists concede that the models are not perfect and that it will be a long road before the models accurately capture all evolutionary processes. Despite this hurdle, they are energized by the prospect of delineating evolution at the sequence level as opposed to being paralyzed by the challenges.

Lateral Gene Transfer: The "Coral of Life"

Biology's confidence in illustrating the relationships among living organisms has been a rollercoaster ride during the past 20 years. One of the most recent challenges has been the realization that DNA is not solely transmitted in a vertical manner to descendents. Multiple studies have shown that horizontal gene transfer (HGT) has played a major role in the flow of genetic information between organisms—especially deep in life's phylogeny. This obviously blurs the phylogenetic picture for life because a basic assumption of the Tree of Life is that information only flows in a one-way vertical direction (from parent to offspring in its broadest sense), not a horizontal direction (from one species to another species). This would be the equivalent of violating the linearity of time in the Universe because you cannot be in two places at a single time. HGT essentially allows two identical pieces of DNA to exist at the same time in complete disregard to evolutionary relatedness.

Does this require that we abandon the concept of vertical transmission and the Tree of Life? Yes and no. As mentioned previously, one alternative view is the Ring of Life. This model assumes that life intermixed so much genetic information shortly after LUCA diverged that there are no dominate traces of vertical inheritance until well after the three of domains of life emerged.

Another alternative not yet mentioned is the Coral of Life (Fig. 1D). This concept is being developed by Peter Gogarten, and like the Ring of Life, allows for genetic information to flow in a horizontal manner from species to species (Fournier et al. 2009). Unlike the Ring of Life, however, the Coral of Life permits a dominant path of vertical inheritance to have occurred for ancient life deep in a phylogeny. This path is thought to be present in the tree by the observation that some genes appear to be *resistant* to HGT. Some genes and their protein products are so entrenched in biochemical and cellular pathways, and protein–protein interactions that have evolved covariantly, that there is no selective advantage, and in all likelihood there would probably be a disadvantage, for a species to acquire a foreign copy of the gene.

These observations have resuscitated the notion of the Tree of Life (or whatever life-like creature it illustrates) and shown that biologists need not be paralyzed by HGT when attempting to understand early life. Now that we have hopefully convinced the reader that *some* phylogeny of life exists, we now discuss how researchers exploit this phylogeny in attempts to understand early life.

Early Life and Its Temperature History

The temperature history of life is a topic that has interested scientists for at least two centuries since Darwin's famous statement regarding a warm little pond and the origins of life. Although this article does not deal with the origins of life per se, there has been an equal interest in the temperature history for early life, in particular the close descendents of LUCA.

All of modern life is categorized into one of four temperature ranges. Heat-loving organisms come in the form of thermophiles (grow optimally $\sim 45°$ to $80°C$) and hyperthermophiles (grow optimally $\geq 80°C$). Cold-loving organisms are called psychrophiles (grow optimally $\leq 15°C$), and middle-loving organisms are called mesophiles (grow optimally $\sim 15°C$ to $45°C$).

Before the mid 1990s, most conclusions about the temperature history of life were based on chemical considerations and the physical behaviors of biomolecules. This changed with the accumulation of DNA sequence information from a broad range of species and the topology of the inferred phylogenies built from this sequence information. For instance, the first comprehensive discussion about thermostability and ancient life was based on the distribution of hyperthermophilic archaea and bacteria in the Tree of Life (Stetter 1996). The grouping of hyperthermophilic species on short branches near the bases of both the bacterial and archaeal domains of the tree parsimoniously suggested that LUCA was a hyperthermophile (Fig. 2).

This conclusion, however, was disputed shortly after it was presented. Some argued for a long-branch attraction artifact in the phylogenetic approach that caused hyperthermophiles to be randomly attracted/grouped instead of grouped because of common ancestry. Others

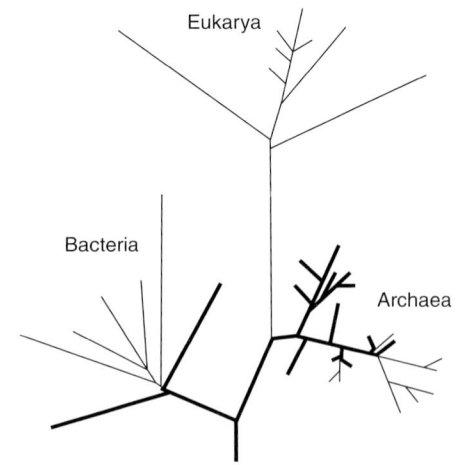

Figure 2. Distribution of modern hyperthermophilic organisms (Stetter 1996). Thick terminal branches lead to hyperthermophiles and thin terminal branches lead to nonhyperthermophiles. Thick internal branches are inferred based on the distribution and relatedness of modern hyperthermophiles.

argued that species sampling was sparse and the distribution of hyperthermophiles was coincidental. Still others argued that all hyperthermophiles require a particular protein to survive (reverse gyrase) and this protein evolved through a fusion of two other nonrelated proteins (Forterre 2002). So, if reverse gyrase is required for hyperthermophiles, and if the reverse gyrase cannot be "spontaneously" evolved from a random sequence, then hyperthermophiles must have evolved from a species that lived at a lower temperature.

A genomic-wide approach to understanding the temperature history of early life exploited the observation that the genomic G+C content of modern organisms correlates to the optimal growth temperature of the host organism itself (Galtier et al. 1999). Higher G+C content equates to a higher growth temperature because Gs and Cs form an extra hydrogen bond between base pairs (bps) compared with an A:T bp. The accumulation of extra hydrogen bonds throughout the genome would therefore make it more stable and resistant to heat denaturation. These researchers used models of molecular sequence evolution to infer the G+C content for gene families believed to have been present in LUCA and inferred to have traversed a mostly vertical descent through the Tree of Life with minimal horizontal gene transfer. The researchers concluded that LUCA did not have a genomic G+C content consistent with a hyperthermophilic life style.

To show how sensitive inferences can be to the use of evolutionary models as input for phylogenetic analysis, Di Giulio analyzed the same genomic dataset as previously discussed but used a different phylogenetic algorithm to infer the G+C content of LUCA (Di Giulio 2000). This analysis resulted in an ancient G+C content for LUCA that is consistent with thermophilic and hyperthermophilic life styles.

The previous contention represents one of multiple examples in which analyses have led to competing conclusions. Table 1 presents a condensed chronological list of studies that have attempted to determine the temperature history of early life. The majority of these studies are strictly computational and were not verified in any experimental manner. The next sections show how phylogenetic analysis can be used to guide laboratory experiments to address the temperature history of early life.

Ancestral Sequence Reconstruction

The recent accumulation of DNA sequence data, combined with advances in evolutionary theory and computational power, have paved the way for innovative approaches to understanding the origins, evolution, and distribution of life and its constituent biomolecules (Pauling and Zuckerkandl 1963; Benner et al. 2002; Gaucher et al. 2004). One approach to understanding ancestral states follows a present-day-backwards strategy, whereby genomic sequences from extant (modern) organisms are incorporated into evolutionary models that estimate the extinct (ancient) character states of genes no longer present on Earth (Fitch 1971; Shih et al. 1993; Benner 1995; Koshi and Goldstein 1996; Schultz et al. 1996; Cunningham 1999; Omland 1999; Pagel 1999; Schultz and Churchill 1999; Chang and Donoghue 2000; Thornton 2004; Hall 2006; Liberles 2007). These inferred ancestral gene sequences act as hypotheses that can be tested in the laboratory through the resurrection of the ancestral proteins themselves. Results from functional assays of the protein products from these ancient genes permit us to accept/reject hypotheses about the sequence themselves, or about their interactions/binding specificities/environments, etc.

Ancestral sequence reconstruction uses standard statistical theory to generate posterior probabilities of different reconstructions given the data at a site from aligned sequences. For each site of the inferred sequence at a phylogenetic node, posterior values for all 20 amino acids are calculated and represent the probability of a particular amino acid occupying a specific site in the protein during its evolutionary history. This posterior probability distribution is calculated from patterns of amino acids in modern sequences as described by a phylogeny, a matrix of amino acid replacement probabilities, amino acid equilibrium (stationary) frequencies, phylogenetic branch lengths,

Table 1. Temperature History of Life

Citation	Taxonomic unit?	Study design and observations	Conclusion
(Stetter 1996)	LUCA	A review of hyperthermophilic archaea and bacteria. In the 16S rRNA-based universal phylogenetic tree, hyperthermophiles are represented in the deepest and shortest lineages.	Hyperthermophile
(Forterre 1996)	LUCA	Reverse gyrase is a hyperthermophile-specific protein formed by the association of a putative topoisomerase and helicase. If reverse gyrase is a prerequisite to life at high temperatures, it suggests that hyperthermophiles descended from less thermophilic organisms that possessed these putative enzymes.	Mesophile or thermophile
(Galtier et al. 1999)	LUCA	A model of sequence evolution, assuming varying G+C content among lineages and unequal substitution rates among sites, was applied to estimate ancestral base compositions of rRNA sequences. The inferred G+C content of the LUCA is incompatible with survival at a high temperature.	Mesophile
(Di Giulio 2000)	LUCA	Reanalysis of the alignment used by Galtier (Science 1999) by maximum parsimony implies that the LUCA may have been a thermophile or hyperthermophile.	Thermophile or hyperthermophile
(Brochier and Philippe 2002)	LUB	Applied the heterotacy method on the rRNA bacterial phylogeny and found that the Planctomycetales are the first branching bacterial group; therefore, concluding the most recent common ancestor of bacteria was not hyperthermophilic.	Mesophile or thermophile
(Gaucher et al. 2003)	LUB	The most probabilistic ancestral sequences of elongation factor Tu (EF-Tu) were reconstructed at nodes in the bacterial evolutionary tree. These resurrected proteins were assayed and their temperature optima of 55°–65°C corresponds to ancient bacteria living as thermophiles.	Thermophile
(Brooks et al. 2004)	LUCA	Inferred amino acid composition of 65 proteins dating to the LUCA by maximum-likelihood using expectation-maximization. The inferred protein sequences were more similar to those found in modern-day thermophilic organisms than mesophilic ones.	Thermophile
(Knauth and Lowe 2003; Knauth 2005)	Ocean	Low oxygen isotopes in diagenetic cherts (3.5–3.2 Ga) in South Africa indicate extremely high ocean temperatures of 55°–85°C. Early thermophilic microbes could have been global and not huddled around hydrothermal vents.	Thermophile

(Continued)

Table 1. Continued

Citation	Taxonomic unit?	Study design and observations	Conclusion
(Iwabata et al. 2005)	LUCA	Studied the thermostabilty of ancestral isocitrate dehydrogenase (ICDH) mutants. The incorporation of ancestral residues into a modern ICDH led to an increase in thermostability.	Hyperthermophile
(Robert and Chaussidon 2006)	Ocean	Study of oxygen isotope ratios of cherts (siliceous sediments) as a measure of the Earth's climate in the Precambrian. The observed silicon isotope variations imply seawater temperature changes from 70°C 3.5 billion years ago to 20°C about 800 million years ago.	Thermophile
(Becerra et al. 2007)	LUCA	A study on the evolution of protein disulfide oxidoreductases (PDO) and its implications to then thermostabilty of the LUCA. The results imply that the LUCA lacked PDO-encoding sequences, and may not have been a thermophile.	Mesophile
(Shimizu et al. 2007)	LUCA	Ancestral glycyl-tRNA synthetases (GlyRS) were deduced and residues were introduced in *Thermus thermophilus* GlyRS. The thermostabilty of these mutants were studied and several were found with higher thermostabilty and activity than wild-type *Thermus*. These results suggest a highly thermophilic protein translation system in the LUCA.	Hyperthermophile
(Gaucher et al. 2008)	LUB	Extensions of earlier work with more than 25 phylogenetically dispersed ancestral EF-Tu's. The resurrected proteins at basal nodes are compatible with thermophilic environments.	Thermophile
(Boussau et al. 2008)	LUCA	A computational analysis of both rRNAs and protein sequences whose results imply that the LUCA was a mesophile. This implies that the two lineages descending from LUCA and leading to the ancestors of Bacteria and Archaea-Eukaryota convergently adapted to high temperatures.	Mesophile
(Glansdorff et al. 2008)	LUCA	Archaea have a uniform membrane lipid composition that is suited to life at extreme conditions (heat and pH); in contrast, bacterial membranes show a high variability in composition. The authors suggest that Archaea emerged from a nonthermophilic LUCA under strong selective pressure for adaptation to high temperature; whereas, bacteria were initially nonthermophilic and adapted by convergent evolution to high temperatures.	Mesophile or thermophile

(LUCA) last universal common ancestor of life; (LUB) last common ancestor of bacteria.

and site-specific replacement rates. The most-probabilistic ancestral sequence (M-PAS) uses the amino acid with the highest posterior probability at each site within the distribution.

RECENT RESULTS

Elongation factor Tu (Bacteria)/1A (Archaea and Eukarya) is an ideal protein family to computationally reconstruct and then resurrect in the laboratory in our attempts to better understand the temperature history of life. There is no evidence that EF genes have been laterally transferred between bacterial lineages, and the thermal stabilities of EFs correlate with the growth temperature of their host organisms. Thus, EFs are optimally stable at temperatures of $15°-45°C$, $45°-80°C$, and $>80°C$ when isolated from mesophiles, thermophiles, and hyperthermophiles, respectively. This relationship is consistent with a correlation coefficient of 0.91 between melting temperatures of proteins and environmental temperatures of their host organisms (Gromiha et al. 1999).

Reconstruction of ancestral EF sequences were computed across two bacterial phylogenies selected from the literature (Battistuzzi et al. 2004; Ciccarelli et al. 2006). Both phylogenies were constructed from the concatenation of numerous gene families and are thus less susceptible to systematic error compared with phylogenies based on single genes. The two phylogenies capture the main competing views for bacterial relationships. One scenario posits that hyperthermophilic lineages occupy basal branches of the bacterial tree, whereas the other places these lineages in a more derived portion of the tree. To accommodate the latter scenario, a phylogeny was selected in which the Firmicute lineage (void of hyperthermophiles) is located at the base of the bacterial tree, although other topologies have been suggested (Brochier and Philippe 2002).

Thermostability of modern and ancestral EF proteins was monitored using circular dichroism spectroscopy. Melting temperatures (T_m) of two modern EFs were determined. The T_m values for EFs from *Escherichia coli* and *T. thermophilus* (HB8) are $42.8°C$ and $76.7°C$. These values highlight the relationship between EF stability and the optimal growth temperature of their respective hosts, $\sim40°C$ and $\sim74°C$ (Williams and da Costa 1992).

T_m values for ancestral EF proteins were determined across the two phylogenies. The thermostability profiles of the ancestral proteins display the same general trend despite the fact that the two phylogenies represent competing hypotheses. Ancestral EF proteins resurrected at basal nodes are compatible with thermophilic environments, whereas ancestral proteins from more derived nodes are compatible with cooler environments. Consistent with this temperature trend is the observation that the node representing the presumed last common ancestor of bacteria (and thus oldest) had the most thermostable protein within each phylogeny ($64.8°C$ and $73.3°C$). The similarity in thermostability ($<9°C$) between these two ancestral proteins is noteworthy because the sequences were identical across only 78% of the amino acid sites.

The environmental temperature of ancient bacteria inferred from resurrected EF proteins can be connected to divergence times of major bacterial lineages to gain a more detailed understanding of temperature trends for Precambrian life (Battistuzzi et al. 2004). Divergence estimates from Battistuzzi et al. (2004) were applied to nodes in the current study. Figure 3 highlights the progressive cooling trend of ancient EF proteins from approximately 3.5 billion to 500 million years ago. This temperature trend is strikingly similar to the temperature trend of the ancient ocean inferred from deposition of oxygen and silicon isotopes (Knauth and Lowe 1978; Knauth and Lowe 2003; Robert and Chaussidon 2006).

Reconstruction of ancestral EF proteins throughout the bacterial domain of life suggests that the organisms that hosted these extinct biomolecules lived in environments that have progressively cooled for approximately 3 billion years. This evidence is predicated on multiple assumptions. For instance, it assumes that ancestral sequence reconstruction recapitulates ancient phenotypes and that phylogenies and divergence dates capture the evolutionary relationships and timing of bacterial divergences.

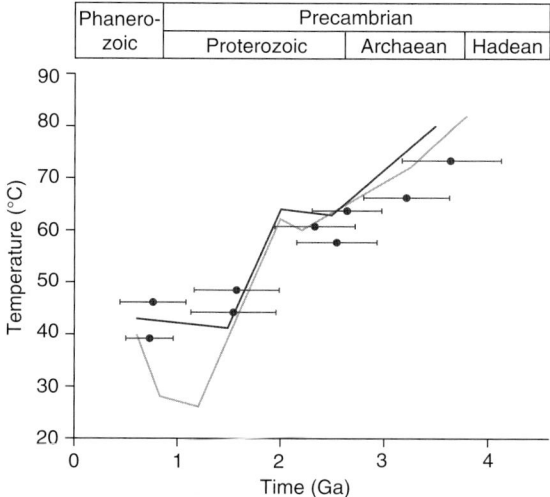

Figure 3. Plot of ancestral EF melting temperatures versus geologic time in billions of years (Ga) (Gaucher et al. 2008). Molecular clock estimates and their confidence intervals (horizontal bars) from Battistuzzi et al., using a 2.3 Ga minimum constraint for the great oxidation event (Battistuzzi et al. 2004). Solid lines are temperature curves of the ancient ocean inferred from maximum $\delta^{18}O$ (light gray [Knauth and Lowe 1978; Knauth and Lowe 2003], dark gray [Robert and Chaussidon 2006]). Although not shown, an analogous trend is seen with $\delta^{30}Si$ isotopes (Robert and Chaussidon 2006).

The inability (short of time travel) to know the true relationships of bacterial lineages and their divergence times should not preclude attempts to understand Precambrian life. Rather, a coherent description of ancient life can be generated when empirical evidence from diverse studies converge on analogous conclusions. For instance, the same paleotemperature trend was observed for ancestral EF proteins regardless of the phylogeny. And for the phylogeny with divergence dates, this trend was substantiated when aligned to the inferred paleotemperature curve of the ancient ocean.

These descriptions are particularly useful when they have predictive value. For instance, the last common ancestor of the mitochondrial bacterium is estimated to have lived 1.66–1.88 Ga based on the T_m's for ancestral EF proteins from the node representing the origins of mitochondria (51.0°C–53.0°C). This is consistent with the origins of mitochondria estimated at 1.8 Ga based on a molecular clock (Hedges et al. 2001), despite the controversial nature of the clock (Graur and Martin 2004) and assuming the last common mitochondrial bacterium lived at a time close to the endosymbiotic event between α-proteobacteria and eukaryotic cells.

Our results suggest early life lived at an environmental temperature similar to today's hot springs. Particular geologic theory and evidence suggests the ancient ocean also had temperatures similar to hot springs (Hoyle 1972; Knauth and Lowe 1978; Knauth and Lowe 2003). As the ocean cooled from 3.5 to 0.5 billion years ago, life may have responded by adapting its range of growth temperatures to correspond to its surrounding environment. This connection assumes early life lived in the ancient ocean, which seems practical based on geologic and biologic constraints such as ocean depth/circulation, land mass exposed to the atmosphere, susceptibility to desiccation, and ultraviolet radiation, among others. Alternatively, it is possible that the inferred paleotemperature trend reflects an ecological trajectory as ancient bacteria transitioned from hot springs/thermal vents to the open ocean.

We note that correlating isotope ratios ($d^{18}O$ and $d^{30}Si$) to ancient ocean temperatures is controversial (Kasting et al. 2006; Jaffres et al. 2007). In particular, the correlation could be invalid if isotope ratios were caused by variation in seawater composition alone. This would translate into a more temperate ancient ocean and be consistent with ancient glaciation events. The similarity, however, in paleotemperature trends inferred from $d^{18}O$, $d^{30}Si$ and ancient EF proteins is striking. Further, the overall trend is compatible with biological evolution. For instance, the thermostability of ancient EFs suggest the origins of cyanobacteria occurred at an environmental temperature approximating 63.7°C. This is consistent with an upper temperature limit of typical cyanobacterial mats in hot springs (~65°C) (Ward et al. 1998).

Overall, the results show that ancient EF thermostability profiles (phenotypes) are robust to uncertainties and potential biases associated with inferring ancestral character states (genotypes). The results also show how ancestral sequence reconstruction can connect physical and natural sciences in our attempts to understand the environmental conditions that hosted early life.

CHALLENGES

Statistical Models of Molecular Sequence Evolution

Despite insightful studies, the field of ancestral sequence reconstruction is encumbered by its inability to know whether inferred sequences truly recapitulate ancestral forms (Williams et al. 2006). Practitioners in the field acknowledge a certain degree of inaccuracy associated with reconstructing ancestral sequences. The concern is not necessarily whether the resurrected form has the exact composition (genotype) of the true ancestral form, but rather that the resurrected form displays the exact behavior (phenotype). A reconstructed sequence can be considered a consensus of a gene distributed throughout a population before species diverge, or before gene duplication. Inaccuracies in a reconstructed sequence can result from sequence variation of the gene itself within an ancient population. Assuming the variants of a homologous gene within a population had the same phenotype at a specific geologic time, it does not necessarily matter which individual genotype is reconstructed.

This assumption is invalid if recombination of individual genotypes generate new phenotypes and if the reconstructed ancestral gene itself represents a consensus of those genotypes. Additional concerns arise if the reconstruction process generates inaccurate sequences because of (1) bias in the evolutionary models used to infer ancestral states or (2) phylogenetic conditions such as long branches and incorrect branching patterns (Felsenstein 1978; Williams et al. 2006; Kelchner and Thomas 2007).

All methods of phylogenetic inference make assumptions about the underlying evolutionary process of their characters and it is these assumptions that determine their relative successes and failures in the estimation of the true phylogeny for a group (Hillis et al. 1992). Much like the manner in which phylogenetic tree building algorithms were developed, tested, and critiqued during the 1990s, we anticipate that ancestral sequence reconstruction algorithms and methods will go through a similar process in the next couple of years now that the reconstruction field is burgeoning. In particular, we anticipate that the development and use of mixture models will play an important role in the development of the field (Gaucher et al. 2002a; Pagel and Meade 2004; Gaucher and Miyamoto 2005).

Experimental Phylogenetics as a Way to Benchmark Ancestral Sequence Reconstruction

Computer simulations of reconstructed ancestral sequences have unequivocally shown the superior performance of the "maximum likelihood" (ML) sequence in terms of accuracy in recovering a true ancestral sequence when it is inferred from tip/leaf/extinct/modern sequences (Huelsenbeck 1995; Yang et al. 1995; Zhang and Nei 1997; Cai et al. 2004; Krishnan et al. 2004; Williams et al. 2006). Although computer simulations of ancestral genotypes and phenotypes are an intriguing approximation for reality, a true benchmark of method performance requires an evaluation of biological sequences and phenotypes measured in the laboratory. As such, it would be useful to use members of the green fluorescent protein family to generate an "experimental phylogeny" (Hillis et al. 1992; Bull et al. 1993). Green fluorescent proteins (GFP) and their varying-colored homologs are widely used as in vivo fluorescent markers and have also been used in experimental paleogenetic studies (Matz et al. 1999; Matz et al. 2002; Ugalde et al. 2004).

Research in our lab is currently generating leaf/tip sequences from an evolved experimental GFP phylogeny that will in turn be used to estimate ancestral genotypes and phenotypes. Because the leaf/tip sequences will be sequentially evolved from nodes on the experimental phylogeny in the laboratory, we will know the true ancestral genotypes and phenotypes. This presents us with the unique opportunity to compare/contrast different approaches attempting to reconstruct ancestral sequences from biologically relevant conditions. Our work represents the first time evolved sequences will be used to benchmark ancestral sequence reconstruction approaches to address issues of ambiguity and bias associated with both reconstructed genotypes and phenotypes.

Sequences at the tips (leaves) of the evolved phylogeny will then be used to computationally reconstruct the inferred ancestral fluorescent sequences at all nodes of the experimental-derived tree. DNA-, codon-, and amino acid-based approaches will be exploited (Yang et al. 1995; Chang et al. 2002; Thornton 2004; Thomson et al. 2005). For each type of data input, we will test different models of sequence evolution and their potential effects on ancestral sequence reconstruction (e.g., transition/transversion ratios, codon tables, amino acid matrices, rate heterogeneity, and others) (Gaucher et al. 2001; Gaucher et al. 2002b; Gaucher and Miyamoto 2005).

RESEARCH DIRECTIONS

We anticipate that our understanding of the temperature history of early life will continue to improve in the coming years. This improvement will not be driven by any single advancement. Rather, a combination of advances in multiple scientific disciplines will enhance our understanding. This is due in large part to the multidisciplinary nature of studying the temperature history of life. For instance, our understanding of taxonomic and evolutionary relationships of bacteria and archaea will greatly enhance our understanding of deep phylogeny, and this in turn will improve our understanding of the environmental conditions that supported these ancient life forms.

More sophisticated models of molecular sequence evolution will help us to better understand ancient life. Such models will improve our ability to accurately construct phylogenetic trees as well as add rigor to ancestral sequence reconstruction methods. The biologists and computational scientists will not be making improvements alone. We anticipate that advances in chemical and geological techniques will also help us define properties of early life.

We are further energized by the prospect of joining evolutionary biology and synthetic biology in our attempts to dissect early life. The next logical extension of molecular reconstruction beyond natural history is to synthetic biology. Synthetic biology means different things to different scientific disciplines (Benner and Sismour 2005; Endy 2005). Surprisingly, however, biologists seem to have taken a backseat to chemists and engineers in the development of this field. It seems apparent that synthetic biology would stand to benefit if "molecular reconstructionists" contributed to its progress. In this way, an evolutionary synthetic biology is formed. A couple of examples come to mind: cellular machines and recombinant genomes.

Cellular machines have a broad range of potentials, from simple expression of heterologous genes for laboratory analysis to the synthesis of minimal artificial cells (Deamer et al. 2002; Martin et al. 2003; Noireaux and Libchaber 2004; Chen et al. 2005). We anticipate that ancestral reconstructed sequences will provide some of the foundation of genetic information for these machines in the future. As a first step, we have shown that ancestral EF proteins can participate in a reconstituted in vitro translation system designed to incorporate unnatural amino acids (unpubl. data). Further, experimental evolution studies of these ancestral genes introduced into laboratory organisms will enhance our biological understanding of adaptive and sequence landscapes, shed light on the transition to protein synthesis by early life, and help elucidate the evolution and adaptation of biochemical pathways. This work will have obvious extensions to natural history and the origins of (early) life.

We also anticipate that the synthesis of recombinant, minimal, and/or ancestral genomes will have a profound effect on our understanding of early life. The Venter Institute, for instance, is in the process of constructing a minimal synthetic *Mycoplasma* genome (Glass et al. 2006; Lartigue et al. 2007; Gibson et al. 2008). As molecular reconstructionists, we would ask why not construct a complete ancestral biochemical pathway (e.g., operon), or even a complete ancestral genome? The ancestral reconstruction field would no longer be confined to single gene analysis. It is also quite possible that our understanding of what constitutes a sustaining minimal genome required to support life will be altered through ancestral reconstructions. In this way, homologous genes performing two different, but related, functions may share a single common ancestor that performed both of these functions, albeit with less efficiency or specificity.

We anticipate that our understanding of the origins of life and its early evolution will be greatly enhanced by advances in molecular evolution techniques in the coming years. Phylogenetic methods and ancestral sequence reconstruction will continue to be combined in innovative ways to contribute to the Origins of Life field.

REFERENCES

Battistuzzi FU, Feijao A, Hedges SB. 2004. A genomic timescale of prokaryote evolution: Insights into the origin of methanogenesis, phototrophy, and the colonization of land. *BMC Evolutionary Biol* **4:** 44.

Becerra A, Delaye L, Lazcano A, Orgel LE. 2007. Protein disulfide oxidoreductases and the evolution of thermophily: Was the last common ancestor a heat-loving microbe? *J Mol Evol* **65:** 296–303.

Benner SA. 1995. Reconstructing Ancient Forms Of Life. *J Cell Biochem:* 200–200.

Benner SA, Sismour AM. 2005. Synthetic biology. *Nat Rev Genet* **6:** 533–543.

Benner SA, Caraco MD, Thomson JM, Gaucher EA. 2002. Planetary biology–paleontological, geological, and molecular histories of life. *Science* **296:** 864–868.

Boussau B, Blanquart S, Necsulea A, Lartillot N, Gouy M. 2008. Parallel adaptations to high temperatures in the Archaean eon. *Nature* **456:** 942–945.

Brochier C, Philippe H. 2002. Phylogeny: A non-hyperthermophilic ancestor for bacteria. *Nature* **417:** 244.

Brooks DJ, Fresco JR, Singh M. 2004. A novel method for estimating ancestral amino acid composition and its application to proteins of the Last Universal Ancestor. *Bioinformatics* **20:** 2251–2257.

Brown JR, Doolittle WF. 1995. Root of the universal tree of life based on ancient aminoacyl-tRNA synthetase gene duplications. *Proc Natl Acad Sci* **92:** 2441–2445.

Brown JR, Robb FT, Weiss R, Doolittle WF. 1997. Evidence for the early divergence of tryptophanyl- and tyrosyl-tRNA synthetases. *J Mol Evol* **45:** 9–16.

Bull JJ, Cunningham CW, Molineux IJ, Badgett MR, Hillis DM. 1993. Experimental molecular evolution of bacteriophage-T7. *Evolution* **47:** 993–1007.

Cai W, Pei J, Grishin NV. 2004. Reconstruction of ancestral protein sequences and its applications. *Bmc Evol Biol* **4:** 33.

Cavalier-Smith T. 2002. The neomuran origin of archaebacteria, the negibacterial root of the universal tree and bacterial megaclassification. *Int J Syst Evol Microbiol* **52:** 7–76.

Cavalier-Smith T. 2006. Rooting the tree of life by transition analyses. *Biol Direct* **1:** 19.

Chang BSW, Donoghue MJ. 2000. Recreating ancestral proteins. *Trends In Ecology & Evolution* **15:** 109–114.

Chang BS, Jonsson K, Kazmi MA, Donoghue MJ, Sakmar TP. 2002. Recreating a functional ancestral archosaur visual pigment. *Mol Biol Evol* **19:** 1483–1489.

Chatton E. 1925. Pansporella perplexa. Réflexions sur la biologie et la phylogénie des protozoaires. *Ann Sci Nat Zool (Ser 10)* **8:** 5–84.

Chen IA, Salehi-Ashtiani K, Szostak JW. 2005. RNA catalysis in model protocell vesicles. *J Am Chem Soc* **127:** 13213–13219.

Ciccarelli FD, Doerks T, von Mering C, Creevey CJ, Snel B, Bork P. 2006. Toward automatic reconstruction of a highly resolved tree of life. *Science* **311:** 1283–1287.

Cunningham CW. 1999. Some limitations of ancestral character-state reconstruction when testing evolutionary hypotheses. *Systematic Biology* **48:** 665–674.

Deamer D, Dworkin JP, Sandford SA, Bernstein MP, Allamandola LJ. 2002. The first cell membranes. *Astrobiology* **2:** 371–381.

Di Giulio M. 2000. The universal ancestor lived in a thermophilic or hyperthermophilic environment. *J Theor Biol* **203:** 203–213.

Endy D. 2005. Foundations for engineering biology. *Nature* **438**: 449–453.

Felsenstein J. 1978. Cases in Which Parsimony or Compatibility Methods Will Be Positively Misleading. *Systematic Zoology* **27**: 401–410.

Fitch W. 1971. Towards defining the course of evolution. Minimum change for a specific tree topology. *Syst Zoology* **20**: 406–416.

Forterre P. 1996. A hot topic: The origin of hyperthermophiles. *Cell* **85**: 789–792.

Forterre P. 2002. A hot story from comparative genomics: Reverse gyrase is the only hyperthermophile-specific protein. *Trends Genet* **18**: 236–237.

Fournier GP, Huang J, Gogarten JP. 2009. Horizontal gene transfer from extinct and extant lineages: Biological innovation and the coral of life. *Philos Trans R Soc Lond B Biol Sci* **364**: 2229–2239.

Galtier N, Tourasse N, Gouy M. 1999. A nonhyperthermophilic common ancestor to extant life forms. *Science* **283**: 220–221.

Gaucher EA, Miyamoto MM. 2005. A call for likelihood phylogenetics even when the process of sequence evolution is heterogeneous. *Mol Phylogenet Evol* **37**: 928–931.

Gaucher EA, Das UK, Miyamoto MM, Benner SA. 2002a. The crystal structure of eEF1A refines the functional predictions of an evolutionary analysis of rate changes among elongation factors. *Mol Biol Evolution* **19**: 569–573.

Gaucher EA, Govindarajan S, Ganesh OK. 2008. Palaeotemperature trend for Precambrian life inferred from resurrected proteins. *Nature* **451**: 704–707.

Gaucher E, Graddy L, Li T, Simmen R, Simmen F, Schreiber D, Liberles D, Janis C, Benner S. 2004. The planetary biology of cytochrome P450 aromatases. *BMC Biology* **2**: 19.

Gaucher EA, Gu X, Miyamoto MM, Benner SA. 2002b. Predicting functional divergence in protein evolution by site-specific rate shifts. *Trends Biochem Sci* **27**: 315–321.

Gaucher EA, Miyamoto MM, Benner SA. 2001. Function-structure analysis of proteins using covarion-based evolutionary approaches: Elongation factors. *Proc Natl Acad Sci* **98**: 548–552.

Gaucher EA, Thomson JM, Burgan MF, Benner SA. 2003. Inferring the palaeoenvironment of ancient bacteria on the basis of resurrected proteins. *Nature* **425**: 285–288.

Gibson DG, Benders GA, Andrews-Pfannkoch C, Denisova EA, Baden-Tillson H, Zaveri J, Stockwell TB, Brownley A, Thomas DW, Algire MA, et al. 2008. Complete chemical synthesis, assembly, and cloning of a *Mycoplasma genitalium* genome. *Science* **319**: 1215–1220.

Glansdorff N, Xu Y, Labedan B. 2008. The last universal common ancestor: Emergence, constitution and genetic legacy of an elusive forerunner. *Biol Direct* **3**: 29.

Glass JI, Assad-Garcia N, Alperovich N, Yooseph S, Lewis MR, Maruf M, Hutchison CA 3rd, Smith HO, Venter JC. 2006. Essential genes of a minimal bacterium. *Proc Natl Acad Sci* **103**: 425–430.

Gogarten JP, Kibak H, Dittrich P, Taiz L, Bowman EJ, Bowman BJ, Manolson MF, Poole RJ, Date T, Oshima T, et al. 1989. Evolution of the vacuolar H+-ATPase: Implications for the origin of eukaryotes. *Proc Natl Acad Sci* **86**: 6661–6665.

Graur D, Martin W. 2004. Reading the entrails of chickens: Molecular timescales of evolution and the illusion of precision. *Trends Genet* **20**: 80–86.

Gribaldo S, Cammarano P. 1998. The root of the universal tree of life inferred from anciently duplicated genes encoding components of the protein-targeting machinery. *J Mol Evol* **47**: 508–516.

Gromiha MM, Oobatake M, Sarai A. 1999. Important amino acid properties for enhanced thermostability from mesophilic to thermophilic proteins. *Biophys Chem* **82**: 51–67.

Hall BG. 2006. Simple and accurate estimation of ancestral protein sequences. *Proc Natl Acad Sci* **103**: 5431–5436.

Hedges SB, Chen H, Kumar S, Wang DY, Thompson AS, Watanabe H. 2001. A genomic timescale for the origin of eukaryotes. *BMC Evol Biol* **1**: 4.

Hillis DM, Bull JJ, White ME, Badgett MR, Molineux IJ. 1992. Experimental phylogenetics—generation of a known phylogeny. *Science* **255**: 589–592.

Hoyle F. 1972. History of Earth. *Q J Roy Astron Soc* **13**: 328–345.

Huelsenbeck JP. 1995. Performance of phylogenetic methods in simulation. *Systematic Biology* **44**: 17–48.

Iwabata H, Watanabe K, Ohkuri T, Yokobori S, Yamagishi A. 2005. Thermostability of ancestral mutants of Caldococcus noboribetus isocitrate dehydrogenase. *FEMS Microbiol Lett* **243**: 393–398.

Iwabe N, Kuma K, Hasegawa M, Osawa S, Miyata T. 1989. Evolutionary relationship of archaebacteria, eubacteria, and eukaryotes inferred from phylogenetic trees of duplicated genes. *Proc Natl Acad Sci* **86**: 9355–9359.

Jaffres JBD, Shields GA, Wallmann K. 2007. The oxygen isotope evolution of seawater: A critical review of a long-standing controversy and an improved geological water cycle model for the past 3.4 billion years. *Earth-Science Rev* **83**: 83–122.

Kasting JF, Howard MT, Wallmann K, Veizer J, Shields G, Jaffres J. 2006. Paleoclimates, ocean depth, and the oxygen isotopic composition of seawater. *Earth and Planetary Science Letts* **252**: 82–93.

Kelchner SA, Thomas MA. 2007. Model use in phylogenetics: Nine key questions. *Trends in Ecology & Evolution* **22**: 87–94.

Knauth LP. 2005. Temperature and salinity history of the Precambrian ocean: Implications for the course of microbial evolution. *Palaeogeogr Palaeocl* **219**: 53–69.

Knauth LP, Lowe DR. 1978. Oxygen isotope geochemistry of cherts from onverwacht group (3.4 billion years), Transvaal, South-Africa, with implications for secular variations in isotopic composition of cherts. *Earth and Planetary Science Lett* **41**: 209–222.

Knauth LP, Lowe DR. 2003. High Archean climatic temperature inferred from oxygen isotope geochemistry of cherts in the 3.5 Ga Swaziland Supergroup, South Africa. *Geol Soc Am Bull* **115**: 566–580.

Koshi JM, Goldstein RA. 1996. Probabilistic reconstruction of ancestral protein sequences. *J Mol Evolution* **42**: 313–320.

Krishnan NM, Seligmann H, Stewart CB, de Koning APJ, Pollock DD. 2004. Ancestral sequence reconstruction in primate mitochondrial DNA: Compositional bias and effect on functional inference. *Mol Biol Evolution* **21**: 1871–1883.

Lartigue C, Glass JI, Alperovich N, Pieper R, Parmar PP, Hutchison CA 3rd, Smith HO, Venter JC. 2007. Genome transplantation in bacteria: Changing one species to another. *Science* **317**: 632–638.

Liberles DA. 2007. *Ancestral Sequence Reconstruction*. Oxford University Press, Oxford.

Martin VJJ, Pitera DJ, Withers ST, Newman JD, Keasling JD. 2003. Engineering a mevalonate pathway in *Escherichia coli* for production of terpenoids. *Nat Biotechnol* **21**: 796–802.

Matz MV, Lukyanov KA, Lukyanov SA. 2002. Family of the green fluorescent protein: Journey to the end of the rainbow. *Bioessays* **24**: 953–959.

Matz MV, Fradkov AF, Labas YA, Savitsky AP, Zaraisky AG, Markelov ML, Lukyanov SA. 1999. Fluorescent proteins from nonbioluminescent Anthozoa species. *Nat Biotechnol* **17**: 969–973.

Noireaux V, Libchaber A. 2004. A vesicle bioreactor as a step toward an artificial cell assembly. *Proc Natl Acad Sci* **101**: 17669–17674.

Omland KE. 1999. The assumptions and challenges of ancestral state reconstructions. *Systematic Biol* **48**: 604–611.

Pagel M. 1999. Inferring the historical patterns of biological evolution. *Nature* **401**: 877–884.

Pagel M, Meade A. 2004. A phylogenetic mixture model for detecting pattern-heterogeneity in gene sequence or character-state data. *Syst Biol* **53**: 571–581.

Pauling L, Zuckerkandl E. 1963. Chemical paleogenetics molecular restoration studies of extinct forms of life. *Acta Chem Scand* **17**: 89.

Rivera MC, Lake JA. 2004. The ring of life provides evidence for a genome fusion origin of eukaryotes. *Nature* **431**: 152–155.

Robert F, Chaussidon M. 2006. A palaeotemperature curve for the Precambrian oceans based on silicon isotopes in cherts. *Nature* **443**: 969–972.

Schultz TR, Churchill GA. 1999. The role of subjectivity in reconstructing ancestral character states: A Bayesian approach to unknown rates, states, and transformation asymmetries. *Systematic Biol* **48**: 651–664.

Schultz TR, Cocroft RB, Churchill GA. 1996. The reconstruction of ancestral character states. *Evolution* **50**: 504–511.

Shih P, Malcolm BA, Rosenberg S, Kirch JF, Wilson AC. 1993. Reconstruction and testing of ancestral proteins. in *molecular evolution: Producing The Biochemical Data*, pp. 576–590.

Shimizu H, Yokobori S, Ohkuri T, Yokogawa T, Nishikawa K, Yamagishi A. 2007. Extremely thermophilic translation system in the common ancestor commonote: ancestral mutants of Glycyl-tRNA synthetase from the extreme thermophile *Thermus thermophilus*. *J Mol Biol* **369**: 1060–1069.

Stetter KO. 1996. Hyperthermophilic procaryotes. *Fems Microbiology Reviews* **18**: 149–158.

Thomson JM, Gaucher EA, Burgan MF, De Kee DW, Li T, Aris JP, Benner SA. 2005. Resurrecting ancestral alcohol dehydrogenases from yeast. *Nat Genet* **37**: 630–635.

Thornton JW. 2004. Resurrecting ancient genes: experimental analysis of extinct molecules. *Nat Rev Genet* **5**: 366–375.

Ugalde JA, Chang BS, Matz MV. 2004. Evolution of coral pigments recreated. *Science* **305**: 1433.

Ward DM, Ferris MJ, Nold SC, Bateson MM. 1998. A natural view of microbial biodiversity within hot spring cyanobacterial mat communities. *Microbiol Mol Biol Rev* **62**: 1353–1370.

Williams RAD, da Costa MS. 1992. The genus Thermus and related microorganisms. in *The prokaryotes* (ed. A. Balows, H.G. Truper, M. Dworkin, W. Harder, K-H. Schleifer), pp. 3745–3753 Springer-Verlag, New York.

Williams PD, Pollock DD, Blackburne BP, Goldstein RA. 2006. Assessing the accuracy of ancestral protein reconstruction methods. *PLoS Comput Biol* **2**: pe69.

Woese CR, Fox GE. 1977. Phylogenetic structure of the prokaryotic domain: the primary kingdoms. *Proc Natl Acad Sci* **74**: 5088–5090.

Woese CR, Goldenfeld N. 2009. How the microbial world saved evolution from the scylla of molecular biology and the charybdis of the modern synthesis. *Microbiol Mol Biol Rev* **73**: 14–21.

Yang Z, Kumar S, Nei M. 1995. A new method of inference of ancestral nucleotide and amino acid sequences. *Genetics* **141**: 1641–1650.

Zhang JZ, Nei M. 1997. Accuracies of ancestral amino acid sequences inferred by the parsimony, likelihood, and distance methods. *J Mol Evolution* **44**: 139–146.

Constructing Partial Models of Cells

Norikazu Ichihashi[1], Tomoaki Matsuura[1], Hiroshi Kita[2], Takeshi Sunami[2], Hiroaki Suzuki[1], and Tetsuya Yomo[1,2,3]

[1]Department of Bioinformatic Engineering, Graduate School of Information Science and Technology, Osaka University, Osaka, Japan
[2]Exploratory Research for Advanced Technology (ERATO), Japan Science and Technology Agency (JST), Japan
[3]Graduate School of Frontier Biosciences, Osaka University, Osaka, Japan

Correspondence: yomo@ist.osaka-u.ac.jp

Understanding the origin of life requires knowledge not only of the origin of biological molecules such as amino acids, nucleotides and their polymers, but also the manner in which those molecules are integrated into the organized systems that characterize cellular life. In this article, we introduce a constructive approach to understand how biological molecules can be arranged to achieve a higher-order biological function: replication of genetic information.

BACKGROUND

What Can We Learn from a Constructive Approach?

In a constructive approach, we aim to reconstitute a biological function, such as genome replication and protein translation, and ultimately fabricate an artificial cell from molecules purified and defined in vitro (Szostak et al. 2001; Deamer 2005). During the process, we can determine what conditions are sufficient to achieve the minimum set of biological functions required for cellular life. For instance, if we can reconstitute a given biological function from a set of defined molecules, we can conclude that the properties of those molecules are sufficient to accomplish that biological function. With regard to the origins of life, this represents a parallel and complementary approach to surveying possible routes from nonliving molecules to extant living systems. If it is difficult for us to reconstruct a biological function, it may have been correspondingly difficult for that function to evolve in a primitive living organism. Knowledge of which functions are difficult to assemble from existing biological molecules, and how such hurdles can be overcome, is expected to provide insights into the origin and evolution of multifunctional extant life.

In the field of synthetic biology, researchers are now constructing artificial networks to understand the "design principles" of biological systems, which is another expression for the "sufficient conditions" concept used here. Most current studies in synthetic biology incorporate modifications of existing cells, but some investigators are constructing artificial networks from defined molecules (Benner and Sismour 2005; Simpson 2006). This is similar to what we refer to here as a constructive approach (Kaneko 2006).

What Should We Construct?

To gain insight into the origins of life, it is important to determine the nonbiological origin of chemical components, such as the amino acids, nucleotides, and lipids discussed in other articles on this topic. These are the small molecules of life that assemble into the proteins, nucleic acids and membranes that are essential for contemporary cells. Primitive versions of such molecules must have given rise to the first forms of cellular life by a process of self-assembly. However, if we observe a mixture of components from disrupted *Escherichia coli* containing all the molecules originally present in the living cells, no spontaneous regeneration of living cells takes place. It follows that molecules *per se* are necessary but not sufficient for life. Molecules and their functions must be coordinated in the correct order according to intrinsic chemical and physical rules, as observed by Schrodinger who famously described life as the "orderly and lawful behavior of matter" (Schrodinger 1944).

The primary aim of the constructive approach to protocellular life is to find sufficient conditions under which biological molecules assemble into systems that display higher-order biological functions, such as translation, replication of genetic information, cell growth, division, and nutrient transport. Some of these functions, including membrane growth (Walde 1994; Hanczyc et al. 2003), membrane growth coordinated with internal replication of genetic information (Chen et al. 2004), membrane transport of nutrients (Chakrabarti et al. 1994; Monnard and Deamer 2001; Fischer et al. 2002; Monnard et al. 2007; Mansy et al. 2008), and coupling of translation and nutrient transport (Noireaux and Libchaber 2004) have been reported previously. Here, we focus on the replication of genetic information, which is a fundamental characteristic of life.

Classifying Self-Replication of Genetic Information

One of the characteristics of life is the possession of genetic information, which replicates by using itself as a template, which we will refer to as "self-replication." Several types of self-replicating systems have been constructed with bioinformational molecules. These include self-replication of DNA by polymerase chain reaction (PCR) (Saiki et al. 1985), self-sustained sequence replication (3SR) (Guatelli et al. 1990), self-replication of RNA by Qβ replicase (Mills et al. 1967; Biebricher et al. 1985; Oberholzer et al. 1995). Other examples include self-replication of peptides (Lee et al. 1996, 1997), low molecular weight compounds (Tjivikua et al. 1990), tetranucleotides (Zielinski and Orgel 1987) and other oligomers (Sievers and von Kiedrowski 1994), and most recently the ligation activity of ribozymes (Lincoln and Joyce 2009).

These self-replication reactions can be classified according to the reaction scheme shown in Figure 1. Type 1 is the self-replication reaction in which the information molecule (DNA or RNA) is replicated by an exogenous enzyme and includes PCR (Saiki et al. 1985), 3SR (Guatelli et al. 1990), and RNA replication (Mills et al. 1967; Biebricher et al. 1985; Oberholzer et al. 1995). Type 2 self-replication does not require any replication enzymes and these reactions include self-ligating ribozymes (Lincoln and Joyce 2009), self-replicating peptides (Lee et al. 1996; Lee et al. 1997), and other low molecular weight compounds (Zielinski and Orgel 1987; Tjivikua et al. 1990; Sievers and von Kiedrowski 1994). In this type of reaction, the information molecule replicates itself in a reaction catalyzed by its own activity, and therefore this is the simplest type of self-replication. Type 3 is a modification of type 1 in which the replication enzyme is supplied internally by synthesis rather than being exogenously added. The replication enzyme encoded in the information molecule is first decoded and then catalyzes replication of the original information molecule. The information molecule serves to provide the information required for protein production and the template for replication.

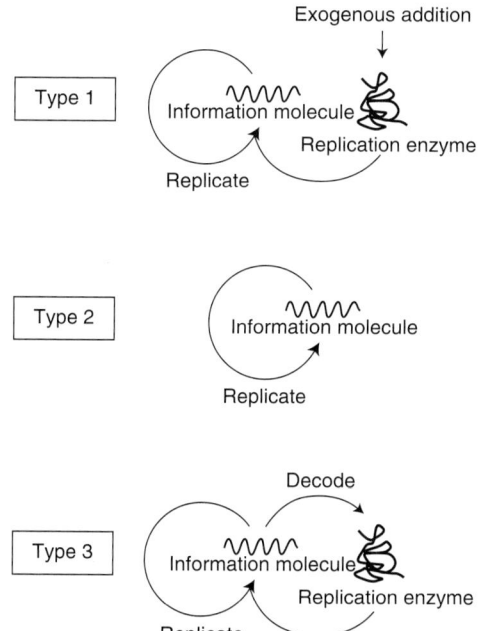

Figure 1. Types of self-replication. Type 1 is the self-replication reaction in which the information molecule (DNA or RNA) is replicated by an exogenous enzyme and includes PCR, 3SR and RNA replication by Qβ replicase. Type 2 self-replication does not require any replication enzymes and these reactions include self-ligating ribozymes, self-replicating peptides. Type 3 is a modification of type 1 in which the replication enzyme is supplied internally by synthesis rather than being exogenously added.

The self-replication of the genome in cells or viruses would also be classified as type 3, because the replication enzyme is translated from the genomic DNA or RNA and then catalyzes replication of the genome. Construction of these types of self-replication systems has been reported previously (Ghadessy et al. 2001; Matsuura et al. 2002). In these reactions, however, translation and replication are separated temporally rather than occurring simultaneously. In the next section, we will describe characteristic features of each type of self-replication from the viewpoint of evolution.

Evolution of Three Types of Self-Replication

If a self-replication process continues for many generations, random mutations can be introduced into the information molecule. This makes it possible for increasingly replicable mutants to appear and dominate the population, as first shown by Bartel and Szostak (1993). There are a number of ways this can occur, including enhancement of template activity (the ability to act as a replication template) and/or catalytic activity that promotes replication. The actual mechanism depends on the type of self-replication. When type 1 self-replication evolves, the template activity is enhanced, as shown previously (Mills et al. 1967). It is notable that although type 1 self-replication requires a replication enzyme, the replication activity does not evolve because the enzyme is not encoded on the information molecule and thus no mutations are introduced to improve the enzyme activity. Therefore, in the evolution of RNA self-replication by Qβ replicase, the template RNA initially encoding three genes were lost during evolution, resulting in a shorter, more rapidly replicating template (Mills et al. 1967).

In type 2 self-replication, a single information molecule has template and catalytic functions, thus both activities are able to evolve (Lincoln and Joyce 2009). In type 3 self-replication, even though the two activities (template activity of DNA or RNA and replication activity of replicase) are encoded on different molecules, both can evolve. This is because the replication catalyst is encoded in the information molecule so that mutations will be introduced. To achieve evolution of the replication enzyme in type 3 self-replication, other conditions are required: encapsulation of the reaction in a compartment with a small number of information molecules. These conditions are required to link information molecules with the encoding replicase so that the translated replication enzyme can interact with its origin information molecule. Without compartmentalization of a low number of information molecules, even if a highly active replicase arose because of mutation, the replicase would amplify the wild-type information rather than its own information containing the mutation (Szostak 1999; Szostak et al. 2001, Matsuura et al. 2002). This requirement for a small number of information molecules was recently shown by an experiment (Sunami et al. 2006) where we encapsulated two types of GFP genes, GFPuv5 and GFPuv2 (GFPuv5:GFPuv2= 0.85:0.15), into a liposome population having a diverse size range together with a cell-free transcription-translation system. GFPuv5 shows eightfold higher fluorescence signal than GFPuv2. Following translation, we collected liposomes showing high GFP fluorescence with a fluorescent-activated cell sorter (FACS) and investigated the ratio of the GFPuv5 gene to the GFPuv2 gene in the collected liposome population. We observed that the ratio was dependent on the liposome volume; within liposomes contained in a large volume (150 fL), the total number of genes per liposome was more than ten and the GFPuv5/GFPuv2 gene ratio was low (Fig. 2). In contrast, liposomes in a small volume (5–10 fL) were found to have nearly one gene per liposome and high GFPuv5/GFPuv2 gene ratio. These observations showed that the gene encoding a highly active protein was selected efficiently in the small liposome, which has a small number of genes.

RECENT EXPERIMENTAL RESULTS

Construction of Type 3 Self-Replication in Liposomes

Primitive life presumably involved type 2 self-replication because of its simplicity. Over time, the replication process would have become more complex by evolutionary selection and approach the type 3 self-replication systems used by extant life. Our purpose was to construct a type 3 self-replication system in which translation and replication occur simultaneously, and to optimize conditions under which type 3 self-replication can function efficiently. The encapsulated system incorporates an information molecule and translation machinery required to decode the information. The information molecule encodes a replication enzyme, which serves to replicate the information molecule.

Figure 3A shows a schematic representation of our type 3 self-replication system. The system consists of a template RNA as an information molecule and a reconstituted cell-free translation system (PURE system) as the decoding machinery, all of which were encapsulated in phospholipid vesicles (liposomes) (Kita et al. 2008). The RNA molecule encodes the catalytic subunit of RNA-dependent RNA polymerase (Qβ replicase), derived from an *E. coli* RNA phage, and has recognition sequences for the Qβ replicase at the termini. During the reaction, the Qβ replicase subunit is first translated from the template RNA and forms active Qβ replicase with EF-Tu and EF-Ts, which are elongation factors for translation and contained in the PURE system. The translated replicase then binds to the original template RNA (plus strand) and synthesizes the complementary RNA strand (minus strand). As the minus strand can also act as a template for the replicase, the RNA strand complementary to the minus strand (i.e., the plus strand) is synthesized in a similar manner. In this way, the information molecule, plus strand RNA, is self-replicated by the self-encoded replication enzyme. Additionally, to monitor

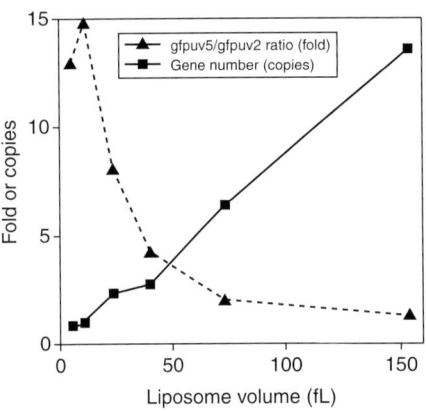

Figure 2. Relationship between liposome volume and efficiency of selection. Two types of GFP genes, GFPuv5 and GFPuv2 (GFPuv5:GFPuv2=0.85:0.15), were encapsulated into a liposome population having a diverse size range together with a cell-free transcription-translation system. GFPuv5 shows eightfold higher fluorescence signal than GFPuv2. Following translation, liposomes showing high GFP fluorescence were collected with a fluorescent-activated cell sorter, and investigated the copy number of total GFP genes and the ratio of the GFPuv5 gene to the GFPuv2 gene in the collected liposome population.

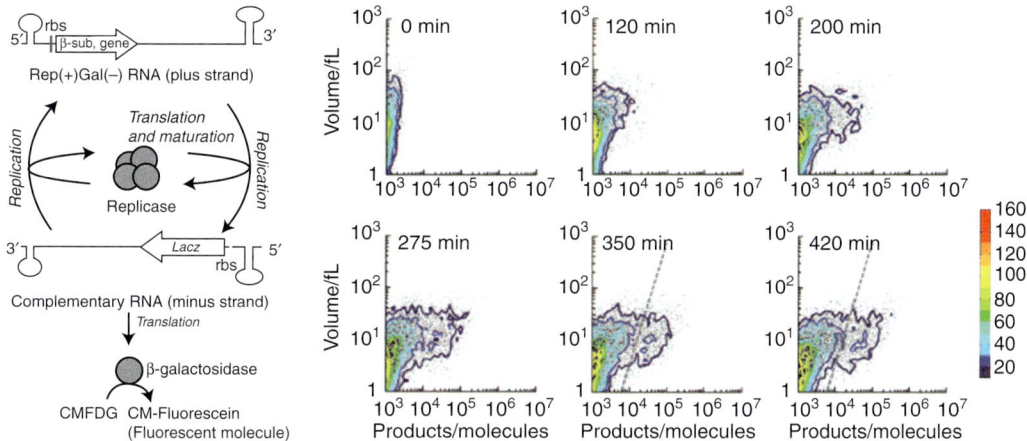

Figure 3. Type 3 self-replication of genetic information in liposomes (A) Schematic representation of the reaction with an additional phenotype that was generated by insertion of the lacZ gene. The Qβ replicase β-subunit was encoded on the plus-strand RNA, and β-galactosidase was encoded on the minus-strand RNA (complement of the plus-strand RNA). Nonfluorescent CMFDG was hydrolyzed by β-galactosidase to yield green fluorescent product, CM-fluorescein. (B) Time course of the reaction analyzed by FACS. The results of 15000 liposomes. The results of FACS analysis of product (horizontal) and internal aqueous volume (vertical) of each liposome are shown. Dots represent the data of individual liposomes. Contour maps are overlaid. The frequency is depicted in color code. At 350 and 420 min, the reacted liposomes were defined as those with a substantial amount of products (right of the dashed lines).

self-replication by fluorescence, we introduced the β-galactosidase sequence into the minus strand. The β-galactosidase is translated after minus strand synthesis and catalyzes hydrolyzation of nonfluorescent 5-chloromethylfluorescein di-β-D-galactopyranoside (CMFDG; Invitrogen, USA) to yield green fluorescent 5-chloromethylfluorescein (CM-fluorescein).

The reaction system consists of 144 gene products (3 rRNAs, 46 tRNAs, 55 ribosomal proteins, 38 proteins), amino acids, other low molecular weight compounds and the template RNA (Table 1). This is a purified reconstituted system in which all of the components and their concentrations are defined. The number of components is amazingly large, yet this is one of the simplest encapsulated systems for carrying out protein translation and RNA replication. With regard to the origin of life, the first living systems would have had functionally identical translation and replication systems, but they must have been simpler and contained machinery for nutrient transport. The complexity of our system implies that extant translation machinery has become highly sophisticated during the evolutionary process.

The self-replication system was encapsulated in lipid vesicles prepared by the freeze-dried empty liposome method (Sato et al. 2005) using the phospholipid mixture, 1-Palmitoyl-2-oleoyl-sn-phosphatidylcholine (POPC):cholesterol: distearoyl phosphatidylethanolamine-polyethylene glycol 5000 (DSPE-PEG5000) at a molar ratio of 58:39:3). The liposomes were multilamellar and the internal volume was found to range from 1 to 100 fL with the most frequent volume about 4 fL. The internal volume was estimated from the fluorescence intensity of the red fluorescent protein, R-phycoerythrin (R-PE), encapsulated as a volume marker and measured by a FACS (Sato et al. 2006; Sunami et al. 2006).

The encapsulated self-replication reaction produces the minus strand, from which β-galactosidase was translated. The translated β-galactosidase hydrolyses the fluorogenic substrate to produce a green fluorescent product. Hence we could monitor the progress of the self-replication reaction by measuring green

Table 1. The structure and composition of a typical functioning liposome

Size of a vesicle	1–100 fL (typically 4 fL) *
Lipid composition	POPC: cholesterol:DSPE-PEG5000 at a molar ratio of 58:39:3
List of proteins and RNAs in the liposome	Template RNA (70 nM), Ribosome (1.2 μM), translational initiation factors: IF1 (2.5 μM), IF2 (0.21 μM), IF3 (0.95 μM), elongation factors: EF-G (3.2 μM), EF-Tu (12 μM), EF-Ts(8.2 μM), releasing factors: RF1 (0.25 μM), RF2 (0.24 μM), RF3 (0.17 μM), RRF (0.48 μM), aminoacyl-tRNA synthetase: AlaRS (725 nM), ArgRS (31 nM), AsnRS (380 nM), AspRS (127 nM), CysRS (24 nM), GlnRS (60 nM), GluRS (233 nM), GlyRS (87 nM), HisRS (8 nM), IleRS (396 nM), LeuRS (42 nM), LysRS (113 nM), MetRS (27 nM), PheRS (676 nM), ProRS (165 nM), SerRS (39 nM), ThrRS (85 nM), TrpRS (28 nM), TyrRS (7 nM), ValRS (17 nM), methionyl-tRNA transformylase (588 nM), creatine kinase (0.47 μM), myokinase (0.93 μM), nucleoside-diphosphate kinase (1.3 μM), pyrophosphatase (0.62 μM), ribosomal protein S1 (4.6 μM), RNasin Plus RNase Inhibitor (1 U/μL), tRNA mix (48 A_{260} units)
List of substrates in the liposome †	ATP (2 mM), GTP (2 mM), CTP (1 mM), UTP (1 mM), 20 amino acids (0.3 mM each), creatine phosphate (20 mM), 10-formyl-5,6,7,8-tetrahydrofolic acid (10 ng/mL)
Other small compounds	Magnesium acetate (13 mM), potassium glutamate (100 mM), spermidine (2 mM), dithiothreitol (1 mM), 2-[4-(2-hydroxyethyl)-1-piperazinyl] ethanesulfonic acid (HEPES) (50 mM, pH 7.6),

*A liposome of a typical size (4 fL) is expected to carry 17 TyrSR molecules (the minimum among protein components), 28 800 EF-Tu molecules (the maximum among protein components), 72 RNA templates and 2 880 ribosomes. It is worth noting that all components that were required for the reactions were likely to be encapsulated into liposomes (>1 fL).

†Our system does not have nutrient transport machinery such as α-hemolysin (Noireaux and Libchaber 2004). It works by consuming internal substrates that have been encapsulated.

fluorescence. FACS analysis showed that the number of liposomes showing green fluorescence increased over time (Fig. 3). We defined the liposomes harboring a substantial amount of products as "reacted liposomes" (liposomes with green fluorescence larger than the dotted line in Fig. 3B). The frequency of the reacted liposome depended on the liposome volume. Statistical analysis showed that the frequency of the reaction occurring per unit volume was constant (0.013 per femtoliter) indicating that the frequency of self-replication was only 5.2% in the case of a typical 4 fL liposome. This implies low efficiency of self-replication in the liposome. The low efficiency is not because of the lack of components because even the smallest liposome at 1 fL is considered to contain all of the components and substrates in our system (Table 1). There are many possible reasons for the low efficiency including degradation of RNAs, inactivation of enzymes, accumulation of inhibitory products and competition between translation and replication. The most plausible possibility for type 3 self-replication is the last one: competition between translation and replication, which could lower reaction efficiency. In the next section, we describe evaluation of the effects of competition for our self-replication system.

Competition between Translation and Replication

Type 3 self-replication is characterized by the dual roles of the information molecule: the information for protein production and template for replication. If the two roles compete, the efficiencies of these reactions in the self-replication reaction become lower than what we expect for the individual activities of translation or replication. Occurrence of this effect arises from the nature of the ribosomes and replication enzyme. In this case, the ribosomes and Qβ replicase were presumed to compete because it was shown that if either is bound to an RNA molecule, the other cannot use the bound RNA as a template (Kolakofsky and Weissmann 1971). The existence of such competition implies that there is an optimum balance between translation and replication. That is, translation of replicase is required for minus strand synthesis, but excess translation by too many ribosomes inhibits minus strand synthesis. To evaluate this competition effect quantitatively, we used a kinetic model to describe part of the self-replication reaction (Fig. 4) (Ichihashi et al. 2008).

The kinetic model contains four components: plus strand RNA, minus strand RNA, RNA replicase (Qβ replicase) and ribosome. It carries out six reactions encompassing binding and dissociation of the plus strand RNA with RNA replicase and ribosome, translation, and minus strand synthesis. The binding and dissociation reactions are assumed to be in equilibrium. The forward reactions favor translation of RNA replicase (decoding processes). The downward reactions tend toward minus strand synthesis (replication processes). The competition effect is represented as the ternary complex of the plus strand with ribosome and replicase (Rep-Rib-P), which is incapable of translation and replication.

The kinetic model has four measurable parameters: dissociation constants for ribosome and replicase, and catalytic constants for translation and replication. Taking advantage of the reconstituted system, we varied the concentration of each component to estimate all four parameters. From the above model and parameters, we could predict an optimum ribosome concentration for minus strand synthesis because of the competition effect (Fig. 5). To examine this prediction experimentally, we varied ribosome concentration in a cell-free translation system, and measured the amount of synthesized minus strand by quantitative PCR after reverse transcription. The experimental results yielded a bell-shaped curve (Fig. 5) and the optimum ribosome concentration was close

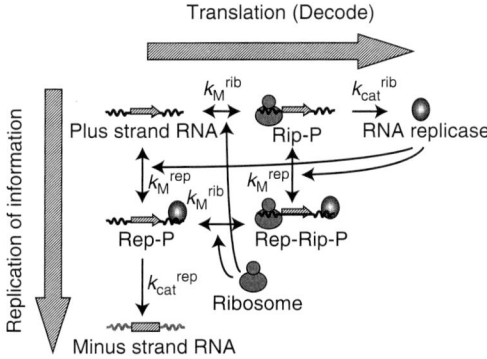

Figure 4. Kinetic model of a part of the type 3 self-replication The kinetic model contains four components: Plus strand RNA, minus strand RNA, RNA replicase (Qβ replicase) and ribosome. It carries out six reactions encompassing binding and dissociation of the plus strand RNA with RNA replicase and ribosome, translation, and minus strand synthesis. The binding and dissociation reactions are assumed to be in equilibrium. The forward reactions favor translation of RNA replicase (decoding processes). The downward reactions tend toward minus strand synthesis (replication processes). The ternary complex of the plus strand with ribosome and replicase (Rep-Rib-P) is incapable of translation and replication. The kinetic model has four measurable parameters: dissociation constants for ribosome (K_M^{rib}) and replicase (K_M^{rep}), and catalytic constants for translation (k_{cat}^{rib}) and replication (k_{cat}^{rep}).

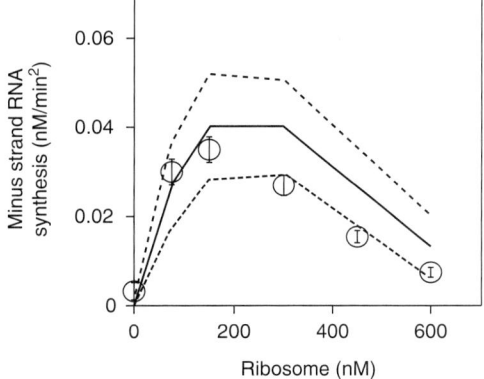

Figure 5. Effects of ribosome on minus strand RNA synthesis Ribosome concentration in a cell-free translation system was varied, and the amount of synthesized minus strand was measured by quantitative PCR after reverse transcription (open circle). Theoretical prediction from the kinetic model and experimentally estimated parameters were shown (solid line) with standard deviation (dotted lines).

to the predicted value, indicating the validity of the kinetic model.

To summarize, we found that our self-replication system showed competition between translation and replication, and we were able to evaluate the effect quantitatively. Such a competition effect in type 3 self-replication is inevitable as long as the dual roles (template and information) are inherent in the information molecule which causes the mutual inhibition of the translation and replication machineries. However, optimizing the ribosome concentration minimizes the inhibitory effect, indicating that the balance between translation and replication is important for efficient type 3 self-replication.

The kinetic analysis provides another implication, with respect to possible size of an evolvable artificial cell. The kinetic analysis revealed that the dissociation constant of the RNA with the replicase was about 20 nM (Ichihashi et al. 2008), indicating that the concentration of RNA should be at more than nanomolar levels for efficient reaction. This requirement limits the possible size of an evolvable artificial cell, considering that the information molecule (RNA in this case) exists in low number for evolution as described above. For example, one information molecule corresponds to approximately 1 nM in a 1 μm cell, but 1 pM in 10 μm cell. Therefore, the size of an evolvable artificial cell including the self-replication system should be 1 μm order for efficient internal RNA replication. This notion implies that the possible size of an evolvable cell would be limited by the affinity of internal components.

CHALLENGES AND FUTURE RESEARCH DIRECTIONS

In this article, we present our recent investigation of a type 3 self-replication system and show the importance of the translation/replication balance. However, with the optimum ribosome concentration, self-replication in liposomes is still inefficient. That is because the system has more than a hundred components and each molecule in the system does not function efficiently, probably because of unexpected interactions such as competition. Further studies are required to determine the conditions under which all components function in a coordinated fashion to achieve efficient self-replication (Pohorille and Deamer 2002).

How do we find these coordinated conditions? One approach is to adopt an evolutionary strategy, mutating the RNA of the self-replication system and selecting mutants showing greater replication. For instance, the selective process could be enabled by selecting for liposomes with higher levels of fluorescence by FACS. This type of evolutionary strategy is also likely to have been adopted by primitive cells, which would need to acquire new functions to replicate efficiently in different environments. The new function or functions acquired by an ancient/primitive cell could sometimes cause a conflict with pre-existing functions. To resolve this conflict, a mutant would evolve such that it would be able to coordinate new functions with pre-existing ones. This conflict resolution process is the same evolutionary strategy that we aim to emulate. Therefore, the construction and improvement of model self-replication systems by evolutionary strategies will provide a deeper understanding of the origin of coordinated biological systems.

REFERENCES

Bartel DP, Szostak JW. 1993. Isolation of new ribozymes from a large pool of random sequences. *Science* **261**: 1411–1418.

Benner SA, Sismour AM. 2005. Synthetic biology. *Nat Rev Genet* **6**: 533–543.

Biebricher CK, Eigen M, Gardiner WC Jr. 1985. Kinetics of RNA replication: competition and selection among self-replicating RNA species. *Biochemistry* **24**: 6550–6560.

Chakrabarti AC, Breaker RR, Joyce GF, Deamer DW. 1994. Production of RNA by a polymerase protein encapsulated within phospholipid vesicles. *J Mol Evol* **39**: 555–559.

Chen IA, Roberts RW, Szostak JW. 2004. The emergence of competition between model protocells. *Science* **305**: 1474–1476.

Deamer D. 2005. A giant step towards artificial life? *Trends Biotechnol* **23**: 336–338.

Fischer A, Franco A, Oberholzer T. 2002. Giant vesicles as microreactors for enzymatic mRNA synthesis. *Chembiochem* **3**: 409–417.

Ghadessy FJ, Ong JL, Holliger P. 2001. Directed evolution of polymerase function by compartmentalized self-replication. *Proc Natl Acad Sci* **98**: 4552–4557.

Guatelli JC, Whitfield KM, Kwoh DY, Barringer KJ, Richman DD, Gingeras TR. 1990. Isothermal, in vitro amplification of nucleic acids by a multienzyme reaction modeled after retroviral replication. *Proc Natl Acad Sci* **87:** 1874–1878.

Hanczyc MM, Fujikawa SM, Szostak JW. 2003. Experimental models of primitive cellular compartments: Encapsulation, growth, and division. *Science* **302:** 618–622.

Ichihashi N, Matsuura T, Kita H, Hosoda K, Sunami T, Tsukada K, Yomo T. 2008. Importance of translation-replication balance for efficient replication by the self-encoded replicase. *Chembiochem* **9:** 3023–3028.

Kaneko K. 2006. *Life: An introduction to complex systems biology.* Springer, Berlin, Germany.

Kita H, Matsuura T, Sunami T, Hosoda K, Ichihashi N, Tsukada K, Urabe I, Yomo T. 2008. Replication of genetic information with self-encoded replicase in liposomes. *Chembiochem* **9:** 2403–2410.

Kolakofsky D, Weissmann C. 1971. Possible mechanism for transition of viral RNA from polysome to replication complex. *Nat New Biol* **231:** 42–46.

Lee DH, Granja JR, Martinez JA, Severin K, Ghadri MR. 1996. A self-replicating peptide. *Nature* **382:** 525–528.

Lee DH, Severin K, Yokobayashi Y, Ghadiri MR. 1997. Emergence of symbiosis in peptide self-replication through a hypercyclic network. *Nature* **390:** 591–594.

Lincoln TA, Joyce GF. 2009. Self-sustained replication of an RNA enzyme. *Science* **323:** 1229–1232.

Noireaux V, Libchaber A. 2004. A vesicle bioreactor as a step toward an artificial cell assembly. *Proc Natl Acad Sci* **101:** 17669–17674.

Mansy SS, Schrum JP, Krishnamurthy M, Tobe S, Treco DA, Szostak JW. 2008. Template-directed synthesis of a genetic polymer in a model protocell. *Nature* **454:** 122–125.

Matsuura T, Yamaguchi M, Ko-Mitamura EP, Shima Y, Urabe I, Yomo T. 2002. Importance of compartment formation for a self-encoding system. *Proc Natl Acad Sci* **99:** 7514–7517.

Mills DR, Peterson RL, Spiegelman S. 1967. An extracellular Darwinian experiment with a self-duplicating nucleic acid molecule. *Proc Natl Acad Sci* **58:** 217–224.

Monnard PA, Deamer DW. 2001. Nutrient uptake by protocells: A liposome model system. *Orig Life Evol Biosph* **31:** 147–155.

Monnard PA, Luptak A, Deamer DW. 2007. Models of primitive cellular life: Polymerases and templates in liposomes. *Philos Trans R Soc Lond B Biol Sci* **362:** 1741–1750.

Oberholzer T, Wick R, Luisi PL, Biebricher CK. 1995. Enzymatic RNA replication in self-reproducing vesicles: an approach to a minimal cell. *Biochem Biophys Res Commun* **207:** 250–257.

Walde P, Wick R, Fresta M, Mangone A, Luisi PL. 1994. Autopoietic self-reproduction of fatty acid vesicles. *J Am Chem Soc* **116:** 11649–11654.

Pohorille A, Deamer D. 2002. Artificial cells: Prospects for biotechnology. *Trends Biotechnol* **20:** 123–128.

Saiki RK, Scharf S, Faloona F, Mullis KB, Horn GT, Erlich HA, Arnheim N. 1985. Enzymatic amplification of β-globin genomic sequences and restriction site analysis for diagnosis of sickle cell anemia. *Science* **230:** 1350–1354.

Sato K, Obinata K, Sugawara T, Urabe I, Yomo T. 2006. Quantification of structural properties of cell-sized individual liposomes by flow cytometry. *J Biosci Bioeng* **102:** 171–178.

Schrodinger E. 1944. *What is life?* Cambridge University Press, Cambridge, UK.

Sievers D, von Kiedrowski G. 1994. Self-replication of complementary nucleotide-based oligomers. *Nature* **369:** 221–224.

Simpson ML. 2006. Cell-free synthetic biology: A bottom-up approach to discovery by design. *Mol Syst Biol* **2:** 69.

Sunami T, Sato K, Matsuura T, Tsukada K, Urabe I, Yomo T. 2006. Femtoliter compartment in liposomes for in vitro selection of proteins. *Anal Biochem* **357:** 128–136.

Szostak JW. 1999. Constraints on the sizes of the earliest cells. In *Size limits of very small microorganisms*, pp. 120–125. National Academy Press.

Szostak JW, Bartel DP, Luisi PL. 2001. Synthesizing life. *Nature* **409:** 387–390.

Tjivikua T, Ballester P, Rebek J. 1990. A self-replicating system. *J Am Chem Soc* **112:** 1249–1250.

Zielinski WS, Orgel LE. 1987. Autocatalytic synthesis of a tetranucleotide analogue. *Nature* **327:** 346–347.

An Origin of Life on Mars

Christopher P. McKay

Space Science Divison, NASA Ames Research Center, Moffett Field, California 94035

Correspondence: chris.mckay@nasa.gov

Evidence of past liquid water on the surface of Mars suggests that this world once had habitable conditions and leads to the question of life. If there was life on Mars, it would be interesting to determine if it represented a separate origin from life on Earth. To determine the biochemistry and genetics of life on Mars requires that we have access to an organism or the biological remains of one—possibly preserved in ancient permafrost. A way to determine if organic material found on Mars represents the remains of an alien biological system could be based on the observation that biological systems select certain organic molecules over others that are chemically similar (e.g., chirality in amino acids).

MARS AND A SECOND GENESIS OF LIFE

Mars today is a cold dry desert world with surface conditions that are not habitable even for the hardiest life forms from Earth. The average surface temperature is $-60°C$ and the atmospheric pressure is near the triple point of water: 120 times lower than sea level pressure on Earth. Even worse for habitability, solar ultraviolet light at wavelengths down to 190 nm penetrates to the surface. Although there is ample evidence for H_2O on Mars, there has been no direct observation of liquid water: only ice, vapor, and geomorphological traces of the action of past liquid water.

In spite of the harshness of the present Martian environment, the Red Planet is a prime target for astrobiology. The motivation for the search for life on Mars comes from the evidence of past water activity. Figure 1 shows an image of Nanedi Vallis, which is perhaps the best example of the long-term flow of liquid water on Mars (Malin and Carr 1999). The canyon is about 2 km across and shows a sinuous pattern consistent with slow erosive fluid flow. The canyon is probably not the actual riverbed. Instead, the bed of the river that carved the canyon is visible in the upper portion of the image. The small size of the riverbed compared to the large canyon indicates that the liquid flowed for a long period of time—although not necessarily continuously. Other fluids suggested as possible geological agents on Mars include wind, glacial ice, lava, and liquefied CO_2. Liquid water, flowing repeatedly and stably on the surface, best explains the features seen in Figure 1.

It is useful to more carefully consider what we are searching for on Mars. Until recently, it was assumed that if there ever were life on Mars, it would necessarily represent a second genesis—a different origin from life on Earth. However, it is now known that many of the meteorites found on Earth have come from Mars (McSween 1984). Furthermore, studies

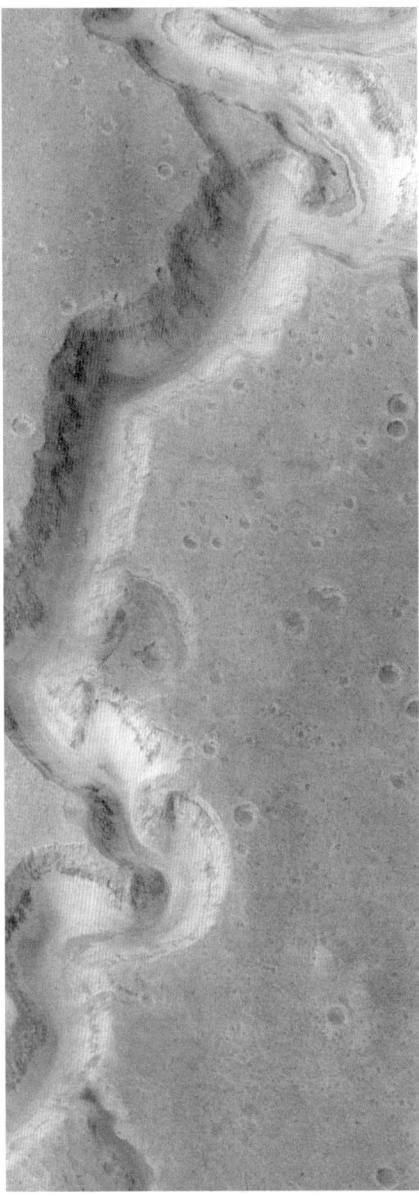

Figure 1. Liquid water on another world. Mars Global Surveyor image showing Nanedi Vallis in the Xanthe Terra region of Mars. Image covers an area 9.8 km by 18.5 km; the canyon is about 2.5 km wide. This image is the best evidence we have of liquid water on Mars. Photo from NASA/Malin Space Sciences.

necessary to consider the possibility that life from Mars was carried to Earth and it is possible that life from Earth could have similarly been carried to Mars—although this path may be less probable. This implies that the discovery of life on Mars does not automatically mean the discovery of a second genesis. To demonstrate a second genesis, it will be necessary to show that Martian life is not related to Earth life.

The search for a second example of life is a key goal for astrobiology. All life on Earth shares common biochemistry and descends from a common ancestor. This prevents us from understanding which aspects of biochemistry and genetics are essential features of life and which are merely particular to the evolutionary history of life on this planet. To develop a more general understanding of life, we need more than one example. Hence, we hope that Mars may have been the site of an independent origin of life.

To determine if life on Mars represents a second type of life requires that we study biological material: Fossils are not enough. Mineralized fossils or tracks of life would be proof of life on Mars, but would not inform us about the relationship of that life to Earth's life. To determine that relationship, we need to study the genetic material and biochemical structure of Martian life—something that can only be done on organisms—alive or dead. We would be convinced of a shared origin of life if Mars had the same chirality, choice of amino acids, genetic code, choice of lipids, and so on. If any or all of these were substantially different, we might conclude a separate origin.

THE ORIGIN OF LIFE: WHAT WE KNOW FROM EARTH THAT APPLIES TO MARS

Everything we know about biology we have learned by studying the single example of life on Earth. The nature of life and its early history inform our search for life on Mars, and elsewhere in the universe. The earliest firm evidence for life on Earth dates back 3.5 billion years ago and is in the form of fossil microbial mats (Tice and Lowe 2004; Allwood et al. 2006). There are also chemical signatures consistent with life in rocks that are 3.8 billion years old (Schidlowski

of the magnetic domains within one of these meteorites by Weiss et al. (2000) have shown that interior temperatures never exceed the survival limits of microorganisms. Thus, it is

1988; Abramov and Mojzsis 2009). The appearance of life on Earth was soon after the end of the late heavy bombardment (Abramov and Mojzsis 2009), suggesting that it appeared as soon as conditions on the surface of the Earth became able to support liquid water environments.

Although the early appearance of life on Earth seems certain, how life originated on Earth remains unclear, with multiple hypotheses competing for attention. Nonetheless, all the scenarios suggested for the origin of life on early Earth can be reviewed for their applicability to early Mars. Figure 2 shows a diagram of the published proposals for the origin of life on Earth (Davis and McKay 1996).

The first divide between the proposals for how life first appeared on the Earth is between those postulating that life arose independently on Earth and those that postulate that life was carried to Earth from elsewhere. If life arose on Earth, the next logical division is between an organic and inorganic nature for that early life. Further divisions within the organic origin of life relate to the nature of the energy source and metabolism for the first organism. How can these ideas be applied to a possible origin of life on Mars? First, what all have in common is a requirement for liquid water environments—the common thread for life today as well as for its origins. Second, all the scenarios listed in Figure 2 could apply to Mars. This is hardly surprising since our understanding of the environment on early Mars suggests that it had a range of environments similar to those on early Earth. Thus, whatever specific conditions and environments led to life on Earth were probably present on Mars as well.

An interesting contrast between Earth and Mars with respect to the origin of life is that the record of early events may be better preserved on Mars than on Earth. The very processes that maintain the habitability of Earth have virtually destroyed the evidence of the first traces of life. Thus, if present on both Earth and Mars sometime before 3.8 billion years ago, the best place to search for physical evidence of the steps that lead to life may be on Mars.

SEARCHING FOR EVIDENCE OF A SECOND GENESIS OF LIFE ON MARS

There are several possible places on Mars where we might find evidence of life that can be used to determine if it emerged as a second genesis. Possible targets include: (1) Life in the surface soil, (2) Life in subsurface liquid water,

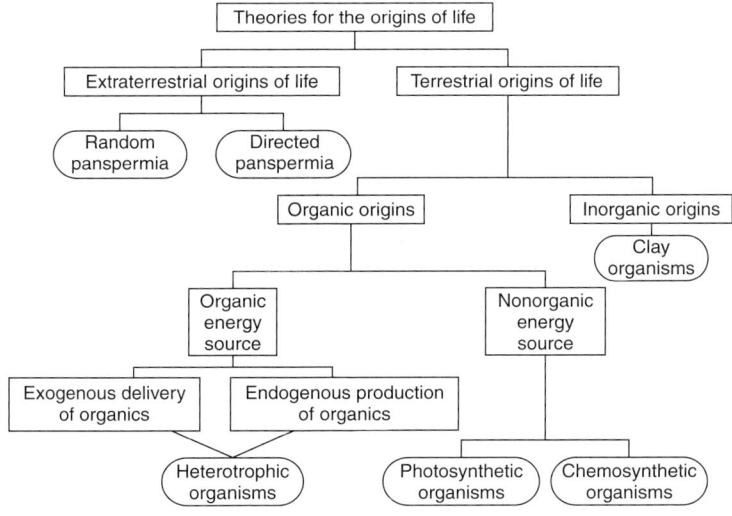

Figure 2. Diagram of the proposed explanations for the origins of life on Earth. Any of these explanations would work on Mars as well as on Earth. From Davis and McKay (1996).

(3) Organisms, probably dead, but preserved in ancient salt or mineral deposits, and (4) Organisms, dead or alive, preserved in ancient ice. Ancient ice is the most promising target known at this time (Smith and McKay 2005), but before discussing this in more detail, it is interesting to first briefly review the case for the other possibilities.

It is unlikely that the surface soil on Mars, such as sampled by the Viking biology experiments, contain life. The general view of the results of the Viking biology experiment is that there is no life present and the reactivity detected was due to chemical processes (Klein 1999). However, the results of the Labeled Release experiment still invoke speculations that the reactions seen were based on biology (Levin 2007; Levin and Straat 1981). If the surface is inhospitable, many have suggested (Boston et al. 1992) that the subsurface may hold liquid water aquifers that support chemosynthetic life. However, there is as yet no direct evidence of subsurface liquid water. On Earth, the oldest biochemical remains of life are found in ancient amber (Cano and Burucki 1995) and salt deposits (Vreeland et al. 2000). However, in both cases, the antiquity of the life found in these deposits is questioned (Willerslev and Cooper 2005). Although amber—a product of trees—is not expected on Mars, if ancient mineral or salt deposits are found, they should be investigated.

Ancient ice presents a known environment on Mars that may contain the frozen remains of life in an accessible form. Smith and McKay (2005) have suggested that the ancient cratered terrain in the southern highlands of Mars near 80°S, 180°W would be an ideal target. The map of Mars (Fig. 3) shows crater distribution, ground ice, and crustal magnetism on Mars. Each green circle represents a crater with a

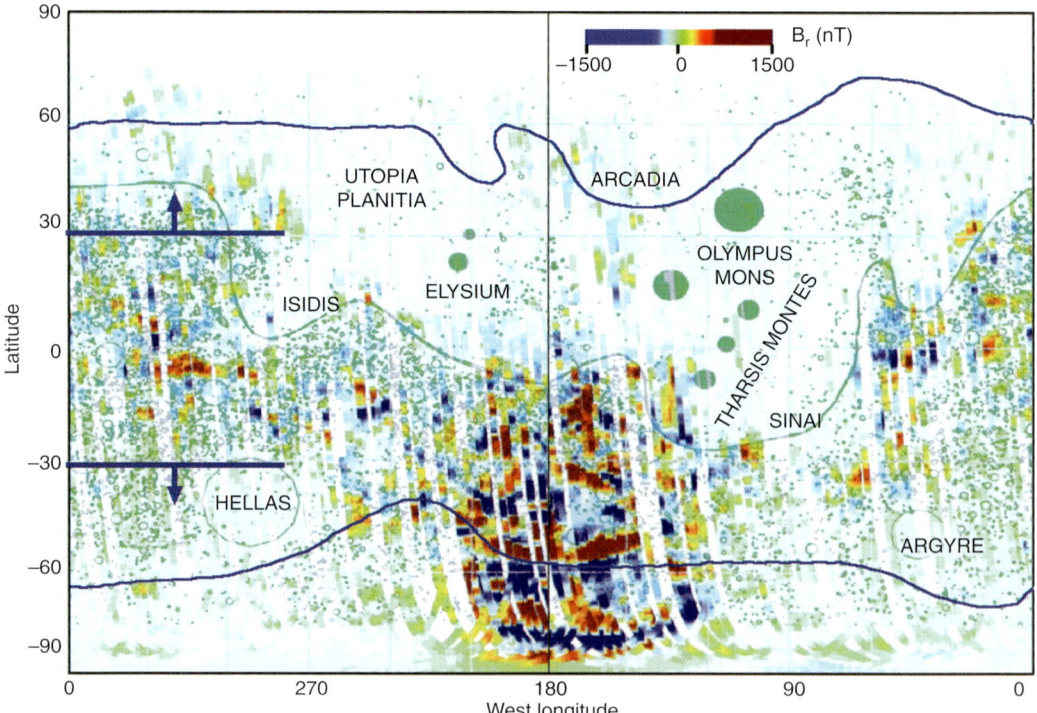

Figure 3. Maps showing crater distribution, ground ice, and crustal magnetism on Mars. The suggested target site for deep drilling to search for evidence of ancient life on Mars is the region between 60° and 80°S at 180°W, where the ground is heavily cratered, crustal magnetism is preserved, and ground ice is present. Figure from Smith and McKay (2005).

diameter greater than 15 km based on the crater distribution in Barlow (1997). The filled green circles are volcanic craters. The boundary between the smooth northern plains and the cratered southern highlands is shown with a green line. The southern regions of Mars are more heavily cratered and therefore considered to be older. The solid blue lines in Figure 4 show the extent of near surface ground ice as determined by the Odyssey mission (Feldman et al. 2002). Ground ice is present near the surface polarward of these lines. Crater morphology indicates deep ground ice poleward of 30° (Squyres and Carr 1986), shown here by dark blue lines and arrows. Also shown on this figure is the crustal magnetism discovered by Acuña et al. (1999). The crustal magnetism is shown as red for positive and blue for negative. Full scale is 1500 nT. The typical strength of Earth's magnetic field at the surface is 50,000 nT. The crustal magnetism is thought to have formed very early in Mars' history, more than 4.5 Gyr ago. The fact of crustal magnetism implies that these surfaces have not been severely heated or shocked. The region between 60° and 80°S at 180°W is heavily cratered, preserves crustal magnetism, and has ground ice present. This is the suggested target site for drilling to find the frozen remains of ancient Martian life.

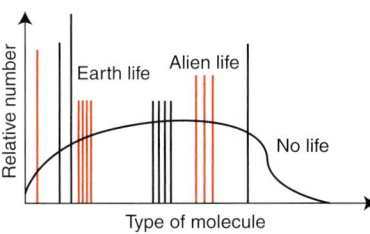

Figure 4. Conceptual comparison of the distribution of molecules in organic matter of biological and nonbiological origin. The ordinate "type of molecule" represents in a general way the size, structure, chirality, and all other features of the molecule. Nonbiological processes produce uniform distributions of organic material, illustrated here by the curve. Biology, in contrast, selects and uses only a few distinct molecules, shown here as spikes (e.g., the 20 L-amino acids on Earth). Alien life might have similar selectivity but based on a different set of molecules (McKay 2004).

DETECTING MARTIAN LIFE

It may be possible that a mission to Mars, landing in the southern highlands and drilling deeply below the surface, would find organic rich sediments that date back to ancient times when conditions on Mars could have supported widespread life. The challenge then would be to determine if this organic material was of biological origin. We know that there is organic material of nonbiological origin in asteroids, comets, and throughout the outer solar system, so organic material from the deep core on Mars could also be nonbiological.

There are two related questions here: First, is this organic material of biological origin, and second, if it is of biological origin, is it evidence of a second genesis of life? If the organic material pulled from a deep ice core on Mars is derived from biochemistry identical to Earth life, then it should be relatively straightforward to show that it is of biological origin. The technology for direct detection of DNA, RNA, ATP, and other key molecules associated with life on Earth has proceeded to the point that it could easily detect the remains of Earth-like life. These molecules would be preserved over the ages by the low temperatures (Kanavarioti and Mancinelli 1990).

The more difficult, but much more interesting case, is that in which the organic material from Mars is of biological origin but does not have the biomolecules associated with Earth life. In this case, our specialized biochemical tools for detecting life are ineffective.

A possible approach to recognizing alien biology has been suggested by McKay (2004) as the "lego" principle. This is the observation that life, like legos, is based on the repeated use of a small number of building blocks. For legos, the building blocks are small plastic interlocking blocks of a few sizes and shapes. Many different structures can be assembled from these blocks. For life on Earth, the building blocks are the 20 amino acids, the bases A, T, C, G, and U in DNA and RNA, the sugars, and a few fatty acids (see, e.g., Lehninger 1975, page 27).

The lego principle can form the basis of a search for biological origins due to the contrast between the selectivity of biology and the non-selectivity of chemistry. A detailed analysis of organic compounds of nonbiological origin would show that a wide range of organic material is present and that their relative concentrations would be determined by their chemical properties. For organic material of biological origin, the distribution would be a series of spikes. In particular, the concentration of chemically identical molecules might be orders of magnitude different in a biological sample, whereas similar in a chemical sample. An example of this is the biological selection of L versus D amino acids for proteins by life on Earth. This concept is presented schematically in Figure 4, which shows the concentration of organic molecules as a function of the type of molecule. The nonbiological distribution is smooth, as seen for example in the Murchison carbonaceous meteorite, whereas the biological distribution is a series of spikes (McKay 2004).

Even if an organism dies, the spiked distribution of organic molecules will persist for some time. However, the chemical distribution in Figure 3 is of higher entropy than the biological distribution, and hence over time, as chemical bonds are broken and reformed, this biological signature will be erased and become indistinguishable from the chemical distribution. Two processes cause this decay: thermal alteration and ionizing radiation. For Mars, the low temperatures in the polar ice ensures that thermal alteration is slow compared to the age of the planet (Kanavarioti and Mancinelli 1990). Ionizing radiation is not temperature dependent. Low but continuous radiation from crustal abundances of U, Th, and K will eventually kill any frozen organism, but will not completely destroy the biological signature of buried frozen organics on Mars (Smith and McKay 2005).

If the organics from Mars show a spiked pattern that is distinct but different from the pattern for Earth life, then this should be both evidence of life and evidence of a second genesis. This is shown schematically in Figure 4 by the red lines. The lego principle thus provides an approach to searching for life that is general enough to detect a second genesis—albeit of carbon-based water-based life. Such life is what we expect to find on Mars and would be a significant first step.

It is not clear how to actually do the measurements illustrated in Figure 4. Currently, the standard approach would be based on GC-MS (gas chromatography-mass spectrometry) with a suitable extraction method. This is, in fact, what was flown to Mars on Viking. However, new methods for organic detection, such as lab-on-a-chip methods, Raman spectroscopy, and UV fluorescence, should also be considered. Analysis of natural soil with low organic content such as the soils of the Atacama Desert (Navarro-Gonzalez et al. 2003, 2006) provide a way to select the most useful methods. Life, or at least its remains, may be out there waiting for us to find it.

It is important to add that even a null result in the search for life on Mars would be informative. It might show that the origin of life is a matter of chance, and the window of opportunity simply was not long enough for life to begin on Mars. Alternatively, it might show that conditions on Mars were too salty, too acid, etc., and inhibited the self-assembly processes required for the origin of life. Finally, we may discover that the origin of life required a fairly specific energy source, mineral, or nutrient (e.g., nitrogen) that was inadequate or absent on Mars. Such insights, even if negative, would help us understand how general the origin of life might be.

The search for life on Mars may be our first chance to discover a second example of life and to investigate the biochemical properties of that life. This possibility is of fundamental importance from both a philosophical and science point of view. Determining where to look and how to search for evidence of a second genesis on Mars is therefore a key task for astrobiology in the next decade.

REFERENCES

Abramov O, Mojzsis SJ. 2009. Microbial habitability of the terrestrial biosphere during the late heavy bombardment. *Nature* **459:** 419–422.

Acuña MH, Connerney JEP, Ness NF, Lin RP, Mitchell D, Carlson CW, McFadden J, Anderson KA, Reme H, Mazelle C, et al. 1999. Global distribution of crustal magnetism discovered by the Mars Global Surveyor MAG/ER experiment. *Science* **284**: 790–793.

Allwood AC, Walter MR, Kamber BS, Burch IW. 2006. Stromatolite reef from the early Archaean era of Australia. *Nature* **414**: 714–718.

Barlow N. 1997. Mars: Impact craters. In *Encyclopedia of planetary sciences* (ed. JH, Shirley, RW Fairbridge), pp. 196–202. Chapman and Hall, London, UK.

Boston PJ, Ivanov MV, McKay CP. 1992. On the possibility of chemosynthetic ecosystems in subsurface habitats on Mars. *Icarus* **95**: 300–308.

Cano RJ, Burucki MK. 1995. Revival and identification of bacterial spores in 25- to 40-million year old Dominican amber. *Science* **268**: 1060–1064.

Davis WL, McKay CP. 1996. Origins of life: A comparison of theories and application to Mars. *Origins Life Evol Biosph* **26**: 61–73.

Feldman WC, Boynton WV, Tokar RL, Prettyman TH, Gasnault O, Squyres SW, Elphic RC, Lawrence DJ, Lawson SL, Maurice S, et al. 2002. Global distribution of neutrons from Mars: Results from Mars Odyssey. *Science* **297**: 75–78.

Kanavarioti A, Mancinelli RL. 1990. Could organic matter have been preserved on Mars for 3.5 billion years? *Icarus* **84**: 196–202.

Klein HP. 1999. Did Viking discover life on Mars? *Origins of Life Evol Biosph* **29**: 1573–0875.

Levin GV. 2007. Possible evidence for panspermia: The labelled release experiment. *Int J Astrobiol* **6**: 95–108.

Levin GV, Straat PA. 1981. A search for a nonbiological explanation of the Viking Labeled Release life detection experiment. *Icarus* **45**: 494–516.

Lehninger AL. 1975. Biochemistry. Worth, New York.

Malin MC, Carr MH. 1999. Groundwater formation on Martian valleys. *Nature* **397**: 589–591.

McKay CP. 2004. What is life—and how do we search for it on other worlds? *PLoS Biol* **2**: 1260–1263.

McSween HY Jr. 1984. SNC Meteorites; are they Martian rocks? *Geology* **12**: 3–6.

Navarro-Gonzalez R, Navarro KF, de la Rosa J, Iniguez E, Molina P, Miranda LD, Morales P, Cienfuego E, Coll P, Raulin F, et al. 2006. The limitations on organic detection in Mars-like soils by thermal volatilization–gas chromatography–MS and their implications for the Viking results. *Proc Nat Acad Sci* **103**: 16089–16094.

Navarro-Gonzalez R, Rainey FA, Molina P, Bagaley DR, Hollen BJ, de la Rosa J, Small AM, Quinn RC, Grunthaner FJ, Ceceres L, et al. 2003. Mars-like soils in the Atacama Desert, Chile and the dry limit of microbial life. *Science* **302**: 1018–1021.

Schidlowski M. 1988. A 3,800-million-year isotopic record of life from carbon in sedimentary rocks. *Nature* **333**: 313–318.

Smith HD, McKay CP. 2005. Drilling in ancient permafrost on Mars for evidence of a second genesis of life. *Planet Space Sci* **53**: 1302–1308.

Squyres SW, Carr MH. 1986. Geomorphic evidence for the distribution of ground ice on Mars. *Science* **231**: 249–252.

Tice MM, Lowe DR. 2004. Photosynthetic microbial mats in the 3416-Myr-old ocean. *Nature* **431**: 549–552.

Vreeland RH, Rosenzweig WD, Powers DW. 2000. Isolation of a 250 million year old bacterium from primary salt crystals. *Nature* **408**: 897–900.

Weiss BP, Kirschvink JL, Baudenbacher FJ, Vali H, Peters NT, Macdonald FA, Wikswo JP. 2000. A low temperature transfer of ALH84001 from Mars to Earth. *Science* **290**: 791–795.

Willerslev E, Cooper A. 2005. Ancient DNA. *Proc R Soc B* **272**: 3–16.

Index

A

Acetal linkage, proto-RNA, 220–221
Amino acid
    curved arrow mechanisms in formation, 72–74
    eutectic enantiomeric excess values, 134–135
    mineral surface binding, 161–162
    monomer polymerization potential, 103–104
    protein synthesis, See Ribosome
    yield in Miller-Urey experiment, 13
Amphiphiles, See Protocell; Vesicle
Apatite, 154
Archaean era
    asteroid impacts, 39–40
    clement conditions on earliest Earth, 37–39
    definition, 36
    mantle paleontology, 43–44
    photosynthesis advent, 41–42
    prospects for study, 44–45
    zircon formation, 40
Atmosphere
    impact degassing, 53–56
    moon formation
        impact, 56–58
        post-impact, 58–60
    origins, 50–52
    prospects for study, 60–62
    secondary atmospheres, 52–53
    volcanic atmospheres, 53

B

Basalt, 47
Belousov-Zhabotinsky reaction, 17
Bilik reaction, 85
Bioenergetics
    energy sources in prebiotic Earth
        carbohydrate chemical energy, 150–151, 153
        chemiosmotic energy conversion to anhydride energy, 149–150
        energy flow, 144–145
        geological electrochemical energy, 148
        photochemical energy, 148
        sulfur chemistry, 148–149
        thermal energy, 146–147, 151
    fundamental considerations from thermodynamics and kinetics, 142–143
    overview, 143–144
    prospects for study
        condensation energy source discovery, 153
        phosphate reactions, 154
        pigments and photosynthesis, 154
Blackmond/brown model, 127–128
Borate, carbohydrate binding and stabilization, 79–84
Breslow, Ronald, 77
Butlerov, Aleksandr, 7, 76

C

Carbohydrate
    chemical energy, 150–151, 153
    curved arrow mechanisms formation, 75–78
    metastability
        improvement
            appendages, 77, 79
            minerals, 79–86
        limitations, 75
    polymerization, 68–69
    prebiotic synthesis driving, 150–151
Carbon, See Cosmic carbon; specific compounds
Carbonaceous meteorite
    abiotic pathways to biomolecules, 100–103
    Antarctica finds, 94–97
    carbon delivery
        molecular evolution
            energetic contingencies, 104–105
            monomer polymerization potential, 103–104
        overview, 26, 89–90
    carbonaceous chondrite characteristics, 90–91
    cosmic history, 97–100
    isotopic ratios, 97–99
    Murchison organic carbon
        classes of compounds, 93
        insoluble organic material, 91–92
        soluble organic compounds, 92
    prospects for study, 105

Index

Carbon bond
    electron pairs, 69–70
    strength, 69
Carbonyl sulfide, prebiotic condensing agent, 152–153
Chatton, E., 281–282
Chemiosmotic energy, conversion to anhydride energy, 149–150
Chirality, *See* Homochirality
Clay, RNA polymerization, 151
Comet, carbon delivery, 26
Cosmic carbon
    compound formation in space environments, 24–25
    extraterrestrial delivery, 26
    gas phase molecules, 26–27
    interstellar ices, 28–29
    interstellar medium phases, 23–24
    inventory, 22–23
    meteorite, *See* Carbonaceous meteorite
    organic molecule cycles, 22–24
    prospects for study, 30–31
    refractory compounds, 27–28
    relevance for early Earth, 29–30
    solar system formation, 25–26
    stability, 29
    young terrestrial planet conditions, 24
Cosmozoa, 8
Crick, Francis, 13, 15
Crystal surface, *See* Mineral surface
Curved arrow mechanisms
    amino acid formation, 72–74
    carbohydrate formation, 75–78
    electron pair movement, 70–71
    nucleobase formation, 74–75

D

Darwin, Charles, 6, 9, 282
D'Herelle, Felix, 8, 10
Diaminopurine, proto-RNA, 214

E

Electron pair
    atomic bonding, 69–70
    curved arrows and movement depiction, 70–71
Electrophilic center, 70
Elongation factors
    ancestral sequence reconstruction, 288–291
    thermostability, 288–289
Enatiomer, *See* Homochirality

Energy flux
    origin of life considerations, 142–143
    prebiotic energy flow, 144
Eschenmoser synthesis, carbohydrate, 75–76
Eve crystal model, conglomerates, 130–131
Extended triple-layer model, ligand exchange, 161

F

Fast ridge axis, 47
Faux amphibolite, 47
Fischer-Tropsch reaction, 145, 183
Formaldehyde, ultraviolet light as energy source for synthesis, 145
Fox, George, 282
Fox, Sidney W., 13
Frank, F.C., 125–127

G

Gabbro, 47
Geological electrochemical energy, prebiotic energy source, 148
Glycerol nucleic acid (GNA), 210, 216
GNA, *See* Glycerol nucleic acid
Goldilocks situation, 47
Graham, Thomas, 6

H

Hadean era
    asteroid impacts, 39–40
    clement conditions on earliest Earth, 37–39
    definition, 36
    geochemical environment, 157–158
    Hudson Bay rock characteristics, 40–41
    mantle paleontology, 43–44
    mineral diversity and distribution, 162–163
    photosynthesis advent, 41–42
    prospects for study, 44–45
    zircon formation, 40
Haldane, John, 10
Harvey, R.B., 10
Herrera, A.L., 8
Heterotrophic theory, evolution, 11–12
Homochirality
    autocatalytic mechanism, 125–127, 137
    chance versus determinism, 124, 128
    enantiomeric excess, 124–126
    enantiomers, 123
    life signature, 123–124
    mechanistic corroboration of Frank model, 127–128

minerals and chiral molecular adsorption, 165–166
phase behavior
  chiral amnesia, 132
  crystal engineering for tuning eutectics, 135–137
  enantioenrichment thermodynamic model, 134
  Eve crystal model for conglomerates, 130–131
  extension to chiral models, 132–134
  modeling, 137–138
  near equilibrium systems, 131–132
  sublime partitioning, 134–135
physical models
  conglomerates, 129–130
  enantiomer partitioning, 130
  eutectic composition, 129
  kinetics versus thermodynamics, 128–129
  phase behavior of chiral solids, 129
  racemic compounds, 130
prospects for study, 137–138
ribosome, 264–265
Hypoxanthine, proto-RNA, 214

I
Infrared Space Observatory (ISO), 28
Insoluble organic material (IOM), Murchison organic carbon, 91–92
IOM, *See* Insoluble organic material
Iron oxide, sources in early Earth, 43
ISO, *See* Infrared Space Observatory
Isochron, 47
Isoguanine, proto-RNA, 214
Isua, 47

K
Kagan model, 127
Kimberlite, formation, 43–44, 47
Komatiite, 47

L
Last universal common ancestor (LUCA), 41, 261, 265, 279
Lateral gene transfer, 283–284
Lewis structure, 70
Lipid carrier model, membrane transport, 200
Lipid head group-gated model, membrane transport, 199–200
Lipman, Charles, 10
Liposome, *See* Vesicle

Lithosphere, 47–48
LUCA, *See* Last universal common ancestor

M
Mantle, paleontology, 43–44, 48
Mars
  astrobiology study rationale, 305–306
  crustal magnetism, 309
  Earth life applicability, 306–307
  life
    detection, 309–310
    evidence searching, 307–309
  water activity, 305–306
Membrane, *See* Protocell; Vesicle
Membrane transport
  influences on cellular processes, 203
  lipid dynamics, 195–196, 200–201
  membrane structure, 193–195
  permeability mechanisms
    lipid carrier model, 200
    lipid head group-gated model, 199–200
    packing defect model, 199
    solubility-diffusion model, 197–198
    transient pore model, 198–199
  prospects for study, 200–203
  protocells, 202–203
  selective permeability and origins of life, 201–202
Metastability
  carbohydrate
    improvement
      appendages, 77, 79
      minerals, 79–86
    limitations, 75
  organic compounds, 68
Meteorite, *See* Carbonaceous meteorite
Methane
  atmospheric stability, 50
  interstellar ice, 28
  methanogenesis, 42
Miller, Stanley C., 12, 146
Miller-Urey experiment, 12–13, 73–74, 145
Mineral surface
  chiral molecular adsorption, 165–166
  Hadean era mineral diversity and distribution, 162–163
  molecular interactions and adsorption, 161–166
  organic synthesis catalysis, 163
  polymerization induction, 166–167
  prospects for study, 167–169
  structure, 159–160
  water interface, 160–161

Index

Molybdate, carbohydrate binding and stabilization, 85–86
Moon
 formation from planetary collision, 56–58
 impact studies, 39
Muller, Hermann J., 9, 13–14
Murchison organic carbon
 classes of compounds, 93
 insoluble organic material, 91–92
 soluble organic compounds, 92
Mush, 48

N

Nucleobase, *See also* Ribonucleotide
 curved arrow mechanisms in formation, 74–75
 novel prebiotic synthesis, 152–153
 polymerization, *See* RNA
Nucleophilic center, 70

O

OB fold, riboproteins, 268
Oken, Lorenz, 6
Oparin, A.L., 9–14, 16, 18
Orgel, Leslie, 207–208
Origin of Species, 6–8
Oró-Orgel synthesis, adenine, 75

P

Packing defect model, membrane transport, 199
PAHs, *See* Polycyclic aromatic hydrocarbons
Pasteur, Louis, 7–8
Peptidyl transferase center, *See* Ribosome
Phosphate, soluble forms, 154
Phosphodiester bond
 linkage in proto-RNA, 218–222
 synthesis driven by anhydrous cycles, 151–152
Phosphoramidate nucleic acids, template replication studies, 253–254
Photochemical energy, prebiotic energy source, 148
Photosynthesis
 advent, 41–42
 early pigments, 148, 154
Phylogenetic tree
 ancestral sequence reconstruction, 285, 287, 290–291
 domains of life, 282
 historical perspective, 280–282
 lateral gene transfer, 283–284
 overview, 279–280
 prospects for study, 291–292
 statistical models of sequence evolution, 290
 temperature history of early life, 284–287
 Tree of Life, 282–283
Polycyclic aromatic hydrocarbons (PAHs)
 early photosynthesis pigments, 154
 formation in space environments, 25
 gas phase molecules, 26–27
 stability, 29
Popper, Karl, 15–16
Protein synthesis, *See* Ribosome
Protocell, *See also* Vesicle
 amphiphiles, 183–184
 assembly, 254–255
 computer simulation, 186
 encapsulated template replication and emergence, 255–256
 membranes
  compartment boundaries, 246–247
  transport, *See* Membrane transport
 partial model construction
  genetic self-replication
   classification of mechanisms, 296–297
   competition between translation and replication, 300–302
   evolution of mechanisms, 297–298
   type 3 self-replication construction in liposomes, 298–300
  goals, 296
  overview, 295–296
  prospects for study, 302
 phosphoramidate nucleic acid template replication studies, 253–254
 physical forces, 186–187
 prebiotic membrane composition and structure, 195
 prospects for study, 186–188, 256–257
 ribozyme-catalyzed RNA replication, 250–253
 simple model, 245–246
 systems-level properties, 185
 vesicles
  division pathways, 249–250
  growth pathways, 247–249
  prebiotic composition, 247
Proto-RNA, *See* RNA
Pyrophosphate bond, thermal energy storage, 147

R

Ribonucleotide, *See also* Nucleobase
 abiotic synthesis, 230–231
 activation and RNA assembly, 109–113
 formation in proto-RNA, 217–218

geochemically plausible synthesis, 118–119
novel self-assembly sequence, 113–118
prospects for study, 119–120
ribozyme synthesis, *See* Ribozyme
Ribosome
  chirality, 264–265
  evolution timeline, 272–273
  genetic self-replication competition with translation, 300–302
  peptidyl transferase center, 262–263
  prospects for study, 273–274
  riboproteins
    age deduction, 266–267
    inferences from structure, 268–269
    recent coevolution of RNA and protein, 270–272
  ribosomal RNA evolution, 269–270
  RNA history, 265–266
  structure and function, 262
  transfer RNA origins, 263–264
Ribozyme
  overview, 229–230
  ribonucleotide synthesis studies, 231–232
  RNA ligase activity
    class I ligase and template-directed RNA polymerization, 235
    cross-replication, 232–233
    molecular evolution via in vitro selection, 234–235
  RNA polymerase activity
    modularity and evolutionary power, 235–238
    polymerization from short RNA genomic elements, 233–234
    protocells, 250–253
  strand displacement mechanisms, 239–242
  strand initiation mechanisms, 238–239
RNA
  base pairing, 207–208
  catalysis, *See* Ribozyme
  inferring events before last universal common ancestor, 265–266
  lipid complexes, 184–185
  proto-RNA
    constraints in search, 209–211
    nucleoside formation, 217–218
    origin theories, 208
    phosphate linkage origins, 218–222
    prospects for study, 222–223
    reinventing challenges, 208–209
    structures of bases, 210–215
    sugar possibilities other than ribose, 215–217
  ribosomal RNA, *See* Ribosome
  self-replication, *See also* Ribozyme
    classification of mechanisms, 296–297
    competition between translation and replication, 300–302
    evolution of mechanisms, 297–298
    type 3 self-replication construction in liposomes, 298–300
  transfer RNA origins, 263–264
RNA ligase, *See* Ribozyme
RNA polymerase, *See* Ribozyme
RNA World, history of study, 14–16

S

Self-replication, *See* Ribozyme; RNA
Serpentine, 48, 81
Shale, 48
Soai reaction, 127–128
Solubility-diffusion model, membrane transport, 197–198
Spontaneous generation, 7–8
Strecker synthesis, amino acids, 72–74
Subduction, 48
Sulfur chemistry, prebiotic energy source, 148–149, 152

T

Temperature, *See* Thermal energy
Thallium, mantle isotope studies, 44
Thermal energy
  geothermal energy and amphiphile synthesis, 151
  prebiotic energy source, 146–147, 151
  pyrophosphate bond, 147
  temperature history of early life, 284–287
Thermostability, elongation factors and ancestral sequence reconstruction, 288–289
α-Threofuranosyl nucleic acid (TNA), 210, 216
Threose, proto-RNA, 216
TNA, *See* α-Threofuranosyl nucleic acid
Transfer RNA, *See* Ribosome
Transient pore model, membrane transport, 198–199
Tree of Life, *See* Phylogenetic tree
Troland, Leonard, 8–9

U

Urey, Harold C., 12

Index

V

Vesicle, *See also* Protocell
- diversity, 180–181
- division pathways, 249–250
- growth pathways, 247–249
- lipid dynamics, 195–196
- membrane structure, 193–195
- mixed composition
  - vesicles, 187–188
- prebiotic composition, 247
- prospects for study, 186–188
- rationale for study, 179–180
- RNA lipid complexes, 184–185
- self assembly of amphiphiles
  - information storage and propagation, 188
  - kinetics, 182–183
  - quantitative trait, 188
  - thermodynamics, 181–182
- self-replication type 3 construction in liposomes, 298–300

W

Wachtershäuser, G., 16
Woese, Carl, 15, 282
Wöhler, Friedrich, 6–7

X

Xanthine, proto-RNA, 214

Z

Zircon, formation, 40